Compressive Sensing
for
Urban Radar

Compressive Sensing
for
Urban Radar

EDITED BY

Moeness Amin

Villanova University, Pennsylvania

CRC Press
Taylor & Francis Group
Boca Raton   London   New York

CRC Press is an imprint of the
Taylor & Francis Group, an **informa** business

CRC Press
Taylor & Francis Group
6000 Broken Sound Parkway NW, Suite 300
Boca Raton, FL 33487-2742

First issued in paperback 2017

© 2015 by Taylor & Francis Group, LLC
CRC Press is an imprint of Taylor & Francis Group, an Informa business

No claim to original U.S. Government works

Version Date: 20140728

ISBN 13: 978-1-4665-9784-6 (hbk)
ISBN 13: 978-1-138-07340-1 (pbk)

Library of Congress Cataloging-in-Publication Data

Compressive sensing for urban radar / edited by Moeness Amin.
    pages cm
  Includes bibliographical references and index.
  ISBN 978-1-4665-9784-6 (hardback)
    1. Through-the-wall radar imaging. 2. Data compression (Telecommunication) 3. Sampling (Statistics) I. Amin, Moeness G.

TK6592.T48C66 2014
621.3848′5--dc23
2014027271

Visit the Taylor & Francis Web site at
http://www.taylorandfrancis.com

and the CRC Press Web site at
http://www.crcpress.com

Contents

Preface

I hope that the in-depth discussions and the scope of the topics covered in this book benefit readers seeking to learn about the applicability and practicality of compressive sensing for urban radar. The book consists of 13 chapters and provides comprehensive coverage of the subtle issues in this area. It presents solutions to many problems that are specific to the sensing of opaque scenes and targets with obstructed lines of sight.

Compressive sensing for urban radar is a hybrid between the two areas of compressive sensing and urban sensing. In essence, it enables reliable imaging of indoor targets using a very small percentage of the entire data volume. This is the first book that focuses on the applications of compressive sensing and sparse reconstructions to urban radar. The latter has been a very active area of research and development. It involves localization of targets inside enclosed structures and sensing of building interiors using electromagnetic modality. Such capabilities are highly desirable for law enforcement, fire and rescue, emergency relief, and military operations. Problems specific to urban radar sensing include near-field, complex multipath and rich scatterings, slow or zero-motions of targets of interest, exterior wall electromagnetic wave blockage and strong signal reflections, shadowing and target obstructions, and detection and classification of human motion both from a biometric and a biomechanical aspect.

With the emergence of compressive sensing and sparse reconstruction, the focus of urban radar has recently shifted toward relaxing constraints on signal sampling schemes in time and space and on addressing logistic difficulties in data acquisition, especially when using ground-based radars. These challenges have traditionally hindered high-resolution imaging by restricting both bandwidth and aperture and by imposing uniformity and bounds on their respective sampling rates.

With a view toward providing timely actionable intelligence in urban environments, it has become important to examine the compressive sensing paradigm and its applicability to through-the-wall radar imaging, given the urban environment–specific problems and challenges. Recent research findings, supported by extensive analysis and experimentations, have revealed performance gains when invoking the sparsity property of urban scenes. Compressive sensing has enabled reduced cost, simplified hardware, and efficient sensing operations of sparse behind-the-wall scenes. As a result, we are now able to reconstruct sparse building interiors with a small fraction of the full data volume previously used for back-projection imaging and target localization. Uniform and Nyquist sampled data are no longer a required strategy for data acquisition; nor are they needed for algorithms imaging or for signal processing.

This book attempts to capture the recent, important contributions to the area of compressive sensing for urban radar. It consists of 13 chapters that address various aspects and have been written by leaders in the field. Some chapters cover stationary scenes, whereas others deal with moving targets. Both ground-based and airborne radars are considered. Analyses and examples provided in the chapters include stepped frequencies, short time pulses, and noise-like transmitted signals, as well as generated radar returns using ray tracing–based simulations, electromagnetic modeling software, and lab experimentations.

Chapter 1 provides an introduction to compressive sensing and presents the various commonly used sparse reconstruction techniques that are applicable to urban radars. Chapter 2 addresses the linear forward model and construction of the dictionary matrix, incorporating major canonical indoor signal reflectors such as walls and dihedrals. Chapter 3 deals with underground objects using ground-penetrating radars and underscores the analogy between through-wall and through-ground compressive radar sensing. Chapter 4 shows how exterior wall mitigation can be effectively achieved with significantly reduced data and presents new techniques more robust to random undersampled time-space and frequency-space radar signals. Chapter 5 models reverberations and multipath clutter from exterior and interior walls and exploits these models, within a sparse reconstruction framework, to reduce ghosts and false alarms. Chapter 6 makes use of prior knowledge on urban scenarios and uses Gaussian mixture models to present information on a wide variety of indoor target objects, achieving high-resolution imaging under sub-Nyquist sampling. Chapter 7 assumes availability of co- and cross-polarization data and casts the urban radar sensing problem as multiple measurement vectors in which there is a common dictionary and same scene supports for different polarimetric observations. Chapter 8 uses change detection and target velocity profiles to induce sparsity and demonstrates successful indoor moving target indications using sparse reconstruction for both sudden and translational motions. Chapter 9 deals with nonstationary radar returns that result from short target motions and incorporates compressive sensing into joint time-frequency signal representations. Chapter 10 presents a sparsity-based algorithm for multiple-target tracking in a time-varying multipath environment and casts the problem of multiple-target tracking as a block support recovery problem. Chapter 11 discusses the three-dimensional wide angle synthetic aperture radar (SAR) imaging of vehicles using airborne radar systems and uses sparsely sampled apertures in lieu of a densely sampled set of points in both azimuth and elevation angles, which are associated with traditional Fourier transform–based methods. Chapter 12 makes a case for compressive sensing multiple-input multiple-output (MIMO) radars in urban settings and demonstrates the advantages of transmitting multiple independent or correlated waveforms utilizing the sparsity of the target returns in angle, range, and Doppler space. Chapter 13 uses noise waveforms and demonstrates how these signals and their system platforms are suitable for realizing the promise of compressive sensing in radar imaging.

I hope you enjoy reading this book in its entirety or the chapters of your interest. I would like to thank all the contributors for their hard work, excellent descriptions of the problems, clear delineations of the approaches, and inclusion of supportive simulation and experiments.

Moeness G. Amin

MATLAB® is a registered trademark of The MathWorks, Inc. For product information, please contact:

The MathWorks, Inc.
3 Apple Hill Drive
Natick, MA 01760-2098 USA
Tel: 508-647-7000
Fax: 508-647-7001
E-mail: info@mathworks.com
Web: www.mathworks.com

Editor

Dr. Moeness G. Amin earned his PhD in electrical engineering from the University of Colorado, Boulder, in 1984. He has been on the faculty of the Department of Electrical and Computer Engineering at Villanova University since 1985. In 2002, he became the director of the Center for Advanced Communications, College of Engineering. Dr. Amin is the chair of the Electrical Cluster of the Franklin Institute Committee on Science and the Arts. He is a fellow of the Institute of Electrical and Electronics Engineers (IEEE), a fellow of the International Society of Optical Engineering, and a fellow of the Institute of Engineering and Technology (IET). He is a recipient of the IEEE Third Millennium Medal, the 2009 Individual Technical Achievement Award from the European Association of Signal Processing, the NATO Scientific Achievement Award, the Chief of Naval Research Challenge Award, the Villanova University Outstanding Faculty Research Award, and the IEEE Philadelphia Section Award.

Dr. Amin has over 700 journal and conference publications in the areas of wireless communications, time-frequency analysis, sensor array processing, waveform design and diversity, interference cancellation in broadband communication platforms, satellite navigations, target localization and tracking, direction finding, channel diversity and equalization, ultrasound imaging, and radar signal processing. He has done extensive research in radar signal processing. Dr. Amin is the editor of the book entitled *Through-the-Wall Radar Imaging* published by CRC Press in 2010. He was the guest editor of the *Journal of the Franklin Institute* September 2008 special issue on advances in indoor radar imaging, of the *IEEE Transactions on Geoscience and Remote Sensing* May 2009 special issue on remote sensing of building interior, of the *EURASIP Journal on Advances in Signal Processing* 2014 special issue on sparse sensing in radar and sonar signal processing, and of the *IEEE Signal Processing Magazine* November 2013 special issue on time-frequency analysis and applications and the July 2014 special issue on recent advances in synthetic aperture radar imaging.

Contributors

Fauzia Ahmad
Center for Advanced
    Communications
Villanova University
Villanova, Pennsylvania

Moeness G. Amin
Center for Advanced
    Communications
Villanova University
Villanova, Pennsylvania

Junhyeong Bae
School of Electrical and Computer
    Engineering
and
Advanced Radar Research Center
The University of Oklahoma
Norman, Oklahoma

Abdesselam Bouzerdoum
School of Electrical, Computer and
    Telecommunications Engineering
University of Wollongong
Wollongong, New South Wales,
    Australia

Phani Chavali
Preston M. Green Department of
    Electrical and Systems
    Engineering
Washington University in St. Louis
St. Louis, Missouri

Jacco de Wit
Department of Radar Technology
Netherlands Organisation for
    Applied Scientific Research
The Hague, the Netherlands

Emre Ertin
Department of Electrical and
    Computer Engineering
The Ohio State University
Columbus, Ohio

Nathan A. Goodman
School of Electrical and Computer
    Engineering
and
Advanced Radar Research Center
The University of Oklahoma
Norman, Oklahoma

Yujie Gu
School of Electrical and Computer
    Engineering
and
Advanced Radar Research Center
The University of Oklahoma
Norman, Oklahoma

Kyle R. Krueger
School of Electrical and Computer
    Engineering
Georgia Institute of Technology
Atlanta, Georgia

Michael Leigsnering
Signal Processing Group
Institute of Telecommunications
Technische Universität Darmstadt
Darmstadt, Germany

Rabinder N. Madan
Champana Scientific, LLC
Annandale, Virginia

James H. McClellan
School of Electrical and Computer
    Engineering
Georgia Institute of Technology
Atlanta, Georgia

Ram M. Narayanan
Department of Electrical Engineering
The Pennsylvania State University
University Park, Pennsylvania

Arye Nehorai
Preston M. Green Department of
    Electrical and Systems
    Engineering
Washington University in St. Louis
St. Louis, Missouri

Irena Orović
Department of Electrical Engineering
University of Montenegro
Podgorica, Montenegro

Athina Petropulu
Department of Electrical
    and Computer Engineering
Rutgers, The State University of
    New Jersey
Piscataway, New Jersey

Muralidhar Rangaswamy
Air Force Research Laboratory
Sensors Directorate
Wright Patterson AFB, Ohio

Waymond R. Scott, Jr.
School of Electrical and Computer
    Engineering
Georgia Institute of Technology
Atlanta, Georgia

Mahesh C. Shastry
Department of Electrical Engineering
The Pennsylvania State University
University Park, Pennsylvania

Ljubiša Stanković
Department of Electrical Engineering
University of Montenegro
Podgorica, Montenegro

Srdjan Stanković
Department of Electrical Engineering
University of Montenegro
Podgorica, Montenegro

Fok Hing Chi Tivive
School of Electrical, Computer and
    Telecommunications Engineering
University of Wollongong
Wollongong, New South Wales,
    Australia

Wim van Rossum
Department of Radar Technology
Netherlands Organisation for
    Applied Scientific Research
The Hague, the Netherlands

Michael B. Wakin
Department of Electrical Engineering
    and Computer Science
Colorado School of Mines
Golden, Colorado

Jack Yang
School of Electrical, Computer and
    Telecommunications Engineering
University of Wollongong
Wollongong, New South Wales,
    Australia

Yao Yu
Department of Electrical
    and Computer Engineering
Rutgers, The State University of
    New Jersey
Piscataway, New Jersey

Yimin D. Zhang
Center for Advanced
    Communications
Villanova University
Villanova, Pennsylvania

Abdelhak M. Zoubir
Signal Processing Group
Institute of Telecommunications
Technische Universität Darmstadt
Darmstadt, Germany

1

Compressive Sensing Fundamentals

Michael B. Wakin

CONTENTS

1.1 Overview

1.1.1 Signal Models and Dimensionality Reduction

As we enter the era of Big Data, the amount we demand from sensing and imaging systems continues to explode. One way this is manifested is in the increasing resolution, or typical number of samples (or pixels, voxels, etc.), in a signal (or image, video, etc.). We will use N to denote the number of samples in a signal of interest. The continuing growth of N places an increasing burden on every stage of the data processing pipeline, from the initial acquisition of the data to the subsequent transmission, storage, and analysis.

In order to control the cost, complexity, and bandwidth of systems for collecting and processing such high-dimensional data, it is critical to exploit *models* that encapsulate prior information regarding the signals of interest. Signal models capture—either implicitly or explicitly—the fact that many N-sample signals actually have far fewer than N degrees of freedom. That is, many signals have an intrinsic information level that can be well described using just K parameters or degrees of freedom, where K might be much smaller than N. This lower information level suggests that it may be possible to reduce the burden of acquiring, processing, and understanding a high-dimensional signal by exploiting that signal's information level K rather than operating at its ambient dimension N. Indeed, there exist a wide variety of techniques for what might be called *dimensionality reduction*: using low-dimensional models to reduce the burden of processing and understanding high-dimensional signals and data sets. Examples include

- *Data compression:* Reducing the number of bits required to store a high-dimensional signal while preserving its critical information.
- *Parameter estimation:* Discerning the specific values of the underlying degrees of freedom of a high-dimensional signal.
- *Feature extraction:* Extracting information-carrying characteristics from a raw high-dimensional signal.
- *Manifold learning:* Identifying the parameterization underlying a collection of high-dimensional signals.

For many signals, the underlying low-dimensional structure can be revealed by applying a transform to the signal—that is, by computing the signal's expansion coefficients in an appropriate basis such as a Fourier basis or wavelet basis [125]. When the basis is well chosen, it is often the case that many of the resulting transform coefficients will be small and only a few (say, K) will be large. In some cases, the locations of the K large coefficients will be fixed from signal to signal or known a priori as part of the model. This corresponds to a low-dimensional *linear* signal model: Geometrically one can envision the set of possible signals as clustering around a K-dimensional linear subspace within the ambient signal space $\mathbb{R}^N$. Linear subspace models are foundational in classical signal processing [108,155]: bandlimited signals cluster along a subspace spanned by the low-frequency Fourier basis functions (i.e., sinusoids), least-squares problems can be solved by projections onto fixed subspaces, optimal subspaces can be learned via principal component analysis (also known as the Karhunen–Loève transform), etc.

In other cases, the locations of the K large coefficients may vary from signal to signal or be unknown a priori. This is known as a *sparse signal model* [82,125]. Sparse signal models are *nonlinear*: Such a model cannot be contained within a single low-dimensional linear subspace, but rather corresponds to a collection of many candidate K-dimensional subspaces within the ambient signal space $\mathbb{R}^N$, each one corresponding to a set of possible locations for the K large coefficients. Consequently, sparse signal models are far more flexible than linear subspace models and can provide economical descriptions of far broader classes of signals. Over the past two decades, a great deal of research in signal processing has been devoted to developing efficient sparsifying transforms and dimensionality reduction techniques that exploit these transforms. The JPEG-2000 standard for image compression, for example, involves computing the wavelet transform of an image and encoding the resulting coefficients; this can be done efficiently because many of them are small [158]. Also, many techniques for noise removal in signals involve applying a sparsifying transform to the noisy signal, keeping the largest transform coefficients (because those likely correspond to signal energy) and throwing away the smallest transform coefficients (because those may be due to noise only) [65].

1.1.2 Motivation for Compressive Sensing

The dimensionality reduction techniques described above all require the N samples of a signal to actually be collected before we can identify and exploit its intrinsic low-dimensional structure. That is, although the signal may depend on just K degrees of freedom, these pieces of information cannot be identified until all N samples of the signal have been collected and, for example, the transform coefficients of the signal have been computed. Informally, this suggests that the sensing process for many signals may be unnecessarily wasteful—after all, sensing high-bandwidth and high-resolution

5	3			7				
6			1	9	5			
	9	8					6	
8				6				3
4			8		3			1
7				2				6
	6					2	8	
			4	1	9			5
				8			7	9

FIGURE 1.1
In a sudoku puzzle, a set of rules allow the missing numbers to be filled in. (From Sudoku layout, http://en.wikipedia.org/wiki/File:Sudoku-by-L2G-20050714.svg.)

signals can require expensive hardware, consume valuable power, etc.—and it raises the question of whether it is possible to incorporate dimensionality reduction into the sensing process itself. For example, could it be possible to design a sensor that intentionally collects fewer samples of a signal, with the expectation that the missing samples can later be reconstructed from the recorded ones?

At an intuitive level, it is not unreasonable to think that low-dimensional models could be useful for *filling in the blanks* when only partial information about a signal is recorded. A sudoku puzzle provides a useful analogy. A sudoku puzzle consists of a 9×9 grid of numbers. The player is allowed to see some of these numbers and must fill in the missing ones (see Figure 1.1). With no additional information, it may seem impossible to determine the missing numbers. However, a correct puzzle solution must obey a certain set of simple rules,* and these rules act as a *model* that constrains the set of possible solutions. A puzzle solver who is skilled in applying these rules is effectively imposing the model in order to resolve the missing information. It is not difficult to think of other examples where models could be useful in dealing with partial obs rv ti ns. (The previous sentence is itself an example.)

Perhaps the most efficient sensor one could imagine would be one that directly senses the K degrees of freedom of the signal. However, since these K degrees of freedom may not be apparent from just K *samples* of the signal (e.g., K pixels from an image), it is common to consider instead sensors that can produce discrete *measurements* that are the result of linear operations on the signal: filtering, modulation, sampling, etc. Transform coefficients of a signal (Fourier coefficients, wavelet coefficients, etc.) can be recorded using

---

* All nine columns, all nine rows, and all nine 3×3 blocks must each use the digits $1, 2, \ldots, 9$ exactly once.

such linear operations. Still, for many signal models it might be impossible to specify *in advance* a measurement protocol that can capture the critical signal information using just K linear measurements. In the case of sparse models, the reason is clear: although the signal information is entirely contained in just K transform coefficients, it is not known a priori which of the N total transform coefficients are the nonzero ones. Any measurement strategy that prescribed measuring a fixed set of K transform coefficients would risk missing critical information about the signal.

Nevertheless, the central idea that underlies compressive sensing (CS) is that it can be possible to record a small number of linear measurements of a signal (slightly more than K, but far fewer than N) and from those measurements reconstruct the complete set of all N samples that a conventional sensor would have recorded [41,44,66]. A critical difference from the doomed strategy described earlier is that the measurements in CS typically do not attempt to access the signal's degrees of freedom directly; in a sparse model, for example, the linear CS measurements typically do not correspond to transform coefficients in the sparsifying basis. Rather, each measurement contains a mix of signal information that must later be decoded.

1.1.3 Compressive Sensing in a Nutshell

There are so many twists on the basic CS setup that it can be difficult to get a sense for what sort of problems fit into CS. In this section, we give a very brief synopsis of a typical CS problem. The rest of this chapter then fills in the details: Section 1.2 discusses sparse models, Section 1.3 discusses compressive measurement protocols, and Section 1.4 discusses the recovery of sparse signals from compressive measurements.

A typical CS problem looks like this: Let f denote a desired collection of N samples of a signal. For convenience, suppose these are arranged into an $N \times 1$ vector; in cases where these are samples of an image or other multidimensional signal, the pixel values can be stacked into the vector f using any ordering rule. We do not attempt to record the N entries in f directly. Instead, we record a smaller number M of linear measurements of f; suppose these are arranged into an $M \times 1$ vector we call y. Since these measurements are linear, we can represent the measurement vector as

$$y = \Phi f,$$

where Φ is an $M \times N$ matrix we refer to as the *measurement matrix*. Typically in CS, Φ is designed with some element of randomness. Also, typically in CS, the number of measurements M must be proportional to the information level K of f according to some model. This model (along with knowledge of Φ) can then be used to recover f from y despite the fact that the system of equations $y = \Phi f$ is underdetermined and therefore has an infinite number of candidate solutions.

What sorts of random measurements can be used in CS? In some cases, it can be appropriate to collect just a random set of *time domain* samples of the signal, for example by recording a random set of, say, 10% of the pixels in an image. In this case, Φ is a binary matrix containing a single randomly positioned 1 in each row. In other cases, it can be appropriate to collect a random set of *frequency domain* measurements of a signal, for example by recording a random set of 10% of the Fourier coefficients of f. In this case, Φ contains a random set of M rows of the $N \times N$ discrete fourier transform (DFT) matrix. In other cases, every measurement in y might be a random linear combination of all of the entries in f. In this case, Φ might be populated with independent and identically distributed (i.i.d.) Gaussian or Rademacher (± 1) random variables. Ultimately the question of which measurement scheme is appropriate depends on which model will be used to recover f. For example, if a signal is sparse in the frequency domain, it would be inappropriate to collect random measurements in the frequency domain, as we have explained in Section 1.1.2. Random matrices populated with i.i.d. Gaussian or Rademacher entries have the attractive feature of *universality*: With high probability they are suitable for measuring signals that are sparse in any fixed basis. This means it is not necessary to know the sparse basis at the time the measurements are collected.

There exist compressive measurement devices that can produce random measurement vectors y directly and without explicitly recording the sample vector f. Several examples are listed in Section 1.3.3 and include devices for imaging and analog-to-digital conversion. When compared with conventional sensors, these devices can enjoy advantages in terms of cost, power consumption, quantity of data to be transmitted, etc. In some other applications, however, one might already have a sample vector f acquired using a conventional sensor and choose to compress f computationally—that is, by multiplying it by a random matrix.*

The appropriate technique for recovering f from y and Φ depends on what low-dimensional model is assumed to capture the information in f. At a high level, all CS recovery algorithms can be interpreted as searching among the candidate solutions to the equation $y = \Phi f$ for the one signal that best matches the low-dimensional model. In the case of a sparse signal model—which is by far the most commonly used model in CS—one might look for the sparsest candidate solution in some known basis. As we describe in Section 1.4, there are a variety of algorithms available for searching for this sparsest candidate solution; some involve convex optimization, while others involve iterative greedy methods. It is known that, under certain assumptions on the random measurement protocol, a K-sparse signal

---

* This could be desirable in situations where conventional signal compression approaches are expensive or difficult to implement. For example, using CS one can compress an ensemble of correlated sparse signals recorded in a sensor network without requiring any communication among the sensors [13].

f can be recovered from a number of measurements M which is proportional merely to $K \log(N/K)$. This number of measurements can be significantly smaller than N and is only greater than the information level K by a logarithmic factor; this logarithmic factor is the price one pays for not knowing the locations of the sparse coefficients in advance. Remarkably, in the absence of noise and assuming f is exactly sparse, this recovery is exact. In the presence of noise or assuming f is nearly sparse, this recovery is provably robust.

1.2 Sparse Modeling

Sparsity is one manifestation of the fact that many high-dimensional signals actually have a small number of degrees of freedom. In sparse models, these degrees of freedom are expressed as nonzero coefficients when the signal of interest is expanded in some basis or dictionary.

1.2.1 Sparsity, Compressibility, and Norms

A real- or complex-valued length-N vector x is said to be *K-sparse* if it contains just K nonzero entries. The ℓ_0 norm can be used to count the number of nonzero elements in x (though it does not meet the formal mathematical definition of a norm). For a K-sparse vector x, $\|x\|_0 = K$. We refer to the set of positions of the nonzero entries of x as the *support* of x and denote this by $\mathrm{supp}(x)$. For any x, $|\mathrm{supp}(x)| = \|x\|_0$.

For any vector $x \in \mathbb{R}^N$ or $\mathbb{C}^N$, we let x_K denote the nearest K-sparse vector to x. This can be obtained simply by keeping the K entries of x with the largest magnitudes and setting all remaining entries to 0. If $\|x\|_0 \leq K$, then $x = x_K$. If the distance from x to x_K is small (but not necessarily zero), x is said to be *compressible*. Vectors whose sorted entries decay like a power law are often considered to be compressible [32].

Other norms besides the ℓ_0 norm can also be useful when discussing and processing sparse signals. The ℓ_1 norm measures the absolute sum of the entries of x:

$$\|x\|_1 = \sum_{n=1}^{N} |x_n|,$$

and the squared ℓ_2 norm measures the sum of squared magnitudes of the entries of x:

$$\|x\|_2 = \sqrt{\sum_{n=1}^{N} |x_n|^2}.$$

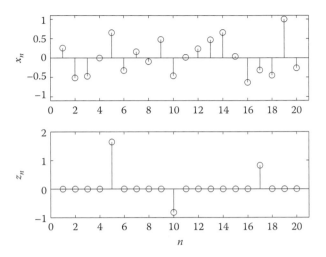

FIGURE 1.2
The length-20 vectors x (top) and z (bottom) have the same ℓ_2 norm ($\|x\|_2 = \|z\|_2 = 2.012$), but vector z has fewer nonzero entries. Consequently, the ℓ_0 and ℓ_1 norms of z are smaller than the ℓ_0 and ℓ_1 norms of x: $\|z\|_0 = 3$ while $\|x\|_0 = 20$, and $\|z\|_1 = 3.285$ while $\|x\|_1 = 7.496$.

Unlike the ℓ_0 norm, the ℓ_1 and ℓ_2 norms both meet the formal mathematical definition of a norm [119] and both are convex functions of x [27].

The ℓ_1 norm has a special connection to sparsity: It tends to be small for sparse signals. More formally, for any $x \in \mathbb{R}^N$ or $\mathbb{C}^N$, the ℓ_1 norm must always obey the bound: $\|x\|_2 \leq \|x\|_1 \leq \sqrt{N}\,\|x\|_2$. For two vectors x and z with the same ℓ_2 norm, the one with fewer nonzero entries will generally have the smaller ℓ_1 norm. This is demonstrated in Figure 1.2.

1.2.2 Sparsity in Orthonormal Bases

Some signals f may themselves be sparse or compressible vectors; that is, they may contain just a few significant entries. One simplified example might be an astronomical image where only a few pixels are illuminated by stars [26]. However, it is common for the sparse structure of a signal to be revealed only when that signal is transformed into an appropriate domain. For example, if f comprised a small number of harmonic tones, then the vector containing the DFT coefficients of f will be sparse.

For any unitary transform, we let Ψ denote an $N \times N$ real- or complex-valued *basis matrix* that implements the inverse transform, taking a coefficient vector $x \in \mathbb{R}^N$ or $\mathbb{C}^N$ and *synthesizing* the corresponding signal $f \in \mathbb{R}^N$ or $\mathbb{C}^N$:

$$f = \Psi x. \tag{1.1}$$

Unitary transforms have the property that $\|x\|_2 = \|\Psi x\|_2$ for all x; this property is equivalent to requiring all columns of Ψ to be orthonormal. The forward transform, taking a signal $f \in \mathbb{R}^N$ or $\mathbb{C}^N$ and *analyzing* it in the basis Ψ to produce the coefficient vector $x \in \mathbb{R}^N$ or $\mathbb{C}^N$, is easy to express thanks to the orthonormality of the columns of Ψ. We have

$$x = \Psi^* f, \tag{1.2}$$

where Ψ^* denotes the transpose of Ψ (or the conjugate transpose if Ψ is complex).

We can interpret the aforementioned equations in terms of the columns of Ψ, which we denote as $\Psi_1, \Psi_2, \ldots, \Psi_N$. Equation 1.1 becomes

$$f = \sum_{n=1}^{N} x_n \Psi_n, \tag{1.3}$$

in which we see f expressed as a linear combination of the basis elements Ψ_n with weights x_n. Equation 1.2 becomes

$$x_n = \Psi_n^* f = \langle f, \Psi_n \rangle, \tag{1.4}$$

for each $n = 1, 2, \ldots, N$, in which we see that each coefficient can be computed as the inner product (also known as the dot product) of f with a basis element.

For some signals f, it is possible to choose a suitable basis Ψ such that the transform coefficients $x = \Psi^* f$ are sparse. This means that f can be written as a weighted sum of just K basis elements (or columns) from Ψ since just K terms in the summation (1.3) will be nonzero. In this event, we say that f is *K-sparse in the basis* Ψ. If $x = \Psi^* f$ is compressible, we say that f is *compressible in the basis* Ψ. If f is itself a sparse vector, one can simply choose $\Psi = \mathcal{I}_N$, the $N \times N$ identity matrix. In this event, f is said to be sparse in the canonical basis, the time domain (if f represents samples of a one-dimensional signal), or the spatial domain (if f corresponds to an N-pixel image). If f contains samples of a uniformly smooth signal or a linear combination of a small number of harmonic tones, one can choose Ψ to be the $N \times N$ DFT matrix, which we denote by $\mathcal{F}_N$ (we assume the DFT is normalized so that each column of $\mathcal{F}_N$ has unit ℓ_2 norm). If f contains samples of a uniformly smooth *or* piecewise smooth signal (one containing a small number of discontinuous or nonsmooth points), one can choose Ψ to be a discrete wavelet transform (DWT) matrix [125]. Note that discontinuities in signals can carry important information. For example, the two-dimensional DWT is commonly used in image processing, and edges—which indicate the boundaries of objects—are responsible for producing most of an image's significant wavelet coefficients.

1.2.3 Sparsity in Nonorthonormal Dictionaries

Some signals exhibit a sparse structure that cannot be easily expressed using
an orthonormal basis.

1.2.3.1 Synthesis Sparsity

In some cases, a signal $f \in \mathbb{R}^N$ or $\mathbb{C}^N$ may be expressible as a weighted
sum of a small number of vectors belonging to some set of interest, but
these vectors may not be necessarily orthonormal. Examples include signals
synthesized using a combination of vectors drawn from multiple orthonormal bases (such as spikes and sines [70]) or signals consisting of pure
tones selected from a grid with finer spacing than the conventional DFT
[76,87].

To accommodate such settings, we can generalize the framework of
Section 1.2.2 by letting Ψ denote an $N \times D$ real- or complex-valued *dictionary matrix* whose columns $\Psi_1, \Psi_2, \ldots, \Psi_D$ can be used to synthesize signals
of interest. For the *spikes and sines* example mentioned earlier, one can construct Ψ by concatenating $\mathcal{I}_N$ and $\mathcal{F}_N$; for the finely gridded tones example,
one can select Ψ as an oversampled DFT matrix. In scenarios involving radar
imaging, a coefficient vector x might encode sparse target positions over a
spatial grid, and a signal f might represent the concatenated responses from
various transceivers over a uniform range of frequencies; in such a case, Ψ
would correspond to a matrix of sampled complex exponentials depending
on the propagation delays and received frequencies [7].

More generally, there also exist numerous techniques for learning an
effective sparsifying dictionary Ψ from a collection of real-world signals
f [6,106,112,115,122,123,151,181]. Some of these techniques ensure that the
learned dictionary elements obey structured models or that the learned dictionary matrix Ψ and its conjugate transpose Ψ^* have efficient computational
implementations (without requiring explicit matrix multiplications).

If $D > N$, an $N \times D$ dictionary Ψ is said to be *overcomplete* or *redundant*.
The columns of Ψ are not assumed to be orthonormal, and if the dictionary
is overcomplete, the columns of Ψ simply cannot all be orthonormal. For
a coefficient vector $x \in \mathbb{R}^D$ or $\mathbb{C}^D$, we can again use (1.1) to represent the
synthesis of f using the coefficients x, and we can make (1.3) compatible with
general dictionaries simply by changing the upper limit of summation from
N to D:

$$f = \sum_{n=1}^{D} x_n \Psi_n. \tag{1.5}$$

If x is a K-sparse vector, then we again see that f can be written as a weighted
sum of just K dictionary elements (or columns) from Ψ since just K terms in
the summation (1.5) will be nonzero.

For a given signal f, however, the question of how to *find* a sparse coefficient vector x that satisfies $f = \Psi x$ can be much more complicated in the case of a general dictionary. The difficulty is typically not in finding just any x that can be used to synthesize f; in fact if Ψ is overcomplete, there exist an infinite number of vectors x that can be used to synthesize a given signal f via (1.1) or (1.5). If Ψ is a special type of dictionary known as a *tight frame* (note that every orthonormal basis is a tight frame), one can in fact still use (1.2) or (1.4) (with a possible need for rescaling) to compute one valid set of coefficients x. The difficulty is that the coefficients produced by (1.2) or (1.4) will not necessarily be sparse; that is, even if f is originally generated using a sparse vector x, these sparse coefficients may not be recoverable simply by computing $\Psi^* f$. A considerable amount of research that predates even CS has focused on how to find, for a given signal f and overcomplete dictionary Ψ, a set of coefficients that can be used to synthesize f and are as sparse as possible [28,67,68, 163,165]. In general, the more correlated the elements of the dictionary are, the harder it is to solve this sparse approximation problem. Insight gleaned from research into this problem has inspired several of the popular techniques for finding sparse solutions x to the CS problem (1.14). Section 1.4 reviews a number of algorithms for solving the CS problem, and so we do not elaborate on how to solve the sparse approximation problem here.

1.2.3.2 Analysis Sparsity

In other settings, the *synthesis* viewpoint may simply be inappropriate for describing the structure of a signal in a concise way. Rather than expecting that a signal be explicitly *built* as a sparse sum of elements from some dictionary, it may be more appropriate to take an *analysis* viewpoint and expect that a signal produce a sparse set of coefficients when it is correlated (using inner products) against the dictionary elements [83]. That is, for a given signal $f \in \mathbb{R}^N$ or $\mathbb{C}^N$ and real- or complex-valued $N \times D$ dictionary matrix Ψ, we may expect that the coefficients $x \in \mathbb{R}^D$ or $\mathbb{C}^D$ computed using (1.2) or (1.4) will be sparse or compressible. When Ψ is an orthonormal basis, the assumptions of synthesis sparsity and analysis sparsity are exactly equivalent. However, when Ψ is overcomplete, these assumptions are different in general. The choice of whether a signal is better described using a synthesis sparsity model or an analysis sparsity model can depend on the context. Synthesis sparsity can be preferable in cases where one views the dictionary elements as *ingredients* that actually comprise the signal; analysis sparsity can be preferable in cases where the dictionary is merely a set of vectors used for *looking at* the signal.

Some example overcomplete dictionaries that can be used for analysis sparsity models include

- The oversampled DFT [36], where one assumes that the zero-padded DFT of the signal f produces a sparse set of coefficients

- The Gabor dictionary [125], where one assumes that the time-frequency analysis of the signal f is sparse
- The curvelet dictionary [35], where one assumes an image f is piece-wise smooth with edges that trace out smooth curves (curvelets provide superior compressible representations compared with wavelets for certain image classes)

One other important example of an analysis sparsity model is the use of *total variation (TV) regularization* to encourage reconstruction of piecewise smooth images [41,121,138,153]. In this setting one lets $\Psi^* = \nabla$, a discrete gradient operator. A quantity known as the *TV norm* is typically used in TV regularization; the TV norm of an image is defined to be the ℓ_1 norm of its gradient. Thus, an image f will tend to have a small TV norm if its gradient $\Psi^*f = \nabla f$ is sparse or compressible; this will occur if the intensity changes in the image are localized along contours such as edges.

We note that one way the study of analysis sparsity has been formalized has been in the context of a proposed *cosparse analysis* model. A signal f is said to be cosparse with respect to some dictionary Ψ if the analysis coefficients $x = \Psi^*f$ contain many zero values. This is closely related to the description of analysis sparsity given earlier, but in the cosparse analysis model the emphasis is particularly on the zeros in x. For a further discussion of this model, the reader is referred to [134].

1.2.4 Extensions of Sparse Models

Much of the work in CS uses the basic sparse models that we have described earlier. However, there are many possible extensions to sparse models, and while a detailed discussion of these extensions is far beyond the scope of this chapter, we highlight a few of the most common.

1.2.4.1 Structured Sparsity Models

As mentioned briefly in Section 1.1.1, K-sparse signal models are nonlinear. Geometrically, the set of all signals f that are K-sparse in an orthonormal basis Ψ consists of a union of $\binom{N}{K}$ different K-dimensional subspaces within the ambient signal space $\mathbb{R}^N$ or $\mathbb{C}^N$. Each subspace is spanned by a different choice of K columns from Ψ. $\left(\text{If } \Psi \text{ is a generic } N \times D \text{ dictionary, the same statements hold true, except the total number of subspaces is } \binom{D}{K}.\right)$

In some cases, it may be possible to refine a sparse model by excluding certain coefficient patterns that are not plausible. For example, the structure of the signal f and the basis Ψ might dictate that the nonzero elements of a coefficient vector x will tend to cluster into a small number of contiguous groups. In cases such as this, it could be reasonable to exclude from consideration any coefficient vectors x whose supports contain isolated indices [11,85]. In some other scenarios, such as when a wavelet basis is chosen for Ψ, it may be that the nonzero elements of a coefficient vector x will tend to cluster along a connected tree structure. (Wavelet transforms produce sets of coefficients that can be naturally organized in a tree structure.) In cases such as this, it could be reasonable to exclude from consideration any coefficient vectors x whose supports correspond to disconnected trees [11].

Geometrically, these types of exclusions reduce the complexity of a sparse model by eliminating certain subspaces (from among the $\binom{N}{K}$ original ones) from consideration. Algorithms can be adopted to exploit the structure of this reduced union of subspaces, and when these algorithms are successfully employed, the CS reconstruction problem can typically be solved using fewer measurements than would be required using a conventional algorithm suitable for unstructured sparsity models. The interested reader is referred to [9,11,21,24,78,86] for more information.

We also note that, as argued in [117,118], certain classes of parametric analog signals (such as spike trains and piecewise polynomial signals) can be viewed as coming from an infinite union of subspaces. These works derive conditions under which a linear sampling strategy can provide an invertible and stable representation of the signals belonging to this union. In a similar vein, a number of papers have focused on sampling signals having what is known as a finite rate of innovation (FROI) [18,75,94,116,126,176]. FROI signals are those that depend only on a finite number of parameters per unit time. Many (but not all) FROI signal classes can be viewed as corresponding to a union of subspaces model. Conventional problems of interest for processing FROI signals include designing sampling kernels to produce digital measurements, estimating the unknown parameters from low-rate measurements, and oversampling such signals to make the estimation process robust to noise. FROI sampling is also closely related to the broad framework for compressively sampling analog signals known generically as Xampling [131,132] and embodied specifically in devices such as the modulated wideband convertor (MWC) [132]. We discuss an application of the MWC in Section 1.3.3.

1.2.4.2 Statistical Sparsity Models

While our focus in this chapter has been on deterministic signal models, it is also possible to formulate probabilistic models in which compressible signals have high likelihood and noncompressible signals do not. One prototypical assumption is that the elements of the coefficient vector x are drawn according to the Laplace distribution [107]; under this assumption, the log-likelihood of a coefficient vector is proportional to its ℓ_1 norm.* Due to this fact, finding the maximum a posterior (MAP) estimate of a coefficient vector can involve solving problems analogous to (1.23). In practice, more complicated hierarchical priors and mixture-of-Gaussian models—and the corresponding signal recovery algorithms—can lead to improved performance [8,52,104,107].

1.2.4.3 Beyond Sparsity

Not all concise models for signal structure fit neatly into the framework of sparse signal representations. To give the reader a sense of perspective, we briefly mention a few alternatives to sparse models.

Some signals could depend explicitly and nonlinearly on a small number of continuous-valued parameters; this differs from a sparse model in that the set of parameters is fixed but the dependence on these parameters is nonlinear. As an example, one might consider an image of a fixed object but where the position and orientation of the camera are degrees of freedom; as another example, one might consider a radar pulse where the frequency and time-of-arrival are degrees of freedom. Geometrically, such signal families correspond not to unions of subspaces but rather low-dimensional *manifolds* embedded in the ambient high-dimensional signal space [69]. For a random measurement matrix Φ with a sufficient number of rows, one can ensure that $\|\Phi f\|_2 \approx \|f\|_2$ holds for all signals f living along a smooth, low-dimensional manifold [12]. This fact is analogous to the restricted isometry property (RIP) for sparse signals (see Section 1.3.1), and so it implies that manifold models can be used as an alternative to sparse models for signal recovery in CS. Algorithms have been proposed for manifold-based signal recovery [21,60,157], but these tend to be less general purpose than algorithms for sparse signal recovery. We also note that the mixture-of-Gaussian models mentioned earlier can be used to construct piecewise linear approximations to signal manifolds and thus it is possible to develop statistical methods for manifold-based signal recovery [52].

Other models can be considered for describing the structure within an ensemble of signals, such as might be acquired using a sensor network. Some models for such ensembles can be expressed in terms of sparsity. For example, one may assume that all signals in the ensemble are sparse in the same basis and that the supports of the coefficient vectors are the same [13,58]. Such models have connections to block sparsity and unions of subspaces [86], and when the signals in the ensemble are stacked into a matrix, certain matrix norms can be proposed that are small when the signals in the ensemble share a common sparse support. Algorithms for recovering a signal

---

* Despite the prevalence of the Laplacian prior assumption, vectors x generated according to this distribution will not have entries that decay rapidly enough to be considered compressible. The reader may refer to [46] for further discussion.

ensemble from compressive measurements (often referred to as the *multiple measurement* vector [MMV] problem) can be formulated by minimizing these matrix norms [58,86,168] or by employing various greedy and iterative strategies [13,166]. Such an approach has been successfully employed in radar imaging to deal with ghosting caused by multipath effects [7]; this work modeled the measurements as a superposition of sparse signals (corresponding to different paths) but with each signal generated from a target coefficient vector with the same sparsity pattern.

Finally, other types of matrix-valued data sets may be well modeled by assuming the data matrix has low rank [37,38,147]. This is equivalent to assuming that all columns of the data matrix live along a common low-dimensional subspace, but when dealing with matrix-valued data, this subspace may not be known in advance and it may not be aligned with the elements of any particular sparse dictionary. Although the technical details differ, low-rank matrix models have many analogous properties to sparse signal models and they also support RIP-like embeddings under certain random measurement protocols [147]. A variety of techniques have been proposed for recovering low-rank matrices from partial information, either from a random set of observed entries of the matrix or from a collection of random (e.g., Gaussian) measurements of the matrix. The most common technique involves a convex optimization program in which the nuclear norm of the matrix is minimized in order to encourage a low-rank solution [38,147]. (Note that the nuclear norm of a matrix equals the ℓ_1 norm of the singular values of the matrix.) Greedy and iterative thresholding algorithms have also been proposed for matrix recovery [29,114].

1.3 Compressive Measurement Protocols

We let $f \in \mathbb{R}^N$ or $\mathbb{C}^N$ denote a vector of samples (or pixels, voxels, etc.) of a signal of interest. Most measurement protocols in CS correspond to collecting a small number of linear measurements of f; often this can be done without explicitly collecting the samples f in the first place. For example, in radar imaging (where, as discussed in Section 1.2.3, f might represent the concatenated responses from various transceivers over a uniform range of frequencies), one could obtain a compressed set of *measurements* of f merely by keeping the responses from some subset of transceivers at some subset of frequencies [7]. In such a scenario, it would not be necessary to ever collect any data from the omitted transceivers or at the omitted frequencies.

When we collect any type of linear measurements of a signal f, we can represent the measurement vector $y \in \mathbb{R}^M$ or $\mathbb{C}^M$ as

$$y = \Phi f + n, \tag{1.6}$$

where Φ is a real- or complex-valued $M \times N$ *measurement matrix* and $n \in \mathbb{R}^M$ or $\mathbb{C}^M$ is a vector of *measurement noise*. Because we are interested in scenarios where the number of measurements M is smaller than the number of samples N, the vector y is often said to contain *compressive measurements* of f. Let us denote the rows of the measurement matrix Φ by the length-N vectors $\Phi_1^*, \Phi_2^*, \ldots, \Phi_M^*$. Examining (1.6), we see that each measurement can be written in terms of an inner product:

$$y_m = \Phi_m^* f + n_m = \langle f, \Phi_m \rangle + n_m, \tag{1.7}$$

for $m = 1, 2, \ldots, M$. Thus, we can think of each row of Φ as a *measurement vector*, which is correlated against f to produce a single measurement.

In this section, we outline several possible measurement strategies and discuss the properties of the corresponding measurement matrices. Of particular interest to us are compressive measurement strategies that will facilitate the recovery of *sparse* signals. To gain some intuition for what makes a matrix suitable for measuring a sparse signal, suppose a signal f was K-sparse in some orthonormal basis Ψ and recall the synthesis and analysis Equations 1.3 and 1.4. If one knew in advance which K basis functions (out of N total) contributed to synthesizing f, a very effective strategy for collecting compressive measurements of f would simply be to set $M = K$ and choose measurement vectors that correspond exactly to the aforementioned K basis functions. In this way (neglecting noise), one would collect measurements that exactly correspond to the nonzero entries of the sparse coefficient vector x.

Unfortunately, in practice one will often not know the positions of the nonzero entries in a sparse representation for a signal of interest. Armed with a limited budget of measurements, then, it would be wasteful to choose measurement vectors that correspond to basis vectors from Ψ: Many of the resulting measurements would simply be zero. For this reason, when a signal f is sparse in some basis (or dictionary) Ψ, effective compressive measurement strategies involve measurement vectors that are very different from the basis vectors that comprise Ψ. These can be obtained in a variety of ways.

1.3.1 Random Gaussian and Subgaussian Matrices

One effective strategy for designing a compressive measurement matrix Φ is simply to populate it with random numbers. Intuitively, a vector of random numbers will, with high probability, be highly uncorrelated with all basis vectors in any fixed sparse basis. The most common example of this is to design the matrix with i.i.d. Gaussian entries, each with mean zero and variance $1/M$. (This variance is chosen merely to ensure the measurements are normalized so as to be compatible with our discussions in what follows; one could use a different normalization but should be aware that this could affect the sensitivity to measurement noise.) An alternative choice is to use a Rademacher distribution, choosing each entry in the matrix Φ independently to be either $1/\sqrt{M}$ or $-1/\sqrt{M}$, with each value chosen with probability $1/2$. Both of these distributions are examples of what are known as subgaussian distributions [175].

Random measurement matrices populated with subgaussian random variables are very powerful for capturing the information in sparse signals. As we will see in Section 1.4, matrices that satisfy a condition known as the restricted isometry property (RIP) can be proven to allow the recovery of sparse signals via efficient algorithms. As we state later, a random measurement matrix populated with subgaussian random variables will satisfy the RIP with high probability.

Definition 1.1 ([43]) *An $M \times N$ measurement matrix Φ is said to satisfy the RIP of order K with respect to the $N \times N$ orthonormal basis (or $N \times D$ dictionary) Ψ if there exists a constant $\delta_K \in (0, 1)$ such that*

$$(1 - \delta_K) \|x\|_2^2 \leq \|\Phi \Psi x\|_2^2 \leq (1 + \delta_K) \|x\|_2^2 \tag{1.8}$$

holds for all coefficient vectors x with $\|x\|_0 \leq K$. The parameter δ_K is known as the isometry constant of order K.

The RIP is essentially a requirement that in the matrix $A := \Phi \Psi$ any submatrix containing K columns will act as an approximate isometry (its K columns will be approximately orthonormal). While the definition of the RIP is itself a deterministic statement, the most efficient constructions of RIP matrices involve randomness. In fact, it has been shown that even checking whether the RIP holds for a given matrix with a specified isometry constant is NP-hard in general [164]. Fortunately, this property can be guaranteed to hold with very high probability under suitable conditions.

Theorem 1.1 ([10,130]) *Let Ψ be an arbitrary fixed $N \times N$ orthonormal basis in $\mathbb{R}^N$ or $\mathbb{C}^N$, and let Φ be an $M \times N$ measurement matrix populated with i.i.d. subgaussian entries having mean zero and variance $1/M$. If*

$$M \geq C_1 \left(K \log \left(\frac{N}{K} \right) + \log \left(\frac{1}{\rho} \right) \right), \tag{1.9}$$

then with probability at least $1 - \rho$, Φ will satisfy the RIP of order K with respect to Ψ with isometry constant δ_K. Here C_1 depends on δ_K and the type of subgaussian distribution; typically C_1 scales quadratically in $1/\delta_K$.

We see that subgaussian random matrices not only can satisfy the RIP but they can do this with a number of measurements that is essentially proportional to the sparsity level of the signal. (As we will see in Section 1.4, recovery of K-sparse signals is typically guaranteed if Φ satisfies the RIP merely of order cK for some small constant c; for example, $c = 2$ in Theorem 1.7.) Thus, in terms of the number of measurements, the penalty that is paid for not knowing the positions of the nonzero coefficients in advance is a constant times a factor that is logarithmic in the signal dimension N. Another remarkable fact about subgaussian matrices is that they are *universal*: for any fixed orthonormal basis Ψ, a randomly generated Φ will satisfy the RIP with respect to Ψ with high probability. Therefore, although it is necessary to specify a sparse basis for use in signal reconstruction, it is not necessary to specify this basis when designing the measurement process.*

Despite their appealing theoretical properties, subgaussian random matrices do have some practical drawbacks. First, aside from the Rademacher distribution and similar choices of discrete distributions, most subgaussian random variables can take arbitrary real values, and it can be difficult to design physical sensing systems that implement the corresponding matrix Φ. (We do note that in some problems it may be acceptable to collect all of the samples f explicitly using a conventional sensor and then compute the measurements $y = \Phi f$ via matrix multiplication in software; see [13] for one such scenario.) Second, typical reconstruction algorithms in CS require the ability to multiply Φ by an arbitrary length-N vector and to multiply Φ^* by an arbitrary length-M vector. For the sort of unstructured matrices that arise when populating Φ with i.i.d. random entries, there may be no fast way to compute these products other than via explicit matrix multiplication. However, for some of the more structured measurement processes we describe later, there can be fast methods for applying the operators Φ and Φ^* in software, without requiring explicit matrix multiplication. We also note that certain structured measurement matrices, which may not themselves be universal, can be converted into universal measurement matrices by randomizing the columns of the matrix using a random sign sequence; see [110] for more details.

Theorem 1.1 applies when Ψ is an $N \times N$ orthonormal basis. When Ψ is a generic $N \times D$ dictionary, it can be difficult for Φ to satisfy the RIP with respect to Ψ unless the columns of Ψ are approximately orthonormal (see [146] for additional discussion). A related condition to the RIP has been proposed, however, for recovery of signals that are sparse in generic dictionaries.

Definition 1.2 ([36]) *Let Ψ be an arbitrary $N \times D$ dictionary. An $M \times N$ measurement matrix Φ is said to satisfy the Ψ-RIP of order K with isometry constant $\delta_K \in (0, 1)$ if*

---

* One of the hallmarks of CS is that the measurement process can be defined independently of the signal; in situations where some feedback from the reconstruction algorithm is possible, however, there can be advantages to adapting the measurement process over time [103,107].

$$(1 - \delta_K) \|\Psi x\|_2^2 \leq \|\Phi \Psi x\|_2^2 \leq (1 + \delta_K) \|\Psi x\|_2^2 \qquad (1.10)$$

holds for all coefficient vectors x with $\|x\|_0 \leq K$.

This property, which is subtly different from the conventional RIP (compare (1.8) with (1.10)), is much easier to satisfy when Ψ has highly correlated columns. The following theorem follows from essentially the same argument used in [10] to prove Theorem 1.1.

Theorem 1.2 *Let Ψ be an arbitrary fixed $N \times D$ dictionary in $\mathbb{R}^N$ or $\mathbb{C}^N$, and let Φ be an $M \times N$ measurement matrix populated with i.i.d. subgaussian entries having mean zero and variance $1/M$. If*

$$M \geq C_1 \left(K \log \left(\frac{D}{K} \right) + \log \left(\frac{1}{\rho} \right) \right), \qquad (1.11)$$

then with probability at least $1-\rho$, Φ will satisfy the Ψ-RIP of order K with isometry constant δ_K. Here C_1 is the same constant that appears in Theorem 1.1.

1.3.2 Random Sampling in an Orthogonal Basis

Another strategy for designing a compressive measurement matrix is to draw the measurement vectors explicitly from some basis. Consider an $N \times N$ orthonormal matrix U with columns $u_1, u_2, \ldots, u_N$ and suppose the measurement vectors $\Phi_1, \Phi_2, \ldots, \Phi_M$ are selected as a random set of M columns from U. We can think of U as a *measurement basis* and the measurements $y = \Phi f + n$ as a random set of M coefficients that are kept after f is transformed into the measurement basis (i.e., y contains a randomly selected, noisy set of entries of $U^* f$). In some cases, these measurements can be computed without explicitly acquiring all entries of f or computing the entire set of coefficients $U^* f$.

One common choice for U is $U = \mathcal{I}_N$, the $N \times N$ identity matrix. In this case, y consists of a randomly selected, noisy set of entries of f. To make this concrete, if we had $M = N/10$ and f were an image, y would correspond to a random set of 10% of the pixel values in the image. Another common choice for U is $U = \mathcal{F}_N$, the $N \times N$ normalized DFT matrix. In this case, y consists of a randomly selected, noisy set of DFT coefficients of f. For both of these choices of U, the resulting matrix Φ and its conjugate transpose Φ^* have efficient implementations (without requiring explicit matrix multiplications). This can help in developing fast algorithms for solving the reconstruction problem.

For this measurement strategy to be effective, the measurement basis U must be sufficiently different from the sparsity basis Ψ. The difference between U and Ψ can be quantified as follows.

Definition 1.3 *For a pair of $N \times N$ orthonormal matrices U and Ψ, the mutual coherence $\mu(U, \Psi)$ between U and Ψ is the maximum absolute correlation between a column of U and a column of Ψ:*

$$\mu(U, \Psi) = \max_{1 \leq m,n \leq N} |\langle u_m, \Psi_n \rangle|. \tag{1.12}$$

The mutual coherence between any two $N \times N$ orthonormal matrices can be no smaller than $1/\sqrt{N}$, and this lower bound is achieved when $U = \mathcal{I}_N$ and $\Psi = \mathcal{F}_N$ (or vice versa). The mutual coherence can be no larger than 1, and this upper bound is achieved when $U = \Psi$. Sampling from a measurement basis that is incoherent with a signal's sparsity basis can be an effective measurement strategy, as the following theorem demonstrates.

Theorem 1.3 ([144,152]) *Let U be an $N \times N$ real or complex orthonormal matrix and suppose the measurement vectors $\Phi_1, \Phi_2, \ldots, \Phi_M$ are selected as a random set of M columns from U. Let Ψ be an arbitrary fixed $N \times N$ orthonormal basis for $\mathbb{R}^N$ or $\mathbb{C}^N$. If*

$$M \geq \frac{C_2 K (\sqrt{N}\mu(U, \Psi))^2 (\log^4(N) + \log(1/\rho))}{\delta_K^2}, \tag{1.13}$$

then with probability at least $1 - \rho$, $\left(1/\sqrt{M}\right)\Phi$ will satisfy the RIP of order K with respect to Ψ with isometry constant $\delta_K < 1/2$. Here C_2 is an absolute constant.

When the coherence between U and Ψ is small, the term $(\sqrt{N}\mu(U, \Psi))^2$ in (1.13) will be approximately 1, and thus the requisite number of measurements M to satisfy the RIP of order K will again scale essentially like the sparsity level K times additional logarithmic factors. Finally, we note that when there are a small number of highly correlated atoms between U and Ψ (but most of the atoms in U are incoherent with most of the atoms in Ψ), it can still be possible to construct efficient measurement matrices by modifying the way that columns are selected from U. The interested reader is referred to [111] for more details.

1.3.3 Measurement Systems

A growing number of hardware systems have been developed for collecting compressive measurements. In most cases, the physics and structure of the measurement process constrain the type of measurements that can be collected, and this imposes a certain type of structure on the measurement matrix Φ. Typically, though, some amount of randomness can also still be incorporated into the sensing matrix. For an extensive review of structured measurement systems and their theoretical properties, the reader is referred to [78,144].

1.3.3.1 One-Dimensional Signals

One of the conceptually simplest strategies for CS measurement involves collecting random samples of a signal in the time domain, as described in Section 1.3.2 with the choice of $U = \mathcal{I}_N$. Such measurements are well suited to capturing signals with a sparse spectrum at average sampling rates well below the Nyquist limit. One architecture for collecting such samples is described in [177]; this architecture involves a customized sample-and-hold circuit fed with a nonuniform clock signal. In this implementation, samples are not quite chosen uniformly at random: The clock signal is generated using a pseudorandom bit sequence (PRBS), and the gap between consecutive samples is restricted to be neither too short nor too long. This device was used to successfully digitize an 800 MHz to 2 GHz band (having 100 MHz total of noncontiguous spectral content) at an average sample rate of just 236 Msps. For other discussions concerning random sampling in the time domain, the reader is referred to [95,142].

Several other architectures have been proposed for collecting compressive measurements by combining elementary operations such as modulation, filtering, multiplexing, and low-rate sampling. The random demodulator (RD) involves multiplying the incoming signal by a PRBS, lowpass filtering the result, and collecting low-rate samples of the output [113]. The resulting Φ matrix has a banded structure and, with high probability, will satisfy the RIP with respect to the sparsity basis $\Psi = \mathcal{F}_N$ [169]. The random-modulation preintegrator (RMPI) is essentially a multichannel implementation of the RD [51,185]. These devices can be used not only to capture spectrally sparse signals but also to capture signals with a sparse time–frequency profile. A hardware implementation [186] of the RMPI has been validated using a $13\times$ sub-Nyquist measurement rate to capture radar pulse parameters in an instantaneous bandwidth spanning 100 MHz to 2.5 GHz. The compressive mutliplexer (CMUX) is another multichannel architecture [159]. The CMUX involves breaking the signal into bands, randomly modulating each band, and adding the bands back together before sampling with a single analog-to-digital converter (ADC). The resulting Φ matrix can be shown to satisfy the RIP for signals with sparse multichannel spectra (see [159] for details). The modulated wideband converter (MWC) is based on an acquisition strategy known as Xampling [131,132]. A hardware prototype described in [132] can be used for digitizing signals with up to a 2 GHz Nyquist rate (having 120 MHz total of noncontiguous spectral content) at an average sample rate of just 280 Msps.

Before moving on to other measurement strategies and signal models, we provide a note of caution concerning the recovery of signals with sparse spectra. In most practical settings where signals with sparse spectra would be encountered (and devices such as those given earlier would be used), the original signal would be an *analog* signal and sparsity would be assumed to exist in that signal's continuous-time Fourier transform. However, this does not necessarily translate into a sparse spectrum when a finite, discrete sample vector f is expanded in the DFT basis: Due to the effects of sampling and time limiting, the original spectrum will often be smeared out, a phenomenon commonly known as *spectral leakage*. The failure of the DFT to efficiently capture the signal's degrees of freedom can be problematic for CS recovery [54]. There have been several strategies proposed for dealing with this problem, including the use of smooth windowing in time to minimize smearing in frequency [169,177], the use of alternative dictionaries Ψ (instead of the DFT) for reconstruction [62,76], and the use of specialized measurement and reconstruction strategies [88,131,132].

Moving on, a different conceptual idea for collecting compressive measurements involves filtering the incoming signal using a filter with randomly chosen impulse response [150,170] and sampling the output at a low rate. The resulting Φ matrix will contain a selection of rows from a random Toeplitz matrix, where the impulse response of the filter appears in each row of the matrix. It has been shown that this Φ matrix can satisfy the RIP with respect to the sparsity basis $\Psi = \mathcal{I}_N$ [109,145], so such a random convolution process can be used to capture signals that are sparse in the time domain. One important application of this fact is in channel sensing [102,154], where the goal is to identify the impulse response of an unknown communication channel. If the channel can be modeled as a linear, time-invariant system with a sparse impulse response (e.g., if there are a few discrete reflectors in a multipath communication environment), then the channel can be identified by sending a randomly generated input signal and taking a small number of samples of the received signal. (Since convolution is commutative, this is just exchanging the roles of the input signal and impulse response; if one is random and known, and the other is sparse and unknown, then the sparse signal can be recovered from a small number of output samples.)

1.3.3.2 Images and Higher-Dimensional Signals

One intriguing application area of CS is in imaging. For an extensive review of possible applications of CS in optical imaging systems, the reader is referred to [179]. Some imaging architectures that predate CS can be viewed through the lens of compressive measurements. For example, tomography, magnetic resonance imaging (MRI), and some microscopy systems effectively measure samples of an image in the two-dimensional Fourier plane [41,93,121]. While these samples may not always be chosen randomly (though they sometimes can be), they are often incomplete: in order to construct an image from the acquired data, it is often necessary to interpolate the missing samples in the Fourier domain. Using the same sort of principles that apply in CS, one can argue that if the unknown image is sparse in the space domain or if it has a sparse gradient (small TV norm), it may be possible to accurately reconstruct the image from a limited number of Fourier-domain samples. The feasibility of using CS techniques to accelerate pediatric MRI has been confirmed in a clinical setting [174]. Random frequency-domain sampling has also been proposed for ground penetrating radars when imaging spaces with a small number of point-like targets [100]. CS techniques for through-the-wall radar imaging are discussed in [7], and additional applications of CS in radar imaging are described in [143].

Other imaging architectures have been proposed exclusively for CS. One proposed architecture [77] for a digital camera involves randomly modulating the incident light field using an array of digital micromirrors in random positions, focusing the light onto a single photodetector, and recording a single measurement at a time. By putting different patterns on the mirrors, different measurements can be collected serially in time; each row of the Φ matrix contains a random binary sequence. A basis U of binary functions known as noiselets [55] can be used as an effective sequence of measurement patterns. Because this device requires only a single light sensing element, it has been nicknamed the *single-pixel camera*, and it can be economical in situations (such as infrared or terahertz imaging [48]) where building a high-resolution sensor array would be expensive. A similar architecture has been proposed for compressive confocal microscopy [183].

A collection of proposed imaging strategies involve the use of randomly designed coded apertures for modulating an image or its Fourier transform (or both) before sampling [128,129,150,160]. Because this modulation (in the Fourier or spatial domain) corresponds to a convolution in the complementary (spatial or Fourier) domain, the resulting Φ matrix will possess a Toeplitz-like structure and is amenable to RIP-based analysis. A random convolution has also been implemented in a CMOS imager [105]. A somewhat related architecture involving a random lens is described in [89].

Several architectures for CS imaging—including the single-pixel camera [140] and coded aperture systems [127]—can be easily extended for acquisition of a moving scene (video) by serially collecting random measurements in time, although the reconstruction process can require a significantly larger amount of computation. In multisignal CS problems such as video acquisition (where each frame could be viewed as a signal) or when CS is employed on an ensemble of signals in a sensor network, the resulting Φ matrix can possess a block diagonal structure. Such matrices can again satisfy the RIP but are most effective only for certain sparsity bases Ψ [81]. When measuring multidimensional signals (such as images [148] or hyperspectral signals [162]) one dimension at a time, the resulting Φ matrix can possess a separability property. The implications of this separability are described in [78].

1.4 Sparse Signal Recovery Algorithms and Guarantees

We now address the problem of how a sparse model can be used to recover a signal f from a vector $y = \Phi f + n$ of compressive measurements. Let us suppose f is K-sparse or compressible in some orthonormal basis Ψ. (We address the case of general dictionaries in Section 1.4.5.) Thus, we can write

$$y = \Phi f + n = \Phi \Psi x + n = Ax + n, \qquad (1.14)$$

where
  y is a length-M measurement vector,
  $A := \Phi \Psi$ is an $M \times N$ matrix,
  x is a K-sparse or compressible vector of length N, and
  n is a length-M vector of measurement noise.

These vectors and matrices can be real- or complex-valued, and our discussion will apply to both cases except where noted. Most CS recovery algorithms can be interpreted as solving for a sparse vector $\widehat{x}$ that satisfies $y \approx A\widehat{x}$ as closely as possible. Once the sparse coefficient vector has been estimated, one can synthesize a signal estimate via multiplication by Ψ: $\widehat{f} = \Psi\widehat{x}$.

1.4.1 Preliminaries

Performance guarantees for many CS recovery algorithms depend on the sparsity basis Ψ and measurement matrix Φ only through the properties of their product: $A = \Phi\Psi$. In Section 1.3, we discussed the condition under which a measurement matrix Φ satisfies the RIP with respect to a sparsity basis Ψ (Definition 1.1). However, this condition is equivalent to requiring merely that

$$(1 - \delta_K) \, \|x\|_2^2 \le \|Ax\|_2^2 \le (1 + \delta_K) \, \|x\|_2^2 \qquad (1.15)$$

holds for all coefficient vectors x with $\|x\|_0 \le K$. Because this requirement depends only on $A = \Phi\Psi$, we refer to this condition simply as A satisfying the RIP (of order K and with isometry constant δ_K).

  A second property of A will be useful for our discussions given later.

Definition 1.4   *The coherence of a matrix A with columns $a_1, a_2, \ldots, a_N$ is the maximum normalized inner product between any two distinct columns of A:*

$$\mu(A) := \max_{1 \le m, n \le N, \, m \ne n} \frac{|\langle a_m, a_n \rangle|}{\|a_m\|_2 \, \|a_n\|_2}. \qquad (1.16)$$

The coherence of a single matrix (Definition 1.4) is not to be confused with the mutual coherence between a pair of orthonormal bases (Definition 1.3), although the two concepts are related: If U and Ψ are orthonormal $N \times N$ matrices, the mutual coherence $\mu(U, \Psi)$ between U and Ψ equals the coherence $\mu([U \ \Psi])$ of the $N \times 2N$ concatenation of the two.

The coherence of an $M \times N$ matrix A can be no smaller than $\sqrt{N - M/M(N - 1)}$ (which scales like $1/\sqrt{M}$ when $M \ll N$) and no larger than 1 [161]. A larger coherence indicates the existence of at least one pair of highly correlated columns in A. Intuitively, such correlations make it difficult to recover the correct sparse solution to Equation 1.14. The reason for this is that when x is sparse, Ax will be a linear combination of a few columns of A. When A has high coherence, it can be difficult to identify the correct columns from the compressive measurements. Effective compressive measurement strategies (see Section 1.3) ensure that $A = \Phi\Psi$ has low coherence. Indeed, coherence is related to the RIP. For example, if the columns of A are normalized, then A will satisfy the RIP of order K as long as $\mu(A) < 1/K$ [31]. However, coherence is generally a less sharp analytical tool when one seeks to minimize the number of measurements M required for efficient signal recovery.

To further discuss the intuition underlying CS, let us assume for the moment that the measurements are noise-free ($n = 0$). Recall that when $M < N$, the system of equations $y = Ax$ has an infinite number of candidate solutions. However, when the coherence of A is sufficiently small (how small depends on the sparsity level of x), one can actually guarantee that $y = Ax$ has a *unique sparse solution*. This is stated formally in the following theorem.

Theorem 1.4 ([82]) *Suppose x is K-sparse and let $y = Ax$. If*

$$\mu(A) < \frac{1}{2K - 1},\tag{1.17}$$

then $\widehat{x} = x$ is the unique solution to $y = A\widehat{x}$ having sparsity K or less.

A related statement holds for the RIP.

Theorem 1.5 *Suppose x is K-sparse and let $y = Ax$. If A satisfies the RIP of order $2K$ with isometry constant $\delta_{2K} < 1$, then $\widehat{x} = x$ is the unique solution to $y = A\widehat{x}$ having sparsity K or less.*

The practical question is how to actually *find* a sparse solution given the matrix A and the measurements $y = Ax$. Theorems 1.4 and 1.5 ensure that if A has sufficiently low coherence or satisfies the RIP of order $2K$ with any isometry constant,* then in principle the original K-sparse coefficient vector x could be recovered from the measurements $y = Ax$ by solving the following problem:

$$\widehat{x} = \arg\min_{x'} \|x'\|_0 \text{ subject to } y = Ax'. \qquad (1.18)$$

Unfortunately, this ℓ_0-minimization problem is known to be NP-hard in general [135,164]. Fortunately, under slightly stronger conditions on A, the correct K-sparse coefficient vector can be recovered using tractable optimization-based or greedy algorithms, of which we outline several in the following sections. For additional surveys and discussion of possible methods for finding sparse solutions to linear inverse problems, the reader is referred to [156,171,187].

1.4.2 Optimization-Based Recovery from Noise-Free Measurements

1.4.2.1 Problem Formulation

As discussed in Section 1.2.1, the ℓ_1 norm tends to be small for sparse signals. The ℓ_1 norm is also a convex function, and so this leads naturally to the formulation of a tractable convex optimization problem for attempting to solve (1.14). In the absence of noise (i.e., when $n = 0$), one can solve[†]

$$\widehat{x} = \arg\min_{x'} \|x'\|_1 \text{ subject to } y = Ax'. \qquad (1.19)$$

This optimization problem is commonly known as *basis pursuit (BP)* [40,53,66].

1.4.2.2 Performance Guarantees

As stated in the following theorem, in the absence of noise, BP can exactly recover sparse vectors if A has sufficiently low coherence.

Theorem 1.6 ([67]) *Suppose A satisfies the coherence bound (1.17). Then for any K-sparse vector x, (1.19) will correctly return $\widehat{x} = x$ given the noise-free measurements $y = Ax$.*

---

[*] The critical fact that underlies the proof of both Theorems 1.4 and 1.5 is that if $\mu(A) < 1/(2K-1)$ or if A satisfies the RIP of order $2K$ with isometry constant $\delta_{2K} < 1$, then any set of $2K$ columns in A will be linearly independent. As long as this condition is satisfied, then solving (1.18) will return the unique and correct K-sparse solution.

[†] In certain cases, the optimal solution for this minimization problem (and others appearing in this chapter) will not be unique. In such an event, one could pick $\widehat{x}$ to be any feasible vector that minimizes the objective function.

Unfortunately, for $A = \Phi\Psi$ to achieve the coherence required in (1.17), one would require a number of measurements M that scales quadratically in the sparsity level K [167]. Fortunately, a smaller requirement on M can be obtained using the RIP.*

Theorem 1.7 ([91]; see also [34]) *Suppose A satisfies the RIP of order $2K$ with isometry constant $\delta_{2K} < 0.4651$. Then for any K-sparse vector x, (1.19) will correctly return $\widehat{x} = x$ given the noise-free measurements $y = Ax$.*

As discussed in Section 1.3, for many random constructions of the measurement matrix Φ, $A = \Phi\Psi$ will satisfy the RIP of order $2K$ and with prescribed isometry constant δ_{2K} with high probability as long as the number of measurements M scales with the sparsity level K times additional logarithmic factors. Combined with Theorem 1.7, this means that exact recovery of K-sparse signals is possible in CS using a number of measurements that scales like $K\log(N/K)$. For an intimate analysis of the recovery problem (including sharp bounds on the number of measurements M) for certain distributions of the random matrix Φ (including i.i.d. Gaussian), the reader is referred to [49,72,152].

In fact, something even stronger is true. Recall that for any vector x we define x_K to be the nearest K-sparse vector to x. When $x_K = x$, x is K-sparse, but when $x_K \approx x$, we say that x is compressible.

Theorem 1.8 ([91]; see also [34]) *Suppose A satisfies the RIP of order $2K$ with isometry constant $\delta_{2K} < 0.4651$. Let $y = Ax$ be noise-free measurements of any vector x. Then the solution $\widehat{x}$ to (1.19) will obey*

$$\|x - \widehat{x}\|_2 \leq C_3 \frac{\|x - x_K\|_1}{\sqrt{K}}, \tag{1.20}$$

where C_3 depends only on δ_{2K}.

As we have discussed, after a sparse coefficient vector has been estimated, one can synthesize a signal estimate via multiplication by Ψ: $\widehat{f} = \Psi\widehat{x}$. If Ψ is an orthonormal basis, the error in the signal domain will equal the error in the coefficient domain:

$$\left\|f - \widehat{f}\right\|_2 = \|\Psi x - \Psi\widehat{x}\|_2 = \|\Psi(x - \widehat{x})\|_2 = \|x - \widehat{x}\|_2. \tag{1.21}$$

Theorem 1.8 is a significant generalization of Theorem 1.7 and implies that approximate recovery of compressible signals (specifically, those close to

---

* While much of our discussion focuses on the recovery guarantees that can be derived using the RIP, in practice the RIP is stronger than what is required for reconstructing many signals; we refer the interested reader to [19,20,64,74,133] for additional discussion.

K-sparse signals) can also be possible in CS using a number of measurements that scales essentially like K. This theorem also provides useful insight in situations where x is sparse but its sparsity level is unknown. As we have previously discussed, depending on how Φ is generated, a matrix $A = \Phi\Psi$ with M rows may satisfy the RIP up to an order K satisfying (up to constants) $K \approx M/\log(N/K)$. Theorem 1.8 states that the recovery error will be small if the sparsity level of x is up to this same order.

In practice, the performance of BP can often be improved by using an iterative reweighting strategy, where the ℓ_1-minimization problem (1.19) is solved multiple times, and entries of x that appear to be significant based on a previous estimate are down-weighted in the ℓ_1 norm used in the next iteration. Further details are available in [45,50,136,180].

1.4.2.3 Computational Considerations

When x is real valued, BP can be cast as a linear program. When x is complex valued, BP can be cast as a second-order cone program (SOCP). Standard techniques from convex optimization, such as the simplex method or interior point methods, can be used for solving these problems [27]. Popular software packages include ℓ_1-MAGIC [39] (for solving a particular set of sparse recovery problems) and CVX [98,99] (for solving general convex optimization problems of modest size). For BP, ℓ_1-MAGIC is restricted to real-valued vectors x, while CVX can be used for real-valued vectors or, if declared as such, complex-valued vectors.

However, it can also be possible to solve the BP problem more efficiently using software designed explicitly to exploit its structure. For real-valued vectors x, the homotopy algorithm [80,139] is one such technique and exploits the similarity between BP and the problem (1.23) as the parameter λ approaches (but does not equal) 0. Another such technique for real-valued vectors utilizes linearized Bregman iterations [184]. Finally, a spectral projected-gradient algorithm (SPGL1) [3,173] and an algorithm based on the alternating direction method (YALL1) [182,189] are available for solving BP in both the real and the complex cases.

1.4.3 Optimization-Based Recovery from Noisy Measurements

1.4.3.1 Problem Formulation

In the presence of noise, one can loosen the equality constraint in (1.19) and solve instead

$$\widehat{x} = \arg\min_{x'} \left\| x' \right\|_1 \text{ subject to } \left\| y - Ax' \right\|_2 \leq \eta, \tag{1.22}$$

where η is a parameter that reflects the anticipated level of measurement noise. This optimization problem is commonly known as *basis pursuit*

de-noising (BPDN) [42,53,172]. Though we focus on BPDN and related problems in this section, other optimization-based strategies such as the Dantzig
selector [33] have been proposed for recovering sparse vectors from noisy
measurements.

1.4.3.2 Performance Guarantees

As stated in the following theorem, BPDN can provide stable recovery of
sparse and compressible signals if A satisfies the RIP.

Theorem 1.9 ([91]; see also [34]) *Suppose A satisfies the RIP of order $2K$ with
isometry constant $\delta_{2K} < 0.4651$. Let $y = Ax + n$ be noisy measurements of any
vector x. If $\eta \geq \|n\|_2$, then the solution $\widehat{x}$ to (1.22) will obey*

$$\|x - \widehat{x}\|_2 \leq C_3 \frac{\|x - x_K\|_1}{\sqrt{K}} + C_4 \eta,$$

where C_3 is as specified in Theorem 1.8 and C_4 depends only on δ_{2K}.

Theorem 1.9 implies both Theorem 1.7 and Theorem 1.8.

1.4.3.3 Computational Considerations

Standard techniques from convex optimization [27,39,98,99] can be used to
solve the BPDN problem (1.22). However, better performance can again be
achieved using algorithms designed explicitly to exploit its structure. Major
work along these lines has included SPGL1 [3,173], an algorithm inspired
by Nesterov (NESTA) [1,16], a formulation of the problem in conic form
(TFOCS) [5,17], the SPArse Modeling Software (SPAMS) toolbox [2,122,123],
and YALL1 [182,189]. Most of these techniques can accommodate both real
and complex matrices and vectors.

For a given value of $\eta > 0$, there typically exists a parameter $\lambda > 0$ such
that the solution returned by (1.22) equals the solution returned by the
unconstrained optimization problem:

$$\widehat{x} = \arg\min_{x'} \frac{1}{2} \|y - Ax'\|_2^2 + \lambda \|x'\|_1. \tag{1.23}$$

This can be viewed as the Lagrangian form of the BPDN problem. The
parameter λ trades off between the complexity of the solution $\widehat{x}$ and the
distance from $A\widehat{x}$ to the measurement vector y. When λ is large, $\widehat{x}$ will
tend to be more sparse; when λ is small, $\|y - A\widehat{x}\|_2$ will tend to be small.
There are a large number of algorithms available for solving (1.23) efficiently.
Some examples include the homotopy algorithm (LARS) [80,124,139], a gradient projection technique (GPSR) [90], a fixed-point continuation method

(FPC) [101], an approach utilizing linearized Bregman iterations [184], a fast iterative shrinking-thresholding algorithm (FISTA) [14], a coordinate descent technique [92], the SPAMS toolbox [2,122,123], and YALL1 [182,189]. Some of these techniques are limited to real vectors and matrices; others, such as YALL1, can accommodate complex matrices and vectors. For more general discussions of developing fast algorithms for solving convex optimization problems, the reader is referred to [15,56,57].

1.4.3.4 Parameter Selection

In general, choosing the best value of the parameter η (for solving (1.22)) or λ (for solving (1.23)) is a difficult problem. Theorem 1.9 gives its strongest guarantee on BPDN performance if η is chosen exactly equal to $\|n\|_2$. In practice, however, the exact value of $\|n\|_2$ may not be known. Given a prior model (such as a probabilistic distribution) for the noise, one could choose η to be slightly larger than the expected value of $\|n\|_2$ so that Theorem 1.9 will provide a guarantee on the recovery performance. However, in some cases, choosing η to be slightly smaller than this level can give better empirical performance.

Similarly, the mapping between η and λ that makes (1.23) equivalent to (1.22) is unknown in general. Homotopy algorithms can be attractive for solving (1.23) because they trace a path of solutions to (1.23) as the parameter λ varies [80,124,139]. If one has a priori knowledge about the sparsity level of the original coefficient vector x, one can use this solution path to choose the value of λ that returns a solution of the anticipated sparsity. Cross-validation [178] and statistical methods such as the GSURE [84] can also be used for selecting λ.

1.4.4 Greedy Methods

Greedy algorithms for CS reconstruction are an alternative to optimization-based recovery methods. Typical greedy algorithms do not explicitly involve *searching* among a candidate set of solutions to the Equation 1.14. Rather, they attempt to explicitly *build up* a sparse solution to the equation.

1.4.4.1 Orthogonal Matching Pursuit (OMP)

Orthogonal matching pursuit (OMP) is a prototypical greedy algorithm for CS recovery [63,141,165,167]. The intuition behind OMP is roughly as follows. Suppose for a moment that the measurements were noiseless (so that $y = Ax$) and that the matrix A was square and orthonormal. In this case, recovery of x would be straightforward: By simply taking $\hat{x} = A^*y$, we could ensure that $\hat{x} = x$ exactly. That is, to recover x_n we would merely compute an inner product between y and the nth column of A. In practice, of course, we are interested in cases where A has size $M \times N$ with $M < N$, and this prevents the columns of A from being orthonormal. However, as we have discussed, if A satisfies the RIP, then small sets of columns of A will be approximately orthonormal, and if x is K-sparse, then y can be written as a weighted sum of just K columns of A. Consequently, although simply computing A^*y will not be sufficient to perfectly recover x, it can provide a crude initial estimate of x. OMP attempts to correct the errors in this estimate caused by minor correlations among the columns of A. Specifically, OMP relies on an iterative process to identify the support of x one element at a time. At each iteration, a residual vector is correlated against the columns of A, and the position with the largest inner product is added to the support estimate. Candidate values for the entries of x on this support are then computed using a least-squares technique, and the residual vector is updated. A key fact is that the residual will always be orthogonal to the previously chosen columns of A, so a new index will always be added to the support at each iteration.

The OMP algorithm is detailed in Algorithm 1.1. The least-squares problem in the update step can be solved by letting $x^{\ell+1} = (A_{\Lambda^{\ell+1}})^{\dagger}y$ on the indices $\Lambda^{\ell+1}$ (and letting $x^{\ell+1} = 0$ elsewhere); here, the superscript † indicates the matrix pseudoinverse, and $A_{\Lambda^{\ell+1}}$ denotes the $M \times (\ell + 1)$ matrix obtained by restricting A to the columns indexed by $\Lambda^{\ell+1}$. The algorithm can be stopped after K iterations if x is known to be K-sparse. However, the correct support may not always be identified in the first K iterations, and additional iterations (but never more than M total) may be helpful. The stopping criterion can also be based on the residual vector r^{ℓ} and its size in comparison to the anticipated noise n. If the measurements are known to be noiseless, OMP can be stopped when $r^{\ell} = 0$. A fast implementation of the OMP algorithm is available in the SPAMS toolbox [2,122,123].

Algorithm 1.1 Orthogonal matching pursuit (OMP)

input data: matrix A, measurements $y = Ax + n$
input parameters: stopping criterion (possibly based on sparsity level K)
initialize: $r^0 = y$, $x^0 = 0$, $\ell = 0$, $\Lambda^0 = \emptyset$
while stopping criterion not met **do**

match:	$h^{\ell} = A^*r^{\ell}$
identify:	$\Lambda^{\ell+1} = \Lambda^{\ell} \cup \{\arg\max_j |h^{\ell}(j)|\}$
	(if multiple maxima exist, choose only one)
update:	$x^{\ell+1} = \arg\min_{z:\ \text{supp}(z) \subseteq \Lambda^{\ell+1}} \|y - Az\|_2$
	$r^{\ell+1} = y - Ax^{\ell+1}$
	$\ell = \ell + 1$

end while
output: $\widehat{x} = x^{\ell}$

As stated in the following theorem, in the absence of noise, OMP can exactly recover sparse vectors if A has sufficiently low coherence.

Theorem 1.10 ([165]) *Suppose the columns of A have unit norm, and suppose A satisfies the coherence bound (1.17). Then for any K-sparse vector x, after K iterations OMP will correctly return $\widehat{x} = x$ given the noise-free measurements $y = Ax$.*

Unfortunately, as noted before, for $A = \Phi\Psi$ to achieve the coherence required in (1.17), one would require a number of measurements M that scales quadratically in the sparsity level K [167]. Using a different type of analysis, it has been shown that with a number of measurements that scales with $K \log(N)$, any fixed K-sparse signal x can be recovered with high probability if Φ is generated from a suitable random distribution (and independent of x) [167]; note that this is slightly different from conclusions such as that of Theorem 1.10, which guarantees recovery of all K-sparse signals using one A. More recently, RIP-based analysis of OMP has led to similar measurement bounds [188] that do hold for all K-sparse signals and also guarantee robustness with respect to measurement noise.

1.4.4.2 Compressive Sampling Matching Pursuit (CoSaMP)

A more recent algorithm known as compressive sampling matching pursuit (CoSaMP) [137] is a refinement of the OMP idea. Like OMP, CoSaMP attempts to identify a sparse support set that leads to a small residual. Unlike OMP, however, CoSaMP constructs this support set multiple elements at a time, and at each iteration elements can be both added to and removed from the estimated support set. In fact, CoSaMP deliberately overestimates the size of the support set and then estimates the signal coefficients and uses these estimates to prune back the support. The CoSaMP algorithm is detailed in Algorithm 1.2. Stopping criteria and useful variations of this algorithm are discussed in [137]. Strategies for choosing the input parameter K—which is used directly in the execution of the CoSaMP algorithm—are also discussed in [137] for cases where the signal's sparsity level is not known a priori.

As shown in the following theorem CoSaMP is amenable to RIP-based analysis that accommodates compressible signals and measurement noise.

Theorem 1.11 ([137]) *Suppose A satisfies the RIP of order 2K with isometry constant $\delta_{2K} < 0.0125$. Let $y = Ax + n$ be noisy measurements of any vector x. Then the solution $\widehat{x}$ obtained by CoSaMP (Algorithm 1.2) with an appropriately chosen sparsity level and with a sufficient number of iterations will obey*

$$\|x - \widehat{x}\|_2 \leq C_5 \left(\frac{\|x - x_K\|_1}{\sqrt{K}} + \|n\|_2 \right),\qquad(1.24)$$

where C_5 depends only on δ_{2K}.

Algorithm 1.2 Compressive sampling matching pursuit (CoSaMP)

---

input data: matrix A, measurements $y = Ax + n$
input parameters: sparsity level K, stopping criterion
initialize: $r^0 = y$, $x^0 = 0$, $\ell = 0$, $\Lambda^0 = \emptyset$
while stopping criterion not met **do**

match:	$h^\ell = A^* r^\ell$
identify:	$\Omega^\ell = $ positions of largest $2K$ entries in h^ℓ
merge:	$\Lambda^\ell = \Omega^\ell \cup \Gamma^\ell$
update:	$\widetilde{x}^\ell = \arg\min_{z:\, \text{supp}(z) \subseteq \Lambda^\ell} \left\| y - Az \right\|_2$
	$\Gamma^{\ell+1} = $ positions of largest K entries in $\widetilde{x}^\ell$
	$x^{\ell+1} = \widetilde{x}^\ell_K$ (keep largest K entries in $\widetilde{x}^\ell$)
	$r^{\ell+1} = y - Ax^{\ell+1}$
	$\ell = \ell + 1$

end while
output: $\widehat{x} = x^\ell$

A closely related algorithm known as subspace pursuit (SP) [59] has steps similar to CoSaMP and has similar performance guarantees when A satisfies the RIP. Stagewise orthogonal matching pursuit (StOMP) [73] is another effective algorithm that involves selecting multiple coefficients at a time.

1.4.4.3 Iterative Hard Thresholding (IHT)

Another method known as iterative hard thresholding (IHT) [22] also involves the iterative refinement of a support estimate. The IHT algorithm is detailed in Algorithm 1.3. In each iteration of IHT, the residual is correlated against the columns of A, these statistics are scaled and added to the

---

Algorithm 1.3 Iterative hard thresholding (IHT)

---

input data: matrix A, measurements $y = Ax + n$
input parameters: sparsity level K, stepsize γ, stopping criterion
initialize: $x^0 = 0$, $\ell = 0$
while stopping criterion not met **do**

update:	$\widetilde{x}^\ell = x^\ell + \gamma A^* (y - Ax^\ell)$
threshold:	$x^{\ell+1} = \widetilde{x}^\ell_K$ (keep largest K entries in $\widetilde{x}^\ell$)
	$\ell = \ell + 1$

end while
output: $\widehat{x} = x^\ell$

---

previous estimate of x, and this estimate is thresholded to preserve only its K largest entries. Each of these steps is very simple, and if A and A^* have efficient implementations (without requiring explicit matrix multiplications) then each iteration of IHT can be performed very quickly. The choice of the stepsize parameter γ is important, however, for the convergence of the algorithm (see [22,25] for more details). IHT is amenable to RIP-based analysis [23,25] that accommodates compressible signals and measurement noise. IHT is similar in spirit to iterative soft thresholding methods that have been proposed for solving (1.23); see [14] and references therein. Approximate message passing (AMP) [71] is a technique similar to iterative thresholding but adds an additional term before thresholding.

1.4.5 Signal Recovery in Nonorthonormal Dictionaries

All of our analyses thus far in Section 1.4 have assumed that the sparsity basis Ψ is an $N \times N$ orthonormal matrix. As discussed in Section 1.2.3, however, nonorthonormal dictionaries Ψ can appear in two different modeling contexts: synthesis sparsity and analysis sparsity. The particular modeling context will determine the appropriate type of reconstruction algorithm.

1.4.5.1 Synthesis Sparsity in Redundant Dictionaries

All of the results presented thus far in Section 1.4 have depended on Ψ only through the properties of the product $A = \Phi\Psi$. In fact, all of these same results hold—with no changes—if we allow Ψ to be a generic $N \times D$ dictionary and consider signals of the form $f = \Psi x$, where x is a sparse or compressible vector. (The only technical difference is that the unknown coefficient vector x will now have length D instead of N, and A will have size $M \times D$ instead of $N \times D$.)

In all of the bounds presented thus far in Section 1.4, a critical condition for successful recovery of a coefficient vector x has been that the columns of A be uncorrelated; this has been formalized using the properties of coherence and the RIP. Unfortunately, ensuring that A has low coherence (or that A satisfies the RIP) can be much more difficult when Ψ is a generic dictionary. If Φ is generated randomly from one of the sort of distributions discussed in Section 1.3, then for $A = \Phi\Psi$ to satisfy the RIP, one essentially needs Ψ to satisfy the RIP in the first place. (But if Ψ does satisfy the RIP, ensuring that A satisfies the RIP is indeed possible [146].) We do note that some work has focused on learning, from a given set of training data, a sparsifying dictionary Ψ in conjunction with an incoherent sensing matrix Φ [79].

Another caveat is that the bounds presented thus far in Section 1.4 have concerned the accuracy to which a coefficient vector x can be recovered. As we have mentioned, after a sparse coefficient vector has been estimated, one can synthesize a signal estimate via multiplication by $\Psi: \widehat{f} = \Psi\widehat{x}$. However, when Ψ is a generic dictionary, the error in the signal domain will typically not equal the error in the coefficient domain. That is, the rightmost equality in (1.21) will not hold. Indeed, it is possible for $\left\|f - \widehat{f}\right\|_2$ to be much larger or much smaller than $\|x - \widehat{x}\|_2$. For all of these reasons, signal recovery using nonorthonormal dictionaries must be approached with caution.

In a slightly different vein, there is emerging research—still assuming synthesis sparsity in a redundant dictionary Ψ—into methods for recovering a signal f without attempting to accurately recover the original coefficient vector x. Extensions of IHT [21,157] and CoSaMP [61] have been proposed toward this end; these algorithms treat Φ and Ψ separately and do not rely merely on the product $A = \Phi\Psi$. Consequently, theoretical performance guarantees for these algorithms do not require that $A = \Phi\Psi$ satisfy the RIP but only that Φ satisfy the Ψ-RIP (recall from Section 1.3.1 that this is often a much easier condition to satisfy). Unfortunately, a practical limitation of these techniques at present is that they rely on projection operations that can be difficult to compute when Ψ has correlated columns.

1.4.5.2 Analysis Sparsity

For signal models where Ψ is a generic dictionary and Ψ^*f is assumed to be sparse or compressible, it is necessary to use a different suite of recovery algorithms than those presented thus far [36,83]. The natural modification of BP is

$$\widehat{f} = \arg\min_{f'} \left\| \Psi^*f' \right\|_1 \text{ subject to } y = \Phi f' \tag{1.25}$$

and the natural modification of BPDN is

$$\widehat{f} = \arg\min_{f'} \left\| \Psi^*f' \right\|_1 \text{ subject to } \left\| y - \Phi f' \right\|_2 \leq \eta. \tag{1.26}$$

We note that these problem formulations are not stated in terms of $A = \Phi\Psi$ but rather treat Φ and Ψ separately. If Ψ is an orthonormal basis, then these formulations are equivalent to the original BP and BPDN formulations. NESTA [1,16], TFOCS [5,17], CVX [98,99], and YALL1 [182,189] are among the available solvers for this ℓ_1-analysis minimization problem, as it is known. Most of these packages can handle both real and complex matrices and vectors. Under certain conditions, ℓ_1-analysis minimization can recover an accurate signal estimate.

Theorem 1.12 ([36]) *Suppose Ψ is a tight frame and suppose Φ satisfies the Ψ-RIP of order 2K with isometry constant $\delta_{2K} < 0.08$. Let $y = \Phi f + n$ be noisy measurements of any signal f. If $\eta \geq \|n\|_2$, then the solution $\widehat{x}$ to (1.26) will obey*

$$\left\| f - \hat{f} \right\|_2 \le C_6 \frac{\left\| \Psi^* f - (\Psi^* f)_K \right\|_1}{\sqrt{K}} + C_7 \eta, \tag{1.27}$$

where

$(\Psi^* f)_K$ *denotes the vector containing the K largest entries of* $(\Psi^* f)_K$ *(and zeros elsewhere), and*

C_6 *and* C_7 *are constants that depend only on* δ_{2K}

We note the presence of the term $\left\| \Psi^* f - (\Psi^* f)_K \right\|_1$ in the bound (1.27); this term will be smallest when the analysis coefficients of f in the dictionary Ψ are sparse or compressible. An analogous term appears in synthesis-based recovery bounds such as (1.20) and (1.24). We also note that an extension of IHT [47] has been proposed as an alternative to ℓ_1-analysis minimization.

TV minimization algorithms for image recovery are typically of the form (1.25) and (1.26), but with $\Psi^* = \nabla$, a discrete gradient operator [41,121,153]. Because ∇^* is not a tight frame, the TV reconstruction problem is more difficult to analyze than conventional CS problems; in particular, Theorem 1.12 does not apply. However, a recent paper [138] has shown a performance guarantee for TV reconstruction analogous to (1.27) (with $\Psi^* = \nabla$) but with additional logarithmic factors on the number of measurements. For the measurement matrix Φ in that paper, subgaussian entries can be used, as can a partial DFT matrix with randomized column signs (see [138] for details). A split Bregman algorithm [97], NESTA [1,16], TFOCS [5,17], ℓ_1-MAGIC [39], and CVX [98,99] are among the available solvers for the TV minimization problem. For additional discussions about the performance of TV-based image reconstruction, wavelet-based image reconstruction, and combinations of the two, the reader may refer to [30,120,149].

For signal recovery under the cosparse analysis model, a variety of algorithms have been proposed and studied. These include ℓ_1-analysis minimization [134], a variant of OMP named greedy analysis pursuit (GAP) [134], and variants of CoSaMP, SP, and IHT named ACoSaMP, ASP, and AIHT, respectively [96]. A property similar to the Ψ-RIP arises in the study of these algorithms.

Acknowledgments

I thank Moeness Amin, Stephen Becker, Mark Davenport, Armin Eftekhari, Marc Rubin, Borhan Sanandaji, and Alejandro Weinstein for suggestions and comments that helped improve this chapter.

References

1. NESTA. A fast and accurate first-order method for sparse recovery. http://www-stat.stanford.edu/~candes/nesta/.
2. SPAMS. SPArse modeling software. http://spams-devel.gforge.inria.fr/.
3. SPGL1. A solver for large-scale sparse reconstruction. http://www.cs.ubc.ca/~mpf/spgl1/.
4. Sudoku layout. http://en.wikipedia.org/wiki/File:Sudoku-by-L2G-200507 14.svg.
5. TFOCS. Templates for first-order conic solvers. http://cvxr.com/tfocs/.
6. M. Aharon, M. Elad, and A. Bruckstein. K-SVD: An algorithm for designing overcomplete dictionaries for sparse representation. *IEEE Transactions on Signal Processing*, 54(11):4311–4322, 2006.
7. M. Amin and F. Ahmad. Compressive sensing for through-the-wall radar imaging. *Journal of Electronic Imaging*, 22(3):030901, 2013.
8. S. D. Babacan, R. Molina, and A. K. Katsaggelos. Bayesian compressive sensing using Laplace priors. *IEEE Transactions on Image Processing*, 19(1):53–63, 2010.
9. F. Bach. Sparsity-inducing norms through submodular functions. In *Advances in Neural Information Processing Systems (NIPS)*, Vancouver, British Columbia, Canada, 2010.
10. R. Baraniuk, M. Davenport, R. DeVore, and M. Wakin. A simple proof of the restricted isometry property for random matrices. *Constructive Approximation*, 28(3):253–263, 2008.
11. R. G. Baraniuk, V. Cevher, M. F. Duarte, and C. Hegde. Model-based compressive sensing. *IEEE Transactions on Information Theory*, 56(4):1982–2001, 2010.
12. R. G. Baraniuk and M. B. Wakin. Random projections of smooth manifolds. *Foundations of Computational Mathematics*, 9(1):51–77, 2009.
13. D. Baron, M. F. Duarte, M. B. Wakin, S. Sarvotham, and R. G. Baraniuk. Distributed compressive sensing. arXiv preprint arXiv:0901.3403, 2009.
14. A. Beck and M. Teboulle. A fast iterative shrinkage-thresholding algorithm for linear inverse problems. *SIAM Journal on Imaging Sciences*, 2(1):183–202, 2009.
15. A. Beck and M. Teboulle. Gradient-based algorithms with applications in signal recovery problems. In D. Palomar and Y. Eldar, editors, *Convex Optimization in Signal Processing and Communications*, pp. 33–88. Cambridge University Press, Cambridge, U.K., 2010.
16. S. Becker, J. Bobin, and E. J. Candès. NESTA: A fast and accurate first-order method for sparse recovery. *SIAM Journal on Imaging Sciences*, 4(1):1–39, 2011.
17. S. R. Becker, E. J. Candès, and M. C. Grant. Templates for convex cone problems with applications to sparse signal recovery. *Mathematical Programming Computation*, 3(3):165–218, 2011.
18. Z. Ben-Haim, T. Michaeli, and Y. Eldar. Performance bounds and design criteria for estimating finite rate of innovation signals. *IEEE Transactions on Information Theory*, 58(8):4993–5015, 2012.
19. J. D. Blanchard, C. Cartis, and J. Tanner. Compressed sensing: How sharp is the restricted isometry property? *SIAM Review*, 53(1):105–125, 2011.

20. J. D. Blanchard, C. Cartis, J. Tanner, and A. Thompson. Phase transitions for greedy sparse approximation algorithms. *Applied and Computational Harmonic Analysis*, 30(2):188–203, 2011.
21. T. Blumensath. Sampling and reconstructing signals from a union of linear subspaces. *IEEE Transactions on Information Theory*, 57(7):4660–4671, 2011.
22. T. Blumensath and M. E. Davies. Iterative thresholding for sparse approximations. *Journal of Fourier Analysis and Applications*, 14(5–6):629–654, 2008.
23. T. Blumensath and M. E. Davies. Iterative hard thresholding for compressed sensing. *Applied and Computational Harmonic Analysis*, 27(3):265–274, 2009.
24. T. Blumensath and M. E. Davies. Sampling theorems for signals from the union of finite-dimensional linear subspaces. *IEEE Transactions on Information Theory*, 55(4):1872–1882, 2009.
25. T. Blumensath, M. E. Davies, and G. Rilling. Greedy algorithms for compressed sensing. In Y. C. Eldar and G. Kutyniok, editors, *Compressed Sensing: Theory and Applications*, pp. 348–393. Cambridge University Press, New York, 2012.
26. J. Bobin, J.-L. Starck, and R. Ottensamer. Compressed sensing in astronomy. *IEEE Journal of Selected Topics in Signal Processing*, 2(5):718–726, 2008.
27. S. P. Boyd and L. Vandenberghe. *Convex Optimization*. Cambridge University Press, Cambridge, U.K., 2004.
28. A. M. Bruckstein, D. L. Donoho, and M. Elad. From sparse solutions of systems of equations to sparse modeling of signals and images. *SIAM Review*, 51(1):34–81, 2009.
29. J.-F. Cai, E. J. Candès, and Z. Shen. A singular value thresholding algorithm for matrix completion. *SIAM Journal on Optimization*, 20(4):1956–1982, 2010.
30. J.-F. Cai, B. Dong, S. Osher, and Z. Shen. Image restoration: Total variation, wavelet frames, and beyond. *Journal of the American Mathematical Society*, 25(4):1033–1089, 2012.
31. T. T. Cai, G. Xu, and J. Zhang. On recovery of sparse signals via $\ell 1$ minimization. *IEEE Transactions on Information Theory*, 55(7):3388–3397, 2009.
32. E. Candès. Compressive sampling. In *Proceedings of the International Congress of Mathematicians*, Madrid, Spain, 2006.
33. E. Candès and T. Tao. The Dantzig selector: Statistical estimation when p is much larger than n. *The Annals of Statistics*, 35(6):2313–2351, 2007.
34. E. J. Candès. The restricted isometry property and its implications for compressed sensing. *Comptes Rendus Mathematique*, 346(9):589–592, 2008.
35. E. J. Candès and D. L. Donoho. New tight frames of curvelets and optimal representations of objects with piecewise C^2 singularities. *Communications on Pure and Applied Mathematics*, 57(2):219–266, 2004.
36. E. J. Candès, Y. C. Eldar, D. Needell, and P. Randall. Compressed sensing with coherent and redundant dictionaries. *Applied and Computational Harmonic Analysis*, 31(1):59–73, 2011.
37. E. J. Candès, X. Li, Y. Ma, and J. Wright. Robust principal component analysis? *Journal of the ACM*, 58(3):11, 2011.
38. E. J. Candès and B. Recht. Exact matrix completion via convex optimization. *Foundations of Computational Mathematics*, 9(6):717–772, 2009.
39. E. J. Candès and J. Romberg. $\ell 1$-MAGIC: Recovery of sparse signals via convex programming. http://users.ece.gatech.edu/~justin/l1magic/.

40. E. J. Candès and J. Romberg. Quantitative robust uncertainty principles and optimally sparse decompositions. *Foundations of Computational Mathematics*, 6(2):227–254, 2006.
41. E. J. Candès, J. Romberg, and T. Tao. Robust uncertainty principles: Exact signal reconstruction from highly incomplete frequency information. *IEEE Transactions on Information Theory*, 52(2):489–509, 2006.
42. E. J. Candès, J. K. Romberg, and T. Tao. Stable signal recovery from incomplete and inaccurate measurements. *Communications on Pure and Applied Mathematics*, 59(8):1207–1223, 2006.
43. E. J. Candès and T. Tao. Decoding by linear programming. *IEEE Transactions on Information Theory*, 51(12):4203–4215, 2005.
44. E. J. Candès and T. Tao. Near-optimal signal recovery from random projections: Universal encoding strategies? *IEEE Transactions on Information Theory*, 52(12):5406–5425, 2006.
45. E. J. Candès, M. B. Wakin, and S. P. Boyd. Enhancing sparsity by reweighted $\ell 1$ minimization. *Journal of Fourier Analysis and Applications*, 14(5-6):877–905, 2008.
46. V. Cevher. Learning with compressible priors. In *Advances in Neural Information Processing Systems (NIPS)*, Vancouver, British Columbia, Canada, 2009.
47. V. Cevher. An ALPS view of sparse recovery. In *IEEE International Conference on Acoustics, Speech and Signal Processing (ICASSP)*, Prague, Czech Republic, pp. 5808–5811, 2011.
48. W. L. Chan, K. Charan, D. Takhar, K. F. Kelly, R. G. Baraniuk, and D. M. Mittleman. A single-pixel terahertz imaging system based on compressed sensing. *Applied Physics Letters*, 93(12):121105–121105, 2008.
49. V. Chandrasekaran, B. Recht, P. A. Parrilo, and A. S. Willsky. The convex geometry of linear inverse problems. *Foundations of Computational Mathematics*, 12(6):805–849, 2012.
50. R. Chartrand and W. Yin. Iteratively reweighted algorithms for compressive sensing. In *IEEE International Conference on Acoustics, Speech and Signal Processing (ICASSP)*, Las Vegas, NV, pp. 3869–3872, 2008.
51. F. Chen, A. P. Chandrakasan, and V. Stojanovic. A signal agnostic compressed sensing acquisition system for wireless and implantable sensors. In *IEEE Custom Integrated Circuits Conference (CICC)*, 2010.
52. M. Chen, J. Silva, J. Paisley, C. Wang, D. Dunson, and L. Carin. Compressive sensing on manifolds using a nonparametric mixture of factor analyzers: Algorithm and performance bounds. *IEEE Transactions on Signal Processing*, 58(12):6140–6155, 2010.
53. S. S. Chen, D. L. Donoho, and M. A. Saunders. Atomic decomposition by basis pursuit. *SIAM Journal on Scientific Computing*, 20(1):33–61, 1998.
54. Y. Chi, L. L. Scharf, A. Pezeshki, and A. R. Calderbank. Sensitivity to basis mismatch in compressed sensing. *IEEE Transactions on Signal Processing*, 59(5):2182–2195, 2011.
55. R. Coifman, F. Geshwind, and Y. Meyer. Noiselets. *Applied and Computational Harmonic Analysis*, 10(1):27–44, 2001.
56. P. L. Combettes and J.-C. Pesquet. Proximal splitting methods in signal processing. In H. H. Bauschke, R. S. Burachik, P. L. Combettes, V. Elser, D. R. Luke, and H. Wolkowicz, editors, *Fixed-Point Algorithms for Inverse Problems in Science and Engineering*, pp. 185–212. Springer, New York, 2011.

57. P. L. Combettes and V. R. Wajs. Signal recovery by proximal forward-backward splitting. *Multiscale Modeling & Simulation*, 4(4):1168–1200, 2005.

58. S. F. Cotter, B. D. Rao, K. Engan, and K. Kreutz-Delgado. Sparse solutions to linear inverse problems with multiple measurement vectors. *IEEE Transactions on Signal Processing*, 53(7):2477–2488, 2005.

59. W. Dai and O. Milenkovic. Subspace pursuit for compressive sensing signal reconstruction. *IEEE Transactions on Information Theory*, 55(5):2230–2249, 2009.

60. M. A. Davenport, C. Hegde, M. F. Duarte, and R. G. Baraniuk. Joint manifolds for data fusion. *IEEE Transactions on Image Processing*, 19(10):2580–2594, 2010.

61. M. A. Davenport, D. Needell, and M. B. Wakin. Signal space CoSaMP for sparse recovery with redundant dictionaries. *IEEE Transactions on Information Theory*, 59(10):6820–6829, 2013.

62. M. A. Davenport and M. B. Wakin. Compressive sensing of analog signals using discrete prolate spheroidal sequences. *Applied and Computational Harmonic Analysis*, 33(3):438–472, 2012.

63. G. M. Davis, S. G. Mallat, and Z. Zhang. Adaptive time-frequency decompositions. *Optical Engineering*, 33(7):2183–2191, 1994.

64. D. Donoho and J. Tanner. Observed universality of phase transitions in high-dimensional geometry, with implications for modern data analysis and signal processing. *Philosophical Transactions of the Royal Society A: Mathematical, Physical and Engineering Sciences*, 367(1906):4273–4293, 2009.

65. D. L. Donoho. De-noising by soft-thresholding. *IEEE Transactions on Information Theory*, 41(3):613–627, 1995.

66. D. L. Donoho. Compressed sensing. *IEEE Transactions on Information Theory*, 52(4):1289–1306, 2006.

67. D. L. Donoho and M. Elad. Optimally sparse representation in general (nonorthogonal) dictionaries via $\ell 1$ minimization. *Proceedings of the National Academy of Sciences*, 100(5):2197–2202, 2003.

68. D. L. Donoho, M. Elad, and V. N. Temlyakov. Stable recovery of sparse overcomplete representations in the presence of noise. *IEEE Transactions on Information Theory*, 52(1):6–18, 2006.

69. D. L. Donoho and C. Grimes. Image manifolds which are isometric to Euclidean space. *Journal of Mathematical Imaging and Vision*, 23(1):5–24, 2005.

70. D. L. Donoho and X. Huo. Uncertainty principles and ideal atomic decomposition. *IEEE Transactions on Information Theory*, 47(7):2845–2862, 2001.

71. D. L. Donoho, A. Maleki, and A. Montanari. Message-passing algorithms for compressed sensing. *Proceedings of the National Academy of Sciences*, 106(45): 18914–18919, 2009.

72. D. L. Donoho and J. Tanner. Precise undersampling theorems. *Proceedings of the IEEE*, 98(6):913–924, 2010.

73. D. L. Donoho, Y. Tsaig, I. Drori, and J.-L. Starck. Sparse solution of underdetermined systems of linear equations by stagewise orthogonal matching pursuit. *IEEE Transactions on Information Theory*, 58(2):1094–1121, 2012.

74. C. Dossal, G. Peyré, and J. Fadili. A numerical exploration of compressed sampling recovery. *Linear Algebra and Its Applications*, 432(7): 1663–1679, 2010.

75. P. Dragotti, M. Vetterli, and T. Blu. Sampling moments and reconstructing signals of finite rate of innovation: Shannon meets Strang-Fix. *IEEE Transactions on Signal Processing*, 55(5):1741–1757, 2007.

76. M. F. Duarte and R. G. Baraniuk. Spectral compressive sensing. *Applied and Computational Harmonic Analysis*, 35(1):111–129, 2013.
77. M. F. Duarte, M. A. Davenport, D. Takhar, J. N. Laska, T. Sun, K. F. Kelly, and R. G. Baraniuk. Single-pixel imaging via compressive sampling. *IEEE Signal Processing Magazine*, 25(2):83–91, 2008.
78. M. F. Duarte and Y. C. Eldar. Structured compressed sensing: From theory to applications. *IEEE Transactions on Signal Processing*, 59(9):4053–4085, 2011.
79. J. M. Duarte-Carvajalino and G. Sapiro. Learning to sense sparse signals: Simultaneous sensing matrix and sparsifying dictionary optimization. *IEEE Transactions on Image Processing*, 18(7):1395–1408, 2009.
80. B. Efron, T. Hastie, I. Johnstone, and R. Tibshirani. Least angle regression. *Annals of Statistics*, 32(2):407–499, 2004.
81. A. Eftekhari, H. L. Yap, C. J. Rozell, and M. B. Wakin. The restricted isometry property for random block diagonal matrices. arXiv preprint arXiv:1210.3395, (in press). http://www.sciencedirect.com/science/article/pii/S1063520314000220
82. M. Elad. *Sparse and Redundant Representations: From Theory to Applications in Signal and Image Processing*. Springer, New York, 2010.
83. M. Elad, P. Milanfar, and R. Rubinstein. Analysis versus synthesis in signal priors. *Inverse Problems*, 23(3):947, 2007.
84. Y. C. Eldar. Generalized sure for exponential families: Applications to regularization. *IEEE Transactions on Signal Processing*, 57(2):471–481, 2009.
85. Y. C. Eldar, P. Kuppinger, and H. Bolcskei. Block-sparse signals: Uncertainty relations and efficient recovery. *IEEE Transactions on Signal Processing*, 58(6):3042–3054, 2010.
86. Y. C. Eldar and M. Mishali. Robust recovery of signals from a structured union of subspaces. *IEEE Transactions on Information Theory*, 55(11):5302–5316, 2009.
87. A. Fannjiang and W. Liao. Coherence pattern-guided compressive sensing with unresolved grids. *SIAM Journal on Imaging Sciences*, 5(1):179–202, 2012.
88. P. Feng and Y. Bresler. Spectrum-blind minimum-rate sampling and reconstruction of multiband signals. In *IEEE International Conference on Acoustics, Speech and Signal Processing (ICASSP)*, Atlanta, GA, pp. 1688–1691, 1996.
89. R. Fergus, A. Torralba, and W. T. Freeman. Random lens imaging. Technical Report MIT-CSAIL-TR-2006-058, MIT Computer Science and Artificial Intelligence Laboratory, 2006.
90. M. A. T. Figueiredo, R. D. Nowak, and S. J. Wright. Gradient projection for sparse reconstruction: Application to compressed sensing and other inverse problems. *IEEE Journal of Selected Topics in Signal Processing*, 1(4):586–597, 2007.
91. S. Foucart. A note on guaranteed sparse recovery via ℓ1-minimization. *Applied and Computational Harmonic Analysis*, 29(1):97–103, 2010.
92. J. Friedman, T. Hastie, and R. Tibshirani. Regularization paths for generalized linear models via coordinate descent. *Journal of Statistical Software*, 33(1):1, 2010.
93. S. Gazit, A. Szameit, Y. C. Eldar, and M. Segev. Super-resolution and reconstruction of sparse sub-wavelength images. *Optics Express*, 17:23920–23946, 2009.
94. K. Gedalyahu, R. Tur, and Y. Eldar. Multichannel sampling of pulse streams at the rate of innovation. *IEEE Transactions on Signal Processing*, 59(4):1491–1504, 2011.
95. A. C. Gilbert, M. J. Strauss, and J. A. Tropp. A tutorial on fast Fourier sampling. *IEEE Signal Processing Magazine*, 25(2):57–66, 2008.

96. R. Giryes, S. Nam, M. Elad, R. Gribonval, and M. E. Davies. Greedy-like algorithms for the cosparse analysis model. *Linear Algebra and its Applications*, 441:22–60, 2013.

97. T. Goldstein and S. Osher. The split Bregman method for ℓ1-regularized problems. *SIAM Journal on Imaging Sciences*, 2(2):323–343, 2009.

98. M. Grant and S. Boyd. CVX: Matlab software for disciplined convex programming. http://cvxr.com/cvx.

99. M. Grant and S. Boyd. Graph implementations for nonsmooth convex programs. In V. Blondel, S. Boyd, and H. Kimura, editors, *Recent Advances in Learning and Control, Lecture Notes in Control and Information Sciences*, pp. 95–110. Springer-Verlag, Berlin, Germany, 2008.

100. A. C. Gurbuz, J. H. McClellan, and W. R. Scott. A compressive sensing data acquisition and imaging method for stepped frequency GPRs. *IEEE Transactions on Signal Processing*, 57(7):2640–2650, 2009.

101. E. T. Hale, W. Yin, and Y. Zhang. Fixed-point continuation for ℓ1-minimization: Methodology and convergence. *SIAM Journal on Optimization*, 19(3):1107–1130, 2008.

102. J. Haupt, W. U. Bajwa, G. Raz, and R. Nowak. Toeplitz compressed sensing matrices with applications to sparse channel estimation. *IEEE Transactions on Information Theory*, 56(11):5862–5875, 2010.

103. J. D. Haupt, R. G. Baraniuk, R. M. Castro, and R. D. Nowak. Compressive distilled sensing: Sparse recovery using adaptivity in compressive measurements. In *Asilomar Conference on Signals, Systems and Computers*, Pacific Grove, CA, pp. 1551–1555. IEEE, 2009.

104. L. He and L. Carin. Exploiting structure in wavelet-based Bayesian compressive sensing. *IEEE Transactions on Signal Processing*, 57(9):3488–3497, 2009.

105. L. Jacques, P. Vandergheynst, A. Bibet, V. Majidzadeh, A. Schmid, and Y. Leblebici. CMOS compressed imaging by random convolution. In *IEEE International Conference on Acoustics, Speech and Signal Processing (ICASSP)*, Taipei, Taiwan, pp. 1113–1116, 2009.

106. R. Jenatton, J. Mairal, F. R. Bach, and G. R. Obozinski. Proximal methods for sparse hierarchical dictionary learning. In *Proceedings of the 27th International Conference on Machine Learning (ICML)*, Haifa, Israel, 2010.

107. S. Ji, Y. Xue, and L. Carin. Bayesian compressive sensing. *IEEE Transactions on Signal Processing*, 56(6):2346–2356, 2008.

108. I. Jolliffe. Principal component analysis. In *Encyclopedia of Statistics in Behavioral Science*. Wiley Online Library, 2005. ISBN 9780470013199.

109. F. Krahmer, S. Mendelson, and H. Rauhut. Suprema of chaos processes and the restricted isometry property. arXiv preprint arXiv:1207.0235, (in press). http://onlinelibrary.wiley.com/doi/10.1002/cpa.21504/abstract

110. F. Krahmer and R. Ward. New and improved Johnson–Lindenstrauss embeddings via the restricted isometry property. *SIAM Journal on Mathematical Analysis*, 43(3):1269–1281, 2011.

111. F. Krahmer and R. Ward. Stable and robust sampling strategies for compressive imaging. *IEEE Transactions on Image Processing*, 23(2):612–622, February 2014.

112. K. Kreutz-Delgado, J. F. Murray, B. D. Rao, K. Engan, T. -W. Lee, and T. J. Sejnowski. Dictionary learning algorithms for sparse representation. *Neural Computation*, 15(2):349–396, 2003.

113. J. N. Laska, S. Kirolos, M. F. Duarte, T. S. Ragheb, R. G. Baraniuk, and Y. Massoud. Theory and implementation of an analog-to-information converter using random demodulation. In *IEEE International Symposium on Circuits and Systems (ISCAS)*, New Orleans, LA, pp. 1959–1962, 2007.

114. K. Lee and Y. Bresler. ADMiRA: Atomic decomposition for minimum rank approximation. *IEEE Transactions on Information Theory*, 56(9):4402–4416, 2010.

115. M. S. Lewicki and T. J. Sejnowski. Learning overcomplete representations. *Neural Computation*, 12(2):337–365, 2000.

116. Y. Lu and M. Do. A geometrical approach to sampling signals with finite rate of innovation. In *IEEE International Conference on Acoustics, Speech and Signal Processing (ICASSP)*, Montreal, QC, Canada, 2004.

117. Y. Lu and M. Do. Sampling signals from a union of subspaces. *IEEE Signal Processing Magazine*, 25(2):41–47, 2008.

118. Y. Lu and M. Do. A theory for sampling signals from a union of sub-spaces. *IEEE Transactions on Signal Processing*, 56(6):2334–2345, 2008.

119. D. G. Luenberger. *Optimization by Vector Space Methods*. John Wiley & Sons, New York, 1968.

120. M. Lustig, D. Donoho, and J. M. Pauly. Sparse MRI: The application of compressed sensing for rapid MR imaging. *Magnetic Resonance in Medicine*, 58(6):1182–1195, 2007.

121. M. Lustig, D. L. Donoho, J. M. Santos, and J. M. Pauly. Compressed sensing MRI. *IEEE Signal Processing Magazine*, 25(2):72–82, 2008.

122. J. Mairal, F. Bach, J. Ponce, and G. Sapiro. Online dictionary learning for sparse coding. In *Proceedings of the 26th Annual International Conference on Machine Learning (ICML)*, Montreal, QC, Canada, pp. 689–696. ACM, 2009.

123. J. Mairal, F. Bach, J. Ponce, and G. Sapiro. Online learning for matrix factorization and sparse coding. *The Journal of Machine Learning Research*, 11:19–60, 2010.

124. D. M. Malioutov, M. Cetin, and A. S. Willsky. Homotopy continuation for sparse signal representation. In *IEEE International Conference on Acoustics, Speech and Signal Processing (ICASSP)*, Philadelphia, PA, 2005.

125. S. Mallat. *A Wavelet Tour of Signal Processing: The Sparse Way*, 3rd edn. Academic Press, Orlando, FL, 2008.

126. I. Maravić and M. Vetterli. Sampling and reconstruction of signals with finite innovation in the presence of noise. *IEEE Transactions on Signal Processing*, 53(8):2788–2805, 2005.

127. R. Marcia and R. M. Willett. Compressive coded aperture video re-construction. In *Proceedings of 2008 16th European Signal Processing Conference (EUSIPCO)*, Lausanne, Switzerland, 2008.

128. R. F. Marcia, Z. T. Harmany, and R. M. Willett. Compressive coded aperture imaging. In *Proceedings of the 2009 IS&T/SPIE Electronic Imaging: Computational Imaging VII*, volume 7246, San Jose, CA, 2009.

129. R. F. Marcia and R. M. Willett. Compressive coded aperture super-resolution image reconstruction. In *IEEE International Conference on Acoustics, Speech and Signal Processing (ICASSP)*, Las Vegas, NV, pp. 833–836, 2008.

130. S. Mendelson, A. Pajor, and N. Tomczak-Jaegermann. Reconstruction and sub-gaussian operators in asymptotic geometric analysis. *Geometric and Functional Analysis*, 17(4):1248–1282, 2007.

131. M. Mishali and Y. C. Eldar. From theory to practice: Sub-Nyquist sampling of sparse wideband analog signals. *IEEE Journal of Selected Topics in Signal Processing*, 4(2):375–391, 2010.

132. M. Mishali, Y. C. Eldar, O. Dounaevsky, and E. Shoshan. Xampling: Analog to digital at sub-Nyquist rates. *IET Circuits, Devices & Systems*, 5(1):8–20, 2011.

133. H. Monajemi, S. Jafarpour, M. Gavish, Stat 330/CME 362 Collaboration, and D. L. Donoho. Deterministic matrices matching the compressed sensing phase transitions of Gaussian random matrices. *Proceedings of the National Academy of Sciences*, 110(4):1181–1186, 2013.

134. S. Nam, M. E. Davies, M. Elad, and R. Gribonval. The cosparse analysis model and algorithms. *Applied and Computational Harmonic Analysis*, 34(1):30–56, 2013.

135. B. K. Natarajan. Sparse approximate solutions to linear systems. *SIAM Journal on Computing*, 24(2):227–234, 1995.

136. D. Needell. Noisy signal recovery via iterative reweighted ℓ1-minimization. In *Asilomar Conference on Signals, Systems and Computers*, Pacific Grove, CA, pp. 113–117, 2009.

137. D. Needell and J. A. Tropp. CoSaMP: Iterative signal recovery from incomplete and inaccurate samples. *Applied and Computational Harmonic Analysis*, 26(3):301–321, 2009.

138. D. Needell and R. Ward. Stable image reconstruction using total variation minimization. *SIAM Journal on Imaging Sciences*, 6(2):1035–1058, 2013.

139. M. R. Osborne, B. Presnell, and B. A. Turlach. On the LASSO and its dual. *Journal of Computational and Graphical Statistics*, 9(2):319–337, 2000.

140. J. Y. Park and M. B. Wakin. A multiscale algorithm for reconstructing videos from streaming compressive measurements. *Journal of Electronic Imaging*, 22(2):021001, 2013.

141. Y. C. Pati, R. Rezaiifar, and P. S. Krishnaprasad. Orthogonal matching pursuit: Recursive function approximation with applications to wavelet decomposition. In *Asilomar Conference on Signals, Systems and Computers*, Pacific Grove, CA, pp. 40–44. IEEE, 1993.

142. S. Pfetsch, T. Ragheb, J. Laska, H. Nejati, A. Gilbert, M. Strauss, R. Baraniuk, and Y. Massoud. On the feasibility of hardware implementation of sub-Nyquist random-sampling based analog-to-information conversion. In *IEEE International Symposium on Circuits and Systems (ISCAS)*, Seattle, WA, pp. 1480–1483. IEEE, 2008.

143. L. C. Potter, E. Ertin, J. T. Parker, and M. Cetin. Sparsity and compressed sensing in radar imaging. *Proceedings of the IEEE*, 98(6):1006–1020, 2010.

144. H. Rauhut. Compressive sensing and structured random matrices. In M. Fornasier, editor, *Theoretical Foundations and Numerical Methods for Sparse Recovery*, pp. 1–92. De Gruyter, Berlin, Germany, 2010.

145. H. Rauhut, J. Romberg, and J. A. Tropp. Restricted isometries for partial random circulant matrices. *Applied and Computational Harmonic Analysis*, 32(2):242–254, 2012.

146. H. Rauhut, K. Schnass, and P. Vandergheynst. Compressed sensing and redundant dictionaries. *IEEE Transactions on Information Theory*, 54(5):2210–2219, 2008.

147. B. Recht, M. Fazel, and P. A. Parrilo. Guaranteed minimum-rank solutions of linear matrix equations via nuclear norm minimization. *SIAM Review*, 52(3):471–501, 2010.

148. R. Robucci, L. K. Chiu, J. Gray, J. Romberg, P. Hasler, and D. Anderson. Compressive sensing on a CMOS separable transform image sensor. In *IEEE International Conference on Acoustics, Speech and Signal Processing (ICASSP)*, pp. 5125–5128, Las Vegas, NV, 2008.

149. J. Romberg. Variational methods for compressive sampling. In *Proc. SPIE 6498*, San Jose, CA, 2007. http://proceedings.spiedigitallibrary.org/proceeding.aspx?articleid=1298718.

150. J. Romberg. Compressive sensing by random convolution. *SIAM Journal on Imaging Sciences*, 2(4):1098–1128, 2009.

151. R. Rubinstein, M. Zibulevsky, and M. Elad. Double sparsity: Learning sparse dictionaries for sparse signal approximation. *IEEE Transactions on Signal Processing*, 58(3):1553–1564, 2010.

152. M. Rudelson and R. Vershynin. On sparse reconstruction from Fourier and Gaussian measurements. *Communications on Pure and Applied Mathematics*, 61(8):1025–1045, 2008.

153. L. I. Rudin, S. Osher, and E. Fatemi. Nonlinear total variation based noise removal algorithms. *Physica D: Nonlinear Phenomena*, 60(1):259–268, 1992.

154. B. M. Sanandaji, T. L. Vincent, and M. B. Wakin. Concentration of measure inequalities for Toeplitz matrices with applications. *IEEE Transactions on Signal Processing*, 61(1):109–117, 2013.

155. L. L. Scharf. The SVD and reduced rank signal processing. *Signal Processing*, 25(2):113–133, 1991.

156. M. Schmidt, G. Fung, and R. Rosales. Optimization methods for ℓ1-regularization. University of British Columbia, Technical Report TR-2009-19, 2009.

157. P. Shah and V. Chandrasekaran. Iterative projections for signal identification on manifolds: Global recovery guarantees. In *Allerton Conference on Communication, Control, and Computing*, Monticello, IL, pp. 760–767, 2011.

158. A. Skodras, C. Christopoulos, and T. Ebrahimi. The JPEG 2000 still image compression standard. *IEEE Signal Processing Magazine*, 18(5):36–58, 2001.

159. J. P. Slavinsky, J. N. Laska, M. A. Davenport, and R. G. Baraniuk. The compressive multiplexer for multi-channel compressive sensing. In *IEEE International Conference on Acoustics, Speech and Signal Processing (ICASSP)*, Prague, Czech Republic, pp. 3980–3983, 2011.

160. A. Stern and B. Javidi. Random projections imaging with extended space-bandwidth product. *IEEE Journal of Display Technology*, 3(3):315–320, 2007.

161. T. Strohmer and R. W. Heath. Grassmannian frames with applications to coding and communication. *Applied and Computational Harmonic Analysis*, 14(3):257–275, 2003.

162. T. Sun and K. Kelly. Compressive sensing hyperspectral imager. In *OSA Computational Optical Sensing and Imaging (COSI)*, San Jose, CA, 2009.

163. V. N. Temlyakov. Greedy approximation. *Acta Numerica*, 17:235–409, 2008.

164. A. M. Tillmann and M. E. Pfetsch. The computational complexity of the restricted isometry property, the nullspace property, and related concepts in compressed sensing. *IEEE Transactions on Information Theory*, 60(2):1248–1259, 2014.

165. J. A. Tropp. Greed is good: Algorithmic results for sparse approximation. *IEEE Transactions on Information Theory*, 50(10):2231–2242, 2004.

166. J. A. Tropp. Algorithms for simultaneous sparse approximation. Part II: Convex relaxation. *Signal Processing*, 86(3):589–602, 2006.
167. J. A. Tropp and A. C. Gilbert. Signal recovery from random measurements via orthogonal matching pursuit. *IEEE Transactions on Information Theory*, 53(12):4655–4666, 2007.
168. J. A. Tropp, A. C. Gilbert, and M. J. Strauss. Algorithms for simultaneous sparse approximation. Part I: Greedy pursuit. *Signal Processing*, 86(3):572–588, 2006.
169. J. A. Tropp, J. N. Laska, M. F. Duarte, J. K. Romberg, and R. G. Baraniuk. Beyond Nyquist: Efficient sampling of sparse bandlimited signals. *IEEE Transactions on Information Theory*, 56(1):520–544, 2010.
170. J. A. Tropp, M. B. Wakin, M. F. Duarte, D. Baron, and R. G. Baraniuk. Random filters for compressive sampling and reconstruction. In *IEEE International Conference on Acoustics, Speech and Signal Processing (ICASSP)*, Toulouse, France, 2006.
171. J. A. Tropp and S. J. Wright. Computational methods for sparse solution of linear inverse problems. *Proceedings of the IEEE*, 98(6):948–958, 2010.
172. Y. Tsaig and D. L. Donoho. Extensions of compressed sensing. *Signal Processing*, 86(3):549–571, 2006.
173. E. Van Den Berg and M. P. Friedlander. Probing the Pareto frontier for basis pursuit solutions. *SIAM Journal on Scientific Computing*, 31(2):890–912, 2008.
174. S. S. Vasanawala, M. T. Alley, B. A. Hargreaves, R. A. Barth, J. M. Pauly, and M. Lustig. Improved pediatric MR imaging with compressed sensing. *Radiology*, 256(2):607–616, 2010.
175. R. Vershynin. Introduction to the non-asymptotic analysis of random matrices. In Y. C. Eldar and G. Kutyniok, editors, *Compressed Sensing: Theory and Applications*, pp. 210–268. Cambridge University Press, New York, 2012.
176. M. Vetterli, P. Marziliano, and T. Blu. Sampling signals with finite rate of innovation. *IEEE Transactions on Signal Processing*, 50(6):1417–1428, 2002.
177. M. Wakin, S. Becker, E. Nakamura, M. Grant, E. Sovero, D. Ching, J. Yoo, J. Romberg, A. Emami-Neyestanak, and E. Candès. A non-uniform sampler for wideband spectrally-sparse environments. *IEEE Journal on Emerging and Selected Topics in Circuits and Systems*, 2(3):516–529, 2012.
178. R. Ward. Compressed sensing with cross validation. *IEEE Transactions on Information Theory*, 55(12):5773–5782, 2009.
179. R. M. Willett, R. F. Marcia, and J. M. Nichols. Compressed sensing for practical optical imaging systems: A tutorial. *Optical Engineering*, 50(7):072601–072601, 2011.
180. D. Wipf and S. Nagarajan. Iterative reweighted $\ell 1$ and $\ell 2$ methods for finding sparse solutions. *IEEE Journal of Selected Topics in Signal Processing*, 4(2):317–329, 2010.
181. Z. J. Xiang, H. Xu, and P. J. Ramadge. Learning sparse representations of high dimensional data on large scale dictionaries. In *Advances in Neural Information Processing Systems (NIPS)*, Vancouver, BC, Canada, 2011.
182. J. Yang and Y. Zhang. Alternating direction algorithms for $\ell 1$-problems in compressive sensing. *SIAM Journal on Scientific Computing*, 33(1):250–278, 2011.
183. P. Ye, J. L. Paredes, G. R. Arce, Y. Wu, C. Chen, and D. W. Prather. Compressive confocal microscopy. In *IEEE International Conference on Acoustics, Speech and Signal Processing (ICASSP)*, Taipei, Taiwan, pp. 429–432, 2009.

184. W. Yin, S. Osher, D. Goldfarb, and J. Darbon. Bregman iterative algorithms for ℓ1-minimization with applications to compressed sensing. *SIAM Journal on Imaging Sciences*, 1(1):143–168, 2008.

185. J. Yoo, S. Becker, M. Monge, M. Loh, E. Candès, and A. Emami-Neyestanak. Design and implementation of a fully integrated compressed-sensing signal acquisition system. In *IEEE International Conference on Acoustics, Speech and Signal Processing (ICASSP)*, Kyoto, Japan, pp. 5325–5328, 2012.

186. J. Yoo, C. Turnes, E. Nakamura, C. Le, S. Becker, E. Sovero, M. Wakin et al. A compressed sensing parameter extraction platform for radar pulse signal acquisition. *IEEE Journal on Emerging and Selected Topics in Circuits and Systems*, 2(3):626–638, 2012.

187. G. -X. Yuan, K. -W. Chang, C. -J. Hsieh, and C. -J. Lin. A comparison of optimization methods and software for large-scale L1-regularized linear classification. *The Journal of Machine Learning Research*, 11:3183–3234, 2010.

188. T. Zhang. Sparse recovery with orthogonal matching pursuit under RIP. *IEEE Transactions on Information Theory*, 57(9):6215–6221, 2011.

189. Y. Zhang, J. Yang, and W. Yin. YALL1 basic solver code. http://yall1.blogs.rice.edu.

2

Overcomplete Dictionary Design for Building Feature Extraction

Wim van Rossum and Jacco de Wit

CONTENTS

2.1 Introduction

There are several scenarios for which it is desirable to obtain the layout of a building interior using outside radar sensing and surveillance. These situations may arise after an earthquake, when a building might become unstable or when the building entrance is blocked in hostage taking or criminal acts by outlaws.

In this chapter, several approaches will be introduced for obtaining building features. Two approaches, point target matched filtering and smashed filtering, are classical Nyquist based. The third approach, overcomplete dictionary (OCD), which uses sparse reconstruction, is the main focus of this chapter. The authors would like to stress that although the sparse reconstruction results shown at the end of the chapter are based on data obtained through noncompressive sensing, the approach can easily be extended to include compressed measurements, as indicated in the theoretical discourse. Also, some concepts will be introduced for completeness but will not be used in obtaining the results shown.

2.1.1 Overview of Through-Wall Radar Mapping

Commercial radar systems for stand-off surveillance of building interiors do exist, but the focus is mostly on detection and imaging of people and objects behind the (first) wall. Obtaining the building layout is generally not a first requirement. Radar technology for stand-off mapping of a *complete building structure* is still under development. Extended surveys of through-wall radar technology and systems can be found in NATO RTO (2011), Huffman and Ericson (2012), and Miller et al. (2012).

Over the years, several approaches have been reported for extracting building structure information from through-wall radar data. One approach is to use a model of the complete building structure to predict radar measurements. In this approach, a detailed electromagnetic simulation is performed for forward prediction of the measurements based on the chosen building layout (Subotic et al. 2008). The layout is updated based on the difference between the predicted measurement vector and the actual measurement vector. These algorithms use the finite element method together with, for instance, simulated annealing (Lavely et al. 2008) or jumped diffusion algorithm (Nikolic et al. 2009), for updating the building layout. The model-based approaches are mentioned here for completeness but are not investigated further in this chapter.

In this chapter, the focus is on feature extraction methods. Feature extraction methods allow detection, classification, and localization of building elemental structures such as walls, ceilings, and corners. Based on the locations of the elemental structures, the building layout can be synthesized.

Different feature extraction methods can be used for extracting building elements, exploiting either focused radar images or raw radar data.

Focused radar images of building interiors may be obtained by point scattering focusing, such as in conventional synthetic aperture radar techniques. Using such techniques, the radar data are matched-filtered with the point spread function of a point target. However, the radar responses to dihedral and trihedral corners, walls, and ceilings are not well represented by such a filter, resulting in classification confusion. Typically, by using point scattering focusing, multipath reflections are also focused, leading to false alarms that clutter the image. Due to the confusion and false alarms, interpretation of focused radar images is difficult without prior knowledge of the building structure (Sévigny et al. 2010, 2012). Furthermore, by using (monostatic) imaging concepts, it is unlikely that walls parallel to the radar line of sight can be detected because they give rise to diffuse scattering in the direction of the radar. To obtain a complete layout, the building needs to be imaged from different sides (Le et al. 2009; Sévigny and DiFilippo 2013).

In literature, there is consensus that building features can be more reliably extracted using an approach exploiting scatterer models (e.g., Marble and Hero 2006; Baranoski et al. 2008; Subotic et al. 2008; Ertin and Moses 2009; Lagunas et al. 2013a). In this approach, the building structural elements are represented by *canonical scatterers* such as spheres, cylinders, corners, and planar surfaces. For these canonical scatterers, models are available describing the amplitude, phase, and polarimetric characteristics of the radar responses (e.g., Potter and Moses 1997). Exploiting such a priori defined scattering models provides robustness with respect to clutter and multipath reflections (assuming these do not match the defined scattering models).

The aforementioned approach may be implemented by defining separate, independent matched filters, each filter tuned to the scattering model of a single canonical scatterer type. The separate filters are applied to the measured radar data sequentially and independently. In Davenport et al. (2007), this approach is referred to as *smashed filter*, to emphasize its similarity to the traditional matched filter while also stressing its compressive nature. The outcomes of the filters are fed to a detection, plot extraction, and classification process to obtain the set of most probable canonical scatterers.

A novel approach is the use of OCDs in which the response in each voxel is assumed to be a weighted superposition of the responses to the different canonical scatterer types. In essence, the different matched filters are applied simultaneously to the radar data. The result is a measurement model in which the output vector is of higher dimension than the input vector. This yields an underdetermined system, and extraction of the building features needs to be performed based on sparse representation algorithms. So, for OCD processing even when the data are Nyquist sampled, due to the simultaneous processing of multiple models, the system becomes underdetermined and sparse representation algorithms are needed.

Point target focusing, smashed filtering, and use of an OCD are three different approaches that use inversion for obtaining building features. These approaches first perform some kind of inversion for obtaining the representation vector. After detection, plot extraction, and classification, the obtained building features should be fed into a mapping algorithm for obtaining the building layout. This approach may be implemented iteratively. In the iterative approach, the scattering models of the canonical scatterers can be updated to include the influence of intervening structures. Thick walls with large permittivity attenuate the signal and change the phase relation of a scatterer behind it. So, both thickness and permittivity of the wall are required for updating the scattering models. If polarimetric measurements are performed, the permittivity can be obtained from the amplitude response of the different polarimetric channels. For instance, when a single polarization channel disappears, this might indicate that the scattering angle with one of the walls or ceilings is equal to the Brewster angle. The Brewster angle is directly related to the permittivity of the material of the wall. The updated scattering models might even include multipath and diffraction effects.

2.1.2 Typical Measurement Geometry

A practical operational concept for through-wall radar building mapping is based on a vertical linear array antenna fixed on the side of a moving vehicle. The radar line of sight is perpendicular to the direction of motion. By driving past the building of interest, a 3D data set is acquired. Resolution is obtained in 3D by applying range compression, synthetic aperture radar techniques, and near-field beamforming. The described measurement geometry is schematically depicted in Figure 2.1.

In this chapter, the term aperture may refer to either the aperture of a physical antenna or a synthetic aperture formed in digital signal processing. The terms parallel and perpendicular are used to indicate the orientation of building structural elements relative to the aperture. A parallel structure is parallel to the aperture, that is, a wall extending in azimuth and elevation. A perpendicular structure is perpendicular to the aperture, that is, a ceiling or floor extending in range and azimuth or a wall extending in range and elevation. A parallel wall and a perpendicular wall are shown in Figure 2.1.

2.1.3 Bases, Frames, and Overcomplete Dictionaries

In linear algebra, an orthonormal basis of a vector space $\mathbf{V}$ with an inner product is a set $\{e_k\}$ of elements of $\mathbf{V}$, which are normalized $(\forall k \, \|e_k\| = 1)$ and satisfy $\sum_k |\langle v, e_k \rangle|^2 = \|v\|^2$ for all $v \in \mathbf{V}$. A frame of this space is a

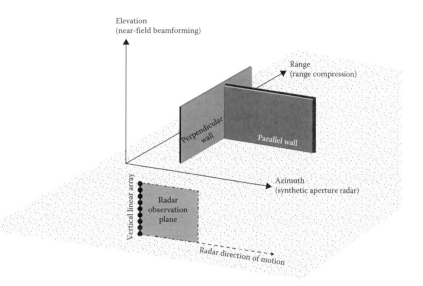

FIGURE 2.1
Schematic overview of the typical radar building mapping measurement geometry.

generalization of a basis to a set of linearly dependent elements. A frame is a set $\{e_k\}$ of elements of **V** that satisfy the frame condition:

$$A\|v\|^2 \le \sum_k |\langle v, e_k \rangle|^2 \le B\|v\|^2 \quad \text{for all } v \in \mathbf{V}. \tag{2.1}$$

Here, $0 < A < B < \infty$. Both A and B are independent of k, they only depend on the set. A frame also spans the vector space **V**. If $A = B$, the frame is called tight; if $A = B = 1$, the frame is called normalized. For a uniform frame, the norm of each element of the set is the same. A uniform normalized tight frame with element norm 1 is an orthonormal basis.

Frames are an overcomplete version of a basis set. They have some redundancy, which is desirable for the robustness of the representation, but at the same time yield an efficiency reduction (larger matrices and longer vectors).

An OCD is a frame of the measurement space. The elements of the set are often referred to as atoms, and this term will be used hereafter. Complex signals, like those obtained in audio recordings, images, and also radar measurements, often include structures that are not well represented by a few atoms in any basis. For instance, in radar measurements, models describing the responses of point targets can be seen as atoms of a basis of the measurement space. A point scatterer is depicted as a single-point target, while a wall will be depicted by a large cluster of point scatterers. Another basis uses the responses of walls as atoms. Using this basis will yield a single entry for a wall but will yield a large number of walls for depicting a single-point scatterer. So, measurements containing both a wall and a point target will yield many atoms contributing to the representation in any of these two bases.

OCDs incorporate multiple patterns. This can increase the sparseness of the vector with the atom coefficients. The use of models of point scatterers together with models of walls will increase the sparseness of an imaging radar output: The wall-point target structure is described by a single wall with a single-point target instead of many point scatterers or many walls. Increase in sparseness results in improvement in compression, in denoising, in pattern recognition, and in the application to inverse problems.

Using an OCD matrix $\mathbf{D} \in \mathcal{R}^{n \times K}$ that contains K atoms for columns, $\{d_j\}_{j=1}^{K}$, a signal $y \in \mathcal{R}^n$ can be represented as a linear combination of these atoms. The representation of y may either be exact:

$$y = \mathbf{D}x, \tag{2.2}$$

or approximate:

$$\|y - \mathbf{D}x\|_p \leq \epsilon. \tag{2.3}$$

Here, the vector $x \in \mathcal{R}^K$ contains the atom coefficients of the signal y and is referred to as the representation of y. In approximation methods, typical norms used for measuring the deviation are the ℓ_p-norms for $p = 1, 2$, and ∞, with $p = 2$ being used most widely.

If $n < K$ and $\mathbf{D}$ is a full-rank matrix, an infinite number of solutions can be found for the representation, therefore constraints on the solution must be set. The solution with the fewest number of nonzero coefficients assumes sparseness of the signal and is a reasonable representation. This sparsest representation is the solution of either

$$\min_{x} \|x\|_0 \text{ subject to } y = \mathbf{D}x, \tag{2.4}$$

or

$$\min_{x} \|x\|_0 \text{ subject to } \|y - \mathbf{D}x\|_2 \leq \epsilon. \tag{2.5}$$

Here, $\|\cdot\|_0$ is the ℓ_0-norm, counting the nonzero entries of a vector. Extraction of the sparsest representation is an NP-hard problem. Algorithms for finding approximating solutions, for instance, based on the ℓ_1-norm, have been extensively investigated.

2.1.4 Layout of This Chapter

In the following section, the three different approaches for extracting building features will be discussed in detail. Special attention will be given to
OCDs and their construction in Section 2.3. The three feature extracting
approaches have been implemented and applied to measured through-wall
radar data. The implementation is based on the backscatter models for
canonical scatterers as explained in Section 2.4. The results obtained are
presented in Section 2.5. Section 2.6 concludes this chapter.

2.2 Building Feature Extraction

The description of the different processing techniques in this section departs
from the assumption that 3D stand-off radar data are available for the building of interest. In the following, fast time is denoted t_f, the transmit antenna
index n_t, the receive antenna index n_r, and the scanning in azimuth or slow
time t_s.

2.2.1 Point Scattering Focusing

In point scattering focusing, the applied matched filters are based on point
scatterers for obtaining the representation. This approach uses a basis of
point scattering models to find a representation based on the ℓ_2 norm. For
each voxel (x, y, z) in the 3D image, a separate matched filter can be obtained.
For instance, consider a frequency modulated continuous wave radar that
transmits a linearly modulated signal:

$$f = f_{start} + \gamma \cdot t_f = f_{start} + \gamma \cdot n_f \cdot \Delta t_f. \tag{2.6}$$

Here, f_{start} is the start frequency and γ the chirp rate. The rightmost part of
(2.6) is the quantized signal that is sampled at Δt_f intervals and the variable
n_f is the sample number in fast time. The matched filter operates in the beat-
frequency domain, that is, after demodulation of the received signals, and
can be written as (ignoring the residual video phase)

$$I(x, y, z) = \sum_{n_s=1}^{N_S} \sum_{n_t=1}^{N_T} \sum_{n_r=1}^{N_R} \sum_{n_f=1}^{N_F} m\left(n_s, n_t, n_r, n_f\right)$$

$$\cdot \exp(-j \cdot 2 \cdot \pi \rightarrow \exp(-j \cdot 2 \cdot \pi)), \tag{2.7}$$

$$R_{n_t} = \sqrt{(x - x_{n_t})^2 + (y - y_{n_t})^2 + (z - z_{n_t})^2}, \qquad (2.8)$$

$$R_{n_r} = \sqrt{(x - x_{n_r})^2 + (y - y_{n_r})^2 + (z - z_{n_r})^2}. \qquad (2.9)$$

where
    I is the intensity of the signal after matched filtering
    m indicates the measurements

Due to the motion of the array, the positions of the transmit and receive elements are functions of the slow-time sample number n_s. The quadruple summation represents over slow time n_s azimuth processing for a moving radar platform or Doppler processing for moving target indication, over the independent transmitters n_t and over the receivers n_r far-field or near-field beamforming (in elevation) depending on the range of the voxel, and over fast time n_f to range compression. Of course, each of the processing steps can be performed more traditionally sequentially (range compression, beamforming, and synthetic aperture radar processing).

Equation 2.7 can be written in matrix form:

$$I_{xyz} = G_{(xyz),(n_s n_t n_r n_f)} m_{n_s n_t n_r n_f}. \qquad (2.10)$$

where
    I_{xyz} is the output vector whose entries represent the intensity at certain
        voxel positions in 3D
    G is the pseudo-inverse of the forward-model matrix
    m is a single vector with all measurements in lexicographic order

The forward-model or measurement matrix A maps the voxel characteristics to the measurements:

$$m_{n_s n_t n_r n_f} = A_{(n_s n_t n_r n_f),(xyz)} I_{xyz}. \qquad (2.11)$$

For instance, A corresponding to (2.7) is given as

$$A_{(n_s n_t n_r n_f),(xyz)} = \exp(j \cdot 2 \cdot \pi \to \exp(j \cdot 2 \cdot \pi)). \qquad (2.12)$$

The synthetic aperture radar processing described here uses a geometrical approach for the migration in the time domain and is known as back projection (Ulander et al. 2003) or diffraction summation (Miller et al. 1987). The algorithm does not take into account the wave equation: the attenuation due to absorption along the way and free-space losses. These methods are simple to implement and very agile because there is no need for a regular measurement grid, only the positions of the transmitters and receivers need to be known and the output grid is user defined. On the other hand, these algorithms require considerable computing power because each voxel is treated independently from all the other voxels. A similar approach is the Kirchhoff migration (Yilmaz and Doherty 1987). It is based on the solution of a scalar wave equation. Kirchhoff migration theory provides a description for obtaining the amplitude and phase along the wave front in an environment with variable propagation velocity and also provides the shape of the wave front. The hyperbola used in back projection is replaced by a more general shape. Wave equation–based migration can also be performed in the frequency domain. The ω-k algorithm, or Stolt migration (Stolt 1978), solves the migration problem using the Fourier transform. This method is fast with low computational complexity, but it is not agile: The algorithm relies on a regular grid for the measurements in slow time and creates a regular output grid.

The aforementioned algorithms rely on sufficiently correct models for the calculation of the matched filters. For walls with unknown thickness and electromagnetic properties, the filtering is distorted and is no longer matched to the signal. Also, position errors of the sensors will yield model errors. Iterative methods can be used to estimate these properties. For instance, in Önhon and Çetin (2009), the motion errors are estimated and used to update the atoms.

After obtaining the 3D image based on point scatterers, traditional detection like constant false alarm rate (CFAR), plot extraction, and classification need to be applied in order to obtain some information about the building elements, such as walls and corners. For instance, the Hough transform can be used on the extracted plots to find linear structures (walls) in the image (Aftanas 2009; Aftanas and Drutarovský 2009).

2.2.2 Smashed Filter Processing

Building interiors may have structures that have recognizable signatures. For example, intersections between walls produce dihedral reflectors, and intersections between multiple walls and ceilings or floors create trihedral reflectors. Both scattering types can provide key anchor points in the building model. Multiple dihedrals might be connected to indicate the presence of a wall that may not be visible due to unfavorable viewing angles. This indicates that the use of canonical scattering types of extended scatterers may yield better and more robust results than the point scatterer–based imaging.

This second approach uses multiple matched filters. Following Davenport et al. (2007), the algorithm can be described as follows:

1. Create different hypotheses for the different models:

$$\mathcal{H}_i : y = \Phi\left(f_i\left(\theta_i\right) + n\right). \tag{2.13}$$

where

    Φ is a projection operator used in compressive sensing
    f is the mapping function of the model i
    θ the state variables
    n is additive white Gaussian noise

2. For each of the hypotheses, $\mathcal{H}_i$ obtains the maximum likelihood estimate of the parameter vector:

$$\widehat{\theta}_i = \arg\min_{\theta_i} \left\| y - \Phi f_i\, (\theta_i)^2_2 \right\|. \tag{2.14}$$

3. Perform maximum likelihood classification:

$$C\,(y) = \arg\max_{i=1,\ldots,P} p\big(y|\widehat{\theta}_i, \mathcal{H}_i\big). \tag{2.15}$$

This labels y with the hypothesis $\mathcal{H}_i$.

When no compressive sensing is applied, this algorithm can be seen as applying different bases for each of the different models. Each model creates its own basis, and the data are projected onto this basis. By applying generalized likelihood testing, the appropriate hypothesis is chosen.

The results of the different matched filters can be used as input into a detection, plot extraction, and classification scheme in order to obtain different building features.

2.2.3 OCD with Sparse Representation

The third approach is based on the use of an OCD. In this approach, the different models are used as atoms of a single, large observation matrix. Even though each of the models creates a basis $\mathbf{A}_i$, the combination of the p different bases creates a frame:

$$\mathbf{A} = \begin{bmatrix} \mathbf{A}_1 & \mathbf{A}_2 & \ldots & \mathbf{A}_p \end{bmatrix}. \tag{2.16}$$

As indicated before, the representation problem now needs some kind of regularization in order to obtain a unique solution. The atom coefficients directly indicate the presence of a building feature at a given position. The detection, location, and classification are performed directly.

The benefit of using all models simultaneously lies in the nonorthogonality of the different models. In the smashed filter approach, careful considerations are required on how to interpret the final output for extended objects. A wall will yield many point target responses when the point scattering filter is used, which may influence the performance of the (CFAR) detection, plot extraction, and classification algorithm. In the OCD approach, there will only be an entry for the wall model and no entries (in the ideal case) for the point scattering model.

2.3 How to Create an OCD

There are two approaches for finding the dictionaries: using known atoms or using adapting-dictionary techniques.

2.3.1 Knowledge-Based Dictionaries

The first approach is simpler and in many cases results in simple and fast algorithms, for example, overcomplete wavelets, curvelets, and short-time Fourier transforms. This approach includes also the use of physical models. Each atom represents a signal based on a physical assumption: for example, the radar response of a point scatter or of a wall. This allows the result to be easily interpreted. Each entry of the representation vector is directly related to a known model. The output yields the desired information. The suitability of the models to describe the signals defines how well the signal is represented.

The models can be based on free-space propagation or purely on geometry as stated before. These models consider mainly the phase of the signal. The polarimetric information (both amplitude and phase) can also be included in the model.

Special care needs to be observed while building the OCD. The mutual coherence of a dictionary $\mathbf{A}$, denoted by $\mu(\mathbf{A})$, is defined as the maximum value of the absolute scalar product between two different normalized atoms, (a_i, a_j) of $\mathbf{A}$:

$$\mu(\mathbf{A}) = \max_{i \neq j} \frac{|a_i \cdot a_j|}{|a_i| \, |a_j|}. \tag{2.17}$$

Here, the dot indicates the scalar product. The mutual coherence of a dictionary yields a measure of the similarity between the atoms. For an orthogonal matrix, the mutual coherence becomes 0. For overcomplete matrices, the mutual coherence becomes larger than 0. If the mutual coherence is equal to 1, there exist two parallel atoms that will cause confusion in the reconstruction of the representation. For a full rank dictionary of size $n \times k$, a lower bound for the mutual coherence has been obtained by Strohmer and Heath (2004):

$$\mu \geq \sqrt{\frac{k - n}{n(k - 1)}}. \tag{2.18}$$

For k much larger than n, the mutual coherence becomes of the order of $1/\sqrt{n}$.

The spark of dictionary $\mathbf{A}$ is the smallest number of columns that form a linearly dependent set (Donoho and Elad 2003). A trivial relation between the spark $\sigma(\mathbf{A})$ and the mutual coherence $\mu(\mathbf{A})$ is

$$\sigma(\mathbf{A}) \geq 1 + \frac{1}{\mu(\mathbf{A})}. \tag{2.19}$$

From Donoho and Elad (2003), for (2.4), a linear representation over m atoms (i.e., $\|x\|_0 = m$) is unique when $m < \sigma(\mathbf{A})/2$. For (2.5), it has been shown that exact uniqueness cannot be guaranteed but an approximate one that allows a bounded deviation can be claimed.

Finding the exact solution of (2.4) and (2.5) is an NP-hard problem, and approximation algorithms for solving these problems have been developed. For instance, matching pursuit (MP) and basis pursuit (BP) use the ℓ_1 norm

$$\min_{x} \|x\|_1 \text{ subject to } y = \mathbf{D}x$$

or   (2.20)

$$\min_{x} \|x\|_1 \text{ subject to } \left\| y - \mathbf{D}x \right\|_2 \leq \epsilon.$$

In Donoho and Elad (2003), the recovery of x by BP and MP will be exact for (2.4) if the representation x satisfies $\|x\|_0 < 1/2\,(1 + (1/\mu(\mathbf{A})))$. Similarly, in other works the stability of the recovery of representation x for (2.5) has been shown. This suggests that representations with nonzero entries of less than $O(\sqrt{n})$ can be successfully recovered by the approximation algorithms.

From the reasoning given earlier, it can be concluded that dictionaries with $\mu(\mathbf{A})$ as small as possible yield the highest probability of recovering the representation vector x.

2.3.2 Adaptive Knowledge-Based Dictionaries

The suitability of the models might be compromised by measurement errors or errors of the environmental model. For instance, the motion of the radar might not be measured with sufficient accuracy or intervening walls might not have been included in the scattering model. Autofocus techniques will change the motion model, and therefore the atoms. Similarly, the atoms will change when intervening walls are taken into account. These new atoms will become specific for the measurements considered and are no longer as general as the pure knowledge–based atoms. The entries in the representation vector will still have the same physical interpretation as the nonadaptive representation. The entries still represent the presence of walls, dihedral, and trihedral corners.

In order to account for and combat motion errors of the radar platform, in Önhon and Çetin (2009) simultaneous optimization is performed for obtaining the representation vector and phase errors. The combined optimization is performed in an iterative way. For each iteration, the regularized optimization is performed first:

$$\hat{x}^t = \arg\min_x \frac{1}{2} \left\| y - \mathbf{D}(\hat{\varphi}^t) x \right\|_2^2 + \lambda \|x\|_1. \tag{2.21}$$

where

t is the iteration number

$\mathbf{D}(\hat{\varphi}^t)$ is the dictionary adapted using the estimated phase error vector $\hat{\varphi}^t$

The phase error is updated for the mth aperture position:

$$\hat{\varphi}_m^{t+1} = \widehat{\Delta\varphi}_m^{t+1} + \hat{\varphi}_m^t, \tag{2.22}$$

$$\widehat{\Delta\varphi}_m^{t+1} = \angle \left\{ \left(\hat{x}^t\right)^H \mathbf{D}_m \left(\hat{\varphi}_m^t\right)^H y_m \right\}, \tag{2.23}$$

where $\angle$ indicates the phase of a complex number.

The dictionary corresponding to the mth position is then updated:

$$\mathbf{D}_m \left(\hat{\varphi}_m^{t+1}\right) = e^{i\widehat{\Delta\varphi}_m^{t+1}} \mathbf{D}_m \left(\hat{\varphi}_m^t\right). \tag{2.24}$$

The updated dictionary is used for the next iteration until convergence (stopping rule) is reached.

2.3.3 Learned Dictionary

The next approach is to completely adapt the dictionary; for instance, from a learning data set. The optimized OCD tries to represent a large diversity of signals with minimal errors and maximal sparseness. The result is independent of how suitable the starting atoms were for representing the signals. But the learning data set needs to be representative of the signals encountered later on. However, when applied to the inverse problem, there might no longer be any correlation with the information content. This approach is well suited for denoising and compression, but might be less suited for physical interpretation of the representation vector.

For all learned dictionaries, the objective is no longer to recover the representation vector from the measurements based on a known dictionary, but to recover the dictionary with the highest cardinality of the representation vectors for the learning data set $\{y_i\}_{i=1}^N$, each known to be represented by a sparse linear combination over the dictionary $\mathbf{D} \in \mathcal{R}^{n \times K}$. Here, n is the length of each measurement, N the total number of different measurements, and K the number of atoms in the dictionary. So, for each measurement y_i, there exists a representation vector x_i, such that

$$y_i = \mathbf{D}x_i \quad \text{and} \quad \|x_i\|_0 \le T. \tag{2.25}$$

Here, T is the known cardinality, with $T \ll n$. Arranging all learning signal columns in a matrix $\mathbf{Y} \in \mathcal{R}^{n \times N}$, and similarly gathering the coefficient vectors as columns of $\mathbf{X} \in \mathcal{R}^{K \times N}$ yields the desired decomposition:

$$\mathbf{Y} = \mathbf{DX}. \tag{2.26}$$

The problem is thus formulated as follows: given the matrix $\mathbf{Y}$, find the factorizing into an arbitrary dictionary $\mathbf{D}$ with normalized columns, and a sparse matrix $\mathbf{X}$ with no more than T nonzero entries in each column. For instance, the K-SVD algorithm can be applied (Aharon et al. 2006) to obtain the desired solution for $\mathbf{D}$.

Learned dictionaries are not used in our work due to the loss of correlation between the physical properties of the building and the atoms obtained.

2.4 Practical Atom Definition

In this section, the practical implementation of the atoms for the matched filters and OCD is discussed. The approach taken here is the use of known atoms based on physical assumptions; each atom is tuned to the backscatter characteristics of a specific building elemental structure.

2.4.1 Starting Points

The main starting point for the definition of atoms is that buildings comprise straight walls intersecting at right angles. Thus, building walls are either approximately parallel to the aperture or approximately perpendicular to the aperture. At the intersections, the walls form dihedral or trihedral corners. By detecting the walls and corners inside a building, the building layout can be synthesized.

Walls, that is, planar surfaces, and corners are canonical scatterers for which models exist describing the amplitude, phase, and polarization characteristics of the backscatter. These specific characteristics can be exploited to extract the canonical scatterers from the radar data. Here, the phase change induced by the different types of scatterers is used for detection, classification, and location. In 3D radar data, canonical scatterers induce different phase changes over the aperture in elevation and azimuth. Considering only specular reflections, the phase change characteristics of planar surfaces, dihedral corners, and trihedral corners are as follows:

- Planar surfaces are formed by walls that are roughly parallel to the aperture. Large planar walls induce a linear phase change over the aperture in azimuth and in elevation.
- Horizontal dihedral corners induce a linear phase change in azimuth and a quadratic phase change in elevation. Horizontal corners are formed by the junction of a wall and the ceiling or floor of a room.
- Vertical dihedral corners induce a quadratic phase change in azimuth and a linear phase change in elevation. Vertical corners are formed by the junction of two walls of a room.
- Trihedral corners induce a quadratic phase change in azimuth as well as in elevation. Trihedral corners are formed by the junction of two walls and the floor or ceiling of a room.

Walls perpendicular to the aperture are difficult to detect since they only give rise to diffuse scattering in the direction of the radar. The presence of perpendicular walls needs to be deduced from the detected room corners.

The ideal phase change for a certain type of scatterer is affected by transmission through walls, but the polynomial degree of the phase change will be preserved. It is noted that, in ground penetrating radar applications, the quadratic range migration curve is observable in a wide variety of soil types. It is shown here that the polynomial degree is indeed preserved after propagation through a single wall based on the geometry presented in Figure 2.2a. The wall with thickness d is assumed to be homogeneous with relative permittivity $\varepsilon_{r,2}$. The medium in which the wall is placed is also assumed to be homogeneous, but with relative permittivity $\varepsilon_{r,1}$.

The radar system is located at (x_r, y_r) and the scatterer is located at (x_s, y_s). The true path of the radar wave propagating through the wall to the

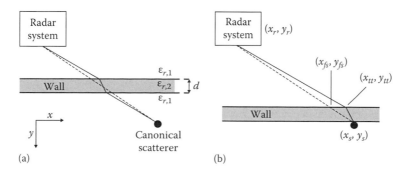

FIGURE 2.2
Schematic geometry of radar wave propagation through a single wall (a) and the re-arranged layout used for the calculation of the path length (b).

scatterer and back to the radar is indicated by the solid line. The dashed line designates the free-space path; that is, the propagation path if the wall would be absent. To determine the length of the *true* propagation path through the wall, the three layers are rearranged as shown in Figure 2.2b. Now, by using the following linear approximation (Johansson and Mast 1994), the diffraction point (x_{tt}, y_{tt}) can be obtained:

$$x_{tt} = x_s + \sqrt{\frac{\varepsilon_{r,1}}{\varepsilon_{r,2}}} \cdot \left(x_{fs} - x_s\right), \tag{2.27}$$

in which (x_{fs}, y_{fs}) is the point where the free-space path intersects the surface of the wall. Once the diffraction point has been obtained, the two-way *true* path length can be determined and the corresponding two-way *true* time delay follows as:

$$\Delta t_{tt} = \frac{2\sqrt{(x_r - x_{tt})^2 + (y_r - y_{tt})^2}}{c} + \sqrt{\varepsilon_{r,2}} \cdot \frac{2\sqrt{(x_{tt-x_s})^2 + d^2}}{c}, \tag{2.28}$$

in which c is the speed of light. The two-way time delay related to the free-space path is given as

$$\Delta t_{fs} = \frac{2\sqrt{(x_r - x_s)^2 + (y_r - y_s)^2}}{c}. \tag{2.29}$$

In Figure 2.3, the two-way difference between the *true* time delay and the free-space time delay is shown, when translated to the corresponding phase error (the frequency is 2.3 GHz). The phase error is given as a function of aperture length, that is, in the figure an aperture of length 0.8 m is located between −0.4 m and +0.4 m. The scatterer is located at 10 m range from the radar aligned with the middle of the aperture. The phase error is given for propagation through a single 15 cm thick wall (solid line) and for propagation through a single 25 cm thick wall (dash-dotted line). The wall is assumed to be dry concrete, that is, $\varepsilon_{r,2} = 8$ (Thajudeen et al. 2011), and the medium is air, that is, $\varepsilon_{r,1} = 1$. A 25 cm thick concrete wall is considered to be worst case; in typical houses and office buildings, concrete walls thicker than 25 cm are not anticipated.

As can be seen in Figure 2.3, the polynomial degree of the phase change is indeed preserved. The deviations from the ideal free-space phase change are small. For example, the required aperture length for 50 cm resolution at 10 m range is 1.40 m, corresponding to an aperture of −0.7 m to +0.7 m in the figure. At the edges of the required aperture, the phase errors are 1.4° for the 15 cm thick wall and 2.3° for the 25 cm thick wall, respectively. These phase errors are assumed negligible for the described feature extraction and representation processes.

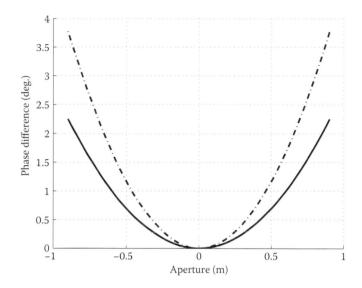

FIGURE 2.3
The two-way phase difference between free-space wave propagation and wave propagation through a single concrete wall of 15 cm thickness (solid) and 25 cm thickness (dash-dotted), respectively.

The two-way offset between the free-space time delay and the *true* time delay is 1.83 ns for the 15 cm thick wall and 3.05 ns for the 25 cm thick wall, leading to range errors of 27 and 46 cm, respectively. These range errors are acceptable considering representation of the major structural elements of a building onto a rather coarse grid, for example, 50 cm grid spacing.

The approximations discussed in this section are valid for representation onto a relatively coarse grid. If high-resolution imagery is desired of objects behind a wall (implying a long aperture), the effects of propagation through the wall need to be taken into account (e.g., Zhang et al. 2011). Representation results are expected to degrade when concrete walls are thicker than 25 cm, when they contain gravel, when they are inhomogeneous or steel-reinforced, etc.

2.4.2 Atom Definition

In this section, the definition of the atoms is discussed. For reasons of clarity, the discussion here is limited to atom definition in the 2D plane. This is sufficient to obtain the building floor plan since it can be deduced from the location of walls and corners in a B-scan. Consequently, only phase changes related to planar walls, dihedral corners, and point-like scatterers are treated. In the 2D plane, trihedral and dihedral corners cannot be distinguished because they give rise to the same quadratic phase change.

Trihedral corners are present at the intersection of two walls and the floor or ceiling of a room. Thus, extraction of trihedral corners provides primarily information on the height of a room. To extract trihedral corners, the representation process needs to be performed on a B-scan and sequentially on a E-scan. When the representation indicates the presence of a dihedral corner for corresponding grid points in both scans, it is likely that a trihedral corner is present. Another option is to define full 3D atoms, but this will significantly increase the dimensions of the representation problem and thus the required processing load.

As mentioned in the previous section, the polynomial degree of the phase change is preserved after propagation through a single wall. The atoms are therefore based on the free-space phase changes; the effect of propagation through one or more walls is neglected. Moreover, the defined atoms are based on specular reflections from walls and corners. This approach mitigates multipath reflections assuming these do not match the defined phase changes.

The application of an atom tuned to a quadratic phase change enables detection and classification of dihedral and trihedral corners. The part of the phase change parabola that is observable in the radar data depends on the orientation of the corner relative to the aperture, as is schematically shown in Figure 2.4. For dihedral corners that are oriented such that one side is perpendicular to the aperture, only half of the phase parabola is observable. Thus, to discriminate left (no. 1) and right (no. 3) dihedral corners, distinction must be made between *forward-looking* and *backward-looking* atoms, each defining one half of the phase parabola. The *forward-looking* phase change (no. 1) is defined as

$$\varphi_i(x) = \frac{4\pi}{\lambda}\sqrt{R_0^2 + x^2}, \quad \text{with} -L \leq x \leq 0, \tag{2.30}$$

where λ is the radar wavelength, R_0 is the shortest distance between the aperture and the scatterer, L is the aperture length, and x denotes azimuth

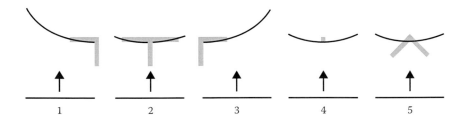

FIGURE 2.4
Schematic picture of five quadratic-phase atoms. For different types of scatterers (gray), the associated quadratic phase change (black curves) over the aperture (horizontal lines) is schematically indicated. The arrows denote the radar line of sight.

position along the aperture. The *backward-looking* phase change (no. 3) is defined as

$$\varphi_i(x) = \frac{4\pi}{\lambda}\sqrt{R_0^2 + x^2}, \quad \text{with} \quad 0 \le x \le L. \tag{2.31}$$

For a double T-shaped corner (no. 2), the full-phase parabola is observable in the radar data. This type of scatterer can be extracted by applying a third atom defining the full-phase parabola. However, to prevent classification confusion, the atoms should be orthogonal. The *forward-looking* and *backward-looking* atoms are orthogonal, but they are both correlated with the full-parabola atom. Consequently, the full-parabola atom is not applied in the smashed filter and the sparse representation with an OCD; it is assumed that a T-shaped corner is present when the representation yields a certain minimum value for both the *forward-looking* and the *backward-looking* atoms at a single point in the representation grid. The ratio between the two values is an indication of the angle of the T-shaped corner with respect to the aperture (Chen et al. 2013a,b).

As indicated in Figure 2.4, point-like scatterers (no. 4) also induce a full-parabolic phase change. By applying the defined *forward-looking* and *backward-looking* atoms, point-like scatterers and T-shaped corners are indistinguishable. Objects inside the building such as furniture may act as point scatterer and lead to false alarms as far as the building map is concerned. This problem can be overcome by first detecting the building inner walls and subsequently by searching for corners only in the range bins where walls have been found (Lagunas et al. 2012, 2013a,b). This approach also reduces the processing load since the search for corners is done on a limited number of grid points. A potential disadvantage of this approach is that building structural elements such as free-standing concrete pillars in office buildings may be missed. Note that point-like scatterers include corners oriented with the open side toward the aperture (no. 5).

Planar walls that are roughly parallel to the radar aperture induce a linear phase change over the aperture. In practice, the exact orientation of walls relative to the aperture is unknown and several atoms tuned to different wall orientations need to be defined (see Figure 2.5). To obtain orthogonal atoms, the difference in angle $\Delta\theta$ between the wall orientations defined in successive atoms is determined by the frequency resolution.

The minimum angular step size $\Delta\theta$, given the frequency resolution, is derived with the aid of Figure 2.5. Within the aperture length, the range to the wall can be written as

$$R_i(x) = R_0 + x \cdot \tan(i \cdot \Delta\theta), \quad \text{with} - L/2 \le x \le L/2, \tag{2.32}$$

in which $i \cdot \Delta\theta$ is the angle between the wall and the aperture, and $i = 0, 1, \ldots, n$. Here, n is the number of linear phase change atoms. The number of

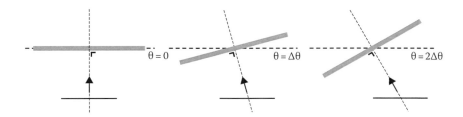

FIGURE 2.5
Schematic picture of three linear-phase atoms. Different atoms need to be defined for different angles between the wall (gray) and the aperture (black lines). The minimum angular step size $\Delta\theta$ is determined by the frequency resolution. The arrows indicate the radar line of sight; note that the specular reflection of the wall is considered.

sensible atoms is limited by the following condition: $n \cdot \Delta\theta \ll 90°$. The (linear) phase change related to (2.32) is

$$\varphi_i(x) = \frac{4\pi}{\lambda} R_i(x),\qquad(2.33)$$

and the corresponding frequency is

$$f_i(x) = \frac{1}{2\pi} \frac{d\varphi_i(x)}{dx} = \frac{2\tan(i \cdot \Delta\theta)}{\lambda}.\qquad(2.34)$$

The angular step size $\Delta\theta$ should be chosen such that the difference in frequency associated with two successive atoms $\mathbf{A}^i$ and $\mathbf{A}^{i+1}$ is larger than the frequency resolution, that is,

$$f_{i+1} - f_i \geq \Delta f = \frac{1}{L},\qquad(2.35)$$

If (2.35) is fulfilled, the atoms are orthogonal because the response of a wall with angle θ and the response of a wall with angle $\theta + \Delta\theta$ are in different frequency bins. By combining (2.34) and (2.35), the following precondition for the minimum angular step size can be derived:

$$\tan(\Delta\theta) \geq \frac{\lambda}{2L}.\qquad(2.36)$$

The defined linear and parabolic atoms are of the following format:

$$\mathbf{A}_i = R_0^\alpha \cdot \exp(j\varphi_i), \quad \text{with } \alpha > 0,\qquad(2.37)$$

in which α is an attenuation parameter, and φ_i is the phase change, defined by either (2.30), (2.31), or (2.33). The attenuation parameter is used to compensate signal attenuation. It may be tuned to compensate free-space losses ($\alpha = 2$) or to compensate attenuation due to propagation through walls ($\alpha > 2$), c.f. the Kirchhoff migration algorithm. Regarding matched filter operation, compensation of propagation losses is not mandatory to maintain the detection probability as a function of range since all cells within a range bin experience the same loss. For sparse representation, the attenuation parameter α plays an important role because sparse representation is an optimization procedure considering all grid points simultaneously. By setting the attenuation parameter to a higher value, points of the representation grid close to the aperture are penalized, enabling detection and classification of canonical scatterers deeper inside the building.

As stated, the atoms need to be orthogonal to prevent classification confusion. However, the atoms within a single model may not be orthogonal, for example, due to oversampling. Also, the quadratic and linear phase change atoms will be correlated since a parabola can be piece-wise approximated by line segments. The level of correlation depends on the measurement geometry and the location of the grid point under test.

2.5 Through-Wall Radar Measurements

The presented approaches have been assessed using measured 3D radar data. For these measurements, the *SAPPHIRE* through-wall radar was used (Smits et al. 2009). SAPPHIRE is designed as a stand-off radar system to sample the observation volume by moving a vertical array parallel to the building (as described in Section 2.1.2). The system is specified to obtain resolution better than 50 cm in 3D at 10 m range. These specifications are chosen such that at least the first rooms, closest to the radar track, are measured with full resolution. The specified resolution allows detection of doorways and separation of walls even in very small rooms. Or in other words, considering the typical sizes of rooms, it is expected that there is at most one building structural element per resolution cell. SAPPHIRE exploits the frequency modulated continuous wave radar principle. The carrier frequency is 2.3 GHz. The four linear polarization pairs are measured (i.e., VV, VH, HV, and HH).

In the next section, the building that has been measured is described in detail. In the succeeding sections, the results obtained with the aforementioned three different feature extraction approaches are presented. All B-scans shown are obtained using 3D free-space back projection. They have varying azimuth extent because of the margins that have to be taken into account for the synthetic aperture length.

2.5.1 Building Layout

The through-wall radar measurements were performed on a three-story building. The layout of the ground floor is schematically shown in Figure 2.6. Note that the radar track is on the azimuth axis. The gray area indicates the building interior. The building consists of a central hallway with rooms of different sizes on either side. The floor and ceilings are concrete. The outer walls of the building comprise concrete pillars with brick walls and windows in between, apart from room no. 1 that has no windows, as annotated in the figure. Here there are just brick walls between the pillars. The walls dividing the rooms are all brick walls. Room no. 3 is split into two by a glass wall with a door. The windows are all fitted with metal venetian blinds. During the radar measurements, the blinds were up, except for rooms 2, 7, 8, and 9.

The rooms on both sides of the hallway are used as storage or offices and contain furniture such as tables, chairs, and metal cupboards. In particular in room no. 4, a storage room, four metal cupboards are placed directly behind the window. Room no. 3 is empty. In the middle of this room, a trihedral radar reflector is placed on a tripod at 1.5 m height as reference.

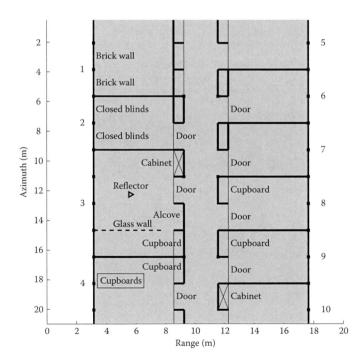

FIGURE 2.6
Schematic view of the building layout. The squares indicate concrete pillars, the thick lines indicate brick walls, and the thin lines indicate wooden or metal doors.

The sides of the hallway consist of concrete pillars and brick walls forming open alcoves, cupboards, or closed spaces. The cupboards are around 70 cm deep, have wooden shelves, and are closed by wooden doors. The cupboards crossed-out in Figure 2.6 are switchboard cabinets with metal doors. The contents of the closed spaces are unknown. The doors of the rooms are all wood.

2.5.2 Point Scattering Focusing

Conventional focusing is based on point-like behavior of scatterers; that is, basically the data are matched filtered using the full-parabola atom with $\alpha = 0$. Therefore, point-like scatterers, corners, and T-shaped corners cannot be distinguished (see Figure 2.4).

2.5.2.1 Reflectivity Maps

In Figure 2.7a, the obtained radar image is shown for VV polarization. In this figure, a B-scan at 1.5 m elevation is presented. The chosen voxel size is 25 cm in three dimensions. The shown B-scan is exactly at the height of the reference reflector placed in room no. 3. Consequently, the response of the reflector is clearly visible; it includes four voxels matching the resolution of 50 cm. At the lower edge of the image, the reflection of the metal cupboards in room no. 4 is visible.

The building façade at 3.5 m range gives rise to an apparent response along the complete azimuth extent of the measurement. In particular, the closed metal blinds in room no. 2 cause high reflections. At the same time

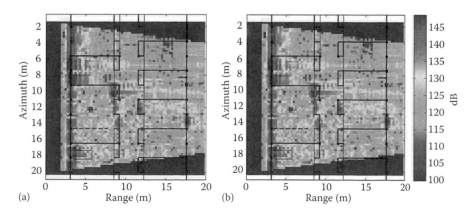

FIGURE 2.7
(See color insert.) B-scan obtained with the matched filter tuned to point-like scatterers, VV polarization (a) and HH polarization (b).

they shield the building interior; behind the closed blinds no inner-building features are discernible. Also, the second parallel wall, around 9 m range, induces a detectable response, apart from the area where it is shielded by the blinds. The switchboard cabinet in room no. 3 has a relatively high reflection due to its metal doors. Generally, the third parallel wall, around 12 m range, is undetectable. This is due to the losses after propagation through two walls. In line with the door of room no. 3, the reflection of the third wall is observable; since the losses due to propagation through the wooden door are relatively low (the door was closed during this measurement). The fourth parallel wall at 17.5 m range is indiscernible aside from a high reflection around 14 m azimuth. This reflection is in line with the alcove in room no. 3, that is, a single brick wall, and the door of room no. 8. Thus, the propagation losses at this particular azimuth location are relatively low. Moreover, the metal blinds before the window of room no. 8 were closed during the measurement. Note that the walls perpendicular to the (synthetic) aperture are indeed undetectable. As explained, this is because perpendicular walls induce only diffuse scattering in the direction of the radar.

The radar image obtained for HH polarization is shown in Figure 2.7b. In the figure, again the B-scan at 1.5 m elevation is presented. The HH-polarized image is similar to the VV-polarized image; that is, it reveals the same building features. That is anticipated because the main building structures are large planar surfaces (i.e., walls) that reflect both polarizations equally well. Considering the wavelength of 12.6 cm, even the closed venetian blinds in room no. 2 form a smooth, impassable surface for both VV and HH polarization. Furthermore, the building structure does not include specific vertically or horizontally oriented elements that could give rise to polarization-dependent backscatter, apart from the metal pipes of the heating system. These vertical pipes are however attached to the concrete pillars of the outer walls, such that a pillar and the attached pipes are in the same resolution cell. Consequently, the pipes are not detectable since the backscatter from the pillars is dominant and independent of polarization.

2.5.2.2 Classified Scatterers

From Figure 2.7, it is clear that through-wall radar images are difficult to (visually) interpret without some knowledge of the actual building layout. As a first step in extracting building features, a fixed-threshold detector is applied to the B-scans. The fixed-threshold detector neglects signal attenuation as function of range, but the integration length used in the back projection increases as function of range to maintain constant resolution, partly compensating the signal losses.

The resulting detections are shown in Figure 2.8. These results confirm that indeed the same building structural elements are highlighted in VV and HH polarization since most detected voxels are detected in both VV and HH

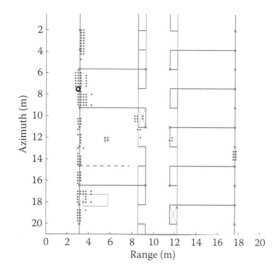

FIGURE 2.8
Detections obtained from the VV-polarized B-scan (**o**). The points denote voxels that are detected in both the VV-polarized B-scan and the HH-polarized B-scan.

polarization. Just one voxel is only detected in VV polarization. Since the VV-polarized and HH-polarized images seem to contain similar information, only results for VV polarization are shown in the remainder of this chapter.

It is apparent that the detections appear in groups. This is due to the chosen voxel size as well as the detection of sidelobes. The chosen voxel size is smaller than the resolution, thus adjacent voxels are correlated. This is clear, considering the response to the reference reflector in room no. 3; it yields four voxels corresponding to a single resolution cell of 50 cm by 50 cm. The metal blinds in room no. 2 give rise to high reflections and correspondingly high sidelobe levels. Here also the sidelobes are detected. The second step of automated feature extraction is therefore plot extraction, that is, clustering of detections.

First, the detections are clustered in range to remove detected sidelobes and correlated voxels. Considering detections in adjacent range bins, only the detection corresponding to the voxel with the highest reflection is kept, the other detections are removed. Subsequently, the remaining detections are clustered in azimuth and the detections within a cluster are aligned, that is, shifted to a single-range bin. The plots resulting after clustering and alignment are shown in Figure 2.9a. The extracted plots are input to the third step of automated feature extraction: plot combination. This third step is needed to translate the extracted plots to building layout information suitable to input to the building mapping process.

The plots obtained with the matched filter approach cannot be used to discriminate different types of scatterers. Consequently, the information to

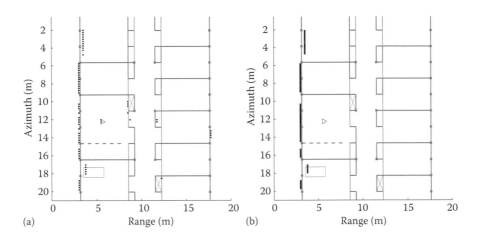

FIGURE 2.9
The plots remaining after clustering (a) and the result of the line extraction using the Hough transform (b), VV polarization.

be gained with plot combination is limited. One possibility is the use of the Hough transform to extract the start and end points of line-like plot structures. Line-like plot structures are due to elongated scattering surfaces, most likely associated with walls. The results obtained with the Hough transform are presented in Figure 2.9b. As can be seen, the building façade is almost completely reconstructed. It can also be seen that the Hough transform approach does not preserve single plots or small clusters of plots. Such small clusters of plots may also contain information about the building structure. Therefore, at this point it seems advisable to present both the uncombined and the combined plots to an analyst.

2.5.3 Smashed Filter Processing

Potentially, the application of several matched filters tuned to different types of scatterers, as in the smashed filter approach, provides additional information for scatterer classification.

For this particular measurement setup, it is known that all walls are either parallel or perpendicular to the radar platform track. Therefore, only a single matched filter tuned to linear phase changes is used, viz., the filter tuned to $\theta = 0$. The attenuation parameter α is also set to zero.

2.5.3.1 Reflectivity Maps

The B-scans obtained with the smashed filter are presented in Figure 2.10. The reflections of the metal blinds of rooms no. 2 and no. 8, the metal doors of the switchboard cabinet in room no. 3, and the metal cupboards in storage room no. 4 appear clearly in all three B-scans. The reflections from these

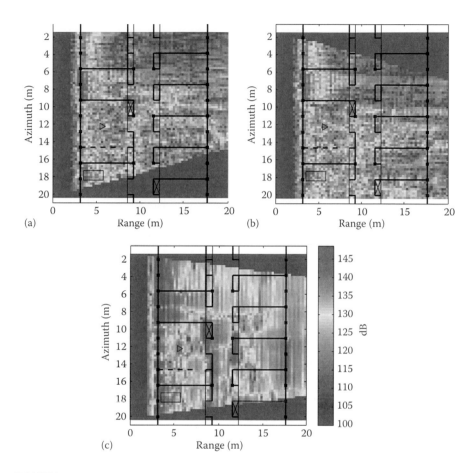

FIGURE 2.10
(See color insert.) B-scans obtained with the *forward-looking* matched filter (a), the *backward-looking* matched filter (b), and the *linear phase* matched filter (c), VV polarization.

metal surfaces are so high that they stand out above the background regardless of the phase reference used in the matched filter. Concentrating on the reflections of the switchboard cabinet in room no. 3, it can be seen that the highest response of the *forward-looking* filter is located exactly in the corner of the room, suggesting strong dihedral-like behavior. The highest response of the *backward-looking* filter is located at the cabinet doors and, at the same time, the response of the corner of the room is low. The response to the reference reflector, on the other hand, is high for both the *forward-looking* and the *backward-looking* filter since this reflector acts as a point scatterer. These results imply that the starting points for the atom definition are sensible. Note that the response to the reference reflector is also apparent in the B-scan obtained with the *linear phase* filter. This is because in broadside view, that is, around its minimum, the parabolic phase change can be approximated by a line segment parallel to the synthetic aperture.

Compared with the application of a single matched filter tuned to point-like scatterers, the smashed filter provides additional information; walls, left corners, and right corners can now be distinguished. To aid visual inspection by an analyst, automated extraction of building features is still desired, for example, to detect coinciding left and right corners implying the presence of a double corner or point scatterer.

2.5.3.2 Classified Scatterers

For detection and plot extraction, the same processing steps are performed as for the matched filter approach described in the previous section. The detections obtained from the different B-scans are shown in Figure 2.11 and the corresponding plots are shown in Figure 2.12.

Now the plots obtained with the separate filters need to be combined. The plot combination is based on the following rules:

- An uninterrupted row of at least four similar corners is inter-preted as a wall. From inspection of the B-scans, it is concluded that planar surfaces inducing high reflections are detected by all three matched filters. Furthermore, considering typical buildings, it is unlikely that walls perpendicular to the aperture are so close together. Consequently, this rule is believed to be sensible.
- Coinciding left and right corners are interpreted as a double T-shaped corner. Corners are seen as *coinciding* if they are less than four voxels (i.e., two resolution cells) apart in any direction. If several plots are coinciding, the left and right corners that are closest are combined into a double corner.
- The remaining plots related to left corners are combined into a single left corner if they are less than four voxels apart. This rule applies to right corners as well.

The results after combining the plots of the different filters are presented in Figure 2.13. The scatterers classified as walls correspond to planar surfaces. These surfaces are not all related to actual building walls. Also, the cupboards in room no. 4 form a planar surface and are thus correctly classified as a wall. The classified double corners are related to point-like scatterers such as the concrete pillars of the building façade and the reference reflector. However, areas of high reflection are also classified as double corners, for example, the metal doors of the switchboard cabinet in room no. 3 and the closed metal blinds in room no. 8. As previously explained, this is because such high reflections stand out against the background regardless of the phase reference used in the matched filter. The corner of room no. 3 next to the switchboard cabinet is correctly classified as a left corner. The other

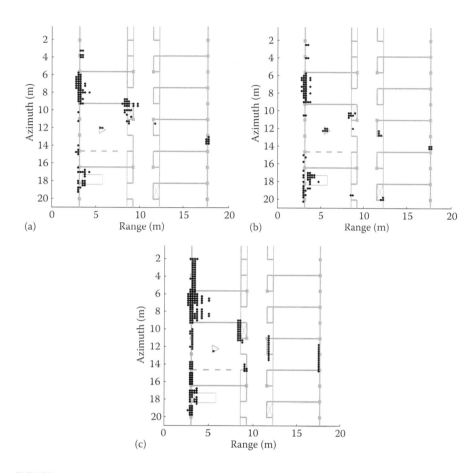

FIGURE 2.11
Detections obtained with the *forward-looking* matched filter (a), the *backward-looking* matched filter (b), and the *linear phase* matched filter (c), VV polarization.

classified single corners seem due to objects giving high reflections and acting like point-like scatterers or possibly classification confusion in packed areas, such as storage room no. 4.

The combined plots provide the building feature information that can be fed to the building mapping process. However, the uncombined plots in Figure 2.12 seem to contain information discarded in the decision process of plot combination. Thus, it seems again worthwhile to present both the uncombined and combined plots to an analyst. One reason is that, by using the listed rules, the combined plots provide a definite classification. In areas where many corners are detected close to each other, the current rules combine these resulting in just a few plots, for instance, in the region around the switchboard cabinet in room no. 3. A trained analyst may conclude that the clouds of corners are more likely due to classification confusion, for

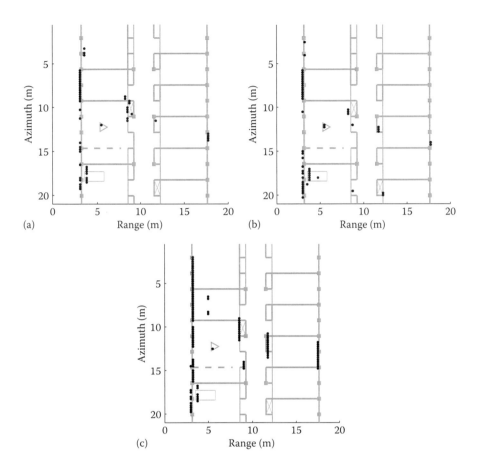

FIGURE 2.12
Plots obtained with the *forward-looking* matched filter (a), the *backward-looking* matched filter (b), and the *linear phase* matched filter (c), VV polarization.

example, because of furniture in the room, or high reflections, than that they are related to building structural elements.

The plot extraction and plot combination rules are based on the physics underlying the radar measurements and constraints given by typical building layouts. Nevertheless, the rules are debatable and other rules could be applied. This is another reason why it might be advisable to examine the uncombined plots. The analyst is able to judge the plot combination outcome and, possibly based on experience, alter the findings.

2.5.4 OCD with Sparse Representation

In this section, the building features are obtained with the use of OCDs. The ℓ_1-regularized least-squares solver used is complex approximate message passing (CAMP). This algorithm will be described in more detail in the next

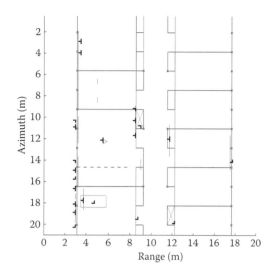

FIGURE 2.13
The combined plots obtained with the smashed filter, VV polarization.

section. The results obtained are similar to the smashed filter approach: For each scattering type, a noisy representation is obtained. The noisy representations are then used as input for detection, plot extraction, and classification algorithms.

2.5.4.1 CAMP Algorithm

CAMP (Anitori et al. 2013 and references therein) is an algorithm based on ℓ_1-regularized least squares, also known as the LASSO or basis pursuit denoising (BPDN). The LASSO is given by

$$x = \arg\min_{x} \frac{1}{2} \|y - \mathbf{D}x\|_2^2 + \lambda \|x\|_1 . \tag{2.38}$$

where
  x and y are complex quantities
  λ is the regularization parameter

The ideal algorithm is given in Algorithm 2.1. The algorithm is denoted ideal due to the knowledge of the true signal, x_0.

  In Algorithm 2.1, $\hat{x}^t$ is the sparse representation of x_0, $\tilde{x}^t$ a nonsparse, noisy estimate of x_0, σ^t is an estimate of the output noise, t denotes the iteration number, δ is the number of measurements divided by the number of samples to be recovered, and ε is the stopping-rule threshold. The soft thresholding function, η, is applied component-wise and is defined as

---

Algorithm 2.1 CAMP algorithm

Input: y, $\mathbf{D}$, τ, x_0, ε
Initialization: $\hat{x}^0 = 0; z^0 = y, t = 0$

Repeat
$t = t + 1$
$\tilde{x}^t = \mathbf{D}^H z^{t-1} + \hat{x}^{t-1}$
$\sigma^t = std\left(\tilde{x}^t - x_0\right)$
$z^t = y - \mathbf{D}\hat{x}^{t-1} + z^{t-1}\frac{1}{2\delta}\left(\left\langle\frac{\partial \eta_R}{\partial x_R}\left(\tilde{x}^t;\tau\sigma^t\right)\right\rangle + \left\langle\frac{\partial \eta_I}{\partial x_I}\left(\tilde{x}^t;\tau\sigma^t\right)\right\rangle\right)$
$\hat{x}^t = \eta\left(\tilde{x}^t;\tau\sigma^t\right)$
Until: $\left\|\hat{x}^t - \hat{x}^{t-1}\right\|_2 < \varepsilon$
Output: $\tilde{x}, \hat{x}, \sigma$

$$\eta\left(x;\lambda\right) \overset{\Delta}{=} \left(|x| - \lambda\right)e^{i\angle x}\mathbf{1}\left(|x| > \lambda\right). \tag{2.39}$$

Here, $\mathbf{1}$ is the indicator function. In Anitori et al. (2013), the connection between τ in CAMP and λ in LASSO is given. In the nonideal case, the output noise needs to be estimated based on the reconstructed result, for example, by considering the median of the output vector.

The CAMP algorithm is used for the noisy output, which allows the use of CFAR detection and also allows the algorithm to become adaptive (Anitori et al. 2013), which solves the problem of choosing the optimum value for the regularization parameter (τ in CAMP or λ in LASSO). In Adaptive-CAMP, the optimal regularization parameter τ is obtained by minimizing the output noise estimate.

2.5.4.2 Reflectivity Maps

The data used as input to Algorithm 2.1 have been resampled down to the Nyquist rate in order to have an underdetermined problem when applying CAMP to obtain the representation vector. Also, only data up to 10 m in range were used in order to keep the computational burden sufficiently low. The output of Algorithm 2.1 is a noisy representation vector for all three different models included in the OCD. These representations can be used as equivalent to reflectivity maps (see Figure 2.14).

2.5.4.3 Classified Scatterers

The rules for obtaining classified scatterers are identical to the rules used for the smashed filter approach (see Section 2.5.4.2). The detections are shown in Figure 2.15; the plots obtained after clustering are shown in Figure 2.16, while the combined plots are shown in Figure 2.17.

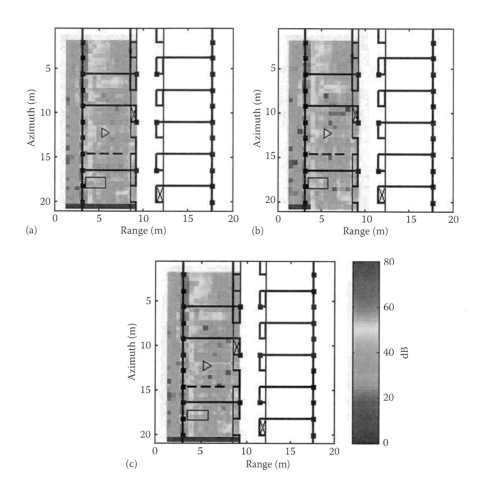

FIGURE 2.14
(See color insert.) Equivalent B-scans obtained with OCD: the *forward-looking* part (a), the *backward-looking* part (b), and the *linear phase* part (c), VV polarization.

The size of the matrix with which the sparse reconstruction is performed is three times larger than the other matrices. This yields some practical limitations on the number of grid points that can be taken into account. The grid considered in the OCD reconstruction is much smaller. This is the reason why only the first and second walls are detected, while in the other results the third and fourth walls have been detected as well.

Although the OCD yields the best reconstruction of the data, there is still some confusion due to coherency between atoms. This coherence is not important when considering the data reconstruction but could be important for building mapping. When a large wall is reconstructed as a row of point scatterers, then the clustering algorithm will yield a long cluster, which in

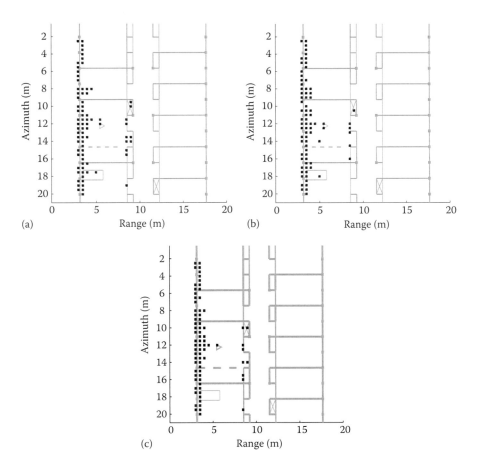

FIGURE 2.15
Detections obtained with OCD: the *forward-looking* part (a), the *backward-looking* part (b), and the *linear phase* part (c), VV polarization.

the end will still be identified as a wall. So the building reconstruction is not compromised. But when the wall is reconstructed as separated point targets, they are not clustered and no wall will be declared. In the latter example, an analyst could still decide to declare a wall based on extra information about buildings in general.

2.6 Conclusion

In this chapter, the use of OCDs for inside-building feature extraction has been investigated. The OCD approach has been compared with more traditional approaches using a single, point scatterer–based model alone

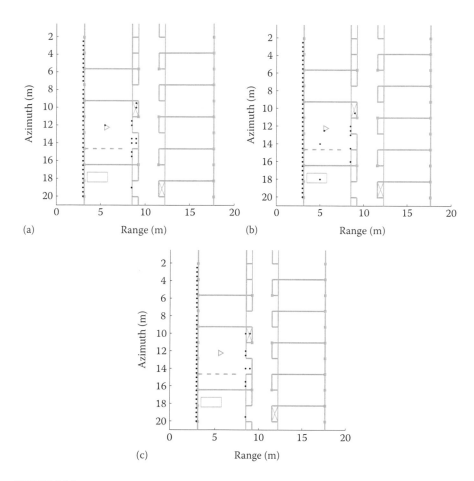

FIGURE 2.16
Plots obtained with OCD: the *forward-looking* part (a), the *backward-looking* part (b), and the *linear phase* part (c), VV polarization.

(conventional synthetic aperture radar approach) or using multiple models independently. Three different models were considered for the smashed filter and OCD: left corner, right corner, and wall. All three approaches have been assessed using measured through-wall radar data.

An advantage of OCD over the other two approaches is the possibility of reconstruction of all models simultaneously. This allows for the regularization of the combined number of scatterers. A drawback of the OCD is the large forward-model matrix. This OCD matrix is N-times larger than its size in the other approaches, N indicates the number of models that are simultaneously reconstructed by the sparse reconstruction algorithm.

In the experiment, the large reflection from the front wall was detected by the three different approaches and with the use of an expert system classified as a large wall. The wall hypothesis is not available in point scattering

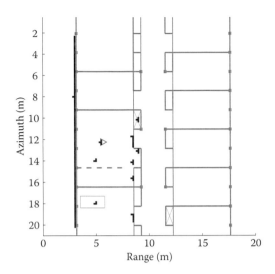

FIGURE 2.17
The combined plots obtained with OCD, VV polarization.

focusing. In this case, the observation that point scatterers associated with a wall lie on a line, can be exploited to locate walls, for example, by using the Hough transform. The distinctions between the smashed filter and OCD approaches are minimal.

By combining the outputs of the three different models, the wall, dihedral corners, and point-like scatterers could be classified. The reconstruction of an inside-building map based on these features was not performed. An automatic reconstruction of the map based on the extracted features is one of the challenges of inside-building mapping. Other challenges include distinguishing between structural features and furniture and detecting humans inside the building.

References

Aftanas, M. Through wall imaging with UWB radar system. PhD dissertation, Technical University of Košice, Košice, Slovakia, 2009.

Aftanas, M. and M. Drutarovský. Imaging of the building contours with through the wall UWB radar system. *J. Radioeng*. 18 (3), 2009, 258–264.

Aharon, M., M. Elad, and A.M. Bruckstein. K-SVD: An algorithm for designing overcomplete dictionaries for sparse representation. *IEEE Trans. Signal Process*. 54 (11), 2006, 4311–4322.

Anitori, L., A. Maleki, M.P.G. Otten, R.G. Baraniuk, and P. Hoogeboom. Design and analysis of compressed sensing radar detectors. *IEEE Trans. Signal Process*. 61 (4), 2013, 813–827.

Baranoski, E.J. Through-wall imaging: Historical perspective and future directions. *J. Franklin Inst.* 345 (6), 2008, 556–569.

Chen, B., T. Jin, Z. Zhou, and B. Lu. Estimation of pose angle for trihedral in ultrawideband virtual aperture radar. *Progr. Electromagnet. Res.* 138, 2013a, 307–325.

Chen, B., T. Jin, Z. Zhou, and B. Lu. Estimation of trihedral pose angle from virtual aperture radar image. In *Proceedings of the 43rd European Microwave Conference (EuRAD)*, Nuremberg, Germany, 2013b, pp. 507–510.

Davenport, M.A., M.F. Duarte, M.B. Wakin et al. The smashed filter for compressive classification and target recognition. In *Proceedings of the Computational Imaging V at SPIE Electronic Imaging*, San Jose, CA, 2007.

Donoho, D.L. and M. Elad. Optimally sparse representation in general (non-orthogonal) dictionaries via l^1 minimization. *PNAS* 100 (5), 2003, 2197–2202.

Ertin, E. and R.L. Moses. Through-the-wall SAR attributed scattering center feature estimation. *IEEE Trans. Geosci. Rem. Sens.* 47 (5), 2009, 1338–1348.

Huffman, C. and L. Ericson. Through-the-wall sensors for law enforcement market survey. Technical Report of National Institute of Justice NLECTC, 2012.

Johansson, E.M. and J.E. Mast. Three-dimensional ground-penetrating radar imaging using synthetic aperture time-domain focusing. *Proc. SPIE* 2275, 1994, 205–214.

Lagunas, E., M.G. Amin, F. Ahmad, and M. Nájar. Sparsity-based radar imaging of building structures. In *Proceedings of the 20th European Signal Processing Conference (EUSIPCO)*, Bucharest, Romania, 2012, pp. 864–868.

Lagunas, E., M.G. Amin, F. Ahmad, and M. Nájar. Determining building interior structures using compressive sensing. *J. Electron. Imag.* 22 (2), 2013a, 021003-1–021003-15.

Lagunas, E., M.G. Amin, F. Ahmad, and M. Nájar. Improved interior wall detection using designated dictionaries in compressive urban sensing problems. *Proc. SPIE* 8717, 2013b, 87170K-1–87170K-7.

Lavely, E.M., Y. Zhang, E.H. Hill, Y. Lai, P. Weichman, and A. Chapman. Theoretical and experimental study of through-wall microwave tomography inverse problems. Special issue, Ed. M.G. Amin. *J. Franklin Inst.* 345 (6), 2008, 592–617.

Le, C., T. Dogaru, L. Nguyen, and M.A. Ressler. Ultrawideband (UWB) radar imaging of building interior: Measurements and predictions. *IEEE Trans. Geosci. Rem. Sens.* 47 (5), 2009, 1409–1420.

Marble, J.A. and A.O. Hero. See through the wall detection and classification of scattering primitives. *Proc. SPIE* 6210, 2006, 62100B-1–62100B-6.

Miller, D., M. Oristaglio, and G. Beylkin. A new slant on seismic imaging: Migration and integral geometry. *Geophysics* 52 (7), 1987, 943–964.

Miller, R.J., W.L. van Rossum, L. Hyde, and J.J.M. de Wit. Radar technology for inside building awareness (RIBA). Technical Report EDA Contract 11.R&T.OP.131, 2012.

NATO RTO. Sensing-through-the-wall technologies. Technical Report RTO-TR-SET-100 AC/323(SET-100)TP/360, 2011.

Nikolic, M.M., M. Ortner, A. Nehorai, and A.R. Djordjevic. An approach to estimating building layouts using radar and jump-diffusion algorithm. *IEEE Trans. Antenn. Propag.* 57 (3), 2009, 768–776.

Önhon, N.Ö. and M. Çetin. A nonquadratic regularization-based technique for joint SAR imaging and model error correction. *Proc. SPIE* 7337, 2009, 73370C-1–73370C-10.

Potter, L.E. and R.L. Moses. Attributed scattering centers for SAR ATR. *IEEE Trans. Image Process.* 6 (1), 1997, 79–91.

Sévigny, P. and D.J. DiFilippo. A multi-look fusion approach to through-wall radar imaging. In *Proceedings of the IEEE Radar Conference*, Ottawa, Ontario, Canada, 2013.

Sévigny, P., D.J. DiFilippo, T. Laneve et al. Concept of operation and preliminary experimental results of the DRDC through-wall SAR system. *Proc. SPIE* 7669, 2010, 766907-1–766907-11.

Sévigny, P., D.J. DiFilippo, T. Laneve, and J. Fournier. Indoor imagery with a 3-D through-wall synthetic aperture radar. *Proc. SPIE* 8361, 2012, 83610K-1–83610K-7.

Smits, F.M.A., J.J.M. de Wit, W.L. van Rossum et al. 3D Mapping of buildings with SAPPHIRE. In *Proceedings of the EMRS DTC Technical Conference*, Edinburgh, U.K., 2009.

Stolt, R.H. Migration by Fourier transform. *Geophysics* 43 (1), 1978, 23–48.

Strohmer, T. and R.W. Heath. Grassmannian frames with applications to coding and communication. *Appl. Comput. Harmon. Anal.* 14 (3), 2004, 257–275.

Subotic, N., E. Keydel, J. Burns et al. Parametric reconstruction of internal building structures via canonical scattering mechanisms. In *Proceedings of the International Conference on Acoustics, Speech, and Signal Processing (ICASSP)*, Las Vegas, NV, 2008, pp. 5189–5192.

Thajudeen, C., A. Hoorfar, F. Ahmad, and T. Dogaru. Measured complex permittivity of walls with different hydration levels and the effect on power estimation of TWRI target returns. *Progr. Electromagnet. Res.* B 30 (2011): 177–199.

Ulander, L., H. Hellsten, and G. Stenstrom. Synthetic aperture radar processing using fast factorized backprojection. *IEEE Trans. Aerosp. Electron. Syst.* 39 (3), 2003, 760–776.

Yilmaz, O. and S.M. Doherty. *Seismic Data Processing. (Investigations in Geophysics, vol. 2).* Society of Exploration Geophysics, Tulsa, OK, 1987.

Zhang, W., A. Hoofar, and C. Thajudeen. Building layout and interior target imaging with SAR using an efficient beamformer. In *Proceedings of the IEEE International Symposium on Antennas and Propagation (APSURSI 2011)*, Spokane, WA, 2011, pp. 2087–2090.

3

Compressive Sensing for Radar Imaging of Underground Targets

Kyle R. Krueger, James H. McClellan, and Waymond R. Scott, Jr.

CONTENTS

ABSTRACT The detection of buried objects is an ongoing and important research topic in application areas such as utility location, treasure hunting, geological exploration, forensics, archeology, snow/ice cover, structural health monitoring, and landmine location. Many of the applications pertain to urban sensing. Different tools have been designed, acquisition systems developed, and detection algorithms created to address this important, wide-ranging problem. In this chapter, we discuss the important aspects of buried target detection with respect to using a ground-penetrating radar

(GPR) applied to locating subterranean landmines with a system that uses compressive sensing (CS) methods. We will show the advantages of using a CS detection algorithm and the considerations that need to be taken to make this type of algorithm efficient enough to solve practical three-dimensional (3D) imaging problems.

3.1 Introduction

GPR has been shown to be an effective tool for imaging a large variety of subterranean targets as they are sensitive to changes in electrical permittivity and conductivity of the subsurface [3,16,33]. Development of GPR systems began in earnest in the early 1970s for applications ranging from detecting buried landmines to locating utilities [30]. Two common transmitter modes for data acquisition for a GPR are time pulse and stepped frequency. While these signals are Fourier transform pairs, the actual acquisition process of each method has its advantages and disadvantages. Time-pulse systems can incorporate automatic gain control (AGC) to adjust the receiver gain vs time/depth in order to increase the signal-to-noise ratio of reflections from deep targets; however, they are susceptible to narrowband interference from communications systems. Stepped frequency GPR systems are much more robust to narrow band noise, but the receiver gain cannot be adjusted for deep targets. Both the time pulse and the stepped frequency systems are hindered by long data acquisition times for many applications, but stepped frequency is generally slower.

Many GPR applications require high spatial resolution, so a large antenna aperture would be required. When trying to locate very shallow buried targets, the desired resolution is on the order of centimeters. The GPR is usually a movable transmit-receive platform that is scanned over a region of interest, for example, some GPRs are on handheld wands. Therefore, it is natural to use a synthetic-aperture approach to acquire coherent data over a large aperture. Synthetic-aperture radar (SAR) has been shown to be effective in GPR applications [14,22,23,31,33,37,38].

SAR is a data acquisition technique that moves a single antenna to different locations along a known path and coherently combines the received samples to create a higher resolution image in the along-track dimension (when the path is linear) [7,15,26]. The system does not make independent decisions at each scan position; instead, it uses the collection of many scans to form the best image. The trade-off with a synthetic aperture is that more measurements are required, which increases the computational complexity of the problem. Data memory is another significant constraint in the GPR problem because many received signals must be collected prior to forming an image.

When the GPR system is used in a mobile environment, the processing would have to be done locally in real time in the device. In other cases, data collection could be done for an entire experiment and then processed offline with a cluster of computers.

The needs for extremely high resolution and data acquisition efficiency have led to investigations of many different inversion techniques when dealing with the GPR problem. An increasingly popular technique that can provide an advantage in both of these areas is CS, which has been studied extensively in the imaging community [5,12,17]. The ability of CS to take a sparse representation of a signal and project it onto a much lower dimensional measurement space can decrease the data acquisition requirements dramatically. CS also produces sparse images, which Gurbuz et al. and Soldovieri et al. have claimed provide an increase to the detection resolution with GPR above that of other commonly used detection algorithms [22,37]. While this has not been directly proven for the general case, the ability to provide sparse solutions does act as an automatic detector and can eliminate the need for image interpretation after the inversion. However, there is a second aspect of CS that is applicable to situations like GPR where a model is available to define the representation matrix Ψ. Often an explicit dictionary matrix, Ψ, is built by enumerating the variables in the expected target response. On the other hand, it might be possible to define a functional representation of Ψ. The dictionary approach can become quite computationally intensive since a 3D imaging problem would require enumeration of a two-dimensional (2D) scan grid, enumeration of a 3D simulated image space, and enumeration of the number of stepped frequencies or time samples. The matrix Ψ scales as $\mathcal{O}(N^6)$ if all variables are generalized to equal discretizations of size N. If a reasonably sized problem has $N = 100$, then 10^{12} complex elements would have to be stored in computer memory. CS can help reduce the number of discretized values that need to be stored for a few variables, but we will show that it is generally not powerful enough. Modifications to the structure and application of Ψ with a functional representation can be used to dramatically reduce the constraints in creating a CS framework that is increasingly more practical for the GPR problem. We envision that this framework can also be adopted in other urban sensing problems, including through-the-wall imaging [1,2].

The remainder of this chapter will be split into four sections. A brief background of the basics of GPR imaging will be given in Section 3.2. The data acquisition system and representation model will be discussed with regards to creating a CS framework for the GPR problem in Section 3.3. Next, some structural modifications will be made to Ψ and the application of Ψ that will result in massive computational advantages in Section 3.4. Finally, the performance of CS in association with applied, real-data experiments will be evaluated and conclusions will be made in Sections 3.5 and 3.6, respectively.

3.2 Background for GPR Imaging

The GPR problem can be described as a parameter-detection problem that relies on model-based inversion. The data flow and processing blocks for this detection system are shown in Figure 3.1. There are two sources of sampling: The environment sensed with a data acquisition system, and a model dictionary matrix, Ψ, constructed using information about the structure of the data acquisition system and the model parameters that need to be extracted. The measurements collected with the data acquisition system and the dictionary Ψ are combined during the inversion step to obtain a 3D image that describes the environment.

The typical data acquisition for GPR can be separated into two categories: time pulse and stepped frequency. The response models for these, as shown in (3.1) and (3.2), are frequency-time pairs, but they have very different characteristics that can effect the CS framework in different ways.

The traditional time-pulse GPR (TPGPR) system sends out a very short electromagnetic pulse from the transmitter (on the order of a nano second). The transmitted pulse reflects off a target, and the reflection is detected by the receiver. The time delay between the received and transmitted pulses is recorded to determine the target location [16]. The data acquisition technique for a GPR includes the creation of a synthetic aperture by moving a sensor to the positions, $l_s = (x_s, y_s, z_s)$. For a 3D image, the scan positions, l_s, would be indexed over a grid in the 2D plane, as shown in Figure 3.2.

In the case where the transmitter and receiver are scanned together, the scalar point target model is

$$r(t, l_s, l_t) = \frac{\rho_1}{S} p(t - \tau(l_s, l_t, c, v)), \qquad (3.1)$$

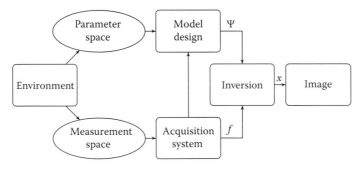

FIGURE 3.1
Detection flow for a GPR system with model-based inversion.

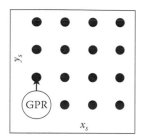

FIGURE 3.2
Two-dimensional scan grid for GPR data acquisition.

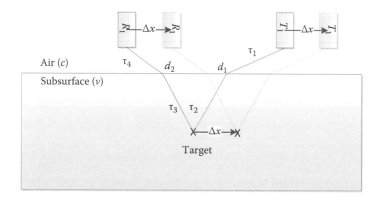

FIGURE 3.3
EM path through multiple mediums.

where

$p(t)$ is a pulse sent from the transmitter at location l_s, which then reflects off a target at location $l_t = (x_t, y_t, z_t)$, with a reflection coefficient ρ_1

S is the electromagnetic spreading function that may or may not be known

τ is a time delay function that uses an approximation of Snell's law to calculate the wave's path through air and at the boundary of a medium with velocity v [24]

An example of the wave path from a bistatic GPR traveling into the ground and reflecting off a target can be seen in Figure 3.3. An example of a simulated time-domain measurement can be seen in Figure 3.4a.

There will also be a reflection from the air–ground interface, and eliminating this ground response is an active research topic in its own right. In particular, some work associated with ground removal for GPR geared toward a CS application was introduced by Tuncer et al., but those details will not be covered in this chapter [40]. To simplify the presentation, the simulations shown in this chapter use a single medium with no velocity changes.

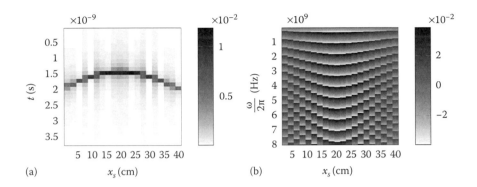

FIGURE 3.4
Simulated measurements from a single-point target in (a) time domain and (b) frequency domain showing the phase.

Furthermore, the subsurface laboratory experiments in Section 3.5.2 were done in an environment with a controlled uniform ground response, so it is possible to subtract out the ground bounce from the measurements since the height of the sensor in the lab is known.

Another way of collecting data with a GPR is to use a stepped frequency GPR (SFGPR). Instead of sending out a short pulse, the GPR sends out a sequence of different constant-frequency sinusoids over a specific bandwidth. The GPR then measures the amplitude and phase at each frequency. For example, the SFGPR system discussed in Section 3.5.2 covers the range 60 MHz to 8 GHz with 401 equally spaced frequencies. These measurements are samples of the Fourier transform of a short pulse, so the phase changes are related to time delay, which can be used to locate the target. An advantage of using stepped frequency over the time–pulse method is that it allows for a wider bandwidth to be covered with equal power because each narrowband sinusoid can be controlled independently [3]. The narrow instantaneous bandwidth also provides resistance to noise and interference [42] because the frequencies can be chosen to avoid communication signals. A disadvantage of using SFGPR is that it can take a long time to acquire data because a different sinusoidal pulse for each stepped frequency must be sent. The target response of a SFGPR,

$$R(\omega, l_s, l_t) = \frac{\rho_2}{S} e^{-j\omega\tau(l_s, l_t, c, v)}, \tag{3.2}$$

has a phase shift instead of a time delay. The response in (3.2) is the frequency domain version of (3.1) if $p(t) = \delta(t)$. An example of the phase of a simulated frequency domain measurement can be seen in Figure 3.4b.

The model dictionary, $\mathbf{\Psi}$, used with a number of different imaging techniques can be created by enumerating (3.2) for all possible parameter discretizations. The first step is to determine which parameters are associated with the measurements and which will be extracted in finding the targets. The highest priority variable in a landmine detection system is target location, l_t. The spreading factor (S) and the strength of the target (ρ_2) can be approximated by a single amplitude, $s(l_t)$, and would not require enumeration. For the remainder of this section, only the SFGPR case will be studied so the frequency ω and sensor locations l_s are the measurement parameters. A very similar approach would be used to make a TPGPR dictionary matrix. The SFGPR model for a single target can be rewritten as

$$R(\omega, l_s; l_t) = s(l_t)e^{-j\omega\tau(d(l_s, l_t), c, v)}$$
$$= s(l_t)\psi(\omega, l_s, l_t), \tag{3.3}$$

where $d(l_s, l_t)$ is a 3D distance function. The measurements are made at a finite set of frequencies ω and a finite number of sensor locations $l_s = (x_s, y_s)$.

One-column vector of the dictionary matrix Ψ is the vector $\psi(l_t)$ whose entries are all the measurements created by evaluating the model $R(\omega, l_s; l_t)$ for a fixed l_t while enumerating all possible triples of the three measurement space parameters, ω, x_s, and y_s. The final dictionary is created by concatenating the target location response vectors as the column space,

$$\Psi = \left[\psi(l_t^1) \mid \psi(l_t^2) \mid \cdots \mid \psi(l_t^{N_{l_t}}) \right], \tag{3.4}$$

where the N_{l_t} values of l_t are obtained by enumerating all possible triples of the target location parameters, x_t, y_t, and z_t. If the number of values for each parameter is N, then Ψ is an $N^3 \times N^3$ matrix. For example, with $N = 100$, the dictionary matrix size would be $10^6 \times 10^6$ with one trillion (10^{12}) entries.

Using the dictionary matrix in (3.4), the response vector can be expressed as

$$f = \sum_{l_t} s(l_t)\psi(l_t) = \Psi x, \tag{3.5}$$

where x is a sparse vector that is only nonzero at the target locations. The indexing of the vector f must follow the enumeration of the triple (ω, x_s, y_s) used for the measurement vectors $\psi(l_t)$. The indexing of x must follow that of the triple (x_t, y_t, z_t) used for the target locations.

The inversion process is done to recover x from the measurements f. However, the size of Ψ and the computational complexity of the inversion algorithms make 3D imaging problematic for real-world applications. To address this issue, some structural properties of Ψ are presented, which can be exploited to simplify the way that the dictionary Ψ is created, stored, and applied in the different inversions.

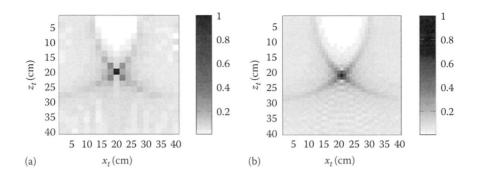

FIGURE 3.5
Standard backprojection of a simulated target with l_t discretization of (a) 2 cm and (b) 1 cm, which is a finer grid.

The most common inversion method for GPR is standard backprojection (BP). In the notation of (3.5), the standard backprojection image is created simply by multiplying the measurements, f, by the adjoint Ψ^*, that is,

$$x_{bp} = \Psi^* f = (\Psi^* \Psi)x. \tag{3.6}$$

The output vector x_{bp} must be reshaped into a 2D array to display the images seen in Figure 3.5. Figure 3.5a shows a single-target image with 2 cm discretization in l_t, and Figure 3.5b shows the same target imaged with 1 cm discretization. Even though there is only one target, it is easy to see that there are many more image pixels containing energy than just the one at the true target location. This is because Ψ is not an orthogonal matrix, and the correlation between two nonidentical columns is not always zero. The coherence of the dictionary, defined by the maximum correlation over all column pairs, will increase for a finer grid where the spacing of l_t decreases, and this fact is important when choosing the discretization of l_t for the dictionary. Figure 3.5 also shows that the actual resolution of the image in terms of detection is not improved by using a finer discretized grid for l_t.

3.3 System Framework with CS

There are many considerations when developing a system framework for a GPR problem that will use CS. The general reason for using CS is to reduce the constraints placed upon the data acquisition and inherently favor sparse solutions. As discussed previously, SFGPR can be hindered by long data acquisition times, especially for array systems that have several transmitting antennas that are not used simultaneously. For instance, if we assume that the GPR has five transmit antennas and requires 100 μs to acquire each frequency, the vehicle could then travel at a maximum speed of approximately

$$\text{max speed} = \frac{\text{aperture spacing}}{\text{freq. duration} \times \text{num. TX antennas} \times \text{num. freqs.}}$$

$$= \frac{0.02}{0.0001 \times 5 \times 400} = 0.1 \text{ m/s,}$$

which is impractically slow for many applications.

Another example comes from the Geneva International Centre for Humanitarian Demining (GICHD) that has published a list of specifications for state-of-the-art GPR landmine detectors [19]. The handheld detectors in the GICHD analysis have optimum sweep speed between 0.2 and 1 m/s, which is the speed at which a GPR wand device can be moved spatially to achieve optimal detection accuracy. While a sweep speed of 1 m/s is probably adequate for a handheld GPR, a sweep speed of 0.2 m/s would be uncomfortably slow to use.

Data acquisition speed is more critical for vehicle-mounted GPR systems as they are used to search larger areas and faster vehicle speed is desirable. In the GICHD equipment catalog, the vehicles with mounted GPRs can travel at speeds between 0.2 and 2 m/s, which are much slower than desired. The data acquisition time is not as problematic for TPGPR, but it is still an issue. For a TPGPR, CS can also provide simplifications to the hardware design, while still decreasing the data acquisition times and achieving acceptable detection accuracy. CS could also be used to decrease the synthetic aperture spacing and/or increase the size of the antenna array for both the handheld and the vehicle-mounted GPR systems. In all of these cases, the objective is to improve the system performance while still forming accurate subsurface images. The rest of this section will address the issues with designing the data flow for the GPR detection problem when using a CS algorithm.

3.3.1 Designing Φ

The first step of the process is designing a sampling scheme to either reduce the sampling time with compressed sensing or to increase the performance of the system without increasing the data acquisition time. In compressed sampling, the compression matrix, Φ, needs to satisfy the restricted isometry property (RIP). If $A = \Phi\Psi$, then

$$1 - \epsilon_{\text{RIP}} \leq \frac{\|Ax\|_2}{\|x\|_2} \leq 1 + \epsilon_{\text{RIP}}, \tag{3.7}$$

where
$\epsilon_{\text{RIP}} > 0$
x is the estimated sparse vector [10]

If A is a random matrix, then the RIP is satisfied with high probability assuming

$$M \gtrsim \mu^2(\mathbf{\Phi}, \mathbf{\Psi}) \log(N)K, \tag{3.8}$$

where
   K is the number of nonzero elements in x
   M is the number of compressed measurements or rows in $\mathbf{\Phi}$
   $N = N_\omega N_{l_s}$ is the number of noncompressed measurements, or the number of rows in $\mathbf{\Psi}$ or columns in $\mathbf{\Phi}$

and

$$\mu(\mathbf{\Phi}, \mathbf{\Psi}) = \max_{\substack{\phi_k \in \text{Rows}(\mathbf{\Phi}) \\ \psi_t \in \text{Cols}(\mathbf{\Psi})}} |\langle \phi_k, \psi_t \rangle| \tag{3.9}$$

is the mutual coherence [9]. The rows of $\mathbf{\Phi}$ are normalized to $\|\phi_k\|_2^2 = N$, and the columns of $\mathbf{\Psi}$ are normalized to one. The desired value for $\mu(\mathbf{\Phi}, \mathbf{\Psi})$ is as close to 1 as possible. For completeness, $D = N_{l_t}$ is the number of parameters, or columns in $\mathbf{\Psi}$.

Three different types of $\mathbf{\Phi}$ are routinely examined: Gaussian (type I), Bernoulli ± 1 (type II), and a random subset of the identity matrix (type III). These $\mathbf{\Phi}$ structures and their naming convention are discussed and analyzed in Gurbuz et al. [23]. For the sampling bound and mutual coherence calculations, the $\mathbf{\Psi}$ matrix was the one used to image Figure 3.5. The sampling parameters are $N_f = N_t = 401$, $N_{l_s} = 20$, and $N_{l_t} = 20 \times 20 = 400$.

3.3.1.1 Time Pulse $\mathbf{\Phi}$

Type I, where $\mathbf{\Phi}$ is a matrix whose entries come from a normal distribution $\mathcal{N}(0, 1)$, is widely used in the literature for CS. Also, since the time pulse measurement is by definition sparse in the time dimension, using a $\mathbf{\Phi}$ that is spread out is beneficial because it is very different from $\mathbf{\Psi}$. Table 3.1 shows the coherence $\mu(\mathbf{\Phi}, \mathbf{\Psi})$ values for type I, II, and III matrices. The value for type I is 4.5, which means that recovery is possible if only approximately 3% of the total measurements are retained.

TABLE 3.1

Mutual Coherence Values for Different $\mathbf{\Phi}$ with TPGPR

Type	Description	$\mu(\mathbf{\Phi}, \mathbf{\Psi})$	$\approx M$
I	$\mathcal{N}(0, 1)$	4.5	240
II	Bernoulli	2.5	80
III	Random sampling	20	5200

Type II matrices have entries taken from a Bernoulli ±1 process. Type II matrices have properties similar to that of type I matrices, in that it is spread out in the dimension where Ψ is sparse. This similarity leads to an even lower value of 2.5 for μ(Φ, Ψ), which corresponds to only requiring approximately 1% of the total measurements to recover a single target.

For type III, Ψ is a random selection matrix, which does just what the name suggests, it takes a random subset of all the measurements. However, type III matrices are not structured to work well with the time pulse system because both matrices Φ and Ψ are sparse in the measurement domain. For example, when a vector of length 50 with a sparsity of one is being sampled, the only way to guarantee that a random selection matrix will obtain any information about the signal is to take all 50 samples. Type III matrices have a much higher value of 20 for μ(Φ, Ψ), which corresponds to a requirement that over half the samples must be taken.

The analysis of the different Φ matrices is important only if they can be effectively applied during the data acquisition process through some type of hardware design, pulse specification, or creative sampling structure. As described in the work by Gurbuz et al., types I and II can be applied using hardware mixers and low-pass filters to allow for the inner product to be taken with a random signal. However, creating Gaussian random pulses at radar rates, generally GHz, is difficult. Creating the random signal vectors for type II is far more reasonable using state machines [23]. Type III matrices are extremely easy to implement by subsampling in either the sensor position domain, as seen in Figure 3.6, where the vertical black lines correspond to l_s positions that are not sampled, or the time domain. This could be ideal for a system where the movement was not uniform, but the sampling time was. As long as the l_s positions are recorded accurately at each sample, the random spatial sampling would be effective.

Random vectors can also be applied in another way, without requiring hardware to perform random inner products. Romberg introduced a method called random convolution, built on the idea that random Toeplitz and circulant matrices possess the same properties that allow for conventional random matrices to fit within the RIP [4,34]. The added advantage of using these structured matrices is that they can be efficiently applied with the fast Fourier transform (FFT) because they are the matrix representations of convolution. Equation 3.1 can be rewritten as a convolution,

$$r(t, l_s, l_t) = \frac{\rho_1}{S} p(t) * \delta(t - \tau(l_s, l_t, c, v)). \tag{3.10}$$

A generic signal can be sent from the receiver, $p(t)$. If $p(t)$ can be constructed to be a known, pseudorandom signal with either a $\mathcal{N}(0, 1)$ distribution or a Bernoulli ±1, the projection of random vectors can be done through the convolution of the transmitted pulse and the target response with no additional

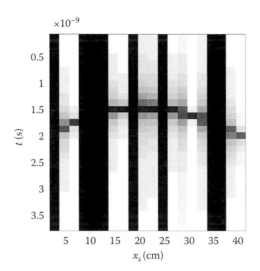

FIGURE 3.6
Time-domain measurement subsampled by omitting sensor positions.

mixing hardware required. The mutual coherence will be similar to that of a type I or type II Φ matrix. Depending on the length of $p(t)$, the resulting measurement response will be, in effect, spread out in time. This will allow for a random sampling step to be performed after convolution to complete the compressive measurements. To recreate the effect of convolution with $p(t)$ and subsequent random sampling for Ψ, all that needs to be done is to apply a Toeplitz matrix, P, to the left of Ψ, followed by the application of a type III Φ matrix. The matrix P is built with a generating vector, p, corresponding to the discretization of $p(t)$.

Using the random convolution method can dramatically reduce the complexity of the hardware in the GPR. In fact, GPRs already exist that use pseudorandom M-sequence pulses as the transmitted signals [36]. Also, it has been shown that pseudorandom sampling in the time domain is an effective strategy for typical GPR systems because it can eliminate the need for using a signal delay line [18,25]. The random sampling could be done in a much shorter sweep time than the conventional sequential sampling while achieving similar accuracy. The use of random convolution and a random sampling GPR make CS seem as though it was created directly with a radar application in mind. The random convolution method could be paired easily with random spatial sampling for additional data acquisition reductions.

3.3.1.2 Stepped Frequency Φ

The same analysis can be done for the SFGPR case as in the previous section for the TPGPR case. However, it is clear from the analysis presented in

TABLE 3.2

Mutual Coherence Values for Different Φ with SFGPR

Type	Description	$\mu(\Phi, \Psi)$	$\approx M$
I	$\mathcal{N}(0, 1)$	3.2	135
II	Bernoulli	3	120
III	Random sampling	1	15

Table 3.2 that all sampling structures for Φ will work well. The major advantage here is that type III matrices, which are by far the easiest to implement in a practical system, actually work the best. These matrices are completely incoherent with the Ψ matrix created for SFGPR, and this is not surprising because Ψ is very similar to a Fourier matrix, which is very dissimilar to an identity matrix. Because of this fact, there is simply no reason in this scenario to use anything other than type III matrices. The dramatic reduction in sampling size seen by using a type III matrix also has the advantage that it can easily and dramatically reduce the data acquisition time for SFGPR, which can discourage its use in practical systems.

3.3.2 CS Inversion

The actual inversion of the CS algorithm is generally done with a convex relaxation of the ℓ_0 minimization, the compressed ℓ_1 minimization (Cℓ_1M). Since the purpose of using these algorithms is to perform inversions on real systems, noise becomes a factor and the basis pursuit method is ineffective. Instead, a measurement constraint inequality should be created to account for the additive noise vector, n,

$$y = \Phi f + n = \Phi \Psi x + n = Ax + n. \tag{3.11}$$

The optimization can allow for noise by using approximate measurement matching,

$$\hat{x} = \min_x \|x\|_1 \qquad s.t. \ \|Ax - y\|_2 < \epsilon_2. \tag{3.12}$$

The optimization in (3.12) is called basis pursuit de-noising (BPDN) [13]. BPDN places an allowable bound on the residual between the estimated signal response and the received signal response. Another important form of de-noising used in these problems is the Dantzig Selector [11], in which the optimization problem from (3.12) becomes

$$\hat{x} = \min_x \|x\|_1 \qquad s.t. \ \|A^*(Ax - y)\|_\infty < \epsilon_d, \tag{3.13}$$

The Dantzig selector (3.13) constrains the size of the residual correlated with the CS matrix, A, instead of just constraining the residual. This formulation has been shown to be an effective de-noising tool in GPR applications [22].

Some common tools used for solving the ℓ_1 minimization only work with real data, such as ℓ_1-MAGIC [8]. The real-only constraint is generally unacceptable for frequency domain data because both A and y are complex. A simple approach to turn complex data into real only is to vertically stack the real part of the data on top of the imaginary part. For instance,

$$A = \begin{bmatrix} \Re(A) \\ \Im(A) \end{bmatrix} \quad \text{and} \quad y = \begin{bmatrix} \Re(y) \\ \Im(y) \end{bmatrix} \tag{3.14}$$

would be sufficient. However, there are other tools, such as CVX and SPGL1, that can deal with complex numbers, so (3.14) is not always necessary [20,41]. In the TPGPR case, complex data is not an issue, unless a complex-valued pulse is used at the transmitter, then similar actions can be taken.

The final consideration is the selection of the ϵ parameters. For ϵ_2 in (3.12), if the noise power, σ^2, is known, then we can use $\epsilon_2 = \sqrt{N}\sigma$. For ϵ_d in (3.13), if the noise power is known, then $\epsilon_d = \sqrt{2\log(N)}\sigma$. However, for practical systems, estimating the noise level accurately can be difficult; if so, there are other techniques that can be used in the selection of the parameters in a data-dependent training mode. The first is using an L-curve, which, in theory, is simple, but can be extremely computationally intense. It is an iterative method that creates a curve of sparsity vs ϵ. For a range of values of ϵ, either (3.12) or (3.13) is solved repeatedly and the sparsity noted. The resulting plot of sparsity vs ϵ will have a distinct *knee* and we select the ϵ value at the knee of the curve to obtain the best solution. An example of an L-curve can be seen in Figure 3.7.

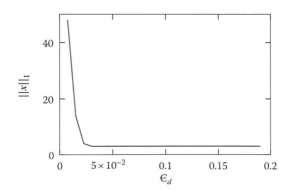

FIGURE 3.7
L-curve for ϵ_d selection.

Algorithm 3.1 CV algorithm for Dantzig selector

Input: Evaluation dictionary, A_E; evaluation measurement vector, y_E;
       validation dictionary, A_V; validation measurement vector, y_V.

Output: The allowable error parameter, ϵ_d.

Initialize $\alpha = 0.99$;

Initialize $\hat{b} = 0$;

Initialize $\epsilon_d = \alpha \left\| A_E^* y_E \right\|_\infty$;

while $\left\| A_V^*(y_V - A_V \hat{b}) \right\|_\infty < \epsilon_d$ **do**

     $\hat{b} = \min_{b} \|b\|_1 \ s.t. \ \left\| A_E^*(A_E b - y_E) \right\|_\infty < \epsilon_d$;

     $\epsilon_d = \left\| A_V^*(y_V - A_V \hat{b}) \right\|_\infty$;

end

A second and much more computationally efficient method is to do cross validation (CV) [6]. The CV method was shown to be effective in CS applications to GPR by Gurbuz et al. [22]. The process involves splitting the measurements into two separate groups. So the compressed measurement vector, y, which is of length M should be split into an estimation set of length $E < M$ and CV set of length $V = M - E$. An example of a CV algorithm used specifically with a Dantzig selector problem for GPR is taken from Gurbuz et al. and seen in Algorithm 3.1 [23]. The algorithm for CV using the quadratic constraint in (3.12) can be found in the original work by Boufounos et al. [6]. The value of α is set to 0.99 to make sure that the data are underfit.

3.3.3 Compressed Orthogonal Matching Pursuit

The algorithms used to compute (3.12) and (3.13) can be rather computationally intensive. An alternative solution for GPR introduced in [21] recommends taking the compressed measurement but instead of using $C\ell_1M$, simply solve a compressed sensing version of orthogonal matching pursuit (COMP). The COMP method does not have the same guaranteed minimum solution that $C\ell_1M$ does, but through experimentation it has been shown to be reasonably competitive in terms of detection accuracy since it also provides sparse solutions. COMP is also much more computationally efficient because it only requires the computation of one adjoint operator per iteration, and the number of iterations will be roughly equal to the sparsity. The experiment done in [21] claims a reduction in computation time by three orders of magnitude when using COMP instead of $C\ell_1M$. The algorithm for computing COMP is given in Algorithm 3.2 and is based on the method by Tropp et al. for signal recovery using OMP [39].

Algorithm 3.2 COMP algorithm

Input: A compressed dictionary, A, where a_t represents the tth column
of A; a compressed measurement vector, y; and a stopping
condition.

Output: An estimated image, Ξ; and vector of indices, λ; an update,
least-squares matrix, Γ; an approximation, θ, of y; and an
update residual, η.

Initialize $\eta = y$;

Initialize Γ to be an empty matrix;

Initialize λ to be an empty matrix;

while *stopping condition is not met* **do**

$\quad t = \arg\max_t |\langle a_t, \eta \rangle|$;

$\quad \lambda = \lambda \cup t$;

$\quad \Gamma = \Gamma \cup a_t$;

$\quad p = \arg\min_p \|\eta - \Gamma p\|_2$;

$\quad \theta = \Gamma p$;

$\quad \eta = y - \theta$;

end

$\Xi(\lambda) = p$

3.3.4 Basic CS Simulation

The same SFGPR simulated measurements from Figure 3.4b are solved using
the $C\ell_1 M$ algorithm in (3.13) with ϵ_d selected using the CV scheme from
Algorithm 3.1 with 50% of the measurements for the evaluation and CV sets.
There is 5 dB SNR present and only six random frequencies are used at each
l_s to create a type III Φ matrix, giving a total compressed measurement size
of $M = 6 \times 20 = 120$. Dropping the amount of frequencies from 400 to 6 would
reduce data acquisition time and Ψ storage by almost two orders of magnitude. The use of only six frequencies would increase the maximum vehicle
speed cited in the example in Section 3.3 from an impractically slow speed of
0.2 m/s to over 10 m/s, which is quite fast for a GPR system. Figure 3.8
shows the resulting sparse solution. Figure 3.8 only has one active pixel,
corresponding to the exact location of the target, and does not have the additional spurious targets due to dictionary coherence seen in the BP solutions
in Figure 3.5. The solution has no need for additional interpretation since
there is only one target shown.

A simple simulation can be run to evaluate the accuracy of CS against
other methods like BP and orthogonal matching pursuit (OMP). OMP provides sparse solutions, but it does not have a guaranteed minimal solution
like ℓ_1 minimization. To evaluate accuracy, a suitable metric is needed. Typically, something like probability of detection or mean-squared error (MSE)

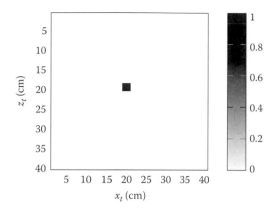

FIGURE 3.8
Simulated $C\ell_1M$ solution of a single target in air; SNR is 5 dB. Six random frequencies out of 400 were measured at 20 sensor positions.

is used in evaluating the result. For this specific application, since the concern is more with the estimated location, than with the amplitude, a measure called the earth mover's distance (EMD) can be used [35]. The EMD takes into account the error between the support, as well as the error in amplitudes. Just like MSE, a lower EMD constitutes a better solution. For example, if a mine was detected but its location was off by 1 cm, the EMD would be lower than if the same mine location was off by 30 cm. However, if the MSE were used, the error would be identical for both cases assuming they had equal amplitudes. A simple one-dimensional (1D) example of EMD vs MSE is given in Figure 3.9.

In order to compare different algorithms, the BP, OMP, COMP, and $C\ell_1M$ methods were used to image a simulated SFGPR three-target scenario with measurements in air while varying the SNR values between −20 dB and

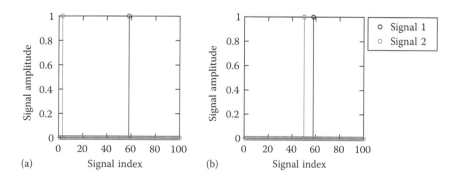

FIGURE 3.9
Signal examples with equal MSE but (a) high EMD and (b) low EMD.

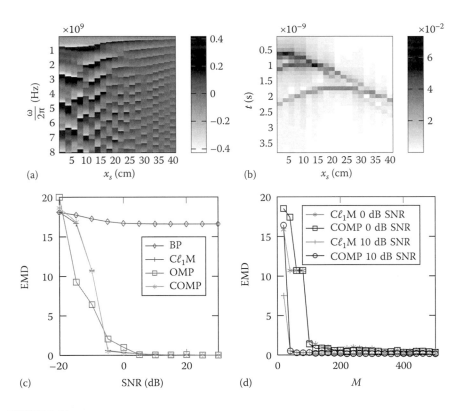

FIGURE 3.10
Figures for three-target simulation: (a) SFGPR measurements, (b) TPGPR measurements, (c) EMD comparison vs SNR for different inversions, and (d) EMD comparison vs M for $C\ell_1M$ methods at two SNRs.

30 dB. $C\ell_1M$ and COMP were run with a number of measurements (M) between 20 and 500, corresponding to randomly selecting 1–20 stepped frequencies for each I_s. The results used for comparison are EMD vs SNR and EMD vs M. $C\ell_1M$ and COMP were also run with M up to 2000 to compare timing.

Each simulation was run 25 times and the median values were stored for accuracy analysis with the results shown in Figure 3.10. Figure 3.10a and b shows the frequency and time measurements, respectively, while Figure 3.10c compares the inversion methods with respect to EMD. Figure 3.10d shows an analysis of the values required for M when the SNR is set to 0 or 10 dB. The EMD is calculated using a fast method from Pele and Werman [32]. It is important to notice that using the EMD tends to favor sparse results, as can be seen from the high floor of the EMD for BP that is nonexistent in the $C\ell_1M$, OMP, or COMP cases. The performance of $C\ell_1M$ compared with COMP in this experiment was first reported in [21]. Comparing the computation time between $C\ell_1M$ and COMP in Figure 3.11 shows

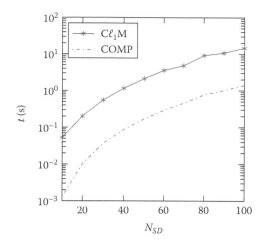

FIGURE 3.11
Computation time for $C\ell_1 M$ and COMP with 97 random frequencies.

that COMP is, in fact, more computationally efficient than even an extremely fast ℓ_1 solver in SPGL1, by about two orders of magnitude. The speed up will be variable as COMP and $C\ell_1 M$ are iterative algorithms. The speed up is not quite as dramatic as first reported in [21] because more efficient ℓ_1 solvers exist than in CVX [20,41].

The preceding synthetic problems have been solved for small 2D images from simulated measurements in order to show the validity of CS for GPR. The creation of a 3D image is much more challenging, so it is often done with 2D slices because the scalability of storing Ψ for a reasonably sized 3D image is unreasonable for most computers. The storage requirements for such problems, with discretization of each variable set at 100, would be $N = N_f N_{l_s} = 10^6$ and $P = N_{l_t} = 10^6$ for a total size of $NP = 10^{12}$ elements. Using CS, if there is a similar reduction of two orders of magnitude from N to M, then the storage requirements become $MP = 10^{10}$. Each element is going to be a complex double of size 64 bits, creating a storage requirement of 80 GB. Since the application for these types of problems is commonly held on small, mobile devices, 80 GB to store a part of the data used for the algorithm is unacceptable. Ways to address this issue are dealt with in the following section.

3.4 Computational Reductions

The validity of CS for GPR applications has been discussed in the previous sections. However, the computational constraints of the inversion algorithm make it undesirable for real-world applications. To address this issue, some structural changes need to be made to the way that the dictionary Ψ is created and applied in the CS inversion.

3.4.1 Shift Invariance Property

A simplification of Ψ is possible because the GPR acquisition system can have the extremely powerful property of spatial shift invariance [27–29]. A graphical example of what the time-domain measurements look like for a target that has been horizontally shifted at the same depth can be seen in Figure 3.12.

The key idea is that the model response (3.2) at the horizontal aperture of sensors will shift in tandem with horizontal shifts in the target positions at a fixed depth. This is true because the distance function $d(l_s, l_t)$ shown in Figure 3.3 does not change with equal horizontal shifts of the sensor and target. However, the computation takes place with discrete grids for the positions l_s and l_t, so the measurement and model grids must also support the shift invariance. While it is not necessary in the general case for the sensor locations l_s to be uniformly spaced, in this section it will be assumed that l_s is uniformly spaced. If l_s is not physically uniformly spaced, it can be interpolated onto a uniformly spaced grid. To show how the shift-invariance property simplifies the computation for the collection of SAR measurements, a 2D example with the target in air will be used. Because the target and sensors are in the air, there is no boundary where the medium changes for the wave, so the wave velocities will be dropped from the equation to keep the notation simpler.

First, rewrite the response vector from (3.5) as a sum of products,

$$r(\omega, x_s) = \sum_{z_t} \sum_{x_t} s(x_t, z_t)e^{-j\omega\tau(x_s, x_t, z_t)}, \tag{3.15}$$

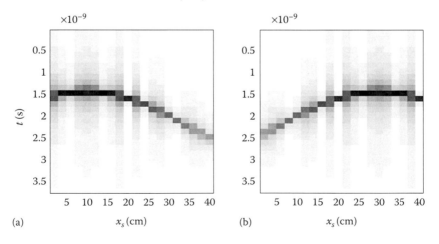

(a)   (b)

FIGURE 3.12
Simulated time-domain measurements of shifted targets when (a) $x_t = 10$ cm and (b) $x_t = 30$ cm.

where $s(x_t, z_t)$ is expected to be sparse when there are few targets. Next we discretize the $x-dir$ dimension as follows: $m\Delta x$ for $m = 1, 2, \ldots, N_{x_s}$ for x_s and $(h + \alpha)\Delta x$ for $h = 1, 2, \ldots, N_{x_t}$ for x_t, where α is a constant such that $0 \leq \alpha < \Delta x$. The $x-dir$ discretization creates a new representation for the response vector,

$$r(\omega, m\Delta x) = \sum_{z_t} \sum_{h=1}^{N_{x_t}} s((h + \alpha)\Delta x, z_t)e^{-j\omega\tau(m\Delta x, h\Delta x, \alpha, z_t)}, \quad (3.16)$$

along with finite limits on the x_t summation. If the time delay τ is examined for a monostatic system measuring in a single medium, we obtain

$$\tau(m\Delta x, h\Delta x, \alpha, z_t) = \left(\frac{1}{c}\right)\sqrt{(m\Delta x - (h + \alpha)\Delta x)^2 + z_t^2}$$

$$= \left(\frac{1}{c}\right)\sqrt{((m - h) - \alpha)^2(\Delta x)^2 + z_t^2}. \quad (3.17)$$

Thus, the time delay depends on the index difference $(m - h)$, and it can be shown that the inner sum in (3.16) is a discrete convolution. The time delay function τ keeps the horizontal shift invariance property even when the system is not monostatic. In general, horizontal shift invariance property, visualized with Figure 3.3, is true when the medium velocity is allowed to change with $z-dir$, but not with $x-dir$ or $y-dir$. In Figure 3.3, if the target, transmitter T_1, and receiver R_1 are all shifted by an equal value, d_1 and d_2 will shift by the same amount, so the time delay τ will be the same as before the shift. Therefore, with a fixed Δx, the exponential in (3.16) is a function of the index difference $(m - h)$:

$$e^{-j\omega\tau(m\Delta x, h\Delta x, \alpha, z_t)} = e^{-j\omega\tau((m-h)\Delta x, \alpha, z_t)}. \quad (3.18)$$

Now we can rewrite the inner sum of (3.16) as a convolution because we have index shift invariance:

$$r(\omega, m\Delta x) = \sum_{z_t} \underbrace{\left(\sum_{h=1}^{N_{x_t}} s((h + \alpha)\Delta x, z_t)e^{-j\omega\tau((m-h)\Delta x, \alpha, z_t)}\right)}_{\text{convolution w.r.t. } m}$$

$$= \sum_{z_t} s((m + \alpha)\Delta x, z_t) \underset{m}{*} e^{-j\omega\tau((m)\Delta x, \alpha, z_t)}$$

$$= \sum_{z_t} s((m + \alpha)\Delta x, z_t) \underset{m}{*} \psi(\omega, m\Delta x, \alpha, z_t). \quad (3.19)$$

The convolution representation shown in (3.19) uses exactly the same dictionary as was created in (3.3), but the discretization for x_t has been replaced by the constant α. Changing the discretization from x_t to α is significant because it drops the storage requirements from N_{x_t} to 1. However, if there is a desire to upsample the image locations, then multiple values of α would be used, and the process would need to be repeated for each α, for example, to get upsampling by a factor of two we could use $\alpha_1 = 0$ and $\alpha_2 = 0.5\Delta x$. With these steps outlined for the forward operator, the adjoint is fairly trivial. Transitioning back to matrix notation, shift invariance leads to a Toeplitz or block-Toeplitz structure in Ψ. For simplicity, consider an example with α_1, α_2, and $N_\alpha = 2$, which could be expanded trivially. The columns correspond to the sensor positions, x_s, the rows correspond to the simulated target locations, x_t, and the entries in the D^* matrix correspond to the distance between x_s and x_t. The distance between x_s and x_t is going to be $\left|m - h - \alpha_i\right|$. A reasonable assumption made here is that $N_{x_t} = N_{x_s}$. When they are not equal, the Toeplitz structure would not be square. For example, with two values of α, the difference matrix is

$$
D^* = \begin{bmatrix}
\alpha_1 & \Delta x - \alpha_1 & \cdots & (N_{x_s}-1)\Delta x - \alpha_1 \\
\alpha_2 & \Delta x - \alpha_2 & \cdots & (N_{x_s}-1)\Delta x - \alpha_2 \\
\Delta x + \alpha_1 & \alpha_1 & \cdots & (N_{x_s}-2)\Delta x - \alpha_1 \\
\Delta x + \alpha_2 & \alpha_2 & \cdots & (N_{x_s}-2)\Delta x - \alpha_2 \\
\vdots & \vdots & \ddots & \vdots \\
(N_{x_t}-1)\Delta x + \alpha_1 & (N_{x_t}-2)\Delta x + \alpha_1 & \cdots & \alpha_1 \\
(N_{x_t}-1)\Delta x + \alpha_2 & (N_{x_t}-2)\Delta x + \alpha_2 & \cdots & \alpha_2
\end{bmatrix}.
$$

$$\tag{3.20}$$

This matrix is not Toeplitz, but with a slight reorganization of the rows, it can become block-Toeplitz with N_α blocks,

$$
D^* = \left[
\begin{array}{cccc}
\alpha_1 & \Delta x - \alpha_1 & \cdots & (N_{x_s}-1)\Delta x - \alpha_1 \\
\Delta x + \alpha_1 & \alpha_1 & \cdots & (N_{x_s}-2)\Delta x - \alpha_1 \\
\vdots & \vdots & \ddots & \vdots \\
(N_{x_t}-1)\Delta x + \alpha_1 & (N_{x_t}-2)\Delta x + \alpha_1 & \cdots & \alpha_1 \\
\hline
\alpha_2 & \Delta x - \alpha_2 & \cdots & (N_{x_s}-1)\Delta x - \alpha_2 \\
\Delta x + \alpha_2 & \alpha_2 & \cdots & (N_{x_s}-2)\Delta x - \alpha_2 \\
\vdots & \vdots & \ddots & \vdots \\
(N_{x_t}-1)\Delta x + \alpha_2 & (N_{x_t}-2)\Delta x + \alpha_2 & \cdots & \alpha_2
\end{array}
\right].
$$

$$\tag{3.21}$$

This difference matrix, D, is built directly into the representation matrix

$$\Psi(\omega_1, D, z_t) = e^{j\omega\tau(D, t_{z_1})},\tag{3.22}$$

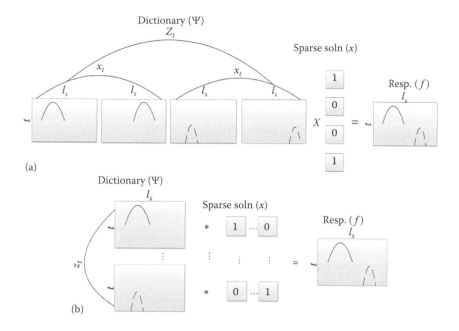

FIGURE 3.13
Dictionary implementation: (a) explicit enumeration with matrix multiplication and (b) exploiting shift invariance by using correlation. (From Krueger, K.R. et al., *proc. SPIE*, 8365, 8365Q, 2012. With permission.)

giving Ψ a block structure with $N_\alpha N_\omega N_{z_t}$ blocks of size $N_{x_s} \times N_{x_t}$. It is well known that a block-Toeplitz matrix can be stored with a single vector for each block.*

Toeplitz matrices have been shown to be an effective way to reduce computational complexity in random sampling matrices for CS [4,34]. A graphical example of the structural changes that would need to be made to the storage and application of the Ψ matrix used for 2D imaging in the time domain can be seen in Figure 3.13. Figure 3.13a shows the traditional dictionary, where every simulated target position l_t is enumerated into the columns of Ψ and the full dictionary matrix is applied using standard matrix–vector multiplication. Figure 3.13b shows a reduced size Ψ with $N_\alpha = 1$, which does not enumerate in the horizontal dimensions and is applied using a convolution operator along the horizontal position, instead of a simple matrix–vector multiply. The particular Ψ structure shown in Figure 3.13b has the added bonus that the computational operations required are $\mathcal{O}(N \log(N))$ by using the FFT along each dimension where the shift invariance can be exploited, instead of $\mathcal{O}(N^2)$. A traditional (explicit) Ψ used to

---

* For the examples given in this section, it will be assumed that the sensor spacing and the simulated target spacing are identical, so $N_\alpha = 1$, $\alpha_1 = 0$.

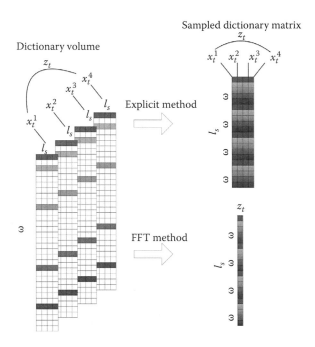

FIGURE 3.14
Data volume for three different Ψ representations.

image three dimensions can be stored and applied in $\mathcal{O}(N^6)$, assuming all measurements and parameters are equally discretized. On the other hand, when Ψ has the Toeplitz structure in both horizontal l_t dimensions equivalent to l_s, taking advantage of the FFT would reduce the storage to $\mathcal{O}(N^4)$ and the computation to $\mathcal{O}(N^4)$. In addition, using a CS inversion would allow for a further reduction in the frequency domain by using $M \ll N$ frequencies. A graphical example of the element reduction in using both CS and exploiting the translational invariance can be seen in Figure 3.14. The takeaway from Figure 3.14 is that each representation can accurately invert a set of measurements, while the method exploiting translational invariance with the FFT is far more efficient. If the example given in Section 3.3.4 is reexamined using these data reduction techniques, then the previous example that required the storage of $MP = 10^{10}$ elements would only require $MP = 10^6$ elements. This reduces the computer storage from approximately 80 GB to 8 MB.

3.4.2 Implementation Specifics for Structure Change

Now that the structure within the dictionary has been identified, the inversion algorithms no longer use a simple matrix–vector multiplication to apply Ψ. The matrix–vector multiplication has to be replaced with a specifically designed function to perform the equivalent of the forward, g_A, and adjoint,

g_A^*, operators of a reduced matrix $\mathbf{\Psi}_\alpha$ to the sparse vector x [27]. It is important to note that $\mathbf{\Psi}_\alpha$ is built by enumerating (3.3), where l_t is only enumerated for $z-dir$ and not $x-dir$ or $y-dir$.

The first step is to make sure that the discretization of l_s and l_t provide the translational invariance discussed in Section 3.4.1. Next, identify how the convolution operation from Figure 3.13b is going to be performed to take advantage of the translational invariance. The FFT can be used to perform circular convolution efficiently, and with the use of a zero-padding operator, Z, linear convolution can be obtained. The FFT can only be performed in the horizontal dimensions if the sampling at each sensor is the same, that is, sampling at the same set of times (for TPGPR), or the same set of frequencies (for SFGPR) for all l_s. Since there is translational invariance in both dimensions of l_s, the zero padding must take place in both dimensions. The simplest thing to do is to add $N_{x_s}/2$ discretizations to the beginning and the end of the x_s dimension, and do the same for $y-dir$. The zero pad will allow for shifts to take place within the desired range of l_s without wrap-around effects from the circular convolution. A slightly more efficient way would be to zero-pad with a number based on the maximal beamwidth of the antenna in space. After zero-padding, the FFT needs to be taken in both $x-dir$ and $y-dir$ of l_s. The same operations must be performed to the corresponding $x-dir$ and $y-dir$ of l_t in the sparse vector x. These are then combined through a series of index multiplications shown specifically in Algorithm 3.3. The summation over z_t should be interpreted as a for loop over depth that adds the results of fast FFT (horizontal) convolution done for each depth. The final step is

---

Algorithm 3.3 Forward function algorithm $y = g_A(\mathbf{\Phi}, \mathbf{\Psi}_\alpha, x)$ for SFGPR

Input: Compression matrix, $\mathbf{\Phi}$; response dictionary, $\mathbf{\Psi}_\alpha$; sparse
      vector, x
Output: compressed measurement vector y
Reshape $\mathbf{\Psi}_\alpha$ and x to have a dimension for every variable (for 3D
imaging $\mathbf{\Psi}_\alpha$ is 4D and x is 3D);
for *all* ω **do**
    $\tilde{f} = 0$;
    for *all* z_t **do**
        $\tilde{x}(k_x, k_y, z_t) = \text{FFT}_x(\text{FFT}_y\{Z(x(x_t, y_t, z_t))\})$;
        $\tilde{\psi}_\alpha(\omega, k_x, k_y, z_t) = \text{FFT}_x(\text{FFT}_y\{Z(\psi_\alpha(\omega, k_x, k_y, z_t))\})$;
        $\tilde{f}(\omega, k_x, k_y) = \tilde{f}(\omega, k_x, k_y) + \tilde{\psi}_\alpha(\omega, k_x, k_y, z_t)\tilde{x}(k_x, k_y, z_t)$;
    end
    $f(\omega, x_s, y_s) = Z^{-1}\{\text{IFFT}_{k_y}(\text{IFFT}_{k_x}\{\tilde{f}(\omega, k_x, k_y)\})\}$
end
Reshape $f(\omega, k_x, k_y,)$ into a vector f;
$y = \mathbf{\Phi}f$;

Algorithm 3.4 Adjoint function algorithm $x = g_A^*(\Phi, \Psi_\alpha, y)$ for SFGPR

Input: Compression matrix, Φ; response dictionary, Ψ_α; compressed
measurement vector y.

Output: Sparse vector, x.

Reshape Ψ_α to have a dimension for every variable (for 3D imaging Ψ
is 4D);

$f = \Phi^* y$;

for *all* z_t **do**

    $\tilde{x} = 0$;

    for *all* ω **do**

        $\tilde{f}(\omega, k_x, k_y) = \text{FFT}_x(\text{FFT}_y\{Z(f(\omega, x_s, y_s))\})$;

        $\tilde{\psi}_\alpha(\omega, k_x, k_y, z_t) = \text{FFT}_x(\text{FFT}_y\{Z(\{\psi_\alpha(\omega, x_t, y_t, z_t)\})\})$;

        $\tilde{x}(k_x, k_y, z_t) = \tilde{x}(k_x, k_y, z_t) + \tilde{\psi}_\alpha^*(\omega, k_x, k_y, z_t)\tilde{f}(\omega, k_x, k_y)$;

    end

    $x(x_t, y_t, z_t) = Z^{-1}\{\text{IFFT}_{k_y}(\text{IFFT}_{k_x}\{\tilde{x}(k_x, k_y, z_t)\})\}$;

end

---

to then compress the measurements with Φ if a compression algorithm is being used. In the case where no compression is required, Φ can be removed, or equivalently set to the identity matrix.

Based on the function to perform the forward operation, a similar function for the adjoint operation can be described. The adjoint matrix Ψ^* is the conjugate transpose, so it possesses Toeplitz subblocks, and fast FFT convolution can be done. In effect, the adjoint can be created by working the forward algorithm backwards. The specifics for how to perform the adjoint operator g_A^* can be found in Algorithm 3.4.

The description of the algorithms can be modified to get a more time-efficient implementation. Many of the steps can be performed offline, for instance, calculating $\tilde{\Psi}_\alpha$ and taking the transform of y. Also, the for loops are easy to vectorize for faster computation when using MATLAB®.

The algorithms described in Algorithms 3.3 and 3.4 describe g_A and g_A^* if a general Φ were to be used in a compression algorithm. However, to use a general Φ requires a complete sampling during data acquisition and eliminates the usefulness of compressive algorithms in terms of data acquisition. Making sure that Φ is a random sampling matrix where the random frequencies, or random time samples, at each l_s are equal, would allow for Φ to be applied to Ψ_α *before the zero pad* and FFT take place. Applying Φ before the zero pad and FFT allows for the compressed sampling to be performed during data acquisition. In other words, if $\Phi\Psi$ has the same shift-invariant property as Ψ, then Φ can be applied before the zero pad and FFT operations.

For the rest of this chapter, the BPDN will be used for $C\ell_1M$ applications, and (3.12) becomes

$$\hat{x} = \min_{x} \|x\|_1 \qquad s.t. \ \|g_A(x) - \tilde{y}\|_2 < \epsilon_2. \qquad (3.23)$$

The reason BPDN is going to be used as opposed to the Dantzig selector that was used in the previous work is simply because the SPGL1 package supports BPDN and is much more computationally efficient than some of the other algorithms like ℓ_1-MAGIC and CVX [8,20,41]. A slight change can be noticed in (3.23). $\tilde{y}$ is used as the compressed measurement vector instead of y, so it is important to remember that the measurements must be zero-padded and passed through the FFT to match the output of g_A.

3.4.3 Simulation Using Functional Dictionary

A simulation was set up to test Algorithms 3.3 and 3.4. The point was to test the accuracy of the functional design and to compare the result to the case when the explicit matrix is used. Unfortunately, an explicit matrix cannot be built directly to image a large enough 3D area, so a quick 2D example is used to compare the explicit method against the functional method directly. A plot showing the timing difference for running a 2D $C\ell_1 M$ problem using SPGL1 with the explicit dictionary and the functional dictionary can be seen in Figure 3.15, with a constant number of 97 compressed frequencies. As the total number of spatial discretizations increases to a value of about

$$N_{SD} = N_{x_s} = N_{x_t} = N_{z_t} = 100,$$

the speed up ratio approaches 15. Since these experiments cannot be done directly in 3D for the explicit model, only a projected ratio of savings for 2D and 3D is seen in Figure 3.16. There will be some scaling factors, and

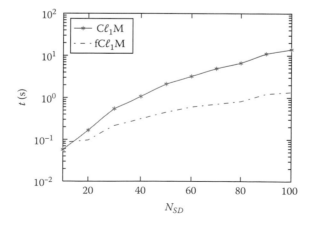

FIGURE 3.15
Timing comparison of functional and explicit $C\ell_1 M$ for different spatial discretizations, N_{SD}.

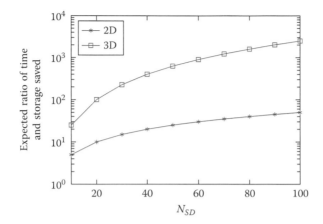

FIGURE 3.16
Approximate ratios of storage and time saved by using the g_A and g_A^* instead of explicit matrix multiplication for different values of N_{SD}.

overhead, so it is not expected that this projected ratio matches the real-time examples directly. The scale up as N_{SD} increases in the 3D case is quite significant, reaching a reduction of about three orders of magnitude when $N_{SD} = 100$.

It is not possible to compare the explicit method with the functional method in 3D, but the method of solving small 2D slices and concatenating them together can be used, as was done in previous work [22]. The point of comparing these two methods is to show why solving the large 3D problem is important. Furthermore, it is important to note that the functional method can be done in 2D as well and would be more time efficient than using the explicit case. This comparison is a relatively small experiment that serves as a proof of concept and gives some simple statistics that predict how the method will work on real data.

The experiment consists of randomly placing two targets in a 3D volume and using $C\ell_1M$ to invert the measurements and create an image. The frequency range used is 0 MHz to 5.02 GHz with $N_\omega = 158$ equally spaced frequencies. The scan positions correspond to a colocated transmitter and receiver. The transmitter locations are uniformly spaced in a 2D square from $y_s = -96$ cm to 90 cm at 6 cm intervals, giving $N_{y_s} = 32$, and similarly from $x_s = -96$ cm to 90 cm at 6 cm intervals, giving $N_{x_s} = 32$. The horizontal target locations are in the range $y_t = -90$ cm to 84 cm at 6 cm intervals, giving $N_{y_t} = 30$; likewise for x_t. The fact that scan positions and the target locations have the same horizontal discretization leads to the simplification that target location shifts will cause a shift in the measurements. Equality is not required, but the ratio of the scan interval to the image location interval must be an integer to get the block-Toeplitz property. The depth locations

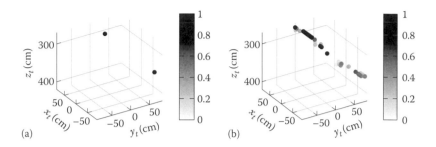

FIGURE 3.17
(a) Full 3D $C\ell_1 M$ solution using FFT method with exact reconstruction and (b) solution using 2D slices. (From Krueger, K.R., 3-D imaging for ground penetrating radar using compressive sensing with block Toeplitz structures, in: *The Seventh IEEE Sensor Array and Multichannel Signal Processing Workshop (SAM)*, Hoboken, NJ, pp. 229–232, June 2012. © 2012 IEEE.)

need not be constrained, but for this experiment, $z-dir$ ranges from 270 to 420 cm deep at 6 cm intervals, giving $N_{z_t} = 26$. The number of compressed measurements, M, used is selected between 0.1% and 2.4% of the total measurements $N = N_\omega N_{I_s} = 158 \times 32^2 \approx 10^5$. The number of discretizations in the target location parameter space is $P = N_{I_t} = 30^2 \times 26 \approx 10^4$. Without using CS or the functional representation, the explicit matrix would be of size $N \times P \approx 10^9$. However, employing CS and the functional representation, N becomes M, $P = 26$, and the total storage requirements for a matrix using 2.4% of the total measurements become $M \times P = 2400 \times 26 \approx 10^5$.

Figure 3.17 shows the detection accuracy advantages that are available when solving the full 3D imaging problem as opposed to imaging the volume with many 2D slices. By not solving the full 3D problem, the synthetic aperture in the extra dimension is not being directly exploited. The full 3D inversion that is made possible by using the functional representations g_A and g_A^* is shown in Figure 3.17a, which is an exact spatial recovery. The sparsity in the 2D slice image from Figure 3.17b shows the resolution issues in x_t. The full 3D inversion provides a much higher resolution image in x_t than using 2D slices. The higher resolution is a by-product of the fact that the full 3D inversion takes advantage of the full 2D array aperture.

3.5 Applied Performance: Laboratory Data

The experimental results in this section are taken from Krueger et al. [28]. This section takes the experiment described in Counts et al. and compares the methods described in Gürbüz et al. with those of Krueger et al. [14,22,27]. The CS algorithm used for these experiments is $C\ell_1 M$.

The frequencies, ω, used in the data collection were $2\pi(60$ MHz$)$ to $2\pi(8.06$ GHz$)$ at $N_\omega = 401$ equally spaced intervals. The first experiment is a target above ground, and the second experiment will be a collection of targets bufried in sand. In order to make a comparison, an explicit dictionary CS method that images the 3D area with a collection of 2D slices and the proposed CS method that images the full 3D area will be used.

The solver that is used to perform the ℓ_1 minimization is SPGL1, which is used because it allows for functional representations of the forward and transpose matrix operators and it is generally faster than ℓ_1-MAGIC or CVX [8,20,41]. The user-chosen parameter that must be set in SPGL1 for BPDN is the tolerance ϵ_2 from (3.23). The tolerance ϵ_2 for the full 3D solver was chosen using CV. Increasing ϵ_2 will increase sparsity, but if it is too large, some targets will be missed.

3.5.1 Air-Target Experiment

The air-target setup can be seen in Figure 3.18a [22]. A metal sphere with a diameter of 2.54 cm is placed on a foam platform and the radar is scanned above it. The 3D time-domain measurements of this experiment are seen in Figure 3.18b, and the target can easily be seen even in the measurements.

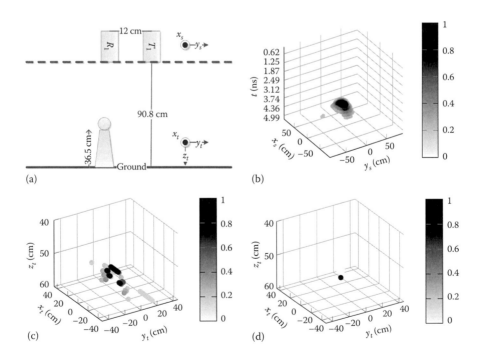

FIGURE 3.18
Air experiment for 1-in. metal sphere (a) setup where the height is measured from the phase center of the antennas, (b) time-domain measurements, (c) solution using 2D slices, and (d) full 3D CS solution using the FFT.

The discretizations used for this problem are as follows: x_s and y_s were both taken from -50 cm to 48 cm at 2 cm increments, x_t and y_t taken from -48 cm to 46 cm at 2 cm increments, and z_t taken from 40 cm to 60 cm with 1 cm increments. This would correspond to an explicit dictionary size of $[379 \times 50^2 \times 48^2 \times 21]$, which is on the order of 10^{10}. Using CS and getting a reduction in frequencies to about 50 reduces the data acquisition time by a factor of about seven. However, the dictionary size is still on the order of 10^9. This means that to use an explicit dictionary this problem must be solved in 2D slices, which does not harness the synthetic aperture in the extra scan dimension and will lower resolution in that spatial dimension. If the functional method is used, there is an elimination of both horizontal spatial image locations in the dictionary for an uncompressed size of $[379 \times 50^2 \times 21]$, which is on the order of 10^7 and a compressed size with 50 frequencies of about 10^6. Using CS and the functional method garner a reduction of three orders of magnitude from the explicit dictionary CS case.

The solution using the explicit dictionary to create 2D slices and build them into a 3D image is seen in Figure 3.18c. The resolution in this image is acceptable in $y-dir$ because this is the dimension where the 2D CS is calculated, but the $x-dir$ has lower resolution because the slice method does not exploit the synthetic aperture in that dimension. Finally, the full 3D solution using the functional representation of the dictionary can be seen in Figure 3.18d. The 3D method is much sparser than the 2D slice solution again because it can take advantage of the synthetic aperture in both scan dimensions, not just one of them.

3.5.2 Subsurface-Target Experiment

The final experiment is one that was performed and documented previously using standard BP in Counts et al. [14]. The acquisition setup for subsurface imaging can be seen in Figure 3.19a, and the ground-truth location of the targets can be seen in Figure 3.19b. A 2D slice CS algorithm was employed for inversion in Gürbüz et al. [22], but we saw in the previous section the issues that can arise from that type of inversion. However, the 2D slice method could also be improved by exploiting the translational invariance if it was desired to still be used. It would be faster and more storage efficient than using the explicit method to calculate each slice because each slice would garner a computational time improvement. The purpose behind inverting this data again is to show a real-world experiment that was previously too computationally intense to invert with full 3D CS, which can now be imaged in 3D with the exploitation of shift invariance. The discretizations used for this problem are as follows: x_s and y_s were both taken from -60 cm to 60 cm at 2 cm increments, x_t and y_t taken from -58 cm to 56 cm at 2 cm increments, and z_t taken from 1 cm to 20 cm with 0.5 cm increments. Figure 3.19c and d shows the top view of the created images using uncompressed BP with 379 frequencies and CS with 100 frequencies. Both of these were computed

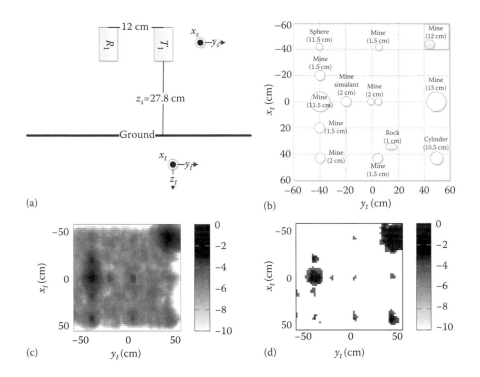

FIGURE 3.19

Underground experiment for multiple objects: (a) sensor setup where z_s is measured from the phase center of the antennas, (b) target locations, the values in parentheses correspond to the depths of the individual targets, (c) solution using BP, and (d) full 3D CS solution using the functional g_A method.

with the functional implementation of the forward and adjoint operators. It is very difficult to distinguish where some of the weaker targets are in the BP image, but it becomes much clearer in the CS image. For example, the mines at $(x_t, y_t) = (-45, 5)$ and $(x_t, y_t) = (0, 50)$, and the cylinder at $(x_t, y_t) = (45, 50)$. The imaging improvement of CS over BP combined with the fact that can be done with much fewer measurements is a substantial reason as to why CS would want to be used over BP. In the work by Gürbüz et al., full 3D inversion for this experiment was impossible with a $C\ell_1M$ version of CS, but now with the modification to the storage and application of Ψ, it can be done.

A problem with imaging the variety of targets in this example is that some of them are much larger than others, so a point target model is not an ideal way to model them with a method that seeks to promote sparsity in the images. By examining Figure 3.19d, each pixel corresponds to a point target, and each point target is counted to determine the sparsity of the image. If a more accurate target model were available for the larger mines, the CS solution would perform much better while requiring fewer measurements by bringing the sparsity down by counting the number of mines as opposed to point targets. This is because the number of measurements required to provide a high probability of accurate recovery is dependent on the sparsity of the image, which can be seen in (3.8).

3.6 Conclusions

CS for a GPR application has been discussed in detail in this chapter. The advantages in the data acquisition are hard to overstate. The landmine detection problem can often be crippled by long data acquisition times, and the use of CS could help eliminate it while possibly providing increased detection capability through sparse results. Even in the situations where data acquisition time is not a hindrance, CS allows for a much smaller number of measurements to be used in the inversion algorithms. The small number of measurements helps with the storage and computational time constraints of the inversion algorithms. A complexity reduction has also been introduced that helps the practicality of the CS framework through exploitation of the shift invariance.

References

1. M. Amin and F. Ahmend. Compressive sensing for through the wall radar imaging. *Journal of Electronic Imaging*, 22(3):021003–021003, 2013.
2. M. G. Amin. *Through-the-Wall Radar Imaging*. CRC Press, Boca Raton, FL, 2011.
3. S. R. J. Axelsson. Analysis of random step frequency radar and comparison with experiments. *IEEE Transactions on Geoscience and Remote Sensing*, 45(4):890–904, 2007.
4. W. U. Bajwa, J. D. Haupt, G. M. Raz, S. J. Wright, and R. D. Nowak. Toeplitz-structured compressed sensing matrices. In *Workshop on Statistical Signal Processing*, Madison, WI, pp. 294–298, August 2007.
5. R. G. Baraniuk. Compressive sensing [lecture notes]. *IEEE Signal Processing Magazine*, 24(July):118–121, 2007.
6. P. Boufounos, M. F. Duarte, and R. G. Baraniuk. Sparse signal reconstruction from noisy compressive measurements using cross validation. In *Workshop on Statistical Signal Processing*, Madison, WI, pp. 299–303, 2007.
7. W. M. Brown and L. J. Porcello. An introduction to synthetic-aperture radar. *IEEE Spectrum*, 6(September):52–62, 1969.
8. E. J. Candès and J. Romberg. ℓ1-magic, October 2005. http://users.ece.gatech.edu/~justin/l1magic/.
9. E. J. Candès and J. Romberg. Sparsity and incoherence in compressive sampling. *Inverse Problems*, (3):1–20, 2007.

10. E. J. Candès, J. Romberg, and T. Tao. Robust uncertainty principles: Exact signal reconstruction from highly incomplete frequency information. *IEEE Transactions on Information Theory*, 52(2):1–41, 2006.

11. E. J. Candès and T. Tao. The Dantzig selector: Statistical estimation when p is much larger than n. *The Annals of Statistics*, 40698:1–37, 2007.

12. E. J. Candès and M. B. Wakin. An introduction to compressive sampling. *IEEE Signal Processing Magazine*, 25(March):21–30, 2008.

13. S. S. Chen, D. L. Donoho, and M. A. Saunders. Atomic decomposition by basis pursuit. *SIAM Journal on Scientific Computing*, 43(1):129–159, 2001.

14. T. Counts, A. C. Gürbüz, W. R. Scott Jr., J. H. McClellan, and K. Kim. Multistatic ground-penetrating radar experiments. *IEEE Transactions on Geoscience and Remote Sensing*, 45(8):2544–2553, August 2007.

15. L. J. Cutrona and E. N. Leith. On the application of coherent optical processing techniques to synthetic-aperture radar. *Proceedings of the IEEE*, 54(8):1026–1032, 1966.

16. D. J. Daniels. Surface-penetrating radar. *Electronics & Communication Engineering Journal*, 8(August):165–182, 1996.

17. D. L. Donoho. Compressed sensing. *IEEE Transactions on Information Theory*, 52(4):1289–1306, April 2006.

18. G. J. Frye and N. S. Nahman. Random sampling oscillography. *IEEE Transactions on Instrumentation and Measurement*, 13(1):8–13, 1964.

19. Geneva International Centre for Humanitarian Demining. *Detectors and Personal Protective Equipment Catalogue*, 2009. http://www.gichd.org/mine-action-resources/documents/detail/publication/detectors-and-personal-protective-equipment-catalogue-2009/#.U1kqMlc0w0w.

20. M. Grant and S. Boyd. Cvx: Matlab software for disciplined convex programming, 2008. http://cvxr.com/cvx/.

21. A. C. Gürbüz. Sparsity enhanced fast subsurface imaging with GPR. In *International Conference on Ground Penetrating Radar (GPR)*, Lecce, pp. 1–5, June 2010. doi: 10.1109/ICGPR.2010.5550130. http://ieeexplore.ieee.org/xpls/abs_all.jsp?arnumber=5550130&tag=1.

22. A. C. Gürbüz, J. H. McClellan, and W. R. Scott Jr. A compressive sensing data acquisition and imaging method for stepped frequency GPRs. *IEEE Transactions on Signal Processing*, 57(7):2640–2650, July 2009.

23. A. C. Gürbüz, J. H. McClellan, and W. R. Scott Jr. Compressive sensing for subsurface imaging using ground penetrating radar. *Signal Processing*, 89(10): 1959–1972, October 2009.

24. E. M. Johansson and J. E. Mast. Three-dimensional ground-penetrating radar imaging using synthetic aperture time-domain focusing. *Proceedings of SPIE*, 2275:205–214, September 1994.

25. M. Kahrs. 50 years of RF and microwave sampling. *IEEE Transactions on Microwave Theory and Techniques*, 51(6):1787–1805, 2003.

26. J. C. Kirk. A discussion of digital processing in synthetic aperture radar. *IEEE Transactions on Aerospace and Electronic Systems*, 11(3):326–337, 1975.

27. K. R. Krueger, J. H. McClellan, and W. R. Scott Jr. 3-D imaging for ground penetrating radar using compressive sensing with block-Toeplitz structures. In *The Seventh IEEE Sensor Array and Multichannel Signal Processing Workshop (SAM)*, Hoboken, NJ, pp. 229–232, June 2012.

28. K. R. Krueger, J. H. McClellan, and W. R. Scott Jr. Dictionary reduction technique for 3D stepped-frequency GPR imaging using compressive sensing and the FFT. *Proceedings of SPIE*, 8365:83650Q–83650Q-9, 2012.
29. K. R. Krueger, J. H. McClellan, and W. R. Scott Jr. Sampling techniques for improved algorithmic efficiency in electromagnetic sensing. In *International Conference on Sampling Theory and Applications*, Bremen, Germany, July 2013.
30. R. M. Lerner. Ground radar system, US Patent 3831173, 1974.
31. P. Millot and A. Berges. Ground based SAR imaging tool for the design of buried mine detectors. In *International Conference on the Detection of Abandoned Land Mines*, October 7–9, 2006, Edinburgh, U.K., Pub. No. 431, 1996.
32. O. Pele and M. Werman. Fast and robust earth mover's distances. In *IEEE 12th International Conference on Computer Vision (ICCV)*, Kyoto, Japan, 2009.
33. L. P. Peters Jr., J. J. Daniels, and J. D. Young. Ground penetrating radar as a subsurface environmental sensing tool. *Proceedings of the IEEE*, 82(12):1802–1822, 1994.
34. J. Romberg. Compressive sensing by random convolution. *SIAM Journal on Imaging Science*, 2(4):1098–1128, December 2009.
35. Y. Rubner, C. Tomasi, and L. J. Guibas. A metric for distributions with applications to image databases. In *International Conference on Computer Vision*, Bombay, India, 1998.
36. J. Sachs. M-sequence ultra-wideband-radar: State of development and applications. In *Proceedings of the International Radar Conference*, Adelaide, Australia, pp. 224–229, 2003.
37. F. Soldovieri, R. Solimene, L. Lo Monte, M. Bavusi, and A. Loperte. Sparse reconstruction from GPR data with applications to rebar detection. *IEEE Transactions on Instrumentation and Measurement*, 60(3):1070–1079, 2011.
38. G. F. Stickley. Synthetic aperture radar for the detection of shallow buried objects. In *International Conference on the Detection of Abandoned Land Mines*, October 7–9, 2006, Edinburgh, U.K., Pub. No. 431, 1996.
39. J. A. Tropp and A. C. Gilbert. Signal recovery from random measurements via orthogonal matching pursuit. *IEEE Transactions on Information Theory*, 53(12):4655–4666, 2007.
40. M. A. C. Tuncer and A. C. Gurbuz. Ground reflection removal in compressive sensing ground penetrating radars. *IEEE Geoscience and Remote Sensing Letters*, 9(1):23–27, 2012.
41. E. van den Berg and M. P. Friedlander. SPGL1: A solver for large-scale sparse reconstruction, June 2007. http://www.cs.ubc.ca/labs/scl/spgl1.
42. J. D. Young, L. Peters Jr., and C. Chen. Characteristic resonance identification techniques for buried targets seen by ground penetrating radar. In *Detection and Identification of Visually Obscured Targets*, pp. 103–162. Taylor & Francis, Boca Raton, FL, 1999.

4

Wall Clutter Mitigations for Compressive Imaging of Building Interiors

Fauzia Ahmad

CONTENTS

ABSTRACT　Exterior wall returns affect the accuracy and fidelity of the imaged scene in urban sensing and through-the-wall radar imaging applications. For reliable imaging of stationary indoor scenes, the front wall reflections need to be properly attenuated. In this chapter, wall mitigations are addressed in the context of compressive sensing for stepped-frequency radar operation. Wall mitigation schemes, originally proposed for imaging using full data volume, maintain their proper performance with few measurements, provided that the same reduced set of frequencies is used at each available antenna position. However, having the same frequency observations across all antennas may not always be feasible. For the more challenging case when different reduced frequencies are employed at different antennas, two alternate methods based on discrete prolate spheroidal sequences (DPSS) and partial sparsity can be applied. The former captures the wall clutter energy at each antenna individually using a DPSS basis and then removes it from the reduced set of measurements. The latter considers the stationary scene reconstruction problem when the support of the image corresponding to the exterior and interior walls is known a priori.

4.1 Introduction

Detection and localization of stationary targets inside enclosed structures using radar are very pertinent to a variety of civil and military applications, including hostage rescue missions, search-and-rescue operations, and surveillance and reconnaissance in urban environments [1–11]. These highly desirable objectives are challenged, among other factors, by the presence of clutter caused by the electromagnetic (EM) scattering from the exterior front wall. Front wall returns in ground-based synthetic aperture radar (SAR) systems are typically stronger than those from targets of interest, such as humans and weapons [12], thus rendering imaging of stationary targets behind walls difficult. The problem is further compounded when the targets are in close vicinity of walls, especially layered walls or hollow cinder block walls. Multiple reflections within the front wall result in wall residuals along the range dimension. These wall reverberations tend to mask and obscure weak and close-by targets [13]. Therefore, stationary targets cannot be generally detected without an effective suppression of front wall clutter, thereby limiting its effects on the imaged scene accuracy and fidelity.

A simple but effective method is based on background subtraction. If the received signals can be approximated as the superposition of the wall and the target reflections, then subtracting the raw complex data without target (reference scene) from that with the target would remove the wall contributions and eliminate its potentially overwhelming signature in the image. Availability of the empty scene, however, is not possible in many applications. For moving targets, Doppler processing [14] or subtraction of data acquired at different times [15,16] alleviates this problem and leads to removal of wall reflections as well as suppression of stationary background. However, when the targets of interest are themselves stationary, one must resort to other means to deal with strong and persistent wall reflections.

For conventional imaging based on beamforming, three main approaches have been proposed to deal with strong wall EM reflections without relying on the background scene data [11,13,17,18]. In the first approach, the parameters of the front wall, such as thickness and dielectric constant, are estimated from the first-wave arrivals [11]. The estimated parameters can be used to model EM wall returns, which are subsequently subtracted from the total radar returns, rendering wall-free signals. Although this scheme is effective, it requires a calibration step, which involves measuring the radar return from a metal plate at the same standoff distance as the front wall under similar, if not identical, operating conditions [19]. The second approach applies a spatial filtering method for wall clutter mitigation [13], which requires measurements from an array aperture that is parallel to the front wall and relies on the wall returns being invariant with changing antenna location. The spatial filter removes the zero spatial frequency component corresponding to the wall return. The third approach recognizes the wall reflections as the strongest component of radar returns, in addition to the invariance of the wall returns across the array aperture [17,18]. By applying singular value decomposition (SVD) to the measured data matrix, the wall returns occupy low-dimensional subspace and can be captured by the singular vectors associated with the dominant singular values. Accordingly, front wall clutter can be effectively removed by projecting the data measurement vectors at each antenna on the wall orthogonal subspace.

Recently, compressive sensing (CS) and sparse reconstruction techniques have been applied, in lieu of beamforming, to reveal the target positions behind walls [20–24]. In so doing, significant savings in acquisition time can be achieved. Further, producing an image of the indoor scene using few observations can be logistically important as some of the data measurements in space and frequency can be difficult or impossible to attain. The application of CS for through-the-wall radar imaging (TWRI) as presented in [20–22] assumed prior and complete removal of front wall EM returns. Without this assumption, strong wall clutter, which extends along the range dimension, reduces the sparsity of the scene and, as such, impedes the application of CS [24,25].

In this chapter, we address wall clutter mitigations in the context of CS for imaging of stationary targets. We examine the effectiveness of the spatial filtering and subspace projection wall mitigation techniques, originally proposed for full data volume, in conjunction with sparse scene reconstruction when only a small subset of measurements is employed. We focus on stepped-frequency operation and consider two cases of reduced frequency measurement distributions over antenna positions in physical or synthetic aperture arrays. In the first case, the same subset of frequencies is used for each antenna. The other case allows the reduced frequencies to differ from one antenna to another, either in a random or preset manner. For the subspace projection and spatial filtering methods, we show that when the same subset of frequency measurements is used at each antenna, those two methods maintain their proper performance as their full data set counterparts. CS techniques for image reconstruction can then be applied with the same reduced measurements but of much higher signal-to-clutter ratio. On the other hand, using different frequencies at different antenna positions would impede the application of either method. This is because the phase returns across the antenna elements would be different, which deprives the wall mitigation algorithms of the spatial invariance and low-dimensional subspace properties of the wall clutter.

In order to overcome the shortcomings of the wall clutter mitigation schemes when a general, nonrestricted reduced data collection scheme is employed, we use a dictionary based on discrete prolate spheroidal sequences (DPSSs) [26] to represent the wall returns, which are then captured by the block sparsity–based approach. This is performed at each available antenna individually. Subtraction of the captured return from the reduced set of measurements at each antenna results in clutter-free data, thereby permitting the application of CS techniques for image reconstruction [27]. Note that, unlike the original wall clutter mitigation methods, this scheme does not require an array aperture to be parallel to the front wall. It can be applied to a single radar unit as well as to significantly reduced array aperture.

Alternately, the idea of partial sparsity can be applied for imaging of stationary indoor scenes using different frequencies from different antennas [28]. Partially sparse scene reconstruction considers the case when the scene being imaged consists of two parts, one of which is sparse and the other is expected to be dense with known support [29,30]. For the problem at hand, the dense part of the image corresponds to the exterior and interior walls. The prior knowledge about the support of the dense part may be available either through building blueprints or from prior surveillance operations.

The chapter is organized as follows. In Section 4.2, we present the through-the-wall signal model and briefly review the wall mitigation techniques of spatial filtering and subspace projection under the full data volume. The performance of these wall mitigation schemes under a reduced set of measurements is discussed in Section 4.3. Section 4.4 deals with the DPSS-based wall clutter suppression scheme, followed by the partial sparsity approach in Section 4.5. Section 4.6 provides concluding remarks.

4.2 Wall Mitigation Techniques of Spatial Filtering and Subspace Projection

4.2.1 Through-the-Wall Signal Model

Consider a homogeneous wall of thickness d and dielectric constant ε located along the x-axis, and the region to be imaged located beyond the wall along the positive y-axis. Assume that an M-element line array of transceivers is located parallel to the wall at a standoff distance y_{off}, as shown in Figure 4.1. Let the mth transceiver, located at $\mathbf{x}_m = (x_m, -y_{off})$, illuminate the scene with a stepped-frequency signal of K frequencies, which are equispaced over the desired bandwidth $f_{K-1} - f_0$,

$$f_k = f_0 + k\Delta f, \quad k = 0, 1, \ldots, K - 1 \tag{4.1}$$

where
  f_0 is the lowest frequency in the desired frequency band
  $\Delta f = (f_{K-1} - f_0)/(K - 1)$ is the frequency step size

The reflections from any targets in the scene are measured only at the same transceiver location. The wall return at the mth transceiver corresponding to

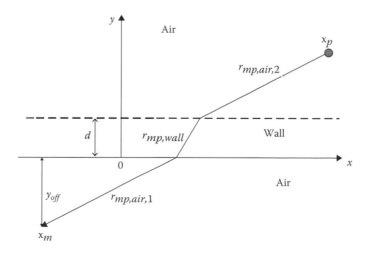

FIGURE 4.1
Scene geometry. (From Lagunas, E. et al., Joint wall mitigation and compressive sensing for indoor image reconstruction, *IEEE Trans. Geosci. Rem. Sens.*, 51 (2), 891, February 2013; Amin, M.G. and Ahmad, F., Compressive sensing for through-the-wall radar imaging, *J. Electron. Imag.*, 22(3), 030901, July 2013. © 2013 IEEE. With permission.)

the kth frequency is given by [31]

$$z_m^w(f_k) = \sum_{l=0}^{L} \sigma_w a_l \exp(-j2\pi f_k \tau_w^{(l)}) \tag{4.2}$$

where
σ_w is the complex reflectivity of the wall
L is the number of wall reverberations
$\tau_w^{(0)}$ is the propagation delay associated with the return from the front face
     of the wall
$\tau_w^{(l)}, l > 0$ are the delays associated with the wall reverberations
a_l is the path loss factor associated with the lth wall return

The decrease in the signal amplitude of the higher-order reverberations is accounted for in the corresponding loss factors a_l. The delay $\tau_w^{(l)}$ is given by

$$\tau_w^{(l)} = \frac{2y_{off}}{c} + l\frac{2d\sqrt{\varepsilon}}{c} \tag{4.3}$$

where c is the speed of light in free space. Assuming the behind-the-wall scene contains P point targets, the target return at the mth transceiver corresponding to the kth frequency can be expressed as

$$z_m^t(f_k) = \sum_{p=0}^{P-1} \sigma_p \exp(-j2\pi f_k \tau_{p,m}) \tag{4.4}$$

where
σ_p is the complex reflectivity of the pth target
$\tau_{p,m}$ is the two-way traveling time between the mth antenna and the pth
     target, given by [7,32]

$$\tau_{p,m} = \frac{2r_{mp,air,1}}{c} + \frac{2r_{mp,wall}\sqrt{\varepsilon}}{c} + \frac{2r_{mp,air,2}}{c} \tag{4.5}$$

The variables $r_{mp,air,1}$, $r_{mp,wall}$, and $r_{mp,air,2}$ in (4.5) represent the traveling distances of the signal before, through, and beyond the wall, respectively, from the mth transceiver to the pth target, as shown in Figure 4.1. Also, the complex amplitude due to free space path loss, wall reflection/transmission coefficients and wall losses, is assumed to be absorbed into the target reflectivity σ_p.

The total baseband signal received by the mth transceiver corresponding to the kth frequency is the superposition of the wall and target returns,

$$z_m(f_k) = z_m^w(f_k) + z_m^t(f_k) = \sum_{l=0}^{L} \sigma_w a_l \exp(-j2\pi f_k \tau_w^{(l)}) + \sum_{p=0}^{P-1} \sigma_p \exp(-j2\pi f_k \tau_{p,m})$$

(4.6)

4.2.2 Wall Clutter Mitigation Techniques

4.2.2.1 Spatial Filtering

From (4.3) and (4.6), we note that $\tau_w^{(l)}$ does not vary with the transceiver location since the array is parallel to the wall. Furthermore, as the wall is homogeneous and assumed to be much larger than the beamwidth of the antenna, each wall reverberation in (4.6) assumes the same value across the array aperture. Unlike $\tau_w^{(l)}$, the time delay $\tau_{p,m}$, given by (4.5), is different for each transceiver location since the signal path from the transceiver to the target is different from one transceiver to the other. For the kth frequency, the received signal is a function of m via the variable $\tau_{p,m}$. Therefore, we can rewrite (4.6) as

$$z_{f_k}(m) = z_{f_k}^w + z_{f_k}^t(m) = \sum_{l=0}^{L} v_{f_k}^l + \sum_{p=0}^{P-1} u_{p,f_k}(m)$$

(4.7)

where

$$v_{f_k}^l = \sigma_w a_l \exp(-j2\pi f_k \tau_w^{(l)})$$

$$u_{p,f_k}(m) = \sigma_p \exp(-j2\pi f_k \tau_{p,m})$$

Thus, separating wall reflections from target reflections amounts to basically separating constant from nonconstant valued signals across the transceivers, which can be performed by applying a proper spatial filter across the array [13].

In its simplest form, the spatial filter, which removes, or significantly attenuates, the zero spatial frequency component, can be implemented as the subtraction of the average of the radar return across the transceivers. That is,

$$\tilde{z}_{f_k}(m) = z_{f_k}(m) - \frac{1}{M} \sum_{m=0}^{M-1} z_{f_k}(m) = z_{f_k}^t(m) - \frac{1}{M} \sum_{m=0}^{M-1} z_{f_k}^t(m)$$

(4.8)

Thus, the subtraction operation removes the zero spatial frequency component corresponding to the wall return and the filter output data will have little or no contribution from the wall reflections.

4.2.2.2 Subspace Projection

The signals received by the M transceivers at the K frequencies are arranged into a $K \times M$ matrix

$$\mathbf{Z} = \begin{bmatrix} \mathbf{z}_0 & \cdots & \mathbf{z}_m & \cdots & \mathbf{z}_{M-1} \end{bmatrix} \tag{4.9}$$

where

$$\mathbf{z}_m = \begin{bmatrix} z_m(f_0) & z_m(f_1) & \cdots & z_m(f_{K-1}) \end{bmatrix}^T = \mathbf{z}_m^w + \mathbf{z}_m^t \tag{4.10}$$

with $z_m(f_k)$ given by (4.6) and $\mathbf{z}_m^w$, $\mathbf{z}_m^t$ representing the wall and target contributions at the mth transceiver, respectively. The eigenstructure of the imaged scene is obtained by performing the SVD of $\mathbf{Z}$,

$$\mathbf{Z} = \mathbf{U}\mathbf{\Lambda}\mathbf{V}^H \tag{4.11}$$

where
   H denotes the Hermitian transpose
   $\mathbf{U}$ and $\mathbf{V}$ are unitary matrices containing the left and right singular vectors, respectively
   $\mathbf{\Lambda}$ is a diagonal matrix

$$\mathbf{\Lambda} = \begin{bmatrix} \lambda_1 & \cdots & 0 \\ \vdots & \ddots & \vdots \\ 0 & \cdots & \lambda_M \\ \vdots & \ddots & \vdots \\ 0 & \cdots & 0 \end{bmatrix} \tag{4.12}$$

and $\lambda_1 \geq \lambda_2 \geq \cdots \geq \lambda_M$ are the singular values. Without loss of generality, the number of frequencies is assumed to exceed the number of antenna locations, that is, $K > M$. The subspace projection method assumes that the wall returns and the target reflections lie in different subspaces. Since the wall reflections are stronger than the target returns, the first J dominant singular vectors of the $\mathbf{Z}$ matrix are used to construct the wall subspace [17],

$$\mathbf{S}_{wall} = \sum_{i=1}^{J} \mathbf{u}_i \mathbf{v}_i^H \tag{4.13}$$

Ideally, a homogeneous wall subspace is spanned by the first singular vector associated with the dominant singular value, that is, $J = 1$. However, factors such as misalignment of antennas and wall heterogeneity will increase the dimensionality of the wall subspace. Methods for determining the dimensionality J of the wall subspace have been reported in [17,33].

The subspace orthogonal to the wall subspace is defined as

$$\mathbf{S}_{wall}^{\perp} = \mathbf{I} - \mathbf{S}_{wall}\mathbf{S}_{wall}^{H} \tag{4.14}$$

where $\mathbf{I}$ is the identity matrix. To mitigate the wall returns, the data matrix $\mathbf{Z}$ is projected on the orthogonal subspace [17,25]

$$\tilde{\mathbf{Z}} = \mathbf{S}_{wall}^{\perp}\mathbf{Z} \tag{4.15}$$

The resulting data matrix has little or no contribution from the front wall.

4.2.3 Scene Reconstruction

An equivalent matrix–vector representation of the wall clutter-free signals in (4.8) and (4.15) can be obtained as follows. Assume that the region being imaged is divided into a finite number of pixels $N_x \times N_y$ in crossrange and downrange. We vectorize the crossrange versus downrange image into an $N_xN_y \times 1$ scene reflectivity vector σ. The qth element of σ takes the value σ_p if the pth point target exists at the qth pixel; otherwise, it is zero. Using (4.4), the wall clutter-free signal vector, $\tilde{\mathbf{z}}_m$, corresponding to the mth transceiver can be expressed in matrix–vector form as

$$\tilde{\mathbf{z}}_m \approx \mathbf{z}_m^t = \mathbf{\Psi}_m\sigma \tag{4.16}$$

where $\mathbf{\Psi}_m$ is a matrix of dimensions $K \times N_xN_y$ with the kth element of its qth column given by

$$[\mathbf{\Psi}_m]_{k,q} = \exp(-j2\pi f_k \tau_{q,m}), \quad k = 0,1,\ldots,K-1, \quad q = 0,1,\ldots,N_xN_y - 1 \tag{4.17}$$

with $\tau_{q,m}$ denoting the two-way traveling time between the qth pixel and the mth transceiver.

Stacking the signals from all M transceiver locations, we obtain the $MK \times 1$ measurement vector $\tilde{\mathbf{z}}$ as [24,25]

$$\tilde{\mathbf{z}} = \mathbf{\Psi}\sigma \tag{4.18}$$

where

$$\tilde{\mathbf{z}} = [\tilde{\mathbf{z}}_0^T \quad \tilde{\mathbf{z}}_1^T \quad \cdots \quad \tilde{\mathbf{z}}_{M-1}^T]^T, \quad \mathbf{\Psi} = [\mathbf{\Psi}_0^T \quad \mathbf{\Psi}_1^T \quad \cdots \quad \mathbf{\Psi}_{M-1}^T]^T \tag{4.19}$$

An image $\hat{\sigma}$ can be reconstructed using delay-and-sum beamforming by premultiplying the wall clutter-free signal $\tilde{z}$ with the adjoint operator Ψ^H. That is,

$$\hat{\sigma} = \Psi^H \tilde{z} \tag{4.20}$$

4.2.4 Illustrative Results

A through-the-wall wideband SAR system was set up in the Radar Imaging Lab at Villanova University. A 67-element line array with an inter-element spacing of 0.0187 m, located along the x-axis, was synthesized parallel to a 0.14-m-thick solid concrete wall of length 3.05 m and at a standoff distance equal to 1.24 m. A stepped-frequency signal covering the 1–3 GHz frequency band with a step size of 2.75 MHz was employed. Thus, at each scan position, the radar collected 728 frequency measurements. A vertical metal dihedral was used as the target and was placed at (0, 4.4)m on the other side of the front wall. The size of each face of the dihedral is 0.39 m by 0.28 m. The back and the side walls of the room were covered with RF-absorbing material to reduce clutter. The empty scene without the dihedral target present was also measured to enable background subtraction for wall clutter removal.

The region to be imaged is chosen to be 4.9 m × 5.4 m centered at (0, 3.7)m and divided into 33 × 73 pixels, respectively. Figure 4.2a depicts the image corresponding to the raw data obtained with beamforming. In this and all subsequent figures in this chapter, we plot the image intensity with the maximum intensity value in each image normalized to 0 dB. The true target position is indicated with a solid red rectangle, while the wall location is indicated by a dashed red rectangle. With the availability of the empty scene measurements, background subtraction generates an image where the target can be easily identified, as shown in Figure 4.2b. Figure 4.2c shows the beamformed image after subspace projection–based wall mitigation approach was applied to the raw data. In all applications of the subspace projection approach in this chapter, the first dominant singular vector of the data matrix $\mathbf{Z}$ is used to construct the wall subspace ($J = 1$). We observe that although the wall return has not been completely suppressed, its shadowing effect has been substantially reduced, allowing the detection of the target. Similar results were obtained with the spatial filtering–based approach [25].

4.3 Spatial Filtering and Subspace Projection under Reduced Data Volume

The data model in (4.6) and the review of the wall clutter mitigation techniques in Section 4.2 involve the full set of measurements made at all

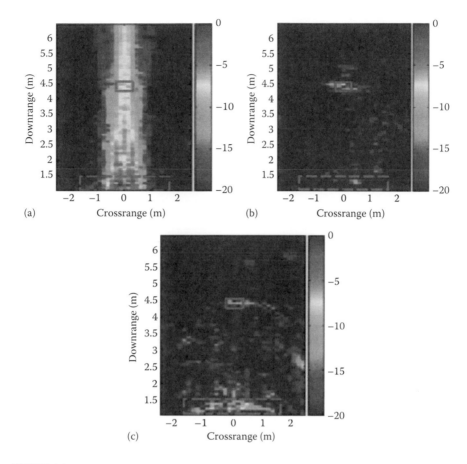

FIGURE 4.2
(a) Beamforming-based imaging result using all of the raw data. (From Lagunas, E. et al., Joint wall mitigation and compressive sensing for indoor image reconstruction, *IEEE Trans. Geosci. Rem. Sens.*, 51 (2), 891, February 2013. © 2013 IEEE. With permission.) **(See color insert.)** (b) Beamforming-based imaging result after background subtraction. (From Lagunas, E. et al., Joint wall mitigation and compressive sensing for indoor image reconstruction, *IEEE Trans. Geosci. Rem. Sens.*, 51(2), 891, February 2013; Amin, M.G. and Ahmad, F., Compressive sensing for through-the-wall radar imaging, *J. Electron. Imag.*, 22 (3), 030901, July 2013. © 2013 IEEE. With permission.) **(See color insert.)** (c) Beamforming-based imaging result after subspace projection based wall mitigation.

M transceiver locations using all K frequencies. Assume only $M_1(<M)$ randomly selected transceiver locations are available for data collection. Let $i_g \in [0, 1, \ldots, M-1]$, for $g = 0, 1, \ldots, M_1 - 1$, be the indices of the employed transceiver locations. Further assume that only $K_1 < K$ frequency measurements are made at each transceiver location. Representing the full frequency measurement at the i_gth transceiver by the $K \times 1$ vector $\mathbf{z}_{i_g}$ defined in

(4.10), the corresponding reduced set of frequency measurements can be expressed as

$$\tilde{z}_{i_g} = \varphi^{(g)} z_{i_g} = \varphi^{(g)} z_{i_g}^w + \varphi^{(g)} z_{i_g}^t = \breve{z}_{i_g}^w + \breve{z}_{i_g}^t \tag{4.21}$$

where $\varphi^{(g)}$ is a $K_1 \times K$ measurement matrix constructed by randomly selecting K_1 rows of a $K \times K$ identity matrix [24,25]. The matrix $\varphi^{(g)}$ determines the reduced set of frequencies corresponding to the i_gth transceiver. Note that the reduced sets of frequencies could either differ from one transceiver to the other (as implied in (4.21)) or be the same for each transceiver ($\varphi^{(g)} = \varphi, g = 0, 1, \ldots, M_1 - 1$).

4.3.1 Wall Mitigations under Reduced Data Volume

Both spatial filtering and subspace projection methods for wall clutter reduction rely on the fact that the wall reflections assume very close, if not equal, values at the different transceiver locations. When the same set of frequencies is used for all employed transceivers, that is, $\varphi^{(g)} = \varphi \ \forall g$, the condition of spatial invariance of the wall reflections is not violated. This permits direct application of the spatial filtering and subspace projection schemes as a preprocessing step to the scene image reconstruction [24,25].

However, use of different sets of reduced frequencies for the various transceivers results in different wall reflection phase returns across the antenna elements. This would deprive the wall mitigation algorithms of the underlying assumption of spatial invariance of the wall clutter, thereby rendering the direct application of the wall mitigation methods ineffective [25].

4.3.2 CS-Based Scene Reconstruction

After the wall clutter mitigation step has been applied to the reduced set of measurements, the image can be reconstructed using sparse reconstruction schemes as follows. Assuming effective suppression of wall clutter and using (4.16) and (4.21), we can express the preprocessed reduced set of frequency measurements at the i_gth transceiver as

$$\tilde{z}_{i_g} \approx \breve{z}_{i_g}^t = \varphi^{(g)} z_{i_g}^t = \varphi^{(g)} \Psi_{i_g} \sigma, \quad g = 0, 1, \ldots, M_1 - 1 \tag{4.22}$$

Considering the preprocessed measurement vectors from all M_1 transceivers, we obtain the $M_1 K_1 \times 1$ measurement vector $\tilde{z}$ as

$$\tilde{z} = \Phi \Psi \sigma \tag{4.23}$$

where

$$\Psi = \begin{bmatrix} \Psi_{i_0}^T & \Psi_{i_1}^T & \cdots & \Psi_{i_{M_1-1}}^T \end{bmatrix}^T \tag{4.24}$$

and $\boldsymbol{\Phi} = \mathrm{bdiag}(\boldsymbol{\varphi}^{(0)}, \boldsymbol{\varphi}^{(1)}, \ldots, \boldsymbol{\varphi}^{(M_1-1)})$ or $\boldsymbol{\Phi} = \mathrm{bdiag}(\boldsymbol{\varphi}, \boldsymbol{\varphi}, \ldots, \boldsymbol{\varphi})$, depending on whether the reduced set of frequencies varies or is the same across the employed transceivers, with $\mathrm{bdiag}(\cdot)$ denoting the block diagonal matrix operation. Given $\tilde{\mathbf{z}}$, we can recover $\boldsymbol{\sigma}$ by solving the following optimization problem:

$$\hat{\boldsymbol{\sigma}} = \arg\min \|\boldsymbol{\sigma}\|_{l_1} \text{ subject to } \tilde{\mathbf{z}} \approx \boldsymbol{\Phi}\boldsymbol{\Psi}\boldsymbol{\sigma} \tag{4.25}$$

The problem in (4.25) can be solved using convex relaxation, greedy pursuit, or combinatorial algorithms [34–39]. In this chapter, we consider orthogonal matching pursuit (OMP), which is a greedy pursuit algorithm and is known to provide a fast and easy to implement solution [38].

4.3.3 Illustrative Results

We consider the same experimental setup as described in Section 4.2.4. For CS, 20% of the frequencies and 51% of the array locations were used, which collectively represent 10.2% of the total data volume. Figure 4.3a depicts the image corresponding to the measured scene obtained with OMP applied directly to the reduced raw data set. The number of iterations of the OMP is usually associated with the level of sparsity of the scene. In this case, the number of OMP iterations was set to 100. We observe that the sparse reconstruction algorithm only reconstructs the wall reverberations and totally misses the target. Since access to the background scene is not available in practice, it is evident from Figure 4.3a that the wall mitigation techniques must be applied, as a preprocessing step, prior to CS in order to detect the targets behind the wall.

First, we consider the case when the same set of reduced frequencies is used for each of the employed transceiver locations. We conducted 100 trials, each with a different random selection of 20% frequencies and 51% array locations. For each trial, the subspace projection method is applied to a $\mathbf{Z}$ matrix of reduced dimension 146×34, followed by scene reconstruction using OMP with 25 iterations. Figure 4.3b shows the corresponding reconstructed image, averaged over 100 trials. It is clear that, even when both spatial and frequency observations are reduced, the joint application of wall clutter mitigation and CS techniques successfully provides front wall clutter suppression and unmasking of the target.

Next, we consider the case when a different randomly chosen set of 20% frequencies is used for different employed transceivers. The corresponding reconstructed image, averaged over 100 trials with 25 iterations of OMP after application of the subspace projection technique, is shown in Figure 4.3c. As expected, the violation of the spatial invariance causes the wall mitigation approach to be unsuccessful, leading to an image reconstruction containing the wall clutter only. Spatial filtering produced similar

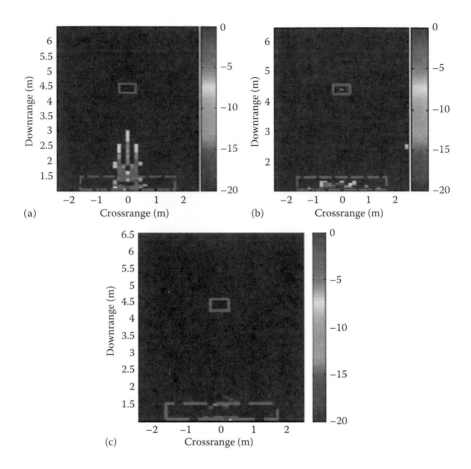

FIGURE 4.3
(See color insert.) (a) CS-based imaging result using all of the raw data. (b) CS-based imaging result using 10.2% data volume with the same frequency set at each antenna, averaged over 100 trials. (c) CS-based imaging result using 10.2% of the data volume with different frequency sets from different antennas, averaged over 100 trials. (From Lagunas, E. et al., Joint wall mitigation and compressive sensing for indoor image reconstruction, *IEEE Trans. Geosci. Rem. Sens.*, 51(2), 891, February 2013. © 2013 IEEE. With permission.)

results for both considered cases of frequency measurement distributions over transceiver locations [25].

4.4 Wall Clutter Mitigation Using DPSSs

As discussed in Section 4.1, having the same frequency observations at all employed antenna locations may not always be possible. In order to overcome the shortcomings of the wall clutter mitigation schemes of spatial filtering and subspace projection in case of the same reduced frequency measurement set for all transceiver locations, an alternate scheme based on discrete prolate spheroidal sequences is considered, which is described next.

4.4.1 Discrete Prolate Spheroidal Sequences

Discrete prolate spheroidal sequences are a collection of index-limited sequences that maximize the energy concentration within a given frequency band [26]. The DPSSs constitute an efficient basis for finite-energy signals that are time-limited with their energy concentrated in a given bandwidth. Since we consider a stepped-frequency signal consisting of K frequencies, we deal with the dual problem to the conventional DPSSs. That is, we are seeking frequency-domain sequences, $s[k]$, confined to the frequency index set $\{0, 1, \ldots, K-1\}$, whose energy is concentrated in a finite time interval $[-\bar{T}, \bar{T})$. Since the unambiguous time interval corresponding to a step size of Δf is $[0, 1/\Delta f)$ or equivalently $[-1/2\Delta f, 1/2\Delta f)$, $\bar{T}$ lies between 0 and $1/2\Delta f$. Let T be the time $\bar{T}$ normalized by $1/\Delta f$ such that $0 < T < 1/2$. Then, exploiting the duality in time and frequency domains, the K-length frequency-domain DPSSs are defined as solutions of [27,40,41]

$$\mathbf{A}\mathbf{s}_i = \lambda_i \mathbf{s}_i, \quad i = 0, 1, \ldots, K-1 \tag{4.26}$$

where

$\mathbf{s}_i$ is a $K \times 1$ vector with elements $s_i[k]$, $k = 0, 1, \ldots, K-1$
λ_i are the eigenvalues of the matrix $\mathbf{A}$

which is given by

$$[\mathbf{A}]_{i,k} = \frac{\sin(2\pi T(i-k))}{\pi(i-k)} \tag{4.27}$$

The DPSSs are orthonormal on the set $\{0, 1, \ldots, K-1\}$ [40,41].

4.4.2 DPSS Basis

Consider the received signal vector $\mathbf{z}_m$ corresponding to the mth antenna location using all K frequencies, as given by (4.10). The time-domain equivalent of each of the M received signals $\{\mathbf{z}_m\}_{m=0}^{M-1}$ is an ensemble of returns concentrated on a number of time intervals within $[-1/2\Delta f, 1/2\Delta f)$. In the case of point targets, each interval would comprise a single resolution cell. However, in practice, since most indoor targets, including walls, are spatially extended, the corresponding returns extend beyond a single resolution cell. For the wall return in particular, the reverberations may not even be separable depending on the wall permittivity, thickness, and the signal bandwidth.

We, therefore, refer to the received signals as *multiduration signals*. We first construct a basis using DPSSs for efficiently capturing the structure of such signals.

We divide the unambiguous time $[-1/2\Delta f, 1/2\Delta f)$ into $N = \lfloor(2/\Delta fD) - 1\rfloor$ overlapping intervals of length D, where D is selected to be a multiple of $1/(K-1)\Delta f$. The nth time interval is centered at $-1/2\Delta f + nD/2$ and has an extent

$$\Delta_n = \left[-\frac{1}{2\Delta f} + \frac{nD}{2} - \frac{D}{2}, -\frac{1}{2\Delta f} + \frac{nD}{2} + \frac{D}{2}\right], \quad n = 1, 2, \ldots, N$$

Note that the choice of nonoverlapping set of intervals would be inadequate since the radar returns from the various scatterers may not lie exactly on the chosen grid. Let $T = D\Delta f/2$ and $t_n = \left(-(1/2\Delta f) + nD/2)\right)\Delta f$. Consider the $K \times K$ matrix $\mathbf{S}_{K,T}$ of K-length frequency-domain DPSSs

$$\mathbf{S}_{K,T} = [\mathbf{s}_0 \quad \mathbf{s}_1 \quad \cdots \quad \mathbf{s}_{K-1}] \tag{4.28}$$

with $\{\mathbf{s}_i\}_{i=0}^{K-1}$ defined in (4.26). Forming the $K \times K$ diagonal matrix $\mathbf{E}_{t_n}$ as

$$\mathbf{E}_{t_n} = \text{diag}(1, \exp(-j2\pi t_n), \ldots, \exp(-j2\pi(K-1)t_n)) \tag{4.29}$$

we can define the time-shifted DPSS basis for Δ_n as $\mathbf{E}_{t_n}\mathbf{S}_{K,T}$. As the first $\lceil 2KT \rceil + 1$ DPSS eigenvalues are close to 1 while the remaining are close to zero, the first $\lceil 2KT \rceil + 1$ time-shifted DPSSs form an efficient signal basis that can capture the energy of the frequency-domain sequences concentrated on the interval Δ_n [40,41]. Therefore, we consider the $K \times (\lceil 2KT \rceil + 1)$ matrix $\mathbf{\Sigma}_n$ comprising the first $\lceil 2KT \rceil + 1$ columns of $\mathbf{E}_{t_n}\mathbf{S}_{K,T}$ as an efficient basis for signals supported on Δ_n. Thus, the $K \times (\lceil 2KT \rceil + 1)N$ DPSS basis $\mathbf{\Sigma}$ for the multiduration signals can be defined as the concatenation of the N time-shifted DPSS bases [27,41]

$$\mathbf{\Sigma} = [\mathbf{\Sigma}_1 \quad \mathbf{\Sigma}_2 \quad \cdots \quad \mathbf{\Sigma}_N] \tag{4.30}$$

Using the DPSS basis $\mathbf{\Sigma}$, the mth received signal $\mathbf{z}_m$ can be expressed as

$$\mathbf{z}_m = \mathbf{z}_m^w + \mathbf{z}_m^t = \mathbf{\Sigma}\boldsymbol{\rho}_m^w + \mathbf{\Sigma}\boldsymbol{\rho}_m^t \tag{4.31}$$

where $\boldsymbol{\rho}_m^w$ and $\boldsymbol{\rho}_m^t$ are the $K \times (\lceil 2KT \rceil + 1)N$-length coefficient vectors corresponding to the wall and target returns, respectively. It is noted that because of the multiduration nature of the radar returns, the wall and target contributions, $\mathbf{z}_m^w, \mathbf{z}_m^t$, can be represented using only the columns of $\mathbf{\Sigma}$ corresponding to the occupied time intervals. Both $\boldsymbol{\rho}_m^w$ and $\boldsymbol{\rho}_m^t$ exhibit a block-sparse structure with the nonzero coefficients occurring in a small number of clusters of size.

Following the data reduction formulation of (4.21), the reduced data counterpart of the model in (4.31) with K_1 frequencies and M_1 transceiver locations can be expressed as

$$\breve{z}_{i_g} = \varphi^{(g)} z_{i_g} = \varphi^{(g)} z_{i_g}^w + \varphi^{(g)} z_{i_g}^t = \breve{z}_{i_g}^w + \breve{z}_{i_g}^t = \varphi^{(g)} \Sigma \rho_{i_g}^w + \varphi^{(g)} \Sigma \rho_{i_g}^t \quad (4.32)$$

with $i_g \in [0, 1, \ldots, M-1]$ for $g = 0, 1, \ldots, M_1 - 1$.

4.4.3 Block-Sparse Reconstruction

The goal is to reconstruct the wall contribution at each employed antenna location individually using the reduced measurement vector $\breve{z}_{i_g}$, which can then be subtracted from $\breve{z}_{i_g}$ to obtain the clutter-free radar return at the i_gth antenna. Because of the block-sparse nature of $\rho_{i_g}^w$ and $\rho_{i_g}^t$, we use the block extension of orthogonal matching pursuit (BOMP) [42] to recover the signal component corresponding to the wall.

The choice of the number of BOMP iterations is critical to this approach. Too small a value may not completely capture the wall reverberations, whereas a large enough value may include the returns from targets located at deeper ranges as part of the wall response reconstruction. In order to sufficiently suppress the wall return without removing targets at deeper ranges, a modified BOMP algorithm, provided in Table 4.1, is used, which employs a larger number of iterations but constrains the reconstructed wall return support to no more than 1.5 m away from the front face of the wall. This constraint on wall clutter support is suggested by EM simulations of various homogeneous and nonhomogeneous walls [43].

TABLE 4.1

Modified BOMP Algorithm

Input: number of iterations I, matrix $\Xi = \varphi^{(g)} \Sigma$ measurements $\breve{z}_{i_g}$, permissible set of wall support indices Ω_w

Initialization: Support set $\Omega_0 = \varphi$, residual error $r_0 = \breve{z}_{i_g}$ iteration index $\bar{i} = 1$

while $\bar{i} \leq I$

1) $\Omega_{\bar{i}} = \Omega_{\bar{i}-1} \cup \left\{ \arg\max_n \left\| \Xi_n^H r_{\bar{i}-1} \right\|_2 \right\}$, where $\Xi_n = \varphi^{(g)} \Sigma_n$

2) $r_{\bar{i}} = \breve{z}_{i_g} - \Xi_{\Omega_{\bar{i}}} \Xi_{\Omega_{\bar{i}}}^\dagger \breve{z}_{i_g}$, where $\Xi_{\Omega_{\bar{i}}}$ denotes the submatrix of Ξ containing only the columns of Ξ corresponding to the set $\Omega_{\bar{i}}$, and the superscript '$\dagger$' denotes Pseudoinverse.

3) $\bar{i} = \bar{i} + 1$

end

Purgation: $\Omega_I' = \Omega_I \cap \Omega_w$

Output: reconstructed signal, $\hat{z}_{i_g}\big|_{\Omega_I'} = \Xi_{\Omega_I'} \Xi_{\Omega_I'}^\dagger \breve{z}_{i_g}$ and $\hat{z}_{i_g}\big|_{(\Omega_I')^c} = 0$, where the superscript 'c' denotes the set complement.

The modified BOMP algorithm will capture the wall contribution only, thereby implying that the output $\hat{\mathbf{z}}_{i_g} \approx \breve{\mathbf{z}}_{i_g}^w$. Thus, the target contribution can be obtained by simply subtracting the reconstructed wall contribution in reduced data domain

$$\breve{\mathbf{z}}_{i_g} - \hat{\mathbf{z}}_{i_g} \approx \breve{\mathbf{z}}_{i_g}^t \tag{4.33}$$

Once the wall clutter has been suppressed individually at each employed antenna location, we can proceed with image formation under reduced data volume by solving the sparse reconstruction problem in (4.25).

4.4.4 Illustrative Results

A stepped-frequency SAR system was used for data measurements in the Radar Imaging Lab at Villanova University. The synthetic linear aperture consisted of 93 uniformly spaced elements, with an inter-element spacing of 0.02 m. The aperture was located parallel to a 0.2-m-thick solid concrete block wall at a standoff distance of 3.13 m. The stepped-frequency signal comprised 641 frequencies from 1 to 3 GHz, with a step size of 3.125 MHz. A vertical metal dihedral, located at -0.29 m in crossrange and 2.05 m away from the other side of the front wall, was used as the target. The side walls were covered with RF-absorbing material while the 0.3-m-thick reinforced concrete back wall was left bare. The distance between the front face of the back wall and the back face of the front wall is 3.76 m.

The scene to be imaged is chosen to be 4 m $\times$ 5.5 m centered at $(0, 4.75)$m and divided into 33×77 pixels. Figure 4.4a depicts the image corresponding to the full raw data set obtained with beamforming. In addition to the wall reverberation, the antenna ringing is clearly visible in Figure 4.4a at downranges prior to the front wall. For CS-based reconstruction, we first randomly selected 20% of the antenna locations with a different set of randomly chosen 20% frequencies at each chosen antenna, which amounts to 4% of the total data volume. We reconstructed the scene after application of the DPSS-based wall suppression scheme 100 times. For each trial, a different random measurement matrix was used to generate the reduced set of measurements, which were then processed for wall clutter mitigation, followed by sparsity-based scene reconstruction. The value of the parameter D was chosen to be 5.5 ns, and the number of iterations for the modified BOMP was selected as 8. Figure 4.4b shows the corresponding reconstructed image averaged over 100 trials. The number of OMP iterations for scene reconstruction was chosen to be 5. We observe from Figure 4.4b that the DPSS-based wall mitigation was successful in removing most of the wall return and the antenna ringing, leading to a *clean* image with the target and back wall clearly visible.

Next, the same set of randomly chosen 20% frequencies was employed for all of the selected 20% antenna locations. We reconstructed the scene

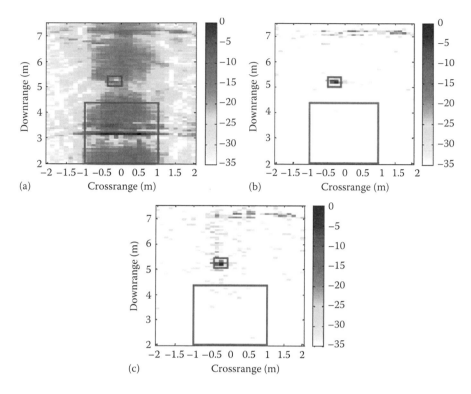

FIGURE 4.4
(a) Beamformed image using full raw data set. (b) Sparse reconstruction result after DPSS-based wall mitigation using 4% of the total data volume with different reduced frequency sets at different employed antennas, averaged over 100 trials. (c) Sparse reconstruction result after DPSS-based wall mitigation using 4% of the total data volume with the same reduced frequency set at each employed antenna, averaged over 100 trials.

after application of the DPSS-based wall suppression scheme 100 times. The number of BOMP and OMP iterations and the value of D were chosen to be kept the same as for Figure 4.4b. Figure 4.4c shows the corresponding sparsity-based reconstruction image. Again, the DPSS-based scheme successfully removed both the antenna ringing and the wall, thereby allowing the subsequent sparse reconstruction to localize the target and the back wall.

4.5 Partially Sparse Reconstruction of Indoor Scenes

In this section, we assume prior knowledge of the building layout, which is exploited under the partial sparsity framework for imaging of stationary targets behind walls.

4.5.1 Partially Sparse Signal Model

Again, consider a stepped-frequency signal of K equispaced frequencies and a monostatic SAR with M antenna positions located at a standoff distance y_{off}, parallel to a homogenous front wall, which is located along the x-axis. The scene behind the front wall is assumed to be composed of P point targets, $N_I - 1$ interior walls, which are parallel to the front wall and to the radar scan direction, and $N_C - 1$ corners corresponding to the junctions of two walls perpendicular to each other. It is noted that, due to the specular nature of the wall reflections, a SAR system located parallel to the front wall will only be able to receive backscattered signals from interior walls, which are parallel to the front wall. The contribution of walls perpendicular to the front wall will be captured primarily through the backscattered signals from the corners [44].

The received signal corresponding to the kth frequency at the mth antenna position, with phase center at $\mathbf{x}_m = (x_m, -y_{off})$ is given by [44]

$$
z_m(f_k) = \sum_{p=0}^{P-1} \sigma_p \exp(-j2\pi f_k \tau_{p,m}) + \sum_{l=0}^{N_I-1} \sigma_{w,l} \exp(-j2\pi f_k \tau_{w,l})
$$

$$
+ \sum_{i=0}^{N_C-1} \Gamma_{i,m} \bar{\sigma}_i \mathrm{sinc}\left(\frac{2\pi f_k \bar{L}_i}{c}\right) \sin(\theta_{i,m} - \bar{\theta}_i) \exp(-j2\pi f_k \tau_{i,m}) \quad (4.34)
$$

where

  $\sigma_p, \sigma_{w,l}, \bar{\sigma}_i$ are the respective complex amplitudes associated with the pth target, lth wall, and the ith corner

  $\tau_{p,m}, \tau_{w,l}, \tau_{i,m}$ are the two-way traveling times from the mth antenna to the pth target, the lth wall, and the ith corner, respectively

  $\bar{L}_i$ is the length

  $\bar{\theta}_i$ is the orientation angle of the ith corner

  $\theta_{i,m}$ is the aspect angle associated with the ith corner and the mth antenna

  $\Gamma_{i,m}$ is an indicator function, which assumes a unit value only when the mth antenna illuminates the concave side of the ith corner

We note that $\sigma_p, \sigma_{w,l}, \bar{\sigma}_i$ contain contributions from free space path loss, attenuation due to propagation through the wall(s), and the reflectivity of the corresponding scatterer. Further, since the scan direction is parallel to the walls, the delay $\tau_{w,l}$ does not depend on the variable m and is a function only of the downrange distance between the lth wall and the array baseline.

Let $\mathbf{z}_m$ represent the received signal vector corresponding to the K frequencies and the mth antenna location. Under the assumption that the building layout is known a priori, the $N_x N_y \times 1$ scene reflectivity vector, $\boldsymbol{\sigma}$, can be expressed as $\boldsymbol{\sigma} = [\boldsymbol{\sigma}_1^T \quad \boldsymbol{\sigma}_2^T]^T$, where the $Q_1 \times 1$ vector $\boldsymbol{\sigma}_1$ is the dense part with known support and the $Q_2 \times 1$ vector $\boldsymbol{\sigma}_2$ is the sparse part with

$Q_2 = N_x N_y - Q_1$. Note that σ_1 corresponds to the walls that are parallel to the antenna baseline. Further, since the wall junctions lie along the parallel walls, the corner locations would correspond to the support of a subset of σ_1, say an $R \times 1$ vector $\bar{\sigma}_1$ with $R < Q_1$. Then, using (4.34), we obtain the matrix–vector form

$$\mathbf{z}_m = \mathbf{\Psi}_{wall,m}\,\sigma_1 + \mathbf{\Psi}_{corner,m}\,\bar{\sigma}_1 + \mathbf{\Psi}_{tgt,m}\,\sigma_2 \qquad (4.35)$$

where $\mathbf{\Psi}_{wall,m}$, $\mathbf{\Psi}_{corner,m}$, $\mathbf{\Psi}_{tgt,m}$ are the dictionaries corresponding to the walls, corner reflectors, and point targets, respectively. The matrix $\mathbf{\Psi}_{tgt,m}$ is of dimension $K \times Q_2$ with its (k, q_2)th element given by

$$[\mathbf{\Psi}_{tgt,m}]_{k,q_2} = \exp(-j2\pi f_k \tau_{q_2,m}) \qquad (4.36)$$

where $\tau_{q_2,m}$ is the two-way traveling time between the mth antenna and the q_2th grid point of the sparse part. The wall dictionary $\mathbf{\Psi}_{wall,m}$ is a $K \times Q_1$ matrix, whose (k, q_1)th element takes the form [28]

$$[\mathbf{\Psi}_{wall,m}]_{k,q_1} = \exp\left(\left(\frac{-j2\pi f_k 2(y_{q_1} + y_{off})}{c}\right)\Im_{q_1,m}\right) \qquad (4.37)$$

In (4.37), y_{q_1} represents the downrange coordinate of the q_1th pixel in the dense part, and $\Im_{q_1,m}$ is an indicator function, which assumes a unit value only when the q_1th grid point lies in front of the mth antenna. The corner dictionary $\mathbf{\Psi}_{corner,m}$ is a $K \times R$ matrix whose (k, r)th element is given by [44]

$$[\mathbf{\Psi}_{corner,m}]_{k,r} = \exp(-j2\pi f_k \tau_{r,m})\Gamma_{r,m}\mathrm{sinc}\left(\left(\frac{2\pi f_k \bar{L}_r}{c}\right)\sin(\theta_{r,m} - \bar{\theta}_r)\right) \qquad (4.38)$$

Equation 4.35 considers the contribution of only one antenna location. Stacking the measurement vectors corresponding to all M antennas, we obtain the linear model

$$\mathbf{z} = \mathbf{\Psi}_{wall}\sigma_1 + \mathbf{\Psi}_{corner}\bar{\sigma}_1 + \mathbf{\Psi}_{tgt}\sigma_2 \qquad (4.39)$$

where

$$\mathbf{z} = \begin{bmatrix} \mathbf{z}_0^T & \mathbf{z}_1^T & \cdots & \mathbf{z}_{M-1}^T \end{bmatrix}^T,$$

$$\mathbf{\Psi}_{corner} = \begin{bmatrix} \mathbf{\Psi}_{corner,0}^T & \mathbf{\Psi}_{corner,1}^T & \cdots & \mathbf{\Psi}_{corner,M-1}^T \end{bmatrix}^T$$

$$\mathbf{\Psi}_{wall} = \begin{bmatrix} \mathbf{\Psi}_{wall,0}^T & \mathbf{\Psi}_{wall,1}^T & \cdots & \mathbf{\Psi}_{wall,M-1}^T \end{bmatrix}^T, \qquad (4.40)$$

$$\mathbf{\Psi}_{tgt} = \begin{bmatrix} \mathbf{\Psi}_{tgt,0}^T & \mathbf{\Psi}_{tgt,1}^T & \cdots & \mathbf{\Psi}_{tgt,M-1}^T \end{bmatrix}^T$$

For reduced data volume, assume that only M_1 randomly selected transceiver locations are available and only K_1 frequency measurements are made at each transceiver location. With the full frequency measurement at the i_gth transceiver given by the $K \times 1$ vector $\mathbf{z}_{i_g}$ in Equation 4.35, the corresponding reduced frequency measurements can be expressed as

$$\breve{\mathbf{z}}_{i_g} = \varphi^{(g)}\mathbf{z}_{i_g} = \varphi^{(g)}\Psi_{wall,i_g}\sigma_1 + \varphi^{(g)}\Psi_{corner,i_g}\bar{\sigma}_1 + \varphi^{(g)}\Psi_{tgt,i_g}\sigma_2 \qquad (4.41)$$

where $i_g \in [0, 1, \ldots, M-1]$ for $g = 0, 1, \ldots, M_1-1$. Concatenating contributions corresponding to all M_1 antennas, we obtain the reduced data counterpart of the linear model in (4.39) as

$$\breve{\mathbf{z}} = \Phi\Psi_{wall}\sigma_1 + \Phi\Psi_{corner}\bar{\sigma}_1 + \Phi\Psi_{tgt}\sigma_2 \qquad (4.42)$$

where the $K_1 M_1 \times KM$ block diagonal matrix Φ is defined in Section 4.3.2.

4.5.2 Sparse Scene Reconstruction

Given the reduced measurement vector $\breve{\mathbf{z}}$ and knowledge of the support of the walls and corners, the goal is to reconstruct the sparse part of the image where the stationary targets of interest are located. Toward this goal, we first need to remove the contributions of the dense part of the scene from $\breve{\mathbf{z}}$. Let $\mathbf{P}_{wall}$ be the orthogonal projection onto the orthogonal complement of the range space of the matrix $\Phi\Psi_{wall}$. If $\Phi\Psi_{wall}$ is a full rank matrix, then $\mathbf{P}_{wall}$ can be expressed as

$$\mathbf{P}_{wall} = \mathbf{I}_{M_1 K_1} - (\Phi\Psi_{wall})(\Phi\Psi_{wall})^{\dagger} \qquad (4.43)$$

where
  $\mathbf{I}_{M_1 K_1}$ is an $M_1 K_1 \times M_1 K_1$ identity matrix
  $(\Phi\Psi_{wall})^{\dagger}$ denotes the pseudoinverse of $\Phi\Psi_{wall}$

On the other hand, if $\Phi\Psi_{wall}$ has a reduced rank, then we have to resort to the SVD of $\Phi\Psi_{wall}$ to obtain the matrix $\mathbf{P}_{wall}$ as

$$\mathbf{P}_{wall} = \mathbf{U}_{wall}\mathbf{U}_{wall}^{H} \qquad (4.44)$$

where $\mathbf{U}_{wall}$ is the matrix consisting of the left singular vectors corresponding to the zero singular values. Applying the projection matrix $\mathbf{P}_{wall}$ to the observation vector $\breve{\mathbf{z}}$, we obtain

$$\breve{\mathbf{z}}_A \equiv \mathbf{P}_{wall}\breve{\mathbf{z}} = \mathbf{P}_{wall}\Phi\Psi_{corner}\bar{\sigma}_1 + \mathbf{P}_{wall}\Phi\Psi_{tgt}\sigma_2 \qquad (4.45)$$

Next, consider the projection matrix $\mathbf{P}_{corner}$ given by

$$\mathbf{P}_{corner}$$

$$= \begin{cases} \mathbf{I}_{M_1 K_1} - (\mathbf{P}_{wall}\boldsymbol{\Phi}\boldsymbol{\Psi}_{corner})(\mathbf{P}_{wall}\boldsymbol{\Phi}\boldsymbol{\Psi}_{corner})^\dagger & \text{if } \mathbf{P}_{wall}\boldsymbol{\Phi}\boldsymbol{\Psi}_{corner} \text{ has full rank} \\ \mathbf{U}_{corner}\mathbf{U}_{corner}^H & \text{otherwise} \end{cases}$$

$$(4.46)$$

where $\mathbf{U}_{corner}$ is the matrix consisting of the left singular vectors corresponding to the zero singular values of the matrix $\mathbf{P}_{wall}\boldsymbol{\Phi}\boldsymbol{\Psi}_{corner}$. Application of $\mathbf{P}_{corner}$ to the measurement vector $\breve{\mathbf{z}}_A$ leads to

$$\breve{\mathbf{z}}_{BA} \equiv \mathbf{P}_{corner}\breve{\mathbf{z}}_A = \mathbf{P}_{corner}\mathbf{P}_{wall}\boldsymbol{\Phi}\boldsymbol{\Psi}_{tgt}\sigma_2 \qquad (4.47)$$

Thus, after the sequential application of the two projection matrices, the measurement vector $\breve{\mathbf{z}}_{BA}$ contains contributions from only the sparse image part, σ_2, which can then be recovered by solving the problem

$$\hat{\sigma}_2 = \arg\min \|\sigma_2\|_{l_1} \text{ subject to } \breve{\mathbf{z}}_{BA} \approx \mathbf{P}_{corner}\mathbf{P}_{wall}\boldsymbol{\Phi}\boldsymbol{\Psi}_{tgt}\sigma_2 \qquad (4.48)$$

The problem in (4.48) belongs to the classical setting of CS and, thus, can be solved using existing sparse reconstruction algorithms [34–39].

4.5.3 Illustrative Results

A simulation was performed in Xpatch®, which is a computational EM code implementing an approximate ray tracing/physical optics computational approach. A computer model of a single story building was created, with overall dimensions of 7 m × 10 m × 2.2 m, containing four humans (labeled 1 through 4) and several furniture items, as shown in Figure 4.5 [45]. The origin of the coordinate system was chosen to be in the center of the building, with the x-axis and the y-axis oriented as shown in Figure 4.5b. The exterior walls were made of 0.2-m-thick bricks and had glass windows and a wooden door. The interior walls were made of 5-cm-thick sheetrock and had a wooden door. The ceiling/roof is flat, made of a 7.5-cm-thick concrete slab. The entire building is placed on top of a dielectric ground plane. The furniture items, namely, a bed, a couch, a bookshelf, a dresser, and a table with four chairs, were made of wood, while the mattress and cushions were made of generic foam/fabric material. Humans 1 through 4 were positioned at various locations in the interior of the building with 45°, 0°, −20°, and 10° azimuthal orientation angles. Note that an orientation angle of 0° corresponds to the human facing along the positive x direction and the positive angles correspond to a counterclockwise rotation in the horizontal plane. Human 3, positioned inside the interior room, was carrying a rifle. The human model was made of a uniform dielectric material with properties

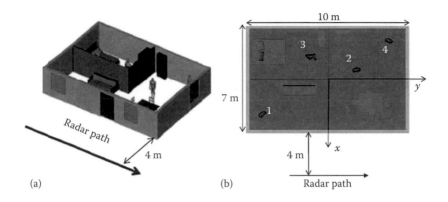

FIGURE 4.5
(a) Three-dimensional view of the scene layout. (b) Top-down view of the scene layout. (From Ahmad, F. et al., A beamforming approach to imaging of stationary indoor scenes under known building layout, in: *Proceedings of the Fifth IEEE International Workshop on Computational Advances in Multi-Sensor Adaptive Processing*, Saint Martin, December 15–18, 2013. © 2013 IEEE. With permission.)

close to those of skin [46]. The rifle is made of metal and wood [47,48]. The dielectric properties of the various materials employed are listed in Table 4.2.

A 6-m-long monostatic synthetic aperture array, with an inter-element spacing of 2.54 cm and located parallel to the front of the building at a stand-off distance of 4 m, was used for data collection. A stepped-frequency signal covering the 0.7–2 GHz frequency band with a step size of 8.79 MHz was

TABLE 4.2

Complex Dielectric Constant for Materials Used in the EM Simulation

Material	ε'	ε''
Brick	3.8	0.24
Concrete	6.8	1.2
Glass	6.4	0
Wood	2.5	0.05
Sheetrock	2.0	0
Foam cushion and fabric	1.4	0
Ground	10	0.6
Human	50	12

Source: Ahmad, F. et al., A beamforming approach to imaging of stationary indoor scenes under known building layout, in: *Proceedings of the Fifth IEEE International Workshop on Computational Advances in Multi-Sensor Adaptive Processing*, Saint Martin, December 15–18, 2013. © 2013 IEEE. With permission.

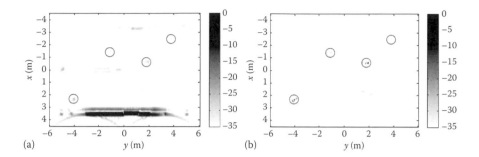

FIGURE 4.6
(a) Beamformed image using full data volume. (b) Scene reconstruction using partial sparsity–based approach. (From Ahmad, F. et al., A beamforming approach to imaging of stationary indoor scenes under known building layout, in: *Proceedings of the Fifth IEEE International Workshop on Computational Advances in Multi-Sensor Adaptive Processing*, Saint Martin, December 15–18, 2013. © 2013 IEEE. With permission.)

employed. Thus, at each of the 239 scan positions, the radar collected 148 frequency measurements over the 1.3 GHz bandwidth.

The region to be imaged was chosen to be 9 m × 12 m centered at the origin and divided into 121 × 161 pixels, respectively. Figure 4.6a shows the image obtained with beamforming using the full data set. Hanning window was applied to the data along the frequency dimension in order to reduce the range sidelobes in the image. The humans in the image are indicated by red circles. We can clearly see the front wall, some of the corners, and humans 1 and 2. Human 3 in the interior room is barely visible due to the additional EM loss as the transmitted signal has to penetrate through both the exterior and the interior walls. Likewise, it is a challenge to detect human 4, who is the farthest away from the front wall.

Next, we reconstructed the scene with the partial sparsity–based approach using 118 randomly selected frequencies (79.7% of 148) and 79 randomly chosen antenna locations (30% of 239), which collectively represent 26% of the total data volume. The dense part of the scene, corresponding to the exterior and the interior walls parallel to the array and corners, consisted of 7,196 pixels, while the sparse part of the scene consisted of the remaining 12,285 pixels. The number of OMP iterations was set to 10. The corresponding reconstruction of the sparse part of the scene is shown in Figure 4.6b, wherein each imaged pixel is the result of averaging 100 trials, with a different random selection for each trial. We observe from Figure 4.6b that the partial sparsity–based scheme was able to detect and localize humans 1 through 3 successfully, while it missed human 4. In addition, some clutter (arising from the left chair and table) and background noise is also visible in the reconstructed image. A similar performance of the partial sparsity–based approach was observed when the same set of reduced frequencies was employed at each selected transceiver location.

4.6 Conclusion

This chapter addressed wall clutter mitigation in the context of CS for stepped-frequency SAR imaging of stationary targets behind walls. First, the performance of two leading methods for combating wall clutter under full data volume, namely, spatial filtering and subspace projection approaches, was investigated under the reduced data volume. Using real data collected in a laboratory environment, we showed that these two methods maintain proper performance when acting on reduced data measurements, provided that the same set of reduced frequencies was used for each employed antenna. Subsequent sparse reconstruction successfully detected and accurately localized the targets. However, when different frequencies were used at different antennas, the aforementioned approaches were unsuccessful in removing the front wall clutter and the sparse reconstruction failed to localize the targets. Second, a DPSS-based wall mitigation method was presented for capturing and removing the wall clutter energy at each antenna individually. Using real data, this method was shown to be flexible in the sense that it permits the use of a different set of frequencies at each antenna location. Third, a partial sparsity–based scene reconstruction approach for imaging of stationary indoor targets was presented. The technique made use of the prior information of the support of the dense part of the scene, corresponding to the building layout, to design projection matrices for removal of reflections from exterior and interior walls and room corners. Using numerical EM data of a single-story building, the effectiveness of the partially sparse reconstruction under both same and different frequency measurement distributions over antennas was demonstrated.

References

1. M.G. Amin (Ed.), *Through-the-Wall Radar Imaging*, CRC Press, Boca Raton, FL, 2011.
2. M.G. Amin (Ed.), Special issue on advances in indoor radar imaging, *Journal of the Franklin Institute*, 345 (6), 556–722, September 2008.
3. M. Amin and K. Sarabandi (Eds.), Special issue on remote sensing of building interior, *IEEE Transactions on Geoscience and Remote Sensing*, 47 (5), 1270–1420, 2009.
4. H. Burchett, Advances in through wall radar for search, rescue and security applications, in *Proceedings of the Institute of Engineering and Technology Conference on Crime and Security*, London, U.K., June 2006, pp. 511–525.
5. C.P. Lai and R.M. Narayanan, Ultrawideband random noise radar design for through-wall surveillance, *IEEE Transactions on Aerospace and Electronic Systems*, 46 (4), 1716–1730, 2010.

6. E.F. Greneker, RADAR flashlight for through-the-wall detection of humans, in *Proceedings of the SPIE: Targets and Backgrounds: Characterization and Representation IV*, Vol. 3375, Orlando, FL, 1998, pp. 280–285.
7. M.G. Amin and F. Ahmad, Wideband synthetic aperture beamforming for through-the-wall imaging, *IEEE Signal Processing Magazine*, 25 (4), 110–113, July 2008.
8. C. Le, T. Dogaru, L. Nguyen, and M.A. Ressler, Ultrawideband (UWB) radar imaging of building interior: Measurements and predictions *IEEE Transactions on Geoscience and Remote Sensing*, 47 (5), 1409–1420, May 2009.
9. F. Soldovieri and R. Solimene, Through-wall imaging via a linear inverse scattering algorithm, *IEEE Geoscience and Remote Sensing Letters*, 4 (4), 513–517, 2007.
10. L.P. Song, C. Yu, and Q.H. Liu, Through-wall imaging (TWI) by radar: 2-D tomographic results and analyses, *IEEE Transactions on Geoscience and Remote Sensing*, 43 (12), 2793–2798, 2005.
11. M. Dehmollaian and K. Sarabandi, Refocusing through building walls using synthetic aperture radar, *IEEE Transactions on Geoscience and Remote Sensing*, 46 (6), 1589–1599, 2008.
12. C. Thajudeen, A. Hoorfar, F. Ahmad, and T. Dogaru, Measured complex permittivity of walls with different hydration levels and the effect on power estimation of TWRI target returns, *Progress in Electromagnetics Research B*, 30, 177–199, 2011.
13. Y.-S. Yoon and M.G. Amin, Spatial filtering for wall-clutter mitigation in through-the-wall radar imaging, *IEEE Transactions on Geoscience and Remote Sensing*, 47 (9), 3192–3208, 2009.
14. P. Setlur, M. Amin, and F. Ahmad, Analysis of micro-Doppler signals using linear FM basis decomposition, in *Proceedings of the SPIE Security and Defense Conference, Radar Sensor Technology*, Vol. 6210, Orlando, FL, May 2006, pp. 62100M.
15. A. Martone, K. Ranney, and R. Innocenti, Automatic through the wall detection of moving targets using low-frequency UWB radar, in *Proceedings of the IEEE International Radar Conference*, Washington D.C., May 2010, pp. 39–43.
16. M.G. Amin and F. Ahmad, Change detection analysis of humans moving behind walls, *IEEE Transactions on Aerospace and Electronic Systems*, 49 (3), 1410–1425, July 2013.
17. F.H.C. Tivive, A. Bouzerdoum, and M.G. Amin, An SVD-based approach for mitigating wall reflections in through-the-wall radar imaging, in *Proceedings of the IEEE Radar Conference*, Kansas City, MO, 2011, pp. 519–524.
18. R. Chandra, A.N. Gaikwad, D. Singh, and M.J. Nigam, An approach to remove the clutter and detect the target for ultra-wideband through wall imaging, *Journal of Geophysics and Engineering*, 5 (4), 412–419, 2008.
19. J. Zhang and Y. Huang, Extraction of dielectric properties of building materials from free-space time-domain measurement, in *Proceedings of the High Frequency Postgraduate Student Colloquium*, Leeds, U.K., September 1999, pp. 127–132.
20. Y.-S. Yoon and M. G. Amin, Compressed sensing technique for high-resolution radar imaging, *Proceedings of SPIE*, 6968, 69681A-1–69681A-10, 2008.
21. Q. Huang, L. Qu, B. Wu, and G. Fang, UWB through-wall imaging based on compressive sensing, *IEEE Transactions on Geoscience and Remote Sensing*, 48 (3), 1408–1415, 2010.

22. M. Leigsnering, C. Debes, and A.M. Zoubir, Compressive sensing in through-the-wall radar imaging, *Proceedings of the IEEE International Conference on Acoustics, Speech and Signal Processing*, Prague, Czech Republic, 2011, pp. 4008–4011.
23. F. Ahmad and M.G. Amin, Through-the-wall human motion indication using sparsity-driven change detection, *IEEE Transactions on Geoscience and Remote Sensing*, 51 (2), 881–890, February 2013.
24. M.G. Amin and F. Ahmad, Compressive sensing for through-the-wall radar imaging, *Journal of Electronic Imaging*, 22 (3), 030901, July 2013.
25. E. Lagunas, M.G. Amin, F. Ahmad, and M. Nájar, Joint wall mitigation and compressive sensing for indoor image reconstruction, *IEEE Transactions on Geoscience and Remote Sensing*, 51 (2), 891–906, February 2013.
26. D. Slepian, Prolate spheroidal wave functions, Fourier analysis, and uncertainty. V—The discrete case, *Bell System Technical Journal*, 57 (5), 1371–1430, 1978.
27. F. Ahmad, J. Qian, and M.G. Amin, Wall mitigation using discrete prolate spheroidal sequences for sparse indoor image reconstruction, in *Proceedings of the 21st European Signal Processing Conference*, Marrakech, Morocco, September 9–13, 2013.
28. F. Ahmad and M.G. Amin, Partially sparse reconstruction of behind-the-wall scenes, *Proceedings of SPIE*, 8365, 83650W, 2012.
29. N. Vaswani and W. Lu, Modified-CS: Modifying compressive sensing for problems with partially known support, *IEEE Transactions on Signal Processing*, 58 (9), 4595–4607, 2010.
30. A.S. Bandeira, K. Scheinberg, and L.N. Vicente, On partially sparse recovery, preprint 11–13, Department of Mathematics, University of Coimbra, Coimbra, Portugal, 2011. Available at: http://www.optimization-online.org/DB_FILE/2011/04/2990.pdf.
31. M. Leigsnering, F. Ahmad, M.G. Amin, and A.M. Zoubir, Multipath exploitation in through-the-wall radar imaging using sparse reconstruction, *IEEE Transactions on Aerospace and Electronic Systems*, 50(2), 2014.
32. F. Ahmad and M.G. Amin, Noncoherent approach to through-the-wall radar localization, *IEEE Transactions on Aerospace and Electronic Systems*, 42 (4), 1405–1419, 2006.
33. F. Tivive, M. Amin, and A. Bouzerdoum, Wall clutter mitigation based on eigen-analysis in through-the-wall radar imaging, in *Proceedings of the 17th International Conference on Digital Signal Processing*, Corfu, Greece, 2011.
34. S. Boyd and L. Vandenberghe, *Convex Optimization*, Cambridge University Press, Cambridge, U.K., 2004.
35. S.S. Chen, D.L. Donoho, and M.A. Saunders, Atomic decomposition by basis pursuit, *SIAM Journal of Scientific Computing*, 20 (1), 33–61, 1999.
36. S. Mallat and Z. Zhang, Matching pursuit with time-frequency dictionaries, *IEEE Transactions on Signal Processing*, 41 (12), 3397–3415, 1993.
37. J.A. Tropp, Greed is good: Algorithmic results for sparse approximation, *IEEE Transactions on Information Theory*, 50 (10), 2231–2242, 2004.
38. J.A. Tropp and A.C. Gilbert, Signal recovery from random measurements via orthogonal matching pursuit, *IEEE Transactions on Information Theory*, 53 (12), 4655–4666, 2007.
39. D. Needell and J.A. Tropp, CoSaMP: Iterative signal recovery from incomplete and inaccurate samples, *Applied and Computational Harmonic Analysis*, 26 (3), 301–321, May 2009.

40. T. Zemen and C. Mecklenbräuker, Time-variant channel estimation using discrete prolate spheroidal sequences, *IEEE Transactions on Signal Processing*, 53 (9), 3597–3607, 2005.
41. M.A. Davenport and M.B. Wakin, Compressive sensing of analog signals using discrete prolate spheroidal sequences, *Applied and Computational Harmonic Analysis*, 33 (3), 438–472, 2012.
42. Y. Eldar, P. Kuppinger, and H. Bolcskei, Block-sparse signals: Uncertainty relations and efficient recovery, *IEEE Transactions on Signal Processing*, 58 (6), 3042–3054, 2010.
43. NAVY SBIR FY08.1 Solicitation, *Radio Frequency (RF) Modeling of Layered Composite Dielectric Building Materials*, pp. 94–95. Available: http://www.acq.osd. mil/osbp/sbir/solicitations/sbir20081/navy081.pdf
44. E. Lagunas, M.G. Amin, F. Ahmad, and M. Najar, Determining building interior structures using compressive sensing, *Journal of Electronic Imaging*, 22 (2), April 2013. doi: 10.1117/1.JEI.22.2.021003.
45. F. Ahmad, M.G. Amin, and T. Dogaru, A beamforming approach to imaging of stationary indoor scenes under known building layout, in *Proceedings of the Fifth IEEE International Workshop on Computational Advances in Multi-Sensor Adaptive Processing*, Saint Martin, December 15–18, 2013.
46. T. Dogaru, L. Nguyen, and C. Le, Computer models of the human body signature for sensing through the wall radar applications, ARL-TR-4290, U.S. Army Research Laboratory, Adelphi, MD, 2007.
47. T. Dogaru and C. Le, Through-the-wall small weapon detection based on polarimetric radar techniques, ARL-TR-5041, U.S. Army Research Lab, Adelphi, MD, 2009.
48. F. Ahmad and M. Amin, Stochastic model based radar waveform design for weapon detection, *IEEE Transactions on Aerospace and Electronic Systems*, 48 (2), 1815–1826, 2012.

5

Compressive Sensing for Urban Multipath Exploitation

Michael Leigsnering and Abdelhak M. Zoubir

CONTENTS

5.1 Introduction

Multipath exploitation has received increasing attention in through-the-wall radar imaging (TWRI) [10,15,31,32,43]. The key idea is to make use of the additional information contained in the signals that reach the receiver through an indirect propagation path. Potentially, multiple benefits arise from the exploitation of multipath, for example, removal of ghost targets, improved signal-to-clutter ratio (SCR), detection of obscured targets, and increased image resolution.

In this chapter, we consider multipath exploitation for stationary and moving targets in an indoor environment based on compressive sensing (CS). Also, we take into account multiple internal reflections within the front wall that have an adverse effect on the imaging results. We consider multipath propagation due to some signal reaching the receiver via an indirect path, that is, an additional specular reflection either at an interior wall or at one of the front wall surfaces. Multipath can be broadly categorized into two types: target multipath and wall reverberations. We refer to the former as an indirect propagation path that involves some interaction with the target of interest. At the image formation step, the additional delay of multipath results in energy being focused at locations other than that of the target—a ghost target. This ghost can be confused with a real target and, thus, complicates the interpretation of the scene [32]. Wall reverberations, however, are multiple reflections within the front wall without any interaction with the targets. Hence, these returns do not give any information about the scene of interest, but result in artifacts in the image located along the range dimension. Therefore, wall reverberations can potentially mask targets, especially those that are close to the wall and should be mitigated [11,35,36,39].

Multipath has been shown to affect radar imaging in urban scenarios [2,10,13]. This has given rise to further work, treating multipath effects that belong roughly to one of two categories, namely, multipath mitigation and multipath exploitation. The former aims at mitigating adverse effects like ghosts and, thereby, obtaining cleaner images [2]. The latter uses multipath to one's advantage by exploiting the additional energy and information contained in the indirect returns [20,32]. We will treat multipath exploitation because it has potentially larger benefits as compared with multipath mitigation.

CS has proven to be a powerful tool to tackle the problem at hand. In view of reduced measurements in the data acquisition, CS has been first applied to TWRI by Yoon and Amin [38]. Subsequently, others showed that as long as the scene is sparse, CS is able to reconstruct very clean and highly resolved images [1,25,38]. However, multipath effects, causing ghost targets, and wall reverberation artifacts render the scene less sparse and, therefore, diminish the benefits and applicability of CS. Hence, multipath propagation should be carefully taken into consideration when formulating the measurement model and reconstruction problem in a CS framework. We will describe methods where each feasible case of multipath propagation is included and superimposed in the received signal model [22,24,26]. The scene is recovered by jointly taking into account multipath. This is essentially done by solving an inverse problem to find the scene that best "explains" all received direct and indirect returns from the targets and the wall. Throughout this chapter, we assume that the front wall properties and the locations of the interior walls are known. A similar approach is used in dealing with wall reverberations. The wall and target returns are jointly modeled and reconstructed in order to separate the two contributions from the received signal.

Related work includes multipath exploitation as developed by Kidera et al. [20], where multipath returns are used to acquire details about shadowed or obscured targets. Setlur et al. [32] laid the foundation of multipath exploitation in TWRI by modeling returns from a room with a known wall layout. They took advantage of the additional energy residing in the ghost targets by first predicting their location and then mapping them back onto the corresponding real targets. Thus, a clean image with improved SCR could be obtained. CS, however, has not been taken into consideration in [32]. Within the CS framework, various methods of wall mitigation for TWRI have been proposed. Classical wall mitigation techniques have been successfully obtained from a CS formulation in [21]. Furthermore, projecting the measurements onto a space orthogonal to the wall, as performed in [3], has been shown to attenuate returns from the front wall. The described approaches have the common feature to deal with the wall returns in a preprocessing step prior to target image formation. We will describe a method where wall mitigation and target image formation is achieved jointly.

The outline of the chapter is as follows. First, we briefly introduce the wideband received signal model for TWRI. In Section 5.35.3, we explain and model the various multipath contributions in the received signal. We show in Section 5.4 how to recover the scene considering stationary and moving targets. Dealing with wall reverberation is delineated in Section 5.5. Each of these sections is accompanied with a short example using simulated as well as experimental data. Finally, we conclude the chapter in Section 5.6.

5.2 Ultrawideband Signal Model

In this section, the signal model for an ultrawideband radar system with one transmitter and N receivers is introduced. Though the model can be easily extended to multiple transmitters, see for example, [29], we restrict ourselves to the single-transmitter case for notational simplicity. In TWRI, there is an interest of imaging both stationary scenes and scenes of moving targets.

A wideband pulsed radar model is developed that can be used to image moving targets [1,29]. Stationary targets are a special case of moving targets, thus they are included in the model. Nevertheless, we will also relate the pulsed radar model to the stepped-frequency radar, which is often used for imaging of stationary scenes [2].

First, we develop a target model, based on the assumption that targets follow a uniform linear motion model in a 2D space. We assume that K wideband pulses are transmitted with a pulse repetition interval (PRI) of T_r. The pulse index $k = 0, \ldots, K - 1$, is referred to as slow time. If the PRI is sufficiently small, target movement of indoor targets should be approximately of constant velocity and slow enough so that they are within a range cell. Using these assumption, one can argue that the pth target at pulse k is located at position

$$z_p(k) = (x_p + v_{xp}kT_r, y_p + v_{yp}kT_r), \quad k = 0, \ldots, K - 1, \tag{5.1}$$

in a Cartesian coordinate system, where (x_p, y_p) are the initial positions and (v_{xp}, v_{yp}) are the respective velocities for $p = 0, \ldots, P - 1$.

Assume that N receivers constitute a line array aperture, located parallel to the x-axis. The array is placed at a standoff distance from a homogeneous wall of thickness d. The single transmitter is also at a certain standoff distance from the wall, which may not be the same as for the receiver array. The transmitter emits a modulated wideband pulse with duration T_p that can be described as $\Re\{s(t)\exp(j2\pi f_c t)\}$, where t is the so-called fast time, $s(t)$ is the pulse in the complex baseband, and f_c is the carrier frequency. The geometry of the scene and radar system is depicted in Figure 5.1. The received signal in the complex baseband corresponding to receiver $n = 0, \ldots, N - 1$, pulse $k = 0, \ldots, K - 1$, and target $p = 0, \ldots, P - 1$, can be expressed as

$$y_{nk}^p(t) = \sigma_p s\left(t - kT_r - \tau_{pn}(k)\right) \cdot \exp\left(-j2\pi f_c\left(kT_r + \tau_{pn}(k)\right)\right), \tag{5.2}$$

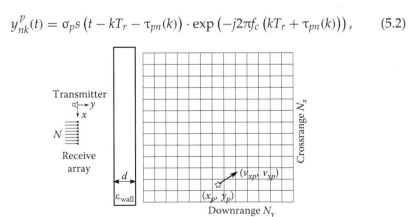

FIGURE 5.1
Geometry of the imaging system.

where

$\tau_{pn}(k)$ is the bistatic two-way delay

σ_p is the reflectivity of the pth point target

As slowly moving targets are considered, the delays can be assumed constant over the wideband pulse duration. Hence, the delay does not depend on the fast time t. However, the delays depend on the pulse number k as the target is moving slowly away from its initial position. For now, it is assumed that no multipath propagation exists; this effect will be treated in Section 5.3. Each receiver collects the superposition of all target responses stemming from the kth pulse. This can be written as

$$y_{nk}(t) = \sum_{p=0}^{P-1} \sigma_p s\left(t - kT_r - \tau_{pn}(k)\right) \cdot \exp\left(-j2\pi f_c\left(kT_r + \tau_{pn}(k)\right)\right). \qquad (5.3)$$

Note that setting the target velocity vector $(v_{xp}, v_{yp}) = (0, 0)$ corresponds to the case of a stationary target p. In this case, the delays do not change with the slow time index $k = 0, \ldots, K - 1$.

On the other hand, the front wall, being a specular reflector, will also contribute to the received signal. The wall response corresponding to receiver n and pulse k can be expressed as

$$y_{nk}^{\text{wall}}(t) = \sigma_{\text{wall}} s\left(t - kT_r - \tau_n^{\text{wall}}\right) \cdot \exp\left(-j2\pi f_c\left(kT_r + \tau_n^{\text{wall}}\right)\right) + y_{nk}^{\text{wall,reverb}}(t), \qquad (5.4)$$

where

σ_{wall} is the complex reflectivity of the wall

τ_n^{wall} is the two-way delay from the transmitter to the wall and back to receiver n

$y_{nk}^{\text{wall,reverb}}(t)$ is the wall reverberation

Wall reverberations stem from multiple reflections within the wall and will be treated in detail in Section 5.5. Note that the delays associated with the wall returns do not vary with the slow time index k as the wall is a stationary reflector.

The target returns (5.3) and the wall returns (5.4) are received simultaneously, resulting in a total received signal for $n = 0, \ldots, N - 1$, $k = 0, \ldots, K - 1$, and for time $t \in \mathbb{R}$

$$y_{nk}^{\text{tot}}(t) = y_{nk}(t) + y_{nk}^{\text{wall}}(t). \qquad (5.5)$$

At first, we deal with the contribution of the target returns only. The model in (5.3) can be discretized in time, velocity, and space and vectorized to obtain a discrete model of the system. The targets are assumed to reside on a discrete grid with size $N_x \times N_y$ and similarly, the velocities are sampled on a discrete grid with size $N_{v_x} \times N_{v_y} = N_v$, (see Figure 5.1). In this 4D space, any possible target can be described, whereas a nonexisting target is represented by a zero reflectivity. Thus, in total there are $N_x N_y N_{v_x} N_{v_y} = P$ possible target states for each path, which can be stacked into a $N_x N_y N_{v_x} N_{v_y} \times 1$ vector σ. Finally, the fast time variable t needs to be sampled in order to obtain a fully discrete model. Hence, for each pulse k, the received signal $y_{nk}(t)$ is uniformly sampled at T time steps with sampling interval T_s. Note that the sampling interval should be chosen to attain the Nyquist rate for the wideband transmit pulse $s(t)$. Subsequently, the samples can be stacked into $T \times 1$ vectors y_{nk}, which are defined as

$$y_{nk} = \Psi_{nk}\sigma, \quad n = 0,\ldots,N-1, k = 0,\ldots,K-1, \tag{5.6}$$

where $\Psi_{nk} \in \mathbb{R}^{T \times P}$ are the dictionary matrices that are obtained by discretizing the right-hand side of (5.2) and are defined as

$$[\Psi_{nk}]_{i,p} = s\left(t_i - kT_r - \tau_{pn}(k)\right) \cdot \exp\left(-j2\pi f_c \left(kT_r + \tau_{pn}(k)\right)\right),$$
$$i = 0,\ldots,T-1, \quad p = 0,\ldots,P-1. \tag{5.7}$$

Performing a last stacking operation, results in a $TNK \times 1$ measurement vector y and a $TNK \times P$ overcomplete dictionary matrix Ψ, obtaining

$$y = \Psi\sigma, \tag{5.8}$$

with

$$\Psi = \left[\Psi_{00}^T \quad \Psi_{10}^T \quad \cdots \quad \Psi_{N-1\,K-1}^T\right]^T. \tag{5.9}$$

Later, we will exploit this linear measurement model for the CS formulation. But let us first examine the relations of this model to the stationary target model and to the stepped-frequency case.

5.2.1 Relation to the Stationary Scene Model

Stationary targets are naturally included in (5.3) as targets with zero velocity. If only interested in a stationary reconstruction of the scene, N_v should be set to one, so that the target state vector σ describes the spatial domain only. In this case, the delays will not depend on the slow time index k, hence one can equivalently use a single pulse. It should be noted that transmitting fewer pulses will lead to a degradation of the SNR.

The stationary model lends itself to a frequency domain formulation. The pulsed radar model in (5.3) can be transformed to the frequency domain (FD), assuming only a single pulse

$$y_{FD,n}(f) = \sum_{p=0}^{N_x N_y - 1} S(f) \sigma_p \exp\left(-j2\pi(f + f_c)\tau_{pn}\right), \quad (5.10)$$

where
  f is the continuous frequency variable
  $S(f)$ is the Fourier transform of the wideband transmit pulse $s(t), t \in \mathbb{R}$

When discretizing into M frequencies, $\{f_m\}_{m=0}^{M-1}$, regularly spaced around the center frequency f_c over the desired bandwidth, (5.10) writes as

$$y_{FD}[m, n] = \sum_{p=0}^{N_x N_y - 1} S(f_m) \sigma_p \exp\left(-j2\pi f_m \tau_{pn}\right). \quad (5.11)$$

This is the same stepped-frequency model used, for example, in [2,24].

This model can be likewise expressed in vector matrix formulation, where the dictionary $\mathbf{\Psi}_{FD} \in \mathbb{R}^{MN \times N_x N_y}$ now contains the delayed pulses transformed to the frequency domain

$$y_{FD} = \mathbf{\Psi}_{FD} \sigma, \quad (5.12)$$

where $y_{FD} = \left[y_{FD}[0,0], y_{FD}[1,0], \ldots, y_{FD}[M-1, N-1]\right]^{\mathsf{T}}$.

5.2.2 Conventional Image Formation

Conventional image formation for TWRI is carried out using backprojection or delay-and-sum beamforming (DSBF) [4,5,33]. This is only applicable for stationary scenes and can easily be applied to the frequency domain model (5.11). The complex image value I_p at the pth grid point (x_p, y_p) is obtained by summing phase shifted copies of the MN signals [4],

$$I_p = \frac{1}{MN} \sum_{n=0}^{N-1} \sum_{m=0}^{M-1} y_{FD}[m, n] \exp\left(j2\pi f_m \tau_{pn}\right), \quad (5.13)$$

where τ_{pn} is the focusing delay for the nth transceiver and the pth grid point. Intuitively, the expected phase shifts are compensated in the received signal and then summed over all frequencies and array elements. This can be equivalently expressed as

$$\hat{\sigma} = \Psi_{\mathrm{FD}}^{\mathrm{H}} y_{\mathrm{FD}}, \qquad (5.14)$$

where
   $(\cdot)^{\mathrm{H}}$ denotes Hermitian transpose
   $\hat{\sigma}$ is an estimate of the target state vector in space, or simply the image

Operation (5.14) is the adjoint of (5.12).

In the case of moving targets, standard backprojection cannot be used as moving targets will be blurred and possibly mislocated. Hence, traditional Doppler processing is used to discriminate the Doppler velocities occurring in the scene. The latter can be done by taking the discrete Fourier transform (DFT) w.r.t. the slow-time dimension k. Hence, K measurement sets are obtained, each corresponding to a particular Doppler velocity cell. Now, conventional DSBF algorithms can be applied to each Doppler cell individually to obtain an image for each discrete velocity. Assuming slowly moving targets, the phase delays are considered constant during a transmit pulse. Using the complete model including stationary and moving targets (5.3), the backprojection can be adjusted to each velocity and finally sum over all N array elements and K pulses. This can be expressed as the adjoint of (5.8)

$$\hat{\sigma} = \Psi^{H} y, \qquad (5.15)$$

where $\hat{\sigma}$ contains vectorized images for each considered target velocity (v_{xp}, v_{yp}). This type of velocity matched beamforming result will be used for later comparison. The resolution is limited by the classical point spread function or Rayleigh resolution [30].

Note that the conventional image formation is usually limited to the case where the full measurement data are available. In the case of missing or undersampled data, DSBF yields a severely degraded image quality. Hence, to fulfill the purpose of reconstructing the scene from few measurements, CS is applied in order to obtain highly resolved images. This will be detailed in the following sections.

5.3 Multipath Propagation Model

In this section, we discuss various cases of multipath propagations, their properties, and modeling. This can further be used to devise a received signal model that can accurately capture all indirect propagation paths. The direct path, that is, the path that is not subject to any secondary reflections, is obstructed and, therefore, influenced by the front wall. The transmitted wave is refracted twice at the front and back interfaces of the exterior building wall between the radar and the scene of interest. The backscattered wave is subject to the same double refraction before reaching the receiver. The front wall is modeled as a homogeneous dielectric slab; more complex or unknown wall structures are not considered here. In the case of a plane wave and homogeneous front walls, these effects can be accurately described by Snell's law [2].

Multipath propagation corresponds to indirect paths that involve reflections at one or more secondary reflectors in addition to diffuse scattering at the target of interest. Depending on the characteristic reflections, multipath can be divided into the following categories [24]:

- *Interior wall target multipath*: Specular reflection at one or more interior walls.
- *Floor/ceiling target multipath*: Specular reflections at the floor and/or ceiling.
- *Wall ringing multipath*: Multiple reflections inside the front wall when the wave travels to/from the target.
- *Target interaction multipath*: The wave traveling along this path interacts with more than one target.
- *Target-independent multipath*: Reflections of the signal from interior walls.

We only deal with interior wall target multipath and wall ringing multipath returns. Target-independent multipaths do not add coherently across the aperture and, as such, can be ignored. Floor/ceiling multipath are not considered as they are usually not present when using antennas with a narrow vertical beamwidth. Note that these multipaths can be treated in the same way as interior wall multipaths. Target interaction multipath is also not considered. The interior wall multipath returns can be further subdivided into the following classes:

- *First-order multipath*: This scattering scenario involves a direct propagation to the target on transmit and one secondary reflection at an interior wall on the way back to the receiver, or vice versa. Hence, the path to the target is different from the path back to the receiver. This is the dominant case of multipath in TWRI. Note that even for a monostatic radar the scattering on the target is of bistatic nature.
- *Second-order multipath*: The signal on the round-trip path is reflected twice at an interior wall. Two cases can be further distinguished:
  - *Quasi-monostatic*: The two reflections occur at the same interior wall, one on the transmit and receive path, respectively. This corresponds to monostatic scattering at the target for a monostatic radar, or at least to a very small bistatic angle, as compared

with first-order multipath when using a bistatic radar with a small
baseline.
  – *Bistatic*: The two specular reflections take place at two different
    walls.
• *Higher-order multipath*: Three or more specular reflections during the
  round-trip path may occur as well.

In the sequel, only first-order multipath and quasi-monostatic second-order
multipath from interior walls are considered. The signal is attenuated at
each secondary wall reflection. Thus, the second- and higher-order multipath returns are usually weak enough to be safely neglected. Assuming
that (quasi-)monostatic reflections from indoor targets are stronger than
bistatic scattering, quasi-monostatic second-order multipath is included in
the model.

By assuming perfect knowledge of the building layout, that is location,
thickness, and permittivity of the front wall, as well as the location of the
interior walls, multipath can be accurately described. Using a geometrical
optics (GO) model, which is valid for plane waves, the exact delays corresponding to each path can be calculated. The derivation follows from
[2,24,32]. We adopt the GO principles for reasons of simplicity, ease of
calculation, and lower computational complexity. Further, it allows us to
model multipath as a finite number of discrete paths, which is later used in
the CS reconstruction method. More general propagation models, such as
FDTD [13] or linear scattering [15], may also be used to describe multipath
responses at the expense of higher computational complexity.

5.3.1 Interior Wall Multipath

Interior wall multipath can be easily described by introducing a virtual
target. This can be illustrated through the monostatic radar example in
Figure 5.2, where the front wall has been ignored in the geometry for

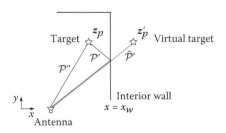

FIGURE 5.2
Multipath propagation via reflection at an internal wall. (From Leigsnering, M. et al., Multipath
exploitation in through-the-wall radar imaging using sparse reconstruction, *IEEE Trans. Aerosp.
Electeon. Syst.*, in press. With permisson. © 2013 IEEE.)

simplicity. The scene consists of a target located at $z_p = [x_p, y_p]^T$ and one interior wall (side wall of the room) parallel to the y-axis and located at $x = x_w$. Now, consider multipath propagation along the path $\mathcal{P}'$ from the target via secondary reflection at the interior wall back to a receiver. As the surface roughness of a building wall is usually much smaller than the wavelengths in TWRI, specular reflection is assumed. Hence, the multipath can be expressed as a direct return path $\tilde{\mathcal{P}}'$ from a virtual target located at $z_p' = [2x_w - x_p, y_p]^T$, whose location can simply be found by reflecting the original target at the wall. From the receiver perspective, the two paths are equivalent in terms of delay and incident angle. Hence, the calculation of a one-way propagation delay for the interior wall multipath can be carried out by a calculation using a direct propagation assumption. Note that the case of the transmit path and reflection from a different wall can be treated similarly.

There is an alternative view of the propagation geometry under specular reflection. Instead of reflecting the target by the interior wall, the antenna can be reflected equivalently. This gives rise to virtual transmitters and receivers when the scene is interrogated. This scenario can be seen as a virtual MIMO configuration. However, there is no control over the virtual transmitters and receivers. The virtual transmitters will transmit exactly the same signal as the corresponding physical transmitter. Likewise, the physical receivers along with its virtual receivers will simultaneously receive the wave and output a superposition of the signals received at their location. Even without control over this virtual array elements, the spatial diversity can be utilized in order to gain more information of the target. This will be explained in the following section.

For the pair of receiver n and target p, the delay is denoted by $\tau_{pn}^{(\mathcal{P}')}$, which is equivalent to $\tau_{pn}^{(\tilde{\mathcal{P}}')}$. When ignoring the front wall, the delay can be simply calculated as the Euclidean distance divided by the propagation speed. If the front wall is present, the double refraction at the two wall interfaces has to be considered. As the two paths $\mathcal{P}'$ and $\tilde{\mathcal{P}}'$ are equivalent, the calculation can be carried out in the same fashion as for the direct path using Snell's law [2]. This type of propagation is also a special case ($l = 0$) of the wall ringing multipath as described in the sequel.

5.3.2 Wall Ringing Multipath

When an electromagnetic wave hits the planar interface between medium A and medium B, it is partly refracted into medium B and partly reflected back to medium A [8]. Consequently, the signal on transit to/from the target may undergo multiple reflections inside the front wall [19]. This effect can be separated into two cases; that is, wall ringing and wall reverberation. The latter describes the case where the wave never reaches the target but only interacts within the front wall. This results in several wall responses with equally spaced delays and exponentially decaying in amplitude. In the

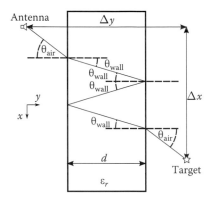

FIGURE 5.3
Wall ringing propagation with $l = 1$ internal bounces. (From Leigsnering, M. et al., Multipath exploitation in through-the-wall radar imaging using sparse reconstruction, *IEEE Trans. Aerosp. Electron. Syst.*, in press. With permission. © 2013 IEEE.)

beamformed image, multiple copies of the wall appear equally spaced along the range direction. The portion of the signal reaching the target may also undergo multiple reflections within the front wall, referred to as wall ringing. This has a similar effect on the imaged targets as wall reverberation has on the imaged wall. A target will be accompanied by multiple decaying copies equally spaced in the range direction. First, consider only the effect of wall ringing; reverberation effects are analyzed in Section 5.5.

Figure 5.3 depicts the front wall along with the incident, reflected, and refracted waves associated with a certain target/receiver pair. The distance between the target and the receiver in crossrange direction, Δx, can be expressed as

$$\Delta x = (\Delta y - d)\tan\theta_{air} + d(1 + 2l)\tan\theta_{wall}, \tag{5.16}$$

where
Δy is the distance between target and array element in downrange direction
θ_{air} and θ_{wall} are the angles in the air and in the wall medium, respectively

The integer l denotes the number of internal reflections within the wall. The case $l = 0$ describes the direct path as derived in [2]. The relationship between the angles of incidence and refraction are governed by Snell's law.

$$\frac{\sin\theta_{air}}{\sin\theta_{wall}} = \sqrt{\varepsilon_r}. \tag{5.17}$$

Equations 5.16 and 5.17 form a nonlinear system of equations that can be solved numerically for the unknown angles, for example, using Newton's method. From the angles follows the calculation of the one-way propagation delay associated with the wall ringing multipath as [2]

$$\tau(\Delta x, \Delta y, l) = \frac{(\Delta y - d)}{c \cos \theta_{\text{air}}} + \frac{\sqrt{\varepsilon_r} d (1 + 2l)}{c \cos \theta_{\text{wall}}}, \tag{5.18}$$

where c is the propagation speed in vacuum. Note that the direction of propagation is not relevant, that is the one-way delay from the transmitter to the target is obtained in exactly the same way. Note also that a frequency independent wall medium has been assumed. For real wall media, like concrete, the permittivity may depend on the frequency, which could lead to a distortion of the wideband pulse.

5.3.3 Bistatic Received Signal Model

The aim of this section is to arrive at a comprehensive bistatic received signal model under multipath propagation. We can exploit the aforementioned multipath mechanisms to find a model that describes the expected return signal. For now, we assume that the measurements contain only the target returns and the front wall scatterings have been removed. This can be achieved by one of the wall mitigation techniques described in Section 5.5. In that section, the case including wall returns as well as wall reverberations will be treated.

Any round-trip path $\mathcal{P}$ can be described by the combination of two one-way paths, namely, the path $\mathcal{P}''$ from the transmitter to the scattering target and the path $\mathcal{P}'$ from the target back to the receiver. As explained earlier, the one-way path $\mathcal{P}'$ can be the direct path or any feasible type of multipath, that is interior wall or wall ringing multipath. Hence, there exist R_1 return paths from a certain target back to the receiver, which will be denoted as $\mathcal{P}'_{r_1}, r_1 = 0, \dots, R_1 - 1$. The same observation holds for the one-way transmit paths, which are denoted by $\mathcal{P}''_{r_2}, r_2 = 0, \dots, R_2 - 1$. Hence, for the round-trip path $\mathcal{P}_r, r = 0, \dots, R - 1$, one can conclude a maximum number of $R \leq R_1 R_2$ paths, which represent all possible combinations of one-way paths. A function can be established to describe these combinations by mapping the index r of the round-trip path to a pair of indices of the one-way paths, $r \mapsto (r_1, r_2)$. It should be noted that $R_1 R_2$ is the maximum possible number of round-trip paths. However, some paths $\mathcal{P}_r$ belong to the class of second-order multipath and are strongly attenuated, and thereby ignored. In what follows, we will consider $\mathcal{P}_0$ as the direct path, that is, the case without any multipath. This model is illustrated in Figure 5.4, which depicts three possible return paths, namely, "direct propagation," "secondary reflection from a side wall," and "wall ringing." Three equivalent transmit paths will be present for the propagation from the transmitter to the scatterer. The combination of three transmit paths and three return paths results in a total of nine round-trip paths, as in Figure 5.5. The paths $\mathcal{P}_1, \mathcal{P}_2, \mathcal{P}_3$, and $\mathcal{P}_6$ correspond

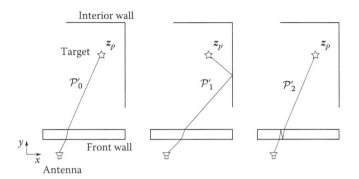

FIGURE 5.4
Example for three possible one-way return paths. (From Leigsnering, M. et al., Multipath exploitation in through-the-wall radar imaging using sparse reconstruction, *IEEE Trans. Aerosp. Electron. Syst.*, in press. With permission. © 2013 IEEE.)

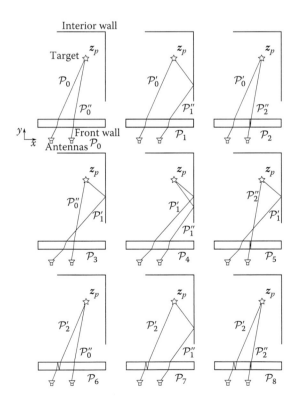

FIGURE 5.5
Round-trip paths between transceiver and target for the partial paths shown in Figure 5.4.

to first-order multipath, P_4 is a quasi-monostatic second-order multipath, which is also included. Round-trip paths P_5 and P_7 are general second-order multipaths that are ignored due to higher attenuation. A round-trip with two wall ringing paths, as P_8 may or may not be included, depending on the attenuation properties of the front wall. Note that the first-order multipaths occur in symmetric pairs in a monostatic setup.

The calculation of the round-trip delays under multipath propagation conditions is dealt with in the sequel. As the round-trip path P_r consists of the one-way paths P'_r and P''_r, the round-trip delay follows as

$$\tau_{pn}^{(P_r)} = \tau_{pn}^{\left(P'_{r_1}\right)} + \tau_{pn}^{\left(P''_{r_2}\right)}. \tag{5.19}$$

The method to obtain the appropriate delays for the indirect one-way paths has been described earlier in this section. For notational simplicity, the round-trip delay between the nth array element and pth target associated with path P_r is denoted as $\tau_{pn}^{(r)}$.

In a similar manner, the complex amplitudes $T_{pn}^{(P_r)} \in \mathbb{C}$ associated with each possible combination of path, transceiver, and target can be calculated. With each reflection and refraction, the traveling wave will suffer some attenuation and possibly a phase shift. For each one-way path, the complex amplitude $T_{pn}^{(\cdot)}$ can be derived from the dielectric properties of the front and sidewalls and the corresponding angles of incidence and refraction. For example, complex amplitude for the direct path $T_{pn}^{(P'_0)}$ is the product of the two transmission coefficients corresponding to the air–wall and wall–air interface under the respective angles of incidence. If an additional secondary reflection at an interior wall occurs, the reflection coefficient at the respective interface has to be taken into account also. A detailed derivation can be found in [24,32]. The total path amplitude is the product of the amplitudes of both one-way paths

$$T_{pn}^{(P_r)} = T_{pn}^{\left(P'_{r_1}\right)} T_{pn}^{\left(P''_{r_2}\right)}.$$

In the sequel, two simplifications are made. First, provided that the incident, refraction, and reflection angles associated with one path do not vary much across the array, one can assume that

$$T_{pn}^{(P_r)} \approx T_p^{(P_r)}, \quad n = 0, \ldots, N-1, \quad p = 0, \ldots, P-1. \tag{5.20}$$

In other words, the complex amplitude for each path depends only on the target position. This approximation generally holds only in far-field conditions where all angles are approximately equal across all target/array

TABLE 5.1

Frequency for Various Paths That the Relative Error of (5.20) Exceeds 10%

Standoff Distance (m)	Direct Path (%)	Left-Side Wall (%)	Back Wall (%)	Back Corner (%)
0.5	21	18	8	5
2.5	1	7	0	0
5	0	0	0	0

element pairs. However, a small numerical example shows that the approximation holds for typical TWRI scenarios. We examined the relative error for a 1.5 m monostatic array and a 4 m by 5 m room at various standoff distances. Table 5.1 illustrates the frequency that the relative error exceeds an acceptable level of 10%. It is evident that the approximation is sufficiently accurate for standoff distances larger than 2.5 m. Only for the very near-field case a significant error level is observed.

Furthermore, as the direct path is typically the strongest path under propagation through walls, all complex amplitudes are normalized w.r.t. the direct path in order to avoid overparameterization:

$$w_p^{(r)} = \frac{T_p^{(\mathcal{P}_r)}}{T_p^{(\mathcal{P}_0)}}, \quad r = 0, \ldots, R-1, \quad p = 0, \ldots, P-1. \tag{5.21}$$

Hence, from (5.19) and (5.20) a complex path weight $w_p^{(r)}$ is obtained for each possible path corresponding to the pth pixel, with the direct path having the weight $w_p^{(0)} = 1$.

Having calculated the focusing delays and path weights, one can formulate a target signal model under multipath propagation conditions. The receiver picks up a superposition of delayed and weighted versions of the transmitted signal corresponding to all possible propagation paths $r = 0, \ldots, R-1$. Hence, (5.3) can be extended to

$$y = \Psi^{(0)}\bar{\sigma}^{(0)} + W^{(1)}\Psi^{(1)}\bar{\sigma}^{(1)} + \cdots + \Psi^{(R-1)}W^{(R-1)}\bar{\sigma}^{(R-1)}, \tag{5.22}$$

where the path weight matrices are given by $W^{(r)} = \mathrm{diag}(w_0^{(r)}, w_1^{(r)}, \ldots, w_{P-1}^{(r)})$ for $r = 0, \ldots, R-1$. The dictionaries $\Psi^{(r)}$ are defined according to (5.7) and (5.9), where $\tau_{pn}(k)$ is replaced by $\tau_{pn}^{(r)}(k)$. An individual target reflectivity vector $\bar{\sigma}^{(r)}$ is assumed for each path, as the phase and amplitude of the reflectivity of a target changes in general with the bistatic angle and aspect angle. Without loss of generality, we can assume the same number of paths for each target in (5.22) as a particular path weight can be set to zero if the corresponding path is not available for this target.

For notational convenience, the path weight can be absorbed into the target reflectivity vectors, as it is only a per-pixel scaling $\sigma^{(r)} = \Psi^{(r)} \bar{\sigma}^{(r)}$. It follows for the measurement model

$$y = \Psi^{(0)} \sigma^{(0)} + \Psi^{(1)} \sigma^{(1)} + \cdots + \Psi^{(R-1)} \sigma^{(R-1)}. \tag{5.23}$$

Equivalently, the measurement model can be expressed for stationary scenes using the stepped-frequency approach

$$y_{\text{FD}} = \Psi^{(0)}_{\text{FD}} \sigma^{(0)} + \Psi^{(1)}_{\text{FD}} \sigma^{(1)} + \cdots + \Psi^{(R-1)}_{\text{FD}} \sigma^{(R-1)}. \tag{5.24}$$

Note that (5.23) and (5.24) are generalizations of the nonmultipath propagation models (5.3) and (5.10), respectively. If the number of propagation paths is set to $R = 1$, the multipath signal models are equivalent to the direct-path models. Using these linear measurement models, CS can be applied to achieve an accurate reconstruction of the scene. Note that, in practice, the number of multipath contributions is limited by the number of large flat surfaces. Thus, for monostatic radar imaging of a single room, one would expect $K = 4$ propagation paths: one direct path and three propagation paths due to the interior walls as multipath. This number increases if second-order multipaths, wall-ringing, or a bistatic operation are considered.

At this point, we would like to emphasize the relation of (5.23) and (5.23) to a MIMO radar formulation. MIMO radar is an emerging concept that has potentially large benefits in radar detection, estimation, and imaging [18,27,40,41]. When considering several switched transmit antennas, the presented model can be seen as a time-multiplexed MIMO radar configuration [29]. Specular reflections at interior walls give rise to virtual transmitters and receivers that extend the aperture of the real array. However, the virtual transmitters, by definition, transmit exactly the same signal as the physical transmitter. Likewise, the signals measured by the physical receivers contain the contributions of the virtual receivers as well. Hence, we have the summation over all possible round-trip paths in the model formulation. This model is very similar to the MIMO setup in [40]; however, the incorporation of multipath is a generalization thereof. It incorporates not only MIMO operation corresponding to direct path propagation, but also multipath propagation in a SIMO setup. Note that if we would be able to resolve and associate the received signals with their propagation paths, the multipath from a single transmitter can be seen as a MIMO system. In this case, the virtual transmitters act as multiple sources.

5.4 Compressive Sensing Reconstruction with Multipath Exploitation

Employing CS, we aim at achieving a high-quality reconstruction of the sparse scene with efficient data acquisition, using only a subset of the full measurements. However, due to multipath propagation, the imaged scene becomes populated by unwanted ghost targets, which reduce the scene sparsity. Within the CS framework, we aim at removing the ghosts, that is, inverting the multipath measurement model and achieving a reconstruction wherein only the true targets remain.

Three different cases are discussed in this chapter: stationary scenes with stepped-frequency measurements, moving/stationary targets using wideband pulsed radar, and stationary scenes with joint wall reconstruction. In this section, we treat the former two cases, assuming that the wall returns have been removed from the measurements. In the following section, the wall returns are taken into consideration by using a wall response model. For each case, the measurement model and the reconstruction problem is adopted to the specific properties. Also, we discuss the choice of the data acquisition process, that is, how the compressive measurements should be acquired.

5.4.1 Stationary Scenes

Let us first turn our attention to the reconstruction of targets in a stationary scene. We assume stepped-frequency measurements and that the wall returns have been properly compensated for. Various data acquisition schemes have been proposed for stepped-frequency radar [6,16,38]. The common feature of these schemes is that they select a random subset of the full measurements. Instead of taking measurement at all array elements and all frequencies, only data at a few array element/frequency pairs are acquired. This can be represented by a binary random measurement matrix $\mathbf{\Phi}_{FD} \in \{0,1\}^{J \times MN}$ acting on the full measurement vector. One can think of $\mathbf{\Phi}_{FD}$ as an $MN \times MN$ identity matrix where all but J rows have been deleted. Consequently, an undersampled measurement vector is obtained from model (5.24) as

$$\bar{y}_{FD} = \mathbf{\Phi}_{FD} y_{FD} = \mathbf{\Phi}_{FD} \left(\mathbf{\Psi}_{FD}^{(0)} \sigma^{(0)} + \mathbf{\Psi}_{FD}^{(1)} \sigma^{(1)} + \cdots + \mathbf{\Psi}_{FD}^{(R-1)} \sigma^{(R-1)} \right). \quad (5.25)$$

Using the reduced data model in (5.25), the image formation process can be cast into a sparse reconstruction problem.

5.4.2 Group Sparse Reconstruction of Stationary Scenes

In practical scenarios, the scatterers are nonisotropic, that is the magnitude and phase of the reflectivities change with aspect angle and bistatic angle.

Thus, the exact relationship between the subimages $\sigma^{(r)}$ corresponding to paths $r = 0, \ldots, R - 1$ is usually unknown. However, it is known that the subimages $\sigma^{(0)}, \sigma^{(1)}, \ldots, \sigma^{(R-1)}$ describe targets at the same locations in the underlying sparse scene. Provided that no propagation paths are blocked, a target that can be observed through one path is also observable through all other paths. Thus, a particular target will populate the same pixel location in any of the subimages. The same reasoning holds for an empty place in the scene, that is if it is empty in one subimage, this should hold in all other subimages. In other words, the support of the R subimages is equal. Even if some paths are blocked or below the noise level for some scatterers, the support is at least approximately equal. This property gives rise to a particular sparse structure of the unknown vector σ. The subimages exhibit a group sparse structure, where the individual groups extend across the paths for each pixel, as illustrated in Figure 5.6. The image vectors $\sigma^{(r)}$ are depicted as image matrices for illustration purposes.

A reconstruction method based on group sparsity has been proposed in [24]. To this end, all unknown vectors in (5.25) can be stacked to form a tall vector

$$\tilde{\sigma} = \left[\left(\sigma^{(0)} \right)^{\mathrm{T}} \left(\sigma^{(1)} \right)^{\mathrm{T}} \cdots \left(\sigma^{(R-1)} \right)^{\mathrm{T}} \right]^{\mathrm{T}} \in \mathbb{C}^{N_x N_y R \times 1}. \qquad (5.26)$$

The reduced measurement vector $\bar{y}_{\mathrm{FD}}$ can then be expressed as

$$\bar{y}_{\mathrm{FD}} = \boldsymbol{\Phi}_{\mathrm{FD}} \tilde{\boldsymbol{\Psi}}_{\mathrm{FD}} \tilde{\sigma}, \qquad (5.27)$$

where the new dictionary matrix has now the form $\tilde{\boldsymbol{\Psi}}_{\mathrm{FD}} = \left[\boldsymbol{\Psi}_{\mathrm{FD}}^{(0)} \, \boldsymbol{\Psi}_{\mathrm{FD}}^{(1)} \cdots \right.$ $\left. \boldsymbol{\Psi}_{\mathrm{FD}}^{(R-1)} \right] \in \mathbb{C}^{MN \times N_x N_y R}$. The unresolvability of multipaths is behind forming the wide matrix $\tilde{\boldsymbol{\Psi}}_{\mathrm{FD}}$, rather than having a block diagonal matrix structure, which is the case in a MIMO configuration.

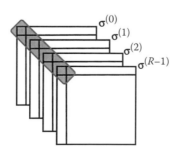

FIGURE 5.6
Group sparse structure for the subimages.

In the next step, model (5.27) and the group sparsity assumption is exploited to reconstruct the unknown reflectivity vectors or images, $\tilde{\sigma}$. It has been shown that a group sparse reconstruction can be obtained by a mixed ℓ_1/ℓ_2-norm regularization [9,12,37,42]. This results in the reconstruction problem

$$\hat{\tilde{\sigma}} = \arg\min_{\tilde{\sigma}} \frac{1}{2} \left\| \bar{y}_{FD} - \mathbf{\Phi}_{FD} \tilde{\mathbf{\Psi}}_{FD} \tilde{\sigma} \right\|_2^2 + \lambda \left\| \tilde{\sigma} \right\|_{2,1}, \tag{5.28}$$

where

$$\left\| \tilde{\sigma} \right\|_{2,1} := \sum_{p=0}^{N_x N_y - 1} \left\| \left[\sigma_p^{(0)}, \sigma_p^{(1)}, \ldots, \sigma_p^{(R-1)} \right]^{\mathrm{T}} \right\|_2$$

$$= \sum_{p=0}^{N_x N_y - 1} \sqrt{\sum_{r=0}^{R-1} \sigma_p^{(r)} \left(\sigma_p^{(r)} \right)^*} \tag{5.29}$$

and λ is the so-called regularization parameter. The convex optimization problem (5.28) can be solved using SparSA [37] or another available scheme [9,12,14]. SparSA uses a sequence of subproblems that can be solved efficiently and converges to the overall optimum solution. The method can be applied to both complex-valued and large-scale problems, for example, when matrix $\tilde{\mathbf{\Psi}}_{FD}$ does not fit into the memory.

By incorporating the multipath model into the CS reconstruction, the number of unknowns has been multiplied with the number of paths. However, the degrees of freedom of the problem can be reduced through the structured sparse constraint on the solution. Hence, the number of measurements does not have to be increased accordingly [12]. Thus, the recovery performance is improved by the group sparse problem formulation.

Once a solution $\hat{\tilde{\sigma}}$ is obtained, the subimages can be combined to form an overall image. Since the phase relationship between the individual subimages depends on the radar cross section (RCS) characteristics of the targets, it is generally unknown. Thus, the best way of combining the subimages is an incoherent accumulation. Effectively, the ℓ_2-norm of each group is formed to obtain the final pixel values for the combined image, expressed as

$$\left[\hat{\sigma}_{GS} \right]_p = \left\| \left[\sigma_p^{(0)}, \sigma_p^{(1)}, \ldots, \sigma_p^{(R-1)} \right]^{\mathrm{T}} \right\|_2, \quad p = 0, \ldots, N_x N_y - 1. \tag{5.30}$$

The incoherent combination of the subimages does not improve the SNR, as spatially white noise will also be accumulated. However, the SCR of the final image can be improved. First, clutter caused by ghost targets is largely suppressed as the multipath returns are accounted for in the model formulation. If residual clutter remains in the reconstruction, caused, for example, by propagation paths or physical effects that were not considered in the model, it is attenuated in the final image. It is expected that residual clutter is spatially nonwhite and independently distributed with respect to the subimages. Hence, after incoherent combination, these residuals are averaged out in the results. Note that the performance of this approach relies on sufficient power in the multipath returns. If the multipath returns are very weak, they cannot contribute to an improvement in the image and should be neglected.

Note that there are two challenges when dealing with practical scenarios. First, in order to achieve good recovery performance, just the significant multipath paths should be included in the model. Neglecting significant paths leads to remaining ghost targets and, thus, increased clutter in the final image. Including many paths results in an unnecessary increase of the number of unknowns and, thereby, the CS reconstruction performance will drop. Hence, the significant paths must be inferred from prior knowledge of the building layout. Second, the precise knowledge of locations of the interior wall is very important. All returns via one particular path are coherently processed. Hence, an error in an interior wall location has the same effect as an error in the array element positions. Further, the target positions, as viewed via different paths, only overlap if the wall locations errors are sufficiently low. Thus, the group sparse property is lost if the errors are too large. A possible solution to the afore-explained issues may be the introduction of additional parameters in the model, which are estimated in the reconstruction process. This, however, increases the numerical complexity. Further, if, for example, the interior wall locations are parameterized, the measurement model becomes nonlinear, rendering the reconstruction problem much harder.

5.4.3 Example

For both simulation and experiments, we assume the same measurement setup and room layout, depicted in Figure 5.7. A 77-element uniform linear monostatic array with an inter-element spacing of 1.9 cm is used for imaging. The origin of the coordinate system is chosen to be at the center of the array. The concrete front wall is located parallel to the array at 2.44 m downrange and has a thickness $d = 20$ cm and relative permittivity $\epsilon = 7.6632$. The left sidewall is at a crossrange of 1.83 m, whereas the back wall resides at 6.37 m downrange. Also, there is a protruding corner on the right at 3.4 m crossrange and 4.57 m downrange. Hence, the approximation in (5.20) is sufficiently accurate. A stepped-frequency signal, consisting of $M = 801$ equally spaced frequency steps covering the 1–3 GHz band, is employed for scene interrogation.

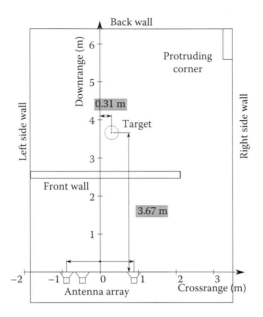

FIGURE 5.7
Measurement setup and room layout. (From Leigsnering, M. et al., Multipath exploitation in through-the-wall radar imaging using sparse reconstruction, *IEEE Trans. Aerosp. Electron. Syst.*, in press. With permission. © 2013 IEEE.)

5.4.3.1 Simulation Results

Two point targets, located at coordinates $(0.31, 3.6)$ m and $(-0.62, 5.2)$ m, are simulated using the target-only model (5.25). We consider $R = 5$ round-trip propagation paths, where the first one-way path is always the direct path and the second one-way path corresponds to direct, back wall multipath, left-side wall multipath, multipath w.r.t. the protruding right corner, and the wall ringing multipath. We apply weights to the paths to account for additional losses due to secondary reflection, which are set to 1, 1, 0.3, 0.5, and 0.4, respectively. For the sake of simplicity, we assume that the weights vary only with the propagation path. White noise with 0 dB SNR is added to the simulated measurements. For comparison, the beamformed image using the full data record is depicted in Figure 5.8a. For the CS results in Figure 5.8, only one-fourth of the array elements and one-eighth of the frequencies have been used for scene reconstruction. The array elements and frequencies are chosen randomly and the results are averaged over 100 Monte Carlo runs.

Figure 5.8b shows the reconstruction result using a conventional CS approach, which does not exploit multipath [38]. Essentially, this corresponds to the approach described earlier when setting the number of paths to one, that is, considering direct propagation only. It is observed that the

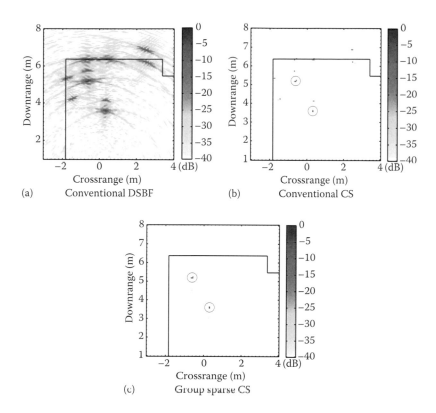

FIGURE 5.8
(See color insert.) Reconstruction results using different algorithms for the simulated scene with two point targets. One-fourth of the array elements and one-eighths of the frequency bins were used for the formation of (b) and (c). Image (a) was created using the full set of measurements. (From Leigsnering, M. et al., Multipath exploitation in through-the-wall radar imaging using sparse reconstruction, *IEEE Trans. Aerosp. Electron. Syst.*, in press. With permission. © 2013 IEEE.)

true targets are reconstructed along with all ghost targets, resulting in a highly cluttered scene. The group sparse reconstruction approach shown in Figure 5.8c provides a superior performance. All ghost targets have been suppressed while the two correct targets remain perfectly visible.

5.4.3.2 Experimental Results

Further, we present experimental results using data collected in a semicontrolled environment at the Radar Imaging Lab, Villanova University. A single aluminum (Al) pipe (61 cm long, 7.6 cm diameter) was placed upright on a 1.2 m high foam pedestal at 3.67 m downrange and 0.31 m crossrange. The left- and right-side walls were covered with RF-absorbing

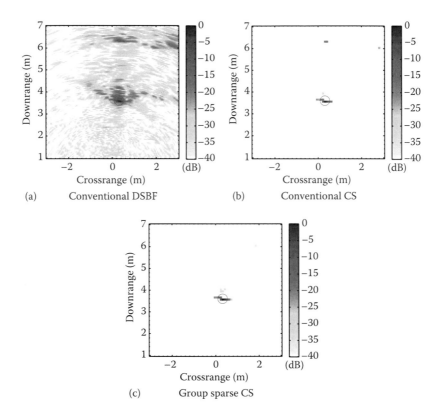

FIGURE 5.9
Imaging results using different algorithms for the Al pipe scene. Conventional DSBF reconstruction using full measurements is shown in (a). CS reconstruction using one-fourth of the array elements and half of the frequency bins is depicted in (b) and (c). (From Leigsnering, M. et al., Multipath exploitation in through-the-wall radar imaging using sparse reconstruction, *IEEE Trans. Aerosp. Electron. Syst.*, in press. With permission. © 2013 IEEE.)

material, but the protruding right corner and the back wall were left uncovered. Background subtraction [2,28] has been performed on the data, as the aim is to focus on target multipath only.

Figure 5.9a depicts the beamformed image using all available data. When comparing with our simulated data, we can conclude that only the multipath ghosts, due to the back wall, and the protruding corner at the back on the right side are visible. Hence, we only consider the direct path and these two multipath propagation cases, that is, $R = 3$, for the group sparse reconstruction. One-fourth of the array elements and one-half of the frequencies are used for CS. The conventional CS reconstruction is shown in Figure 5.9b, where the multipath ghost from the back wall is still visible. The group sparse reconstruction approach with multipath exploitation is capable of suppressing this ghost, see Figure 5.9c.

5.4.4 Moving Targets

When dealing with moving targets, we add additional dimensions both in the measurements and the image. Clearly, the unknown vector must include the target velocities. All possible target states are discretized on two spatial and two velocity dimensions, as described in Section 5.2. Likewise, we gain another dimension in the measured signal, that is, the slow-time index k. In order to resolve target movements, K pulses are transmitted and coherently processed. The subtle changes in the round-trip delay results in changes of the received phase of the signal, which, in turn, serves to distinguish the velocities of the targets.

In order to achieve the maximum benefit from CS, the signal is undersampled in all three dimensions, that is, fast time, slow time, and array elements. For the latter, most savings are achieved by simple omission of some elements, leading to a sparse array. Randomly undersampling pulses does not lead to benefits in terms of time or cost savings. The total duration of transmit/receive does not change if the first and the last pulses are kept. However, reducing the number of pulses leads to savings in energy, which may be relevant in portable applications. It should be noted that this is traded-off against a decreased SNR in the received signal. Various methods have been applied to compressively sample a time domain signal, that is, the fast time in the problem at hand. A random mixing scheme is adopted, where each pulse is correlated with a set of random signals and only the corresponding correlation result is sampled. For a detailed discussion of this scheme, the reader is referred to [17,29]. The compressively sampled signal can be expressed as

$$\bar{y} = \mathbf{\Phi} y = \mathbf{\Phi} \left(\mathbf{\Psi}^{(0)} \sigma^{(0)} + \mathbf{\Psi}^{(1)} \sigma^{(1)} + \cdots + \mathbf{\Psi}^{(R-1)} \sigma^{(R-1)} \right), \tag{5.31}$$

where $\mathbf{\Phi} \in \mathbb{R}^{J \times NKT}$ represents the undersampling operation or measurement matrix. With these considerations, reducing the number of samples along array elements to N_d, along slow time to K_d and along fast time to T_d is achieved by a measurement matrix constructed by

$$\mathbf{\Phi} = \left(\mathbf{\Phi}_1 \otimes I_{T_d K_d} \right) \left(\mathbf{\Phi}_2 \otimes I_{T_d N} \right) \text{diag} \left(\mathbf{\Phi}_3^{(0)}, \ldots, \mathbf{\Phi}_3^{(NK-1)} \right), \tag{5.32}$$

where
$\otimes$ denotes Kronecker's product
I_a is an identity matrix of dimension a

Hence, the total number of measurements is given by $J = T_d N_d K_d$. The matrices $\mathbf{\Phi}_1 \in \mathbb{R}^{N_d \times N}$ and $\mathbf{\Phi}_2 \in \mathbb{R}^{K_d \times K}$ consist of randomly chosen rows from an identity matrix to achieve the aforementioned goals of measurement reduction. Random mixing in the time domain is achieved by Gaussian random matrices $\mathbf{\Phi}_3^{(i)} \in \mathbb{R}^{T_d \times T}$ with entries drawn from a standard normal distribution. Other random matrices, for example, drawn from a Bernoulli distribution, can also be considered to achieve a good trade-off between ease of implementation and performance, see [17]. In order to achieve the reduced complexity of the receiver and the data reduction, the downsampling operation has to be implemented in hardware. The considered scheme lends itself to a hardware implementation using microwave mixers and lowpass filters [29].

5.4.5 Group Sparse Reconstruction of Stationary/Nonstationary Scenes

Next, the complete target information, that is, location and velocity, shall be reconstructed using CS principles [23]. Following the group sparse reconstruction approach in the stationary case, a high-dimensional model is constructed in order to account for all propagation paths. Model (5.31) can be rewritten as

$$\bar{y} = \boldsymbol{\Phi}\tilde{\boldsymbol{\Psi}}\tilde{\sigma}, \tag{5.33}$$

where $\tilde{\boldsymbol{\Psi}} = \begin{bmatrix} \boldsymbol{\Psi}^{(0)} & \boldsymbol{\Psi}^{(1)} \cdots \boldsymbol{\Psi}^{(R-1)} \end{bmatrix} \in \mathbb{C}^{TNK \times N_x N_y N_v R}$ is the concatenated over-complete dictionary for all possible paths. The unknown vectors in (5.33) are stacked into one tall vector $\tilde{\sigma}$ as in (5.26).

Given the reduced measurements $\bar{y}$ in (5.33), we aim at recovering the target state information $\tilde{\sigma}$ using CS reconstruction. If no multipath propagation is present, this can be achieved by standard ℓ_1-minimization as considered in [29]. However, this method is suboptimal in the presence of multipath. As detailed in the previous section, we need to take the prior information of the various subimages $\sigma^{(r)}, r = 0, \ldots, R - 1$ into account. Again, as each target is assumed to be visible from all paths, or at least from more than one path, those subimages should share approximately the same support. Thus, the reconstruction algorithm should consider the structured sparsity of the target state information σ. Note that the apparent Doppler speed, that is, the Doppler shift of the carrier, for a particular target may differ when observed through different paths. This, however, is incorporated in the model, as delays $\tau_{pn}^{(r)}(k)$ depend on the slow time and are all calculated based on the same coordinate system. In this way, the reconstruction benefits from additional diversity in the received signal due to different Doppler shifts from the same target.

The recovery can be again posed as a mixed ℓ_1/ℓ_2-norm minimization problem

$$\hat{\sigma} = \arg\min_{\tilde{\sigma}} \left\| \bar{y} - \boldsymbol{\Phi}\tilde{\boldsymbol{\Psi}}\tilde{\sigma} \right\|_2 + \lambda \left\| \tilde{\sigma} \right\|_{1,2}, \tag{5.34}$$

where

$$\|\tilde{\sigma}\|_{1,2} = \sum_{p=0}^{N_x N_y N_v - 1} \left\| \left[\sigma_p^{(0)}, \sigma_p^{(1)}, \ldots, \sigma_p^{(R-1)} \right]^T \right\|_2 \qquad (5.35)$$

and λ is a regularization parameter.

The partial results are combined in an incoherent fashion using the ℓ_2-norm as in (5.30). The final recovery result contains the information of the location and the translatory motion of all targets. Stationary targets in the pulse-Doppler radar scenario are included in the zero crossrange and downrange velocity case.

5.4.6 Example

5.4.6.1 Simulation Results

Simulations were performed for a wideband real aperture pulse-Doppler radar with one transmitter and a uniform linear array with $N = 11$ receivers. A modulated Gaussian pulse, centered around $f_c = 2$ GHz, with a relative bandwidth of 50% is transmitted. The PRI is set to 10 ms and $K = 15$ pulses are transmitted and processed coherently. At the receiving side, $T = 150$ fast-time samples in the relevant interval, covering the target and multi-path returns, are taken at a sampling rate of $f_s = 4$ GHz. The receive array with element spacing of 10 cm is centered around the transmitter and is located 3 m from the wall. The wall, which is parallel to the array, is modeled with $d = 20$ cm thickness and relative permittivity $\epsilon_r = 7.66$. The imaged region extends 6 m in crossrange and 4 m in downrange and is centered around a point in the broadside direction of the array at 4 m downrange. Two side walls are considered at ± 2 m that cause three different multipaths each. These are in total 4 first-order multipaths and 2 second-order quasi-monostatic multipaths, which are all considered to be 6 dB weaker than the direct path. Hence, in total there are $R = 7$ paths that are considered in the received signal. We do not consider any wall returns or reverberations as they can be usually gated out efficiently in bistatic radars [29]. The scene of interest is spatially discretized into an $N_x \times N_y = 32 \times 32$ pixel grid. The target velocities are discretized on an $N_{v_x} \times N_{v_y} = 5 \times 7$ crossrange by downrange grid, spanning target velocity components of ± 0.9 m/s. The results are shown on a 40 dB scale, and the individual images for a corresponding velocity pair are shown side by side.

We consider a scenario where two strong stationary reflectors block the line of sight to two moving targets. The two stationary targets reside at coordinates $(0.5,3.2)$ m and $(-1.5,3.2)$ m and there is a moving target 1 m behind each stationary target. The moving targets are 8 dB weaker than the stationary targets and posses velocities $(0.45,0)$ m/s and $(0,0.3)$ m/s. As we assume that the line of sight to the moving targets is blocked by a stationary

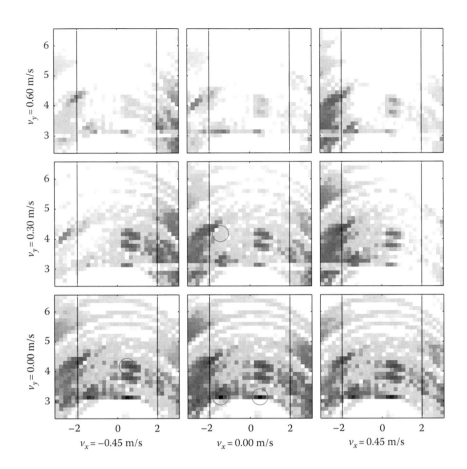

FIGURE 5.10
(See color insert.) Delay-and-sum beamforming result using full data.

scatterer, we only receive their quasi-monostatic multipath signal. However, the stationary targets are visible for all $R=7$ possible paths. Hence, the cumulated received power from the moving targets is 20 dB weaker than from the stationary targets.

We first show the beamforming results using full measurements in Figure 5.10, where the beamformers for each image have been matched to the corresponding velocity pair according to (5.15). The image appears very cluttered due to multipath responses and the moving targets cannot be discerned. This is to be expected as the standard beamforming algorithm only takes the direct propagation path into account. As the information of the moving targets is only contained in the multipath, we cannot gain any knowledge about the moving targets without multipath exploitation.

Next, we show in Figure 5.11 a group sparse reconstruction using 7.8% of the full Nyquist measurements averaged over 10 Monte Carlo runs.

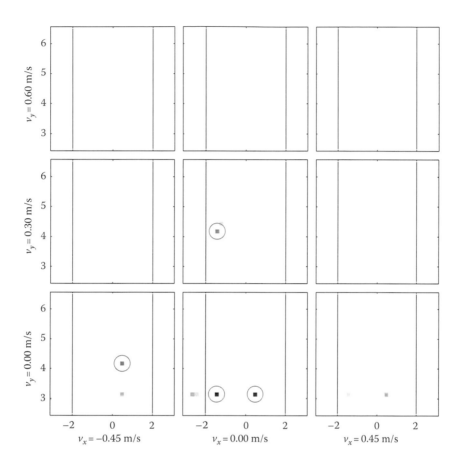

FIGURE 5.11
(See color insert.) CS reconstruction using 7.8% of the measurements.

The downsampling parameters of (5.32) are set to $T_d = 20$, $N_d = 8$, and $K_d = 12$, performing linear measurements, using a Gaussian random mixing matrix in the fast time. It is evident that both the location and the velocities of the targets have been correctly recovered. Even with only 2 out of 7 paths available for the moving targets, they are accurately located. The ghost targets have been largely suppressed and only a few weak clutter pixels remain. The clutter can mostly be attributed to the high correlation in the dictionary for neighboring crossrange velocities, leading to a certain "leakage." Overall, the CS reconstruction features a very clean image with high resolution and accuracy.

5.4.6.2 Experimental Results

We show experimental results for a wideband real aperture pulse-Doppler radar with one transmitter and a uniform linear array with $N = 8$ receivers.

The data have been recorded at the Radar Imaging Lab, Villanova University, in a controlled lab setup. A modulated Gaussian pulse is transmitted, covering the frequency range of 1.5–4.5 GHz. We recorded 768 fast-time samples at sampling rate $f_s = 7.68$ GHz and gated out the early and late returns to clean the data, resulting in $T = 153$ samples. The transmitter was placed 61 cm away from a side wall and the receive array (element spacing 6 cm) was placed on the other side of the transmitter at a distance (to the first element) of 29.2 cm on the same baseline. We did not include a wall in the measurements as it can be usually gated out easily in bistatic pulse radar measurements [29]. A total number of $R = 4$ possible propagation paths is expected, the direct path, two first-order multipaths and one second-order multipath via the side wall. Again, the scene of interest is spatially discretized into a $N_x \times N_y = 32 \times 32$ pixel grid. The target velocities are discretized on a $N_{v_x} \times N_{v_y} = 5 \times 7$ grid spanning target velocity components of ± 0.9 m/s. A scenario with a human walking diagonally toward the radar was recorded. All of the propagation paths are expected to be observed for the human.

Group sparse reconstruction using 20% of the full Nyquist measurements is shown in Figure 5.12. The downsampling parameters of (5.32) are set to $T_d = 15$, $N_d = 5$, and $K_d = 50$, performing linear measurements, using a Gaussian random mixing matrix in the fast time. The human as a moving target is recovered with approximately correct location and velocities. At least the direction of the movement is consistent with the experiment. There is some leakage in the neighboring velocity cell, probably owing to high coherence in the measurement matrix or to the nature of human motion. Additionally, some residual clutter in the stationary image can be observed. This might be explained by the reflections of some stationary objects that were present in the lab.

5.5 Compressive Sensing Reconstruction with the Wall Included

Until this point, the direct returns from the front wall have not been considered. However, in TWRI, the front wall is a very strong stationary reflector, which can potentially mask targets within the room, especially when monostatic radar is used. Due to reverberations within the front wall, the returns may even extend far beyond the wall itself and leak into the room. In a CS framework, this represents an additional challenge. The wall as an extended and strong reflector occupies a large number of pixels in the scene. Hence, the vector of interest is rendered less sparse, which adversely affects the reconstruction performance. Usually, when not mitigated, the front wall returns make conventional CS approaches fail [21]. Hence, various schemes for wall removal or wall clutter mitigation have been developed.

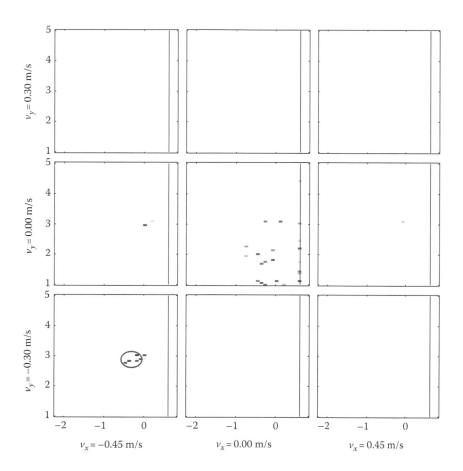

FIGURE 5.12
CS reconstruction of the walking human using 20% of the measurements.

One widely used approach is background subtraction [2,28], where background data are coherently subtracted from the target scene. Provided that background data are available, this method works well in sparse reconstruction [25,38]. If background data are not available, one may resort to change detection. Assuming slowly moving targets, two measurements are taken within a short time interval and subtracted subsequently [1,6,7]. The drawback of this method is that it will not work with stationary targets. For purely stationary scenes, other methods have been proposed in the literature. Spatial filtering [39] and subspace projection [36] approaches have been developed for full data measurements. These methods have also been shown to perform well within the CS framework [21]. Further approaches include modeling the wall returns with subsequent subtraction [11] or time gating [29].

In this section, a conceptually different approach is described [24]. Instead of dealing with wall clutter mitigation and target scene reconstruction separately, both are achieved at the same time. Two signal models are used, one is adapted to target returns and the other is matched to wall returns. Based on these models, the wall and the target scene are recovered jointly. The target and wall returns are captured in different parts of the reconstructed signal and, hence, they are separated in the process. As mentioned earlier, in the case of moving targets, the wall return issue can be dealt with efficiently. In the remainder of this section, we will focus on the challenging problem of a purely stationary scene using stepped-frequency measurements.

5.5.1 Wall Reverberation Model

As discussed in Section 5.2, the wall returns consist of the direct returns and the reverberation. The wall return model (5.4) shall now be described in detail. Wall reverberation can be treated similarly to wall ringing target multipath, but the wave never interacts with the targets, it solely reflects inside the front wall.

The return from each internal reflection within the wall can be considered as an additive contribution equivalent to the first part of (5.4). However, the delays associated with each reverberation return will be different as the internal reflections cause additional delays. At the same time, the amplitude will decay as part of the energy passes the wall and due to attenuation in the wall medium. In bistatic setups, the amplitude of the reverberation returns are usually very weak and can be dealt with by simple time gating [29]. In the sequel, we will focus on monostatic operation modes where significant reverberation returns are expected.

In a monostatic radar, the wall returns arrive with the same time delay at all transceivers. Hence, only a single delay is needed for the direct return and for each of the reverberation returns. Assuming a number of total wall returns R_w, the corresponding delays $\tau_w^{(r)}, r = 0, 1, \ldots, R_w - 1$, can be calculated according to (5.18). There is an amplitude, $\sigma_w^{(r)}$, associated with each delay. It depends on the attenuation and reflection laws governing the wall reverberation [8]. Superimposing all wall returns yields a wall reverberation model as

$$y_w[m, n] = \sum_{r=0}^{R_w - 1} \sigma_w^{(r)} \exp\left(-j2\pi f_m \tau_w^{(r)}\right). \tag{5.36}$$

Keeping the convention that $\tau_w^{(0)}$ describes the direct path, all $\tau_w^{(r)}, r > 0$, correspond to contribution of the wall reverberations. The decrease in amplitude for the higher-order reverberation paths is accounted for in the wall reflectivities $\sigma_w^{(r)} \in \mathbb{C}$. Note that even in a monostatic setup only up to three wall reverberation responses are typically observed [34]. Due to the strong attenuation in wall materials, higher-order reverberations usually drop below the noise floor. Also, most wall materials exhibit a frequency dependent attenuation behavior, which has been neglected here for simplicity. Furthermore, note that bistatic wall reverberations can be modeled in a very similar fashion. However, as the angle of reflection at the wall depends on the receiver position, the delays and the attenuation generally depend on the sensor index n.

For ease of notation and implementation, a modified wall model is used for the CS reconstruction. The wall can be equivalently modeled as a composition of small wall segments that correspond to the grid of the target image. This gives the additional flexibility of modeling returns that do not originate from a perfectly homogeneous wall. Further, despite assuming knowledge of the front wall parameters, these parameters are usually estimates that contain errors. The proposed model has sufficient degrees of freedom to represent the wall returns when dealing with errors in the wall parameters. This is a potential advantage over modeling and coherent subtraction of the wall returns [11], as the coherent subtraction may fail even for small deviations of the parameters.

The vectorized wall reverberation model, similar to (5.24), can be expressed as

$$y_{\text{FD},w} = \mathbf{\Psi}_w^{(0)} \sigma_w^{(0)} + \mathbf{\Psi}_w^{(1)} \sigma_w^{(1)} + \cdots + \mathbf{\Psi}_w^{(R_w-1)} \sigma_w^{(R_w-1)}, \qquad (5.37)$$

where

$\sigma_w^{(r)}$ are the vectorized reflectivities of the wall segments

$\mathbf{\Psi}_w^{(r)}$ contains the phase information for path r

The matrices $\mathbf{\Psi}_w^{(r)}$ are the same as $\mathbf{\Psi}_{\text{FD}}^{(r)}$. However, only the contributions of specular reflections are retained. This can be described by a masking operation,

$$\mathbf{\Psi}_w^{(r)} = \mathbf{M} \circ \mathbf{\Psi}_{\text{FD}'}^{(r)} \qquad (5.38)$$

where "$\circ$" denotes the element-wise (or Schur) product and $\mathbf{M} \in \{0,1\}^{MN \times N_x N_y}$ is a binary matrix. An element $[\mathbf{M}]_{pn}$ is equal to one if the pth wall segment is visible to the nth array element traveling along the rth path and zero otherwise. In the monostatic case, the contribution from all specular targets that are not directly in front of the antenna should be masked out by $\mathbf{M}$. Any wall segment that is not exactly in the broadside direction of the array is not visible, as the transmitted wave will be reflected away from the transceiver. We note that in the bistatic case, the geometry of the propagation path changes with the reverberation order, r, and the considered receive element [29]. Hence, different masking operations need to be assumed for each path r. Specular reflection model (5.38) typically results in wall dictionary matrices, $\mathbf{\Psi}_w^{(r)}$, that contain many zeros [3].

For the CS reconstruction, analogous to (5.25), the downsampled superposition of target and wall contributions in the measured signal is observed:

$$\bar{y}_{\mathrm{FD}} = \bar{y}_t + \bar{y}_w = \mathbf{\Phi}_{\mathrm{FD}} y_t + \mathbf{\Phi}_{\mathrm{FD}} y_w, \tag{5.39}$$

where the target contribution $\bar{y}_t$ is defined in (5.25). For the wall response, the downsampled measurements from (5.37) result in

$$\bar{y}_w = \mathbf{\Phi}_{\mathrm{FD}} \left(\mathbf{\Psi}_w^{(0)} \sigma_w^{(0)} + \mathbf{\Psi}_w^{(1)} \sigma_w^{(1)} + \cdots + \mathbf{\Psi}_w^{(R_w-1)} \sigma_w^{(R_w-1)} \right). \tag{5.40}$$

5.5.2 Separate Reconstruction

Making use of the models that are matched to describe target and wall returns, a sparse reconstruction of the scene is sought. First, we describe a rather simplistic approach, that is, to reconstruct the wall and target images separately.

For CS multipath exploitation, method (5.28), as described in Section 5.4, can be applied. This method is applied twice, first to reconstruct the target scene model (5.25) and then to reconstruct the wall using model (5.40). Hence, from the measurements $\bar{y}_{\mathrm{FD}}$ two images, $\hat{\sigma}$ and $\hat{\sigma}_w$, can be reconstructed to describe the targets and the wall, respectively. Note that the two reconstructions are independent in the sense that no information from the wall image is used to form the target image and vice versa.

5.5.3 Joint Group Sparse Reconstruction

A more sophisticated reconstruction approach makes use of the knowledge that a superposition of the wall and target returns is received. As mentioned earlier, the wall and target images shall be reconstructed jointly [24]. To this end, a similar group sparse approach as described in Section 5.4.2 is used. First, the wall and target models (5.25) and (5.40) are combined, resulting in

$$\bar{y}_{\mathrm{FD}} = \mathbf{\Phi}_{\mathrm{FD}} \tilde{\mathbf{\Psi}}_j \tilde{\sigma}_j. \tag{5.41}$$

The vector $\tilde{\sigma}_j \in \mathbb{C}^{N_x N_y (R + R_w) \times 1}$ is obtained by stacking the various vectors of the scene of interest, that is,

$$\tilde{\sigma}_j = \left[\left(\sigma^{(0)} \right)^{\mathrm{T}} \cdots \left(\sigma^{(R-1)} \right)^{\mathrm{T}} \left(\sigma_w^{(0)} \right)^{\mathrm{T}} \cdots \left(\sigma_w^{(R_w-1)} \right)^{\mathrm{T}} \right]^{\mathrm{T}}. \tag{5.42}$$

Then, the new measurement matrix has the form

$$\tilde{\Psi}_j = \left[\Psi_{\mathrm{FD}}^{(0)} \; \Psi_{\mathrm{FD}}^{(1)} \; \cdots \; \Psi_{\mathrm{FD}}^{(R-1)} \; \Psi_{w}^{(0)} \; \Psi_{w}^{(1)} \; \cdots \; \Psi_{w}^{(R_w-1)} \right] \in \mathbb{C}^{MN \times N_x N_y (R+R_w)}.$$
$$(5.43)$$

From the aforementioned high-dimensional joint model in (5.41) to (5.43), the reconstruction problem is posed using a group sparse regularization term. The convex optimization problem is virtually the same as in (5.28)

$$\hat{\tilde{s}}_j = \arg\min_{\tilde{s}_j} \frac{1}{2} \left\| \bar{y}_{\mathrm{FD}} - \Phi_{\mathrm{FD}} \tilde{\Psi}_j \tilde{s}_j \right\|_2^2 + \lambda \rho_j(\tilde{s}_j), \qquad (5.44)$$

where

$$\rho_j(\tilde{s}_j) := \sum_{p=0}^{N_x N_y - 1} \left\| \left[\sigma_p^{(0)}, \sigma_p^{(1)}, \dots, \sigma_p^{(R-1)} \right]^{\mathsf{T}} \right\|_2$$

$$+ \sum_{p=0}^{N_x N_y - 1} \left\| \left[\sigma_{w,p}^{(0)}, \sigma_{w,p}^{(1)}, \dots, \sigma_{w,p}^{(R_w-1)} \right]^{\mathsf{T}} \right\|_2. \qquad (5.45)$$

Note that the regularizer in (5.45) is different from the case where only targets are considered. This is still a group sparse problem for targets, as reflected in the first part of the regularizer. The subimages of the wall should follow the same behavior as all reverberations originate from the same physical wall. However, no interrelations between wall and target subimages are desired. Consequently, both are regularized by separate terms in order to achieve the goal of separating their respective return signals. In our formulation of the reconstruction problem, we use a single regularization parameter λ to adjust the sparsity of the solution. Introducing separate parameters for the two parts in (5.45) may improve the reconstruction results. However, it aggravates the problem of selecting proper values of the regularization parameters, which is critical for the performance. Methods based on cross-validation have been proposed to find good estimates of the regularization parameter [16].

Further note that in (5.41), the measurement or dictionary matrix is extended to include both target and wall atoms for all possible paths. Using this dictionary, the wall and target contributions in $\bar{y}$ can be expressed in a sparse fashion. However, as the wall and target measurement matrices are very similar, considerable mutual coherence between their columns is observed. This might adversely affect the reconstruction as the wall and target contributions may not get fully separated. A more detailed study of the mutual coherence of the two models can be found in [24].

In order to obtain a final image, the results from (5.44) have to be accumulated. As introduced in Section 5.4, a noncoherent combination is used again. However, two separate images, that is, the wall and the target image, shall be obtained. Hence, the noncoherent combination is performed separately for the target image $\hat{\bar{\sigma}}_{GS}$ and the wall image $\hat{\bar{\sigma}}_{GS,w}$.

5.5.4 Example

Finally, some example results for reconstructing the target image along with the wall are presented. We used the same radar setup and room geometry as in the stationary target case. The returns from the front wall, including reverberations, are taken into consideration this time.

5.5.4.1 Simulation Results

First, the received signals are simulated with a total of $R_w = 4$ propagation paths for the wall model. The direct wall responses is assumed to be 6 dB stronger than the target returns. Subsequently, wall returns are attenuated by 8 dB for each reverberation within the wall. Additive white Gaussian noise and 0 dB SNR is considered for the simulated measurements. The beamforming result using full measurements is depicted in Figure 5.13a. One can clearly see that the first two wall returns lie at the inner and outer surface of the front wall. However, the higher-order returns appear inside the room and potentially mask the targets at those locations.

One-fourth of the array elements and one-fourth of the frequencies are used for CS reconstruction. The corresponding results, averaged over 100 Monte Carlo runs, are displayed in Figure 5.13b through e. Separate reconstruction of the wall and target images with multipath exploitation is shown in Figure 5.13b and c. The targets are reconstructed and the ghost targets are well suppressed. However, the wall response shows up to its full extent in the target image as it is treated as a target response. This result may be considerably improved by employing a wall removal scheme prior to reconstruction. Various CS-based wall mitigation approaches have been discussed in [21]. Employing joint group sparse CS reconstruction as described in Section 5.5.3, a very clean reconstruction of the two targets is achieved, (see Figure 5.13d.) With this reconstruction approach, the ghost targets and the wall returns are well suppressed.

The reconstruction of the wall is very similar for separate and joint group sparse CS reconstruction, refer to Figure. 5.13c and e. The wall response appears in isolated pixels, roughly aligned in two lines. Note that the first reverberation, corresponding to the return from the back face of the wall, is treated as a valid target and not as a multipath return. Hence, it becomes visible in the reconstructed wall images by a second line of pixels.

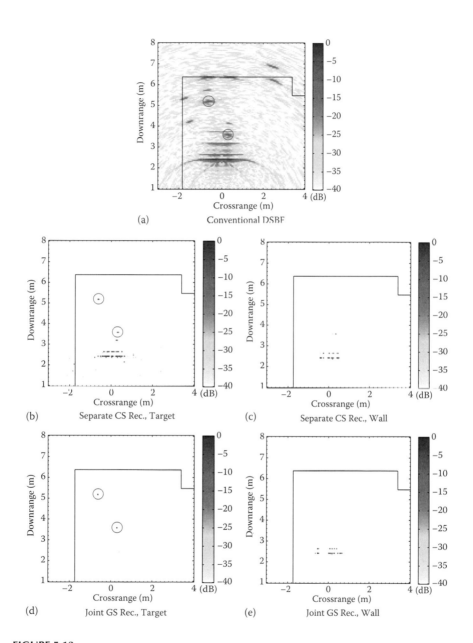

FIGURE 5.13
Image formation results using different algorithms for the simulated scene including the wall response. One-fourth of both the array elements and frequency bins were used for sparse reconstruction. (From Leigsnering, M. et al., Multipath exploitation in through-the-wall radar imaging using sparse reconstruction, *IEEE Trans. Aerosp. Electron. Syst.*, in press. With permission. © 2013 IEEE.)

5.5.4.2 Experimental Results

Finally, we present reconstruction results from experimental data. We use the previously described measurements of the scene with a single aluminum pipe, but without using background subtraction. The raw data suffer from antenna mismatch, and we used a Hamming window across the frequencies to mitigate this effect. Additionally, we gated out all time samples that correspond to returns in front of the wall and far behind the back wall to further clean up the data. These samples cannot contain any target or wall information and, hence, may only distort the reconstructed images.

Figure 5.9a depicts the beamformed image using all available data. In contrast to the simulation setup, the wall response is about 15 dB stronger than the target. The wall responses from the front face and the back face of the wall are clearly visible. Furthermore, one can recognize multipath propagation via the back wall of the room. Thus, we consider only the direct path and one multipath via the back wall in the CS reconstruction algorithms. One-fourth of the array elements and half of the frequency bins are retained before reconstruction. In Figure 5.14b and c, the separate CS reconstructions of the target and wall scenes are shown, whereas the results for the joint group sparse CS method are depicted in Figure 5.14d and e. The latter algorithm obtains a clearer image of the targets with less clutter pixels beyond the front wall. However, neither method is able to separate the wall and target images in a satisfactory way. Also, the wall image reconstruction performance is rather weak as it is not possible to clearly identify the front and back faces of the wall.

The limited capability to separate the wall and the target responses can be attributed to two issues. First, there is residual sidelobe leakage, owing to the very strong antenna mismatch. This signal part is not modeled and, hence, impairs the CS reconstruction. Second, the front wall is constructed by stacking solid concrete blocks without any mortar or plaster, resulting in a wall that has air gaps and no smooth surface. Therefore, the assumptions of homogeneity and specular reflection for the front wall are violated. This leads to a model mismatch in the wall model, which possibly results in a strong "leakage" of the wall into the target image.

5.6 Conclusion

We gave an overview of several compressive sensing methods dealing with multipath propagation effects in TWRI scenarios. A comprehensive model for specular multipath as well as multiple internal reflections inside the front wall was described. This model lends itself to a natural inclusion of multipath returns in the received signal. Employing CS, the model can be

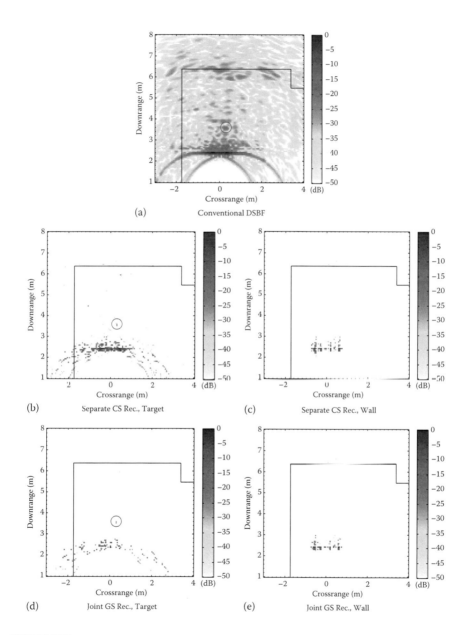

FIGURE 5.14

Image formation results using different algorithms for the Al pipe scene without background subtraction. One-fourth of the array elements and half of the frequency bins were used for the CS reconstruction. (From Leigsnering, M. et al., Multipath exploitation in through-the-wall radar imaging using sparse reconstruction, *IEEE Trans. Aerosp. Electron. Syst.*, in press. With permission. © 2013 IEEE.)

inverted in order to reconstruct the sparse ground truth while exploiting the additional energy in multipath returns.

We have treated different cases: First, we considered stationary scenes, without front wall returns. We have shown that ghost targets due to specular multipath and wall ringing can be suppressed, leading to cleaner images. Second, we presented an extension to stationary and moving targets. Using a similar approach, the benefits of multipath exploiting were applied to scenes with nonstationary targets. Finally, we included the front wall returns in the propagation model of the stationary scene. Exploiting this knowledge, the response from the front wall and the targets can be separated. Hence, a cleaner target image with less wall clutter can be obtained, along with a reconstruction of the wall itself. We illustrated the performance of all three cases using numerical simulations and experimental data.

Acknowledgments

The authors thank Dr. Moeness Amin and Dr. Fauzia Ahmad from the Center of Advanced Communications at Villanova University, Villanova, PA, for fruitful discussions and for providing the experimental data.

References

1. F. Ahmad and M.G. Amin. Through-the-wall human motion indication using sparsity-driven change detection. *IEEE Transactions on Geoscience and Remote Sensing*, 51(2):881–890, February 2013.
2. F. Ahmad and M.G. Amin. Multi-location wideband synthetic aperture imaging for urban sensing applications. *Journal of the Franklin Institute*, 345(6):618–639, September 2008.
3. F. Ahmad and M.G. Amin. Partially sparse reconstruction of behind-the-wall scenes. In *Proceedings SPIE Symposium on Defense, Security, and Sensing, Compressive Sensing Conference*, Vol. 8365, Baltimore, MD, April 2012.
4. F. Ahmad, M.G. Amin, and S.A. Kassam. A beamforming approach to stepped-frequency synthetic aperture through-the-wall radar imaging. In *Proceedings of the IEEE First International Workshop on Computational Advances in Multi-Sensor Adaptive Processing*, Puerto Vallarta, Mexico, December 2005.
5. F. Ahmad, G.J. Frazer, S.A. Kassam, and M.G. Amin. Design and implementation of near-field, wideband synthetic aperture beamformers. *IEEE Transactions on Aerospace and Electronic Systems*, 40(1):206–220, January 2004.
6. M. Amin, F. Ahmad, and Wenji Zhang. A compressive sensing approach to moving target indication for urban sensing. In *IEEE Radar Conference (RADAR)*, pp. 509–512, Kansas City, MO, May 2011.

7. M.G. Amin and F. Ahmad. Change detection analysis of humans moving behind walls. *IEEE Transactions on Aerospace and Electronic Systems*, 49(3):1410–1425, July 2013.

8. C.A. Balanis. *Advanced Engineering Electromagnetics*. Wiley, New York, 1989.

9. R.G. Baraniuk, V. Cevher, M.F. Duarte, and Chinmay Hegde. Model-based compressive sensing. *IEEE Transactions on Information Theory*, 56:1982–2001, April 2010.

10. R.J. Burkholder. Electromagnetic models for exploiting multi-path propagation in through-wall radar imaging. In *International Conference on Electromagnetics in Advanced Applications*, pp. 572–575, Torino, Italy, September 2009.

11. M. Dehmollaian and K. Sarabandi. Refocusing through building walls using synthetic aperture radar. *IEEE Transactions on Geoscience and Remote Sensing*, 46(6):1589–1599, 2008.

12. W. Deng, W. Yin, and Y. Zhang. Group sparse optimization by alternating direction method. *Proceedings of SPIE 8858*, Wavelets and Sparsity XV, 88580R, September 26, 2013.

13. T. Dogaru and C. Le. SAR images of rooms and buildings based on FDTD computer models. *IEEE Transactions on Geoscience and Remote Sensing*, 47(5):1388–1401, May 2009.

14. Y.C. Eldar, P. Kuppinger, and H. Bolcskei. Block-sparse signals: Uncertainty relations and efficient recovery. *IEEE Transactions on Signal Processing*, 58(6):3042–3054, June 2010.

15. G. Gennarelli and F. Soldovieri. A linear inverse scattering algorithm for radar imaging in multipath environments. *IEEE Geoscience and Remote Sensing Letters*, 10(5):1085–1089, September 2013.

16. A.C. Gurbuz, J.H. McClellan, and W.R. Scott. A compressive sensing data acquisition and imaging method for stepped frequency GPRs. *IEEE Transactions on Signal Processing*, 57(7):2640–2650, July 2009.

17. A.C. Gurbuz, J.H. McClellan, and W.R. Scott. Compressive sensing for subsurface imaging using ground penetrating radar. *Signal Processing*, 89(10):1959–1972, October 2009.

18. A.M. Haimovich, R.S. Blum, and L.J. Cimini. MIMO radar with widely separated antennas. *IEEE Signal Processing Magazine*, 25(1):116–129, 2008.

19. A. Karousos, G. Koutitas, and C. Tzaras. Transmission and reflection coefficients in time-domain for a dielectric slab for UWB signals. In *IEEE Vehicular Technology Conference*, pp 455–458, Singapore, May 2008.

20. S. Kidera, T. Sakamoto, and T. Sato. Extended imaging algorithm based on aperture synthesis with double-scattered waves for UWB radars. *IEEE Transactions on Geoscience and Remote Sensing*, 49(12):5128–5139, December 2011.

21. E. Lagunas, M. G. Amin, F. Ahmad, and M. Najar. Joint wall mitigation and compressive sensing for indoor image reconstruction. *IEEE Transactions on Geoscience and Remote Sensing*, 51(2):891–906, February 2013.

22. M. Leigsnering, F. Ahmad, M.G. Amin, and A.M. Zoubir. Compressive sensing based specular multipath exploitation for through-the-wall radar imaging. In *IEEE International Conference on Acoustics, Speech, and Signal Processing (ICASSP)*, Vancouver, British Columbia, Canada, May 2013.

23. M. Leigsnering, F. Ahmad, M.G. Amin, and A.M. Zoubir. General MIMO framework for multipath exploitation in through-the-wall radar imaging.

In *International Workshop on Compressed Sensing Applied to Radar (CoSeRa)*, Bonn, Germany, September 2013.

24. M. Leigsnering, F. Ahmad, M.G. Amin, and A.M. Zoubir. Multipath exploitation in through-the-wall radar imaging using sparse reconstruction. *IEEE Transactions on Aerospace and Electronic Systems*, 50(2), April 2014. http://ieee-aess.org/publications/upcoming-transactions/all.

25. M. Leigsnering, C. Debes, and A.M. Zoubir. Compressive sensing in through-the-wall radar imaging. In *IEEE International Conference on Acoustics, Speech and Signal Processing (ICASSP)*, Prague, Czech Republic, May 2011.

26. M. Leigsnering, F. Ahmad, M.G. Amin, and A.M. Zoubir. CS based wall ringing and reverberation mitigation for through-the-wall radar imaging. In *IEEE Radar Conference (RADAR)*, Ottawa, Ontario Canada, April 2013.

27. J. Li and P. Stoica. MIMO radar with colocated antennas. *IEEE Signal Processing Magazine*, 24(5):106–114, September 2007.

28. J. Moulton, S. Kassam, F. Ahmad, M. Amin, and K. Yemelyanov. Target and change detection in synthetic aperture radar sensing of urban structures. In *IEEE Radar Conference (RADAR)*, Rome, Italy, May 2008.

29. J. Qian, F. Ahmad, and M.G. Amin. Joint localization of stationary and moving targets behind walls using sparse scene recovery. *Journal of Electronic Imaging*, 22(2):021002, June 2013.

30. M. A. Richards, J.A. Scheer, and W.A. Holm, editors. *Principles of Modern Radar: Basic Principles*. SciTech Publishing, Raleigh, NC, 2010.

31. P. Setlur, G. Alli, and L. Nuzzo. Multipath exploitation in through-wall radar imaging via point spread functions. *IEEE Transactions on Image Processing*, PP(99):1–1, 2013.

32. P. Setlur, M. Amin, and F. Ahmad. Multipath model and exploitation in through-the-wall and urban radar sensing. *IEEE Transactions on Geoscience and Remote Sensing*, 49(10):4021–4034, October 2011.

33. M. Soumekh. *Synthetic Aperture Radar Signal Processing with MATLAB Algorithms*. John Wiley and Sons, New York, 1999.

34. C. Thajudeen, A. Hoorfar, F. Ahmad, and T. Dogaru. Measured complex permittivity of walls with different hydration levels and the effect on power estimation of TWRI target returns. *Progress in Electromagnetic Research B*, 30: 177–199, 2011.

35. F.H.C. Tivive, M.G. Amin, and A. Bouzerdoum. Wall clutter mitigation based on eigen-analysis in through-the-wall radar imaging. In *Proceedings 17th International Conference on Digital Signal Processing (DSP)*, pp 1–8, Corfu, Greece, July 2011.

36. F.H.C. Tivive, A. Bouzerdoum, and M.G. Amin. An SVD-based approach for mitigating wall reflections in through-the-wall radar imaging. In *IEEE Radar Conference (RADAR)*, pp. 519–524, Kansas City, MO, May 2011.

37. S.J. Wright, R.D. Nowak, and M.A.T. Figueiredo. Sparse reconstruction by separable approximation. *IEEE Transactions on Signal Processing*, 57(7):2479–2493, July 2009.

38. Y.-S. Yoon and M.G. Amin. Compressed sensing technique for high-resolution radar imaging. In *Proceedings of SPIE Signal Processing, Sensor Fusion, and Target Recognition XVII*, volume 6968, p. 69681A, Orlando, FL, March 2008.

39. Y.-S. Yoon and M.G. Amin. Spatial filtering for wall-clutter mitigation in through-the-wall radar imaging. *IEEE Transactions on Geoscience and Remote Sensing*, 47(9):3192 –3208, September 2009.

40. Y. Yu, A.P. Petropulu, and H.V. Poor. MIMO radar using compressive sampling. *IEEE Journal of Selected Topics in Signal Processing*, 4(1):146–163, February 2010.
41. Y. Yu, A.P. Petropulu, and H.V. Poor. Power allocation for CS-based colocated MIMO radar systems. In *IEEE Sensor Array and Multichannel Signal Processing Workshop (SAM)*, pp. 217–220, Hoboken, NJ, June 2012.
42. M. Yuan and Y. Lin. Model selection and estimation in regression with grouped variables. *Journal of the Royal Statistical Society, Series B*, 68(1):49–67, December 2006.
43. W. Zheng, Z. Zhao, and Z.-P. Nie. Application of TRM in the UWB through wall radar. *Progress in Electromagnetics Research*, 87:279–296, 2008.

6

Measurement Kernel Design for HRR Imaging of Urban Objects

Nathan A. Goodman, Yujie Gu, and Junhyeong Bae

CONTENTS

ABSTRACT Targets of interest in urban applications often include relatively small objects such as chairs, tables, doors, and potential contraband such as handheld weapons. Therefore, radar imaging and classification of these objects can impose extremely high bandwidth and aperture requirements. Regarding the radar bandwidth, the typical manner of obtaining wideband waveforms is to implement swept or stepped waveforms that are instantaneously narrowband, but cover a wide bandwidth over time. On the other hand, some operational scenarios require relatively fast data collection and instantaneously wideband waveforms, which necessitate either expensive high-speed analog-to-digital converters or compressive, sub-Nyquist sampling. In this chapter, we investigate sub-Nyquist sampling of instantaneously wideband waveforms. Our objective is to optimize analog compression kernels for the underlying goals of imaging and/or recognition of small objects of interest in urban scenarios. We use Gaussian mixture models to represent prior information about a wide variety of target objects while also admitting (1) gradient-based optimization of the compression kernels and (2) injection of prior knowledge of the urban scenario. The models are trained using finite-difference time-domain (FDTD)-generated target signatures. Moreover, interfering objects such as walls between the radar and target are also incorporated into the optimization. Simulated performance of optimized kernels is compared with the performance of random-based compression and with Nyquist sampling of reduced-bandwidth waveforms.

6.1 Introduction

Targets of interest in many traditional radar applications are large vehicles such as trucks, tanks, and aircraft. These vehicles are several meters across, even at their narrowest points, such that waveform bandwidths on the order of tens of MHz are sufficient for resolving multiple range bins on the target. Objects encountered in urban applications, however, include relatively small objects such as chairs, tables, and handheld objects and weapons. These items are often less than 1 m in their longest dimension and can be as small as a few centimeters across their narrowest dimensions. To obtain multiple range resolution cells on such targets, waveform bandwidths must be on the order of GHz. For example, consider a rifle that is approximately 1 m in length and a few centimeters wide. When viewed from the side, the radar system must have at least a few GHz of bandwidth before being able to resolve multiple range bins on the target. When viewed from an angle of 30° relative to the side, approximately 500 MHz of bandwidth is required in order for the radar to resolve more than one range bin on the rifle.

In order to achieve such high bandwidths, the usual implementation calls for either a linear frequency modulated (LFM) waveform or a stepped frequency waveform. Both of these waveforms ease requirements for high-rate analog-to-digital converters (ADC) by encoding frequency in the time dimension and using stretch processing (LFM case) [1] or sequential sampling of frequency-domain coefficients (stepped frequency). In other words, the LFM and stepped frequency waveforms are instantaneously narrow-band, but trace out a wide bandwidth over time. Therefore, the tradeoff for high bandwidth with these waveforms is increased acquisition time and potentially severe blind zones, which may be unacceptable for various urban scenarios. For example, consider transmission of an LFM waveform with bandwidth of 1 GHz. With stretch processing, this waveform could be digitized at a sample rate of 100 MHz, but only if the LFM pulse width is at least 10 times longer than the range swath of interest. Assuming even a relatively modest range swath of 20 m, the required pulse width works out to approximately 1.3 μs, which corresponds to a blind range of 200 m. In many urban applications, the scene of interest is much closer than 200 m and would be obscured by the radar's blind zone.

Another option is to employ instantaneously wideband waveforms such as wideband impulses or phase-coded waveforms, but Nyquist sampling of these waveforms requires analog-to-digital (A/D) conversion at the full waveform bandwidth. Unfortunately, current GHz rate ADCs are expensive and consume much more power than ADCs with sampling rates in the low 100s of MHz.

This chapter presents a compressed sensing (CS) approach to easing the radar system issues described above for wideband radar in urban environments. CS techniques offer the potential for achieving high bandwidth in pulsed radar systems while sampling at sub-Nyquist rates. The signal incident on the compressive radar receiver contains the information inherent in a wideband radar signal, and CS principles indicate that this information can be retained through sub-Nyquist sampling if the signal of interest has a sparse representation. However, we take a slightly different approach to both the choice of the compressive measurement kernels and to the subsequent signal processing. We inject prior knowledge into the problem via training data for a set of urban objects. Some objects are ordinary objects such as a chair while others are potential contraband such as assault rifles. The desired high-range resolution (HRR) profiles of these targets are compressible rather than strictly sparse. We use training data in the form of HRR templates at different orientation angles to learn a Gaussian mixture approximation to the probability density function (pdf) of the signals of interest. We also present a method of using this Gaussian mixture density to optimize the CS measurement kernels and to enable minimum mean-squared error (MMSE) inference of the HRR profiles from the compressively sampled data. The MMSE estimator can be expressed in closed form for a Gaussian mixture pdf, and all results presented in this chapter are generated through MMSE estimation rather than sparse reconstruction methods that are more closely associated with CS. Therefore, this chapter focuses on sub-Nyquist methods of acquiring data and pairing those methods with prior pdfs that model compressible signals as a mixture of low-rank Gaussian components.

In this chapter, we first describe the general structure of the compressive receiver that is assumed for this work, including relevant constraints on the allowable measurement kernels. We describe the assumed signal and target models and fit a Gaussian mixture pdf to a training library of HRR profiles obtained via electromagnetic modeling. Next, a method of optimizing the receiver's measurement kernel under a mutual information metric is described and applied to the Gaussian mixture pdf. Finally, the HRR imaging performance of optimized kernels is quantified and compared with random measurement kernels and with Nyquist-sampled narrowband waveforms.

6.2 Sub-Nyquist Sampling Implementations, Models, and Constraints

The sub-Nyquist sampling structure used in this chapter is the basic building block of several sub-Nyquist structures presented in the literature [2,3]. Generally speaking, these structures multiply a noise-corrupted incoming analog signal $f(t)$ with another analog signal $\phi(t)$ that serves as the measurement kernel, as shown in Figure 6.1. The multiplication imprints a time-varying modulation on the desired signal that creates multiple overlapping shifts of the signal in the frequency domain. After multiplication, the resulting signal is passed through an integrator or lowpass filter that completes the projection operation before being sampled by a reduced-rate ADC.

When the measurement kernel is a pseudo-random binary sequence, this structure is often referred to as a random demodulator [2]. When the desired signal is split into multiple branches and the sampling structure of Figure 6.1 is implemented in each branch with a different periodic measurement kernel, the structure has been referred to as the modulated wideband converter [3]. The signal can even be split into multiple branches and multiplied with different subcarrier tones in each branch, in which case the receiver acquires data over a subset of narrow bands within the full signal bandwidth [4].

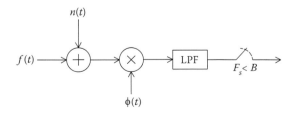

FIGURE 6.1
Single-branch sub-Nyquist sampling structure.

6.2.1 Implementation of Sub-Nyquist Sampling

A more specific implementation diagram is shown in Figure 6.2, where the measurement kernel is assumed to be arbitrary and generated with a digital-to-analog converter (DAC), which is controlled by an field programmable gate array (FPGA). The FPGA performs data handling for the ADC, computes the sensing kernel optimization that will be described later, loads the kernel into the DAC, and may perform the imaging computations. The up- and down-conversion stages as depicted have a flexible interpretation as being either an RF-to-intermediate frequency (IF) stage or as an RF to baseband stage. Furthermore, the receiver might have a second structure operating on a quadrature-demodulated copy of the signal or in general could have multiple branches operating in parallel [5]. Because the measurement kernel has bandwidth similar to the signal being acquired, the analog multiplier must be capable of handling two wideband signals. A simple microwave mixer may not suffice because mixers often have an oscillator port designed for a narrowband tone at a specific power level. Furthermore, the output of the analog multiplier will have bandwidth higher than either of the two input signals.

As an example of the bandwidths and some system considerations involved, consider a scenario where the output of the down-conversion stage is an IF signal centered at 1.5 GHz with bandwidth of 1 GHz (for hardware purposes, the IF center frequency must be larger than the signal bandwidth, so it is not feasible to directly shift the IF signal to the interval from 0 to 1 GHz). Let the frequency spectrum of this IF signal be as depicted in Figure 6.3a. In general, the measurement kernel will have some continuous spectrum, but for a simplified discussion here we assume that the measurement kernel consists of three tones as shown in Figure 6.3b. When the signal and measurement kernel are multiplied in the time domain, their

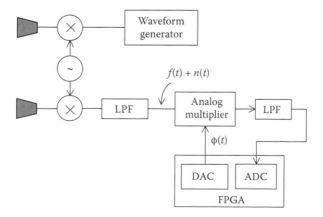

FIGURE 6.2
Implementation of a compressive RF receiver with arbitrary measurement kernel.

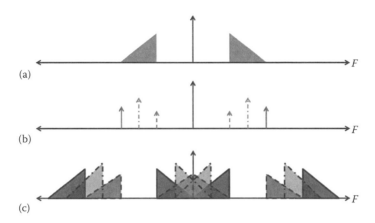

FIGURE 6.3
(See color insert.) Frequency-domain convolution of signal and measurement kernel. (a) Frequency spectrum of original signal, (b) frequency spectrum of the measurement kernel, and (c) frequency spectrum of the time-domain product of the signal and measurement kernel.

frequency spectra are convolved. From convolution, each of the tones in the measurement kernel produces a shifted, scaled version of the desired signal as shown in Figure 6.3c. The replicas in Figure 6.3c are color-coded and outlined with the same line type as the corresponding tone that produced the replica from Figure 6.3b. The replicas are also scaled according to the relative strength of that tone. After time-domain multiplication, the resulting baseband signal has a (two-sided) bandwidth of 2 GHz, which is the sum of the individual bandwidths of the signal and measurement kernel. Therefore, we see that the analog multiplier component in the receiver must be capable of accepting two arbitrary, wideband signals and producing an even wider bandwidth output signal.

After multiplication, the resulting signal is passed through a lowpass filter and sampled. For sampling at sub-Nyquist rates, the lowpass filter will have a cutoff frequency much lower than the combined baseband spectrum shown in Figure 6.3c. From the example numbers presented earlier, the original signal had a bandwidth of 1 GHz while the baseband portion of the postmultiplication signal has a bandwidth of 2 GHz (spanning from -1 to $+1$ GHz). If the intended sampling rate is, say, 100 MHz, then the lowpass filter should remove all frequencies above 50 MHz. Not only will most of the signal energy be removed in the process, but one can see that different pieces of the original spectrum will remain, depending on their shift and scale from the convolution process. The goal is that information from all parts of the original spectrum should be retained in the form of a carefully encoded aliasing behavior. Therefore, the measurement kernel bandwidth should be approximately the same as the original signal bandwidth, otherwise only a small part of the original bandwidth will be retained. For example, if the measurement kernel were to consist of only the center tone in Figure 6.3b, then only a small slice from the center of the original spectrum would be captured. As shown, the highest frequency tone is responsible for shifting the highest frequencies of the original signal into the sampling bandwidth while the lowest frequency tone shifts the lowest frequencies of the original signal into the sampling bandwidth. Expanding this understanding to an arbitrary kernel and multiple samples, we observe that each data sample will be the result of a unique encoding (i.e., unique superposition of spectral shifts and scales) of the original spectrum within the sampling bandwidth, such that sparsity or other prior structural information can be used to effectively reconstruct the original signal.

6.2.2 Power and Cost Benefits

The compressive receiver structure shown in Figure 6.2 introduces additional hardware operations that are unnecessary in a traditional receiver design. Therefore, it is important to consider whether the compressive structure produces system benefits compared with implementing a high-rate Nyquist receiver. The final conclusion to this analysis depends on specific operating frequencies and bandwidths, required effective number of bits in the ADC, available power, and the available cost budget. Initial calculations show that, in general, it may be possible to save significant power and cost with the compressive implementation. These savings are due to the fact that high-bandwidth DACs are less expensive and consume less power than ADCs of the same bandwidth. For example, consider a radar system with instantaneous bandwidth of 1 GHz and assume an intermediate center frequency of 500 MHz such that the signal spectrum spans the interval from 0 to 1 GHz. (As mentioned earlier, practical hardware factors make it difficult to center the signal's spectrum in this way, but the assumption makes for a simpler analysis and favors the traditional, noncompressive system.) Direct sampling of this signal requires sampling at a rate of at least 2 Gigasamples per second (GSPS). A compressive receiver requires a DAC, also operating at 2 GSPS, an analog multiplier, and a reduced rate A/D converter. Assuming a compression ratio (CR) of 10, the compressive receiver's ADC would operate at 200 Megasamples per second (MSPS). Having used commercially available components at the time of writing, we estimate that the extra DAC of the compressive receiver would add about 1 W to power requirements and additional cost of about $100. The analog multiplier is estimated to add about 0.7 W to the power requirements and additional costs of less than $50. On the other hand, a 200-MSPS, 12-bit ADC reduces power consumption by nearly 5 W and cost by nearly $2000 compared with the power and expense of a 2-GSPS, 12-bit ADC. Factored together, the compressive system could save as much as 3 W in power consumption and over $1500 in costs. Although we have not considered data throughput or real-time processing requirements, this analysis demonstrates the potential benefit of sub-Nyquist sampling.

Moreover, as ADC technology improves, the absolute numbers of the analysis will change, but we can expect the benefit to persist for wideband radar systems operating on the "knee" of the performance/cost curve of current ADC technology. As long as DACs can produce wideband signals for lower cost and power than an ADC operating at the same rate, CS opens a new and interesting degree of freedom in the system design space.

6.2.3 Measurement Model with Preprojection Additive Noise

The sub-Nyquist measurement model used in this chapter can be derived from Figure 6.1. Let a finite-duration, approximately band-limited signal be observed by the receiving antenna of a radar receiver. For the purposes of mathematical derivations and simulations presented in this chapter, we assume a complex baseband model such that the frequency spectrum of the information-bearing signal $f(t)$ is essentially contained within the interval $-B/2 < F < B/2$. As soon as the signal reaches the antenna terminals, transmission line, or other electronic component, the signal will be corrupted by additive receiver noise. In addition, the noise term $n(t)$ could also include external interference contributions such as reflections from ground clutter or external jamming. We model the noise as a zero-mean, complex-valued, wide-sense stationary Gaussian random process with autocorrelation function $C_{nn}(\Delta t)$.

The noise-corrupted signal is multiplied with the measurement kernel $\phi(t)$ such that the product, $r(t) = \phi(t)\left[f(t) + n(t)\right]$, is applied to the input of the lowpass filter. Denoting the filter's impulse response as $h(t)$, the filter output is

$$y(t) = \int \phi(\gamma)\left[f(\gamma) + n(\gamma)\right]h(t - \gamma)d\gamma. \tag{6.1}$$

Next, the output of the lowpass filter is sampled by an ADC with sampling interval $T_s = 1/F_s$. If M data samples are captured starting at time $t = T_i$, then the mth sample time is $t_m = T_i + (m - 1)T_s$ for $0 \leq m \leq M - 1$. Denoting the mth sample as $y_m = y(t_m)$, we have

$$y_m = y(t_m) = \int \phi(\gamma)\left[f(\gamma) + n(\gamma)\right]h(T_i + (m - 1)T_s - \gamma)\,d\gamma. \tag{6.2}$$

There are a number of interesting observations that can be made regarding the measurement model described in (6.2). The first observation derives from the fact that the lowpass filter is causal and has a finite-duration impulse response. Letting the impulse response time duration be T_h and assuming that the impulse response begins at $t = 0$, the limits of integration can be expressed as

$$y_m = y(t_m) = \int_{t_m - T_h}^{t_m} \phi(\gamma)\left[f(\gamma) + n(\gamma)\right]h(T_i + (m-1)T_s - \gamma)\,d\gamma. \qquad (6.3)$$

The consequence of (6.3) is that the mth data sample is a result of only the most recent T_h seconds of the analog multiplier output, not the multiplier output over the full duration of the input signal. The impulse response introduces a sliding window effect on the product of the noise signal and the measurement kernel due to the convolution operation. If the impulse response duration T_h is greater than the sampling interval T_s, then part of the contribution to adjacent samples will be identical. If $T_h \leq T_s$, then every data sample comes from a unique, nonoverlapping interval of the input signal.

The second observation is that the relative scaling of the measurement kernel $\phi(t)$ has no impact on signal-to-noise ratio (SNR). Because noise is added prior to multiplication with the measurement kernel, the kernel's overall amplitude affects both the desired signal and the additive noise in equal measure. Third, there is nothing to be gained by repeated or correlated measurement kernels. In some physical applications of CS where fields or signals can be manipulated prior to carrying those signals on a transmission line, the correct model is for noise to be added after signal compression. In this case, it can be beneficial to repeat the same measurement multiple times because repeated measurements of the signal with independent additive noise results in an effective increase in SNR. Unfortunately, this option is not available in the preprojection noise model because repeated measurements would produce the same value of both the signal and noise contributions, yielding an identical measurement. Consequently, measurement kernels corresponding to different data samples should be orthogonal in order to maximize their usefulness. Defining the mth measurement kernel as

$$\phi_m(t) = \begin{cases} \phi(t) & t_m - T_h \leq t \leq t_m \\ 0 & \text{else} \end{cases}, \qquad (6.4)$$

the mth measurement can be expressed as

$$y_m = y(t_m) = \int_{-\infty}^{\infty} \phi_m(\gamma)\left[f(\gamma) + n(\gamma)\right]h(T_i + (m-1)T_s - \gamma)\,d\gamma. \qquad (6.5)$$

For orthogonal kernels, we require

$$\int_{-\infty}^{\infty} \phi_m(t)\phi_l^*(t)\,dt = 0 \qquad (6.6)$$

for $m \neq l$. When $T_h \leq T_s$, different measurement kernels are nonoverlapping in time and satisfaction of (6.6) is guaranteed. When $T_h > T_s$, the kernels are overlapping and identical in some nonzero time interval, and (6.6) cannot be met. Therefore, the lowpass filter should be chosen such that the duration of its impulse response is less than the sub-Nyquist sampling interval. Finally, combining these two observations regarding the relative scale and orthogonal nature of the measurement kernels, we require the measurement kernels to be orthonormal, such that

$$\int_{-\infty}^{\infty} \phi_m(t)\phi_l^*(t)dt = \begin{cases} 1 & m = l \\ 0 & m \neq l \end{cases}. \tag{6.7}$$

The advantage to normalizing the kernels in this way is that if the additive noise is spectrally white such that $C_{nn}(\Delta t) = \sigma_n^2 \delta(\Delta t)$, then the output noise samples will also be uncorrelated with the same average noise power as the input random process. Although orthonormal representation is convenient for developing the mathematics of sensing kernel optimization, in practice the actual scaling of measurement kernels will depend on the dynamic range and optimum operating parameters of the hardware used to implement the compression.

6.2.4 Matrix–Vector Measurement Model

For ease of representation and mathematical manipulation, the data samples are stacked into an $M \times 1$ data vector

$$\mathbf{y} = \begin{bmatrix} y_1 & y_2 & \cdots & y_M \end{bmatrix}^\dagger, \tag{6.8}$$

where $[\cdot]^\dagger$ denotes a matrix transpose operation. In addition, the integral operator of (6.5) can be approximated as a vector dot product between discrete representations of the signal and measurement kernel. For purposes of matrix–vector notation, the mth measurement kernel in vector form is defined to incorporate the effect of the lowpass filter. Therefore, the mth measurement kernel is a row vector $\boldsymbol{\phi}_m$ obtained through a discrete representation of $\phi_m(t)h(T_i + (m-1)T_s - t)$. Likewise, the signal and noise are represented as vectors through discrete representations of $f(t)$ and $n(t)$ to obtain $\mathbf{f}$ and $\mathbf{n}$, respectively. The mth measurement is then

$$y_m = \boldsymbol{\phi}_m(\mathbf{f} + \mathbf{n}). \tag{6.9}$$

In order for (6.9) to be an accurate approximation to the continuous-time expression in (6.5), the measurement kernel, signal, and noise should be represented at a rate significantly higher than the Nyquist sampling frequency. Moreover, since the bandwidth of the product $\phi_m(t)f(t)$ will be higher than the bandwidth of either the kernel or signal in isolation, the sampled representations should be at a rate sufficient to avoid aliasing of the sampled product signal. We define the length of the vector representations of the measurement kernel, signal, and noise to be Q samples, such that ϕ_m is $1 \times Q$, while $\mathbf{f}$ and $\mathbf{n}$ are $Q \times 1$.

Finally, the full measurement vector is represented as

$$\mathbf{y} = \boldsymbol{\Phi}\,(\mathbf{f} + \mathbf{n}), \tag{6.10}$$

where $\boldsymbol{\Phi}$ is an $M \times Q$ matrix according to

$$\boldsymbol{\Phi} = \begin{bmatrix} \phi_1 \\ \phi_2 \\ \vdots \\ \phi_M \end{bmatrix}. \tag{6.11}$$

The orthonormal kernel requirement defined in (6.7) is now manifested as orthonormal rows of the sensing matrix $\boldsymbol{\Phi}$, and the sliding window structure of (6.3) means that $\boldsymbol{\Phi}$ will have a block-diagonal-like structure. This block-diagonal structure is defined by the relative values of T_h and T_s. An example of how the sensing matrix might appear for $T_h = T_s$ is shown in Figure 6.4.

Another representation equivalent to (6.10) is

$$\mathbf{y} = \boldsymbol{\Phi}\mathbf{f} + \boldsymbol{\Phi}\mathbf{n} = \boldsymbol{\Phi}\mathbf{f} + \tilde{\mathbf{n}}. \tag{6.12}$$

Defining the covariance matrix of the input noise vector $\mathbf{n}$ as $\mathbf{C}_{nn} = \mathrm{E}\,[\mathbf{nn}^*]$ (the matrix version of the autocorrelation function, $C_{nn}\,(\Delta t)$, of the noise

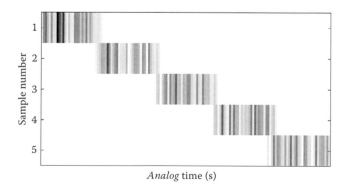

FIGURE 6.4
Example sensing matrix with temporal sampling structure consistent with hardware implementation.

random process), the covariance matrix of $\tilde{\mathbf{n}}$ is

$$\mathbf{C}_{\tilde{n}\tilde{n}} = E\left[\tilde{\mathbf{n}}\tilde{\mathbf{n}}^*\right] = \boldsymbol{\Phi}\mathbf{C}_{nn}\boldsymbol{\Phi}^*. \tag{6.13}$$

where E[.] denotes statistical expectation.

If the input noise is white, then $\mathbf{C}_{nn} = \sigma_n^2 \mathbf{I}_Q$ and due to the orthonormal structure of the sensing matrix, $\mathbf{C}_{\tilde{n}\tilde{n}} = \sigma_n^2 \mathbf{I}_M$. The model in (6.12) appears to be consistent with the postprojection noise model typically used in the CS literature; however, this interpretation can be misleading because the statistics of the postprojection noise vector $\tilde{\mathbf{n}}$ depend on both the input noise covariance matrix and on the structure of the sensing matrix. If one begins directly with the model in (6.12), it is easy to arrive at the incorrect conclusion that the noise statistics are independent of the sensing matrix. Even in the special case of white noise and orthonormal sensing matrix where both the preprojection and postprojection noise covariance matrices are equally scaled identity matrices, the model in (6.12) can be misleading because it appears as though SNR can be increased by amplifying the sensing matrix. On the other hand, (6.13) clearly shows that amplifying the sensing matrix will produce a proportional increase in the noise.

6.3 Radar Target and Received Signal Models

The preceding section of this chapter described mathematical models for compressive sampling of a noisy fast-time signal. In this section, we describe the radar waveform, its interaction with an object that reflects fields back to the receiver, and representations that are compatible with the sampling model described in the previous section.

6.3.1 Linear Target Model

Let the radar transmit a waveform with baseband representation $\psi(t)$. The waveform transmission begins at time $t=0$ and ends at time $t=T_p$ for a total radar pulse width of T_p seconds. The waveform propagates to an object, interacts with that target, and is reflected back toward the radar. The backscattered signal is modeled as the output of a linear system where the target has an impulse response with baseband representation $x(t)$. It is assumed that $x(t)$ is referenced to zero propagation delay; therefore, if the nearest edge of the target is R meters away, the target impulse response becomes $\alpha x(t - \tau)$ where $\tau = 2R/c$, α is a complex-valued global scaling factor that accounts for antenna gain, propagation loss, and phase rotation of the carrier frequency, and c is the velocity of the propagating wave. The signal that arrives at the radar is then

$$f(t) = \psi(t) * \alpha x (t - \tau)$$

$$= \int_{t-T_p}^{t} \alpha x (\gamma - \tau) \psi (t - \gamma) d\gamma, \tag{6.14}$$

where the limits of integration have been derived from the waveform's finite duration. The linear target model describes the received signal as a superposition of delayed transmit waveforms—that is, the waveform that is delayed by γ seconds is scaled by a target- and range-dependent scale factor $\alpha x(\gamma - \tau)$. Superposition of these delayed, scaled waveforms over all delays γ produces the total received waveform.

Approximating the integral in (6.14) with a discrete summation yields

$$f(t) \approx \alpha T_0 \sum_{p=0}^{P-1} x (pT_0 - \tau) \psi (t - pT_0), \tag{6.15}$$

where

T_0 is the width of the bins in the discrete approximation to the integral
P is the number of samples needed to represent the full duration of the waveform at sampling interval T_0

Recalling that the matrix–vector measurement model requires a discrete representation for $f(t)$, we set the sampling grid for $f(t)$ equal to the bin locations for the discrete approximation to (6.15). Let the qth sample of $f(t)$ occur at time $t_q = qT_0$ for $0 \leq q \leq Q - 1$, then substituting qT_0 into (6.15) yields

$$f (t_q = qT_0) \approx \alpha T_0 \sum_{p=0}^{P-1} x(pT_0 - \tau) \psi(qT_0 - pT_0)$$

$$= \alpha T_0 \sum_{p=0}^{P-1} x(pT_0 - \tau) \psi((q - p) T_0). \tag{6.16}$$

Exploring the initial sample values of $f(t)$, we see that

$$f(t_0) = \alpha T_0 x (-\tau) \psi(0)$$
$$f(t_1) = \alpha T_0 (x (-\tau) \psi(T_0) + x (T_0 - \tau) \psi (0)), \tag{6.17}$$

and so on. In other words, the sampled version of $f(t)$ is obtained through a discrete convolution of the sampled transmit waveform and the sampled target impulse response. Allowing the length-D sampled version of the delayed target impulse response to be $\mathbf{x}_\tau$, the discrete convolution can be expressed as

$$\mathbf{f} = \mathbf{\Psi}\mathbf{x}_\tau, \tag{6.18}$$

where the circulant waveform matrix is

$$\mathbf{\Psi} = \begin{bmatrix} \psi(0) & 0 & \cdots & 0 \\ \psi(T_0) & \psi(0) & \ddots & \vdots \\ \vdots & \psi(T_0) & \ddots & 0 \\ \psi((P-1)T_0) & \vdots & \ddots & \psi(0) \\ 0 & \psi((P-1)T_0) & \ddots & \psi(T_0) \\ \vdots & 0 & \ddots & \vdots \\ 0 & 0 & \cdots & \psi((P-1)T_0) \end{bmatrix}. \tag{6.19}$$

Substituting (6.18) into (6.10), the combined radar, target, and compressive measurement linear model yields

$$\mathbf{y} = \mathbf{\Phi}(\mathbf{\Psi}\mathbf{x}_\tau + \mathbf{n}). \tag{6.20}$$

The linear model is the basis for matched filtering and correlation-based radar imaging and facilitates the interpretation of an HRR profile as a map of scattering strength versus time delay for varying range bins on the target.

The discretized target impulse response $\mathbf{x}_\tau$ is the signal that we desire to recover in the HRR imaging application. We can think of the received signal $\mathbf{f}$ as being represented in the *delayed waveform basis*, where the overcomplete basis is a set of delayed transmit waveforms defined on a finely spaced grid of relative delays (with spacing T_0). Because this fine grid comes from the need to approximate analog signals at sampling rates well above Nyquist, the grid spacing is much smaller than the canonical $1/B$ resolution limit. There may be another representation basis in which the target has a more sparse representation; however, an accurate estimate of the HRR profile $\mathbf{x}_\tau$ is the desired output of the system. Furthermore, electromagnetic modeling of various targets and varying orientation provides a direct library of training data for $\mathbf{x}_\tau$, enabling both the definition of prior distributions for $\mathbf{x}_\tau$ and Bayesian MMSE reconstruction. Finally, the use of fine grid spacing will introduce strong mutual coherence between the columns of $\mathbf{\Psi}$; however, this factor is not critical because we are using closed-form MMSE reconstruction with a prior pdf rather than sparse reconstruction where coherence factors are important.

6.3.2 Compression Ratio

The input signal $f(t)$ was defined to have essential bandwidth B and a finite duration. If the duration is defined as T_f, then the number of Nyquist samples required to represent $f(t)$ is $L \approx BT_f$, also known as the essential degrees of freedom of the signal [6]. Compression ratio (CR) is defined in this chapter as the ratio of the signal's essential degrees of freedom to the length of the measurement vector $\mathbf{y}$; hence, $CR = L/M$. For sub-Nyquist sampling of radar pulses reflected from a target, the time duration T_f of the received signal comprises not only the time duration of the target but must also include the time it takes for the end of the pulse to return from the far end of the target. Therefore, the time duration of the received signal is $T_f = T_p + T_x$, which results in different sample lengths M for the same target and same CR, depending on the duration of the transmit pulse. This effect will be outlined in more detail when describing the simulation parameters for results presented later in the chapter.

It is important to note that for an accurate representation of the compression process, the length (defined earlier as Q) of the vector representation of the analog signals should be much larger than the essential degrees of freedom of the input signal, the measurement kernel, and their product. Therefore, accurate matrix–vector representation requires that $Q \gg L$.

6.3.3 Signal-to-Noise Ratio

A disadvantage of sub-Nyquist sampling and radiofrequency compressive sensing in general is a reduction in SNR compared with a fully sampled system. Depending on application and method of compression, the SNR penalty can be viewed as either a loss in collected signal energy [7] due to fewer collected samples or as a noise folding effect [8]. In general, the average SNR loss is equal to the CR of the system. It will be shown that at high SNR, the gains in resolution obtained by transmitting additional bandwidth can outweigh the SNR loss induced by subsequent sub-Nyquist sampling.

6.4 Information-Based Measurement Kernel Optimization

The typical kernel used for CS measurement is a random kernel due to the high likelihood of being able to reconstruct the signal [9] for any sparsifying basis. Although random kernels are flexible and robust in this way, they do not exploit any prior information about the signals being captured other than the fundamental assumption that they are sparse in some basis. In contrast, we consider the scenario where stronger prior information about signals of interest is available and can be used to design the measurement kernels for improved performance with respect to an exploitation task. Because the task considered here is reconstruction of the target impulse response, and because of the inherent relationship between minimum mean-squared error and Shannon information [10], we present an information-based approach to measurement kernel design.

6.4.1 Task-Specific Information

Our goal is to reconstruct the HRR profile of an object and to improve reconstruction performance through optimized compressive measurements. Once reconstructed, the profile can be used to recognize the object or to tag the object for additional scrutiny and observation. Our optimization metric for the measurement kernels is called task-specific information (TSI) [11,12]. TSI is defined as the mutual information [13] between a task-specific source variable and the final measurements. In a detection problem, for example, the source variable is a binary-valued indicator variable defining the presence or absence of a target. Because the maximum entropy of a binary random variable is one bit (when the two hypotheses are equally likely), the maximum task-specific mutual information that can be captured by the measurements for any detection task is also one bit. For a parameter estimation problem, the source variable is the parameter being estimated, which has entropy and maximum TSI determined by the parameter's prior distribution. In an HRR reconstruction problem, the source variable is the profile being reconstructed, x_τ. Thus, the source variable is a continuous-valued random vector with prior pdf $p(x_\tau)$, and TSI is the Shannon information between the measurements, y, and the source variable, x_τ. In terms of entropy values, TSI is defined as

$$I\left(y; x_\tau\right) = h(y) - h\left(y|x_\tau\right). \tag{6.21}$$

Because the measurements y depend on the sensing matrix Φ, the sensing kernel design problem is to design Φ to maximize (6.21), subject to constraints on physical implementation. These physical constraints include orthonormal restrictions as described earlier. In addition, some receiver architectures will require enforcement of the sequential sampling structure as depicted in Figure 6.4.

In the following discussion, we describe our approach to sensing kernel design based on Gaussian mixture distributions. The objective function in (6.21) obviously depends on both the prior distribution of x_τ and the measurement model of (6.20). One method for designing Φ is to implement a brute-force global search. In this approach, (6.21) is evaluated for every allowable sensing matrix, and the sensing matrix with the largest TSI is selected. However, closed-form evaluation of (6.21) is not possible except in a few special cases, thereby requiring (6.21) to be evaluated numerically for every possible Φ. The large dimensionality of Φ makes this approach impractical. A gradient-based search process would allow for optimization of Φ

without trying every possible alternative, but if the gradient calculation must be numerically evaluated, the same dimensionality issue persists. Therefore, we will seek a gradient-based optimization strategy, but one that is capable of calculating the gradient without relying on any numerical evaluations of expected values in high-dimensional spaces.

To enable our approach, we use two approximations. First, we use a Gaussian mixture approximation to the pdf of $\mathbf{x}_\tau$. Gaussian mixtures are useful for several reasons. First, they can be used to approximate a wide variety of other distributions, including distributions for sparse or compressible signals. Second, the accuracy of the approximation depends on the number of components in the mixture; therefore, it is possible to trade between accuracy and computational complexity. Finally, a closed-form MMSE estimator exists for linear measurements of a source signal described by a Gaussian mixture. But even with the Gaussian mixture approximation, it is still not possible to obtain a closed-form expression for the gradient of TSI with respect to the sensing matrix. Therefore, we will also be forced to approximate the entropy expression by performing a Taylor expansion of the logarithm of the Gaussian mixture distribution.

6.4.2 Approximate Gradient of TSI for a Gaussian Mixture Model

Throughout the rest of the chapter, we drop the explicit reminder that the target impulse response depends on target delay τ by using $\mathbf{x}$ rather than $\mathbf{x}_\tau$. In practice, the target impulse response depends on many parameters including the target's delay and its pose or orientation with respect to the radar. These unknown parameters will be used later when we map the training library of target impulse responses to the Gaussian mixture distribution used for sensing kernel optimization. The derivation assumes a complex baseband signal model.

Let the pdf of $\mathbf{x}$ be modeled by a K-component mixture distribution given by

$$p(\mathbf{x}) = \sum_{k=1}^{K} P_k p^{(k)}(\mathbf{x}), \tag{6.22}$$

where the kth component in the mixture is a zero-mean complex Gaussian random vector with pdf

$$p^{(k)}(\mathbf{x}) = \frac{1}{\pi^D \left| \mathbf{C}_{xx}^{(k)} \right|} \exp\left(-\mathbf{x}^* \left(\mathbf{C}_{xx}^{(k)} \right)^{-1} \mathbf{x} \right) = \mathcal{CN}\left(0, \mathbf{C}_{xx}^{(k)} \right) \tag{6.23}$$

and $\mathbf{x}^*$ denotes the conjugate transpose of $\mathbf{x}$. The zero-mean assumption is valid for the radar imaging application because the global phase of the target is typically unknown. Small uncertainty in the target's range—on the order of a wavelength at the carrier frequency—introduce full 2π uncertainty in the global propagation phase of the carrier. For a complex random variable with uniformly distributed phase, the mean is zero. For the zero-mean case, the Gaussian mixture is defined by K covariance matrices (the $\mathbf{C}_{xx}^{(k)}$s) and K probability weights. For a valid pdf, we must have

$$\sum_{k=1}^{K} P_k = 1. \tag{6.24}$$

Let the noise be zero-mean, complex additive Gaussian noise with pdf

$$p\left(\mathbf{n}\right) = \mathcal{CN}\left(0, \mathbf{C}_{nn}\right). \tag{6.25}$$

As seen in (6.20), the model that relates the signal $\mathbf{x}$ to the measurement $\mathbf{y}$ is linear. Given this linear relationship and the additive noise, the measurement is also described by a Gaussian mixture according to

$$p\left(\mathbf{y}\right) = \sum_{k=1}^{K} P_k p^{(k)}\left(\mathbf{y}\right), \tag{6.26}$$

where

$$p^{(k)}\left(\mathbf{y}\right) = \mathcal{CN}\left(0, \mathbf{C}_{yy}^{(k)}\right) \tag{6.27}$$

and

$$\mathbf{C}_{yy}^{(k)} = \mathbf{\Phi}\left(\mathbf{\Psi}\mathbf{C}_{xx}^{(k)}\mathbf{\Psi}^* + \mathbf{C}_{nn}\right)\mathbf{\Phi}^*. \tag{6.28}$$

The differential entropy of $\mathbf{y}$ is defined by

$$h(\mathbf{y}) = -\int p(\mathbf{y})\log p(\mathbf{y})\,d\mathbf{y}, \tag{6.29}$$

which can only be evaluated in closed form for a few cases (not including Gaussian mixtures). Therefore, we first perform a zero-order Taylor expansion of the logarithm. Because the component Gaussians that comprise the distribution are all zero mean, an appropriate expansion point is $\mathbf{y}_0 = \mathrm{E}\left[\mathbf{y}\right] = \mathbf{0}$, which yields

$$\log\left[\sum_{k=1}^{K} P_k p^{(k)}\left(\mathbf{y}\right)\right] \approx \log\left[\sum_{k=1}^{K} P_k p^{(k)}\left(\mathbf{y} = \mathbf{0}\right)\right]. \tag{6.30}$$

Substituting (6.26) and (6.30) into (6.29) yields

$$h\left(\mathbf{y}\right) = -\int \sum_{i=1}^{K} P_i p^{(i)}\left(\mathbf{y}\right) \log \left[\sum_{k=1}^{K} P_k p^{(k)}\left(\mathbf{y} = \mathbf{0}\right)\right] d\mathbf{y}$$

$$= -\sum_{i=1}^{K} P_i \int p^{(i)}\left(\mathbf{y}\right) \log \left[\sum_{k=1}^{K} P_k p^{(k)}\left(\mathbf{y} = \mathbf{0}\right)\right] d\mathbf{y}. \qquad (6.31)$$

The logarithm term in (6.31) has been evaluated at the expansion point of the Taylor series; therefore, it no longer has dependence on $\mathbf{y}$ and can be pulled out of the integration. Likewise, the logarithm term has no dependence on the summation index and can be pulled out of the summation. These steps yield

$$h\left(\mathbf{y}\right) \approx -\log \left[\sum_{k=1}^{K} P_k p^{(k)}\left(\mathbf{y} = \mathbf{0}\right)\right] \sum_{i=1}^{K} P_i \int p^{(i)}\left(\mathbf{y}\right) d\mathbf{y}$$

$$= -\log \left[\sum_{k=1}^{K} P_k p^{(k)}\left(\mathbf{y} = \mathbf{0}\right)\right] \sum_{i=1}^{K} P_i$$

$$= -\log \left[\sum_{k=1}^{K} P_k p^{(k)}\left(\mathbf{y} = \mathbf{0}\right)\right], \qquad (6.32)$$

Finally, noting that

$$p^{(k)}\left(\mathbf{y} = \mathbf{0}\right) = \frac{1}{\pi^M \left| \mathbf{C}_{yy}^{(k)} \right|}. \qquad (6.33)$$

The approximate expression for the entropy of $\mathbf{y}$ is

$$h(\mathbf{y}) \approx -\log \left[\sum_{k=1}^{K} \frac{P_k}{\pi^M \left| \mathbf{C}_{yy}^{(k)} \right|}\right]. \qquad (6.34)$$

The second term in the TSI expression of (6.21) is the entropy of $\mathbf{y}$ conditioned on $\mathbf{x}$. But given $\mathbf{x}$, the only remaining random term is the additive Gaussian noise with covariance matrix $\mathbf{C}_{nn}$. The entropy of a Gaussian random vector is known; hence, the second entropy term in (6.21) is

$$h(\mathbf{y}|\mathbf{x}) = h(\mathbf{\Phi}\mathbf{n}) = \log \left| \pi e \mathbf{\Phi} \mathbf{C}_{nn} \mathbf{\Phi}^* \right|. \qquad (6.35)$$

Combining (6.34) and (6.35) into an approximate TSI expression yields

$$I(\mathbf{y};\mathbf{x}) = h(\mathbf{y}) - h(\mathbf{y}|\mathbf{x})$$

$$\approx -\log\left[\sum_{k=1}^{K} \frac{P_k}{\pi^M \left|\mathbf{C}_{yy}^{(k)}\right|}\right] - \log\left|\pi e \boldsymbol{\Phi} \mathbf{C}_{nn} \boldsymbol{\Phi}^*\right|. \tag{6.36}$$

The next step in the optimization procedure is to compute the gradient of (6.36) with respect to the sensing matrix. The key point here is that the covariance matrices for the components of the mixture distribution describing $\mathbf{y}$ have dependence on $\boldsymbol{\Phi}$ according to (6.28). Substituting (6.28) into (6.36), applying the gradient operator with respect to $\boldsymbol{\Phi}$, and using properties for the derivative of a determinant found in [14] yields

$$\nabla_{\boldsymbol{\Phi}}\{I(\mathbf{y};\mathbf{x})\} = \nabla_{\boldsymbol{\Phi}}\left\{-\log\left[\sum_{k=1}^{K} \frac{P_k}{\pi^M \left|\boldsymbol{\Phi}\left(\boldsymbol{\Psi}\mathbf{C}_{xx}^{(k)}\boldsymbol{\Psi}^* + \mathbf{C}_{nn}\right)\boldsymbol{\Phi}^*\right|}\right]\right.$$

$$\left. - \log\left|\pi e \boldsymbol{\Phi} \mathbf{C}_{nn} \boldsymbol{\Phi}^*\right|\right\}$$

$$= -\frac{1}{\sum_{k=1}^{K} \frac{P_k}{\pi^M \left|\boldsymbol{\Phi}\left(\boldsymbol{\Psi}\mathbf{C}_{xx}^{(k)}\boldsymbol{\Psi}^* + \mathbf{C}_{nn}\right)\boldsymbol{\Phi}^*\right|}}$$

$$\times \nabla_{\boldsymbol{\Phi}}\left\{\sum_{k=1}^{K} P_k \pi^{-M} \left|\boldsymbol{\Phi}\left(\boldsymbol{\Psi}\mathbf{C}_{xx}^{(k)}\boldsymbol{\Psi}^* + \mathbf{C}_{nn}\right)\boldsymbol{\Phi}^*\right|^{-1}\right\}$$

$$- \left(\boldsymbol{\Phi}\mathbf{C}_{nn}\boldsymbol{\Phi}^*\right)^{-1}\boldsymbol{\Phi}\mathbf{C}_{nn}. \tag{6.37}$$

A final gradient of the determinant yields

$$\nabla_{\boldsymbol{\Phi}}\{I(\mathbf{y};\mathbf{x})\}$$

$$= \frac{\sum_{k=1}^{K} P_k \left|\boldsymbol{\Phi}\left(\boldsymbol{\Psi}\mathbf{C}_{xx}^{(k)}\boldsymbol{\Psi}^* + \mathbf{C}_{nn}\right)\boldsymbol{\Phi}^*\right|^{-1}\left(\boldsymbol{\Phi}\left(\boldsymbol{\Psi}\mathbf{C}_{xx}^{(k)}\boldsymbol{\Psi}^* + \mathbf{C}_{nn}\right)\boldsymbol{\Phi}^*\right)^{-1}\boldsymbol{\Phi}\left(\boldsymbol{\Psi}\mathbf{C}_{xx}^{(k)}\boldsymbol{\Psi}^* + \mathbf{C}_{nn}\right)^*}{\sum_{k=1}^{K} P_k \left|\boldsymbol{\Phi}\left(\boldsymbol{\Psi}\mathbf{C}_{xx}^{(k)}\boldsymbol{\Psi}^* + \mathbf{C}_{nn}\right)\boldsymbol{\Phi}^*\right|^{-1}}$$

$$- \left(\boldsymbol{\Phi}\mathbf{C}_{nn}\boldsymbol{\Phi}^*\right)^{-1}\boldsymbol{\Phi}\mathbf{C}_{nn}, \tag{6.38}$$

which is the final gradient approximation used in this work to search for an optimized sensing matrix.

In the special case where the noise is white, then $\mathbf{C}_{nn} = \sigma_n^2 \mathbf{I}$, and under the orthonormal rows assumption, the last term in (6.38) can be simplified to

$$\left(\mathbf{\Phi}\mathbf{C}_{nn}\mathbf{\Phi}^*\right)^{-1}\mathbf{\Phi}\mathbf{C}_{nn} = \frac{1}{\sigma_n^2}\left(\mathbf{\Phi}\mathbf{\Phi}^*\right)^{-1}\mathbf{\Phi}\sigma_n^2 = \mathbf{\Phi}. \tag{6.39}$$

The white noise scenario helps understand the role of the two terms in (6.38). The first term gives the gradient its structure and directs the search process in the right direction. However, a consequence of the preprojection noise model is that performance cannot be improved simply by scaling the sensing matrix. Therefore, the second term removes that part of the gradient that is in the same direction as $\mathbf{\Phi}$—in other words, the second term regularizes the gradient to prevent the gradient search process from seeking a solution that is simply a scaled version of the current sensing matrix.

6.4.3 Application to HRR Imaging

In the previous section, we derived a matrix gradient that will enable a search for the TSI-optimized sensing matrix when the input signal can be described by a Gaussian mixture. We now explain how the gradient can be used to improve performance in an urban radar imaging problem. Let there be a set of targets of interest for a given application. For the urban radar application, likely targets of interest include handheld weapons, explosives, and other contraband. For these targets, it is possible to model their time-domain responses to a wideband pulse using electromagnetic modeling software. If the wideband pulse used in the modeling has higher bandwidth than the transmit radar pulse, then the modeled time-domain response is effectively an impulse response for the target at the particular orientation defined in the simulation. The process can be repeated for different orientation angles in order to build a library of target impulse response templates. This library of templates is the same type of library sometimes used for automatic target recognition [15,16]; furthermore, the templates are sometimes averaged together in order to perform classification based on Gaussian approximations to the library [17]. In this chapter, the template library is used to train the Gaussian mixture prior, which is used for both the sensing kernel optimization and the MMSE-based reconstruction.

One approach to learning a Gaussian mixture earlier is to use expectation maximization [18] or other strategy to learning Gaussian mixture parameters directly from the full set of training data. In our case, however, we train the Gaussian mixture components by grouping together similar parameter values such as orientation angles and target range. Target impulse responses for similar orientation angles and ranges have similar structure and can be approximated with a rank-deficient Gaussian distribution. Consider the target impulse response $\mathbf{x}(\theta)$ where the dependence of the impulse response on the azimuth orientation parameter θ is explicitly noted. As modeled earlier in this chapter, $\mathbf{x}(\theta)$ is a length-D vector that defines a particular point in space for a particular value of θ. As θ varies, the impulse response changes and the tip of $\mathbf{x}(\theta)$ moves throughout D-dimensional space. For a single underlying parameter such as θ, the tip of $\mathbf{x}(\theta)$ traces out a one-dimensional nonlinear manifold in D-dimensional space. For two underlying parameters, the tip of $\mathbf{x}(\theta)$ would lie on a two-dimensional manifold in D-dimensional space, and so on. Our approach is to divide the continuous parameter space into a set of nonoverlapping, adjacent intervals. For example, if θ represents an azimuth pose angle, we define the kth sector as containing templates in the library corresponding to $\theta_{k-1} < \theta \le \theta_k$. The first Gaussian component represents the first target with orientation from θ_0 to θ_1, the second component from θ_1 to θ_2, and so on. In other words, different Gaussian components are assigned to represent adjacent pieces of the low-dimensional nonlinear manifold describing the target, and the Gaussian mixture model approach can be interpreted as a piecewise fit to the manifold [19].

Let $\mathcal{S}_{l,k}$ be the set of all D-dimensional target impulse response templates for the lth target type such that $\theta_{k-1} < \theta \le \theta_k$. A Gaussian component in the mixture is defined by calculating a covariance matrix from elements in the set. Therefore, the covariance matrix for the kth parameter interval of the lth target type is

$$\mathbf{C}_{xx}^{(l,k)} = \frac{1}{\mathcal{K}_l} \sum_{\mathbf{x}(\theta) \in \mathcal{S}_{l,k}} \mathbf{x}(\theta)\mathbf{x}^*(\theta), \tag{6.40}$$

where $\mathcal{K}_l$ is the cardinality of the set $\mathcal{S}_{l,k}$. In other words, the sample covariance matrix computed from all target templates within a parameter interval defines the Gaussian component approximating that piece of the target manifold. The overall carrier phase is still assumed unknown; therefore, the components in the mixture are zero mean and the Gaussian distribution for the kth parameter grouping of the lth target is $\mathcal{CN}\left(\mathbf{0}, \mathbf{C}_{xx}^{(l,k)}\right)$. A benefit of forming components according to locally grouped parameter values is that the mixture probabilities can be derived from prior probabilities on the individual targets and their parameterizations. In many cases, the mixture probabilities will all be equal, indicating noninformative prior information on the target type and its orientation, range, etc. But in other cases where prior information may be available from earlier reports and/or previous radar surveys, this prior information can be exploited by the compressive measurement design by mapping the prior information to mixture probabilities. This approach also admits potential adaptive sensing strategies where the process begins with noninformative priors, and gradually adapts the sensing kernel for subsequent measurements based on previously obtained data. Once the Gaussian components are calculated for each target, the components from all targets are combined into a single mixture distribution.

6.4.4 MMSE HRR Estimation

Given the Gaussian mixture prior distribution of the target impulse response **x**, known waveform matrix **Ψ**, known measurement matrix **Φ**, and compressive measurements **y**, it is possible to compute the MMSE estimate of **x** in closed form. From [20], the MMSE estimate is

$$\hat{\mathbf{x}} = \mathrm{E}\left[\mathbf{x}|\mathbf{y}\right] = \sum_{k=1}^{K} P_{k|y} \mathbf{C}_{xx}^{(k)} (\mathbf{\Phi}\mathbf{\Psi})^* \left(\mathbf{C}_{yy}^{(k)}\right)^{-1} \mathbf{y}, \qquad (6.41)$$

where

$$P_{k|y} = \frac{P_k p^{(k)}(\mathbf{y})}{p(\mathbf{y})} \qquad (6.42)$$

is the posterior probability of the *k*th mixture component after observing **y**. In practice, the denominator of (6.42) does not need to be computed because it is a scaling factor common to all posterior probabilities. Instead, the numerator of (6.42) is calculated for all mixture components, and then the posterior probabilities are all scaled by the same factor such that they sum to one. The zero-mean assumption for the Gaussian components has also been used to simplify (6.41).

6.5 Simulation Results

In this section, we use high-fidelity target models generated with numerical electromagnetic software to assess the performance benefit of our TSI-based approach to kernel optimization for urban target imaging. The high-fidelity target models are used to generate training data for learning the Gaussian mixture distributions necessary for kernel optimization and MMSE imaging. Monte Carlo trials are then performed for varying prior distributions on the target parameters and for various waveform and compression kernel combinations.

6.5.1 Training Data and Gaussian Mixture Calculation

We used the XFdtd software from Remcom, State College, PA to generate high-fidelity target impulse responses for several objects. These objects are either commonly found or of particular interest in urban applications: an AK-47 rifle, and M-16 rifle, a drawer cabinet, and an armchair. The results were generated based on publicly available CAD models for these objects.

The target templates were generated from the CAD models with a transmit pulse having 2 GHz bandwidth (i.e., FDTD timestep of 0.5 ns), and repeated at 0.1° increments in azimuth pose angle from 0° to 90°. Therefore, we obtained 901 templates per target. The resulting templates were $D = 25$ samples long, which corresponds to an impulse response duration of 12.5 ns, or a range swath of 1.875 m. The approximate width, length, and height of the four objects were

AK-47: 5 cm by 122 cm by 28 cm
M-16: 6 cm by 88 cm by 23 cm
Cabinet: 46 cm by 87 cm by 64 cm
Armchair: 56 cm by 58 cm by 83 cm

The objects were grouped into mixture components according to 1° intervals of the azimuth orientation parameter. For example, the 11 templates corresponding to azimuth angles from 0° to 1° were averaged to obtain the covariance matrix for the first Gaussian component of each target. The 11 templates corresponding to azimuth angles from 1° to 2° were averaged to obtain the second covariance matrix, and so on. Note that the templates at the endpoint of each angular sector were shared between components. Using the procedure, we obtained 90 components per target, for a total of $K = 360$ components in the full Gaussian mixture model. The covariance matrices for each component have dimension 25 by 25. In all simulated scenarios, the four objects were present with equal probability. However, the probability distribution of the object's orientation varied across different cases.

6.5.2 Waveforms and Compression Ratio

Not only do we compare optimized sensing kernels to random sensing kernels, but we also compare how these sensing kernels interact with different waveforms. Due to different modulation strategies, it is possible to generate waveforms with the same bandwidth but different pulse widths. Traditionally, modulated waveforms are used to increase the pulse width and energy of a transmit waveform without sacrificing bandwidth.

We compare the performance of sub-Nyquist sampling paired with four different waveforms. The first waveform is an unmodulated simple pulse with bandwidth of 500 MHz. The pulse width of this waveform is the reciprocal of the bandwidth, $T_p = 2$ ns. The total observation time at the receiver is the sum of the pulse width and the impulse response duration, which total 14.5 ns. At the Nyquist sample rate of 500 MHz, the number of samples is approximately eight. At a CR of five, the number of compressive samples is 1.6, which we round up to two.

The second waveform is a linear FM pulse with bandwidth of 500 MHz and duration of $T_p = 38$ ns. Therefore, the chirp rate is 13,150 MHz/us.

TABLE 6.1

Waveform Parameters Used in Performance Simulations

Waveform	Bandwidth (MHz)	Pulse Width (ns)	Samples at 500 MHz	Samples at 100 MHz
Wideband pulse	500	2	~8	~2
LFM pulse	500	38	25	5
Phased-coded pulse	500	38	25	5
Narrowband pulse	100	10	N/A	2

The total observation time is 50.5 ns, and the number of samples at the Nyquist rate of 500 MHz would be approximately 25. At a CR of five, the number of compressive samples is five. The third waveform is a phase-coded waveform with 500 MHz bandwidth and approximately the same duration as the LFM waveform. Therefore, the number of samples at Nyquist is, again, approximately 25 and the number of compressive samples is five.

Finally, we consider a simple, unmodulated waveform with bandwidth of only 100 MHz. We use this waveform as a reference in order to simulate sampling at the Nyquist rate of 100 MHz, which is the same sampling rate as we use to capture the 500 MHz waveforms at a CR of five. The performance of this "narrowband" waveform should be taken as a reference to determine the value of transmitting more bandwidth that necessitates sub-Nyquist sampling. If the compressive approaches do not outperform the narrowband baseline, then the value of Nyquist undersampling of higher-bandwidth waveforms becomes nonexistent. The narrowband waveform has pulse duration that is the reciprocal of its bandwidth, $T_p = 10$ ns, and the number of Nyquist samples for this baseline reference is two. The waveform parameters are summarized in Table 6.1.

6.5.3 Signal Examples

First, we show some qualitative results demonstrating the kernels and test signals as they progress through the system. In these examples, the baseband signals have been shifted to an intermediate frequency (IF) in order to plot real signals. Figure 6.5 shows a sample target impulse response taken from the M-16 rifle at zero degrees azimuth (viewed from the end of the barrel). The envelope of the impulse response carries information about the target's structure. Figure 6.6 shows the reflected signal due to the LFM waveform. Because the pulse width of the LFM waveform is much longer than the duration of the impulse response, the reflected signal is much longer than the original impulse response. The signal in Figure 6.6 is the result of multiple superimposed waveform reflections at various delays and scale factors according to the different scattering points on the target.

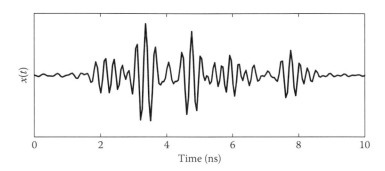

FIGURE 6.5
Target impulse response for the M-16 rifle viewed along its full length.

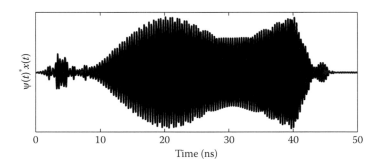

FIGURE 6.6
Reflected signal after the target interacts with the LFM waveform.

As seen in the sampling structure of Figure 6.1, the next step in the process is to multiply the received waveform by the measurement kernel. The result of this multiplication for a TSI-optimized measurement kernel is shown in Figure 6.7. At this point, one can easily see the multiple timescales that are present due to convolution of the two signals' frequency spectra. The high-frequency components still need to be removed by the lowpass filter in order to smooth the signal for final sampling.

Figure 6.8 shows the output of the final lowpass filter, which is modeled in the time domain with a rectangular impulse response. Therefore, it has high-frequency-domain sidelobes, and some residual high-frequency terms remain. In general, the impulse response of the hardware filter must be carefully characterized and incorporated into the CS mathematical model used for reconstruction or other processing. The locations of the five compressive samples are also indicated in Figure 6.8 with circles. For a compressive sample rate of 100 MHz, the final samples are spaced 10 ns apart.

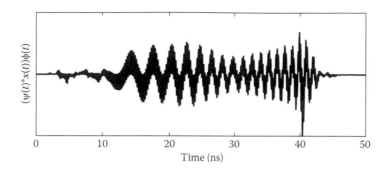

FIGURE 6.7
Product of the reflected signal and the measurement kernel (analog multiplier output).

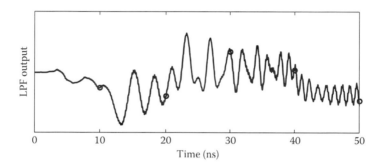

FIGURE 6.8
Lowpass filter output with locations of the five compressive samples indicated.

6.5.4 Quantitative Performance Results

Figures 6.9 through 6.12 show the results of our performance simulations comparing narrowband transmission with Nyquist sampling to wideband transmission with compressive sampling. We also compare random compression kernels to kernels optimized using the TSI/Gaussian mixture approach. The working assumption in performing these simulations is that the receiver's sampling rate is held constant at 100 MHz. We assume that sampling at a higher rate is unacceptable due to increased cost and power consumption. Because sampling rate is held constant, Nyquist versus sub-Nyquist sampling is controlled through the bandwidth of the transmit waveform. For Nyquist sampling, the transmit bandwidth cannot exceed the sampling rate; hence, the Nyquist reference is the 100 MHz unmodulated narrowband pulse. For sub-Nyquist sampling, the transmit bandwidth is increased to 500 MHz. Therefore, some form of compression must be implemented in the receiver, and we compare random compression to our optimized approach. As described earlier, there are three wideband transmit waveforms: an unmodulated pulse, an LFM pulse, and a phase-coded pulse.

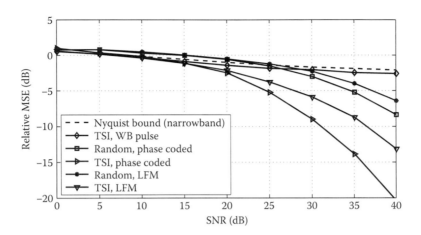

FIGURE 6.9
Reconstruction performance in free space without sequential sampling structure enforced and prior knowledge of object orientation between 0° and 90°.

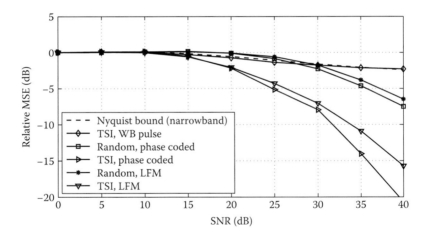

FIGURE 6.10
(See color insert.) Reconstruction performance in free space without sequential sampling structure enforced and prior knowledge of object orientation between 20° and 50°.

Performance results are generated through Monte Carlo trials. For each trial, one of the target objects is selected from an equi-probable distribution. Once the target type is selected, the target's azimuth orientation angle is randomly generated from a uniform distribution. There are two different cases for the uniform distribution representing different levels of prior knowledge. In the weakest case of prior knowledge, the target orientation is generated from a uniform distribution from 0° to 90°. In the stronger case, the distribution is uniform from 20° to 50°. Once the target orientation is generated

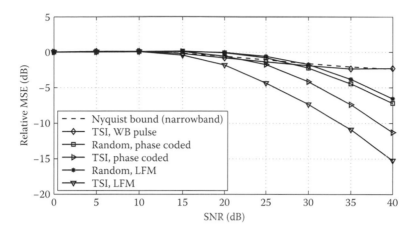

FIGURE 6.11
(See color insert.) Reconstruction performance in free space with prior knowledge of object orientation between 20° and 50°. The sequential sampling structure has now been enforced.

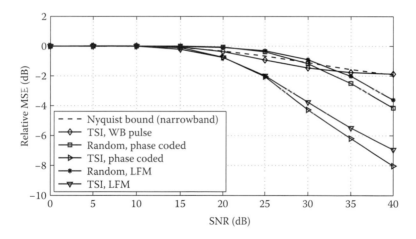

FIGURE 6.12
(See color insert.) Reconstruction performance for an object located behind and in close proximity to a wall with uncertain thickness and permittivity. Prior pdf on orientation window is 20°–50°.

for a particular trial, the truth impulse response is obtained by interpolating between the nearest two templates in the target template library.

After generating the truth target response, data are generated according to (6.20) for each waveform and for random versus optimized sensing kernels. For random compression, a new realization of the measurement kernel is generated for each Monte Carlo trial. For optimized compression, the kernel is calculated before the Monte Carlo simulation starts and held constant throughout the simulation. After obtaining the data **y**, the MMSE estimate of the target profile is calculated using (6.41), and the squared error is calculated and retained for later averaging. All MMSE performance results reported here are normalized to the energy of the target realization; therefore, the worst-case performance in the low SNR regime is a relative MMSE of 0 dB. Figure 6.9 shows the performance results quantified as relative mean-squared reconstruction error versus SNR. Noise power is normalized to one, and SNR is controlled via relative scaling of the waveform matrix. For Figure 6.9, available prior knowledge indicated a target orientation between 0° and 90°. The object was located in free space (objects behind a wall will be considered shortly), and the sequential sampling structure exemplified in Figure 6.4 has not yet been enforced.

As expected, the narrowband waveform with Nyquist sampling has insufficient resolution for resolving any features of the object, and reconstruction is poor even for high SNR. For the wideband waveforms with compressive sampling, reconstruction error improves at high SNR. Once SNR is sufficiently high, the resolution inherent in the wider transmit bandwidth can be exploited, even by a sub-Nyquist receiver. Some dependence on the type of transmit waveform is noted, but the biggest difference in the compressive results is that the exploitation of prior knowledge in the form of optimized measurement kernels provides at least 10 dB savings in transmit power compared with random kernels at the same reconstruction performance.

Figure 6.10 shows the results when prior knowledge is strengthened to an azimuth orientation between 20° and 50°, while Figure 6.11 shows performance with the same 20°–50° range, but sequential sampling structure applied. Sequential sampling structure is enforced via a masking operation during the gradient search for the optimized sensing matrix. The sequential sampling structure places stricter requirements on the structure of the sensing matrix; therefore, we should expect that performance will decrease. Indeed, we observe a performance decrease in comparing Figure 6.11 (with structure) with Figure 6.10 (without structure). The enforced structure has particularly affected the performance of the phase-coded waveform with optimized compression. However, even with the additional structure enforced, the optimized sensing kernel for both waveforms significantly outperforms random compression and the Nyquist-sampled narrowband waveform.

Finally, in Figure 6.12 we include a simulation where a wall has been included between the radar and the object being imaged. To model the wall, we allowed the unknown thickness to vary between 15 and 20 cm, and the wall's unknown relative permittivity to vary from 4 to 10. We then discretized the wall parameters on a finely spaced two-dimensional grid and computed an effective impulse response and transmission coefficients using theory for a dielectric slab [21]. The reflection impulse responses were used to compute an average interference covariance matrix to be included in the noise covariance matrix $\mathbf{C}_{nn}$ in the following manner. Defining $\mathbf{x}_w$ as the impulse response of the wall reflections, we computed

$$\mathbf{C}_{ww} = \mathrm{E}\left[\mathbf{x}_w \mathbf{x}_w^*\right]$$

as the covariance matrix of the expected wall impulse response where the expectation was performed by averaging over the wall parameters. The reflected signal due to the wall was modeled as $\mathbf{\Psi}\mathbf{x}_w$; hence, the additional interference covariance term due to the presence of the wall was

$$\mathbf{C}_{nn}^{(wall)} = \mathbf{\Psi}\mathbf{C}_{ww}\mathbf{\Psi}^*. \tag{6.43}$$

For measurement kernel optimization and MMSE reconstruction, (6.43) is included in the interference term and the transmission coefficient (less than one) is factored into the strength of the received signal from the object. Therefore, the wall produces a signal loss from the desired object, as well as a signal-dependent interference term. In generating signal realizations for the Monte Carlo trials, the wall parameters were randomly selected, and the wall behavior was calculated from those parameters.

We see in Figure 6.12 that, as expected, the overall performance has degraded. Some of the performance degradation with respect to earlier results can be explained by the signal loss of 3–5 dB incurred by propagating to the target through the wall and back. Additional loss is due to interference from the wall. Regardless, the qualitative interpretation of the results is essentially the same as when the target was in free space.

6.6 Conclusions

Radar compressive sensing can provide important benefits in urban applications, especially in operational scenarios where performance is limited by resolution rather than SNR. In particular, small objects of interest in urban applications may place burdensome requirements on the sampling rates of the receiver's ADC. In such cases, it may be cost-effective and performance-effective to transmit more bandwidth than can be supported by traditional Nyquist sampling. The consequence of transmitting more bandwidth is that CS methods for sub-Nyquist sampling must be used.

We have developed an approach for measurement kernel optimization in HRR imaging applications. The information-based approach uses prior knowledge about targets of interest and the imaging task in order to optimize sub-Nyquist compression kernels. The optimization procedure is based on Gaussian mixture priors, which can be learned from target training data.

We have demonstrated the potential utility of measurement kernel optimization via simulation-based quantitative performance analysis in the form of reconstruction MSE.

Available prior information and the ability to describe prior distributions of objects of interest will depend on a mission's concept of operations. Of particular importance is the way that a high-resolution compressive radar might be used in conjunction with other sensors that may be capable of providing prior information to the compressive system. Also of interest is the potential for a compressive system to begin with weak priors and generic measurement kernels, and then to fine-tune the measurement kernels as information is extracted from previous measurements. Although exact future implementations of optimized compressive kernel design are yet unknown, we believe the results presented here provide justification and a potential methodology for considering CS kernel design in future systems.

Acknowledgments

We thank Remcom for providing their XFdtd software. We also gratefully acknowledge that this work was supported in part by the Defense Advanced Research Projects Agency (DARPA) via grant #N66001-10-1-4079.

References

1. M.A. Richards. *Fundamentals of Radar Signal Processing*. (New York: McGraw-Hill, 2005).
2. J.A. Tropp, J.N. Laska, M.F. Duarte et al. Beyond Nyquist: Efficient sampling of sparse bandlimited signals. *IEEE Trans. Inform. Theory* 56(1), (2010): 520–544.
3. M. Mishali and Y.C. Eldar. From theory to practice: Sub-Nyquist sampling of sparse wideband analog signals. *IEEE J. Sel. Top. Signal Process.* 4(2), (2010): 375–391.
4. O. Bar-Ilan and Y.C. Eldar. Sub-Nyquist radar via Doppler focusing. *IEEE Trans. Signal Process.* 62(7), (2014): 1796–1811.
5. B. Pollock and N.A. Goodman. Detection performance of multibranch and multichannel compressive receivers. *Proceedings of the 2012 IEEE 7th Sensor Array and Multichannel Signal Processing Workshop*, Hoboken, NJ, 2012, pp. 341–344.
6. H.J. Landau and H.O. Pollak. Prolate spheroidal wave functions, Fourier analysis and uncertainty – III: The dimension of the space of essentially time- and bandlimited signals. *Bell Sys. Tech. J.* 41 (1962): 1295–1336.
7. B. Pollock and N.A. Goodman. An examination of the effects of sub-Nyquist sampling on SNR. *Proceedings 2012 SPIE Defense, Security, and Sensing: Compressive Sensing I*, Baltimore, MD, 2012.

8. E. Arias-Castro and Y.C. Eldar. Noise folding in compressed sensing. *IEEE Signal Process. Lett.* 18(8), (2011): 478–481.
9. E.J. Candes and T. Tao. Near-optimal signal recovery from random projections: Universal encoding strategies? *IEEE Trans. Inform. Theory* 52(12), (2006): 5406–5425.
10. D. Guo, S. Shamai, and S. Verdu. Mutual information and minimum mean-square error in Gaussian channels. *IEEE Trans. Inform. Theory* 51(4), (2005): 1261–1282.
11. M.A. Neifeld, A. Ashok, and P.K. Baheti. Task-specific information for imaging system analysis. *J. Opt. Soc. Am.* 24(12), (2007): B25–B41.
12. H. Kim and N.A. Goodman. Waveform design by task-specific information. *Proceedings of the 2010 IEEE Radar Conference*, Washington, DC, 2010, pp. 848–852.
13. T.M. Cover and J.A. Thomas. *Elements of Information Theory.* (New York: John Wiley & Sons, 1991).
14. K.B. Petersen and M.S. Pedersen. *The Matrix Cookbook.* (Lyngby, Denmark: Technical University of Denmark, 2012). (Online: www.imm.dtu.dk/pubdb/views/edoc_download.php/3274/pdf/imm3274.pdf).
15. P. Tait. *Introduction to Radar Target Recognition.* (London, U.K.: Institution of Electrical Engineers, 2005).
16. K.M. Pasala and J.A. Malas. HRR radar signature database validation for ATR: An information theoretic approach. *IEEE Trans. Aerosp. Electron. Syst.* 47(2), (2011): 1045–1059.
17. S.P. Jacobs and J.A. O'Sullivan. Automatic target recognition using sequences of high resolution radar range-profiles. *IEEE Trans. Aerosp. Electron. Syst.* 36(2), (2000): 364–382.
18. A.P. Dempster, N.M. Laird, and D.B. Rubin. Maximum likelihood from incomplete data via the EM algorithm. *J. Roy. Stat. Soc. Ser B* 39(1), (1977): 1–38.
19. G. Yu, G. Saprio, and S. Mallat. Solving inverse problems with piecewise linear estimators: From Gaussian mixture models to structured sparsity. *IEEE Trans. Image Processing* 21(5), (2012): 2481–2499.
20. J.T. Flam, S. Chatterjee, K. Kansanen et al. On MMSE estimation: A linear model under Gaussian mixture statistics. *IEEE Trans. Signal Process.* 60(7), (2012): 3840–3845.
21. D.K. Cheng. *Field and Wave Electromagnetics*, 2nd edn. (Reading, MA: Addison-Wesley, 1989).

7

Compressive Sensing for Multipolarization through-the-Wall Radar Imaging

Abdesselam Bouzerdoum, Jack Yang, and Fok Hing Chi Tivive

CONTENTS

ABSTRACT Discrimination of targets can be improved significantly by analyzing the polarization of scattered electromagnetic waves. In radar imaging, the target image can be enhanced by combining measurements from different polarizations. In this chapter, we propose a joint image formation and fusion approach for multipolarization through-the-wall radar imaging, using compressive sensing (CS). The measurements from different polarization channels are processed jointly using the multiple measurement vector (MMV) model to produce several images of the scene, each corresponding to a polarization channel. Furthermore, the measurement vectors are fused together to form a composite measurement vector, which yields a composite image of the scene. The advantage of fusing the measurement vectors before image formation is that the measurement noise is reduced and the target information is enhanced, which leads to a more informative composite image. The MMV model enforces the same sparsity support for all formed images by reinforcing target information across channels and attenuating noise. Experimental results are presented using simulated and real data.

Analysis and comparison of experimental results demonstrate the effectiveness of the proposed through-the-wall radar imaging approach, especially in the presence of high-measurement noise.

7.1 Introduction

Through-the-wall radar imaging (TWRI) is emerging as a viable technology to generate high-resolution images behind walls or inside enclosed building structures. TWRI systems employ electromagnetic (EM) waves that can penetrate opaque materials, such as walls and doors, to detect, recognize, and track targets inside a building. The technology has numerous civilian and military applications, for example, search-and-rescue, law enforcement, and urban surveillance and reconnaissance [3,4,6,7]. There are, however, increasing demands on TWRI systems to produce high-resolution images that can effectively discriminate the targets of interest from clutter without increasing the data acquisition and processing time.

The discrimination of the targets can be enhanced significantly by analyzing the polarization of the EM waves scattered by objects in the scene [11,19]. Several approaches for target detection and classification have been reported based on multiple-polarization signals. In [13], two statistical detectors were proposed for joint target detection and fusion of multipolarization radar images. In [15], a method for through-the-wall detection of small weapons was developed by exploiting target polarization signature. It was found that the ratio of the co- to cross-polarization return can be used to distinguish a human carrying a weapon from a human without a weapon. In [20], target segmentation and classification was achieved using features extracted from multipolarization images. In [26], a human detection method was proposed using a fully polarimetric scattering model of the human body.

The aforementioned studies, however, were not concerned with the problem of image formation from multiple polarizations. The problem of imaging with multiple polarizations is that the target may exhibit different responses when interrogated by different polarized signals. Therefore, the main concern of a practical full-polarization TWRI system is to effectively combine the received radar data from all polarimetric channels so as to produce an image with low background clutter and high target reflections. In [22], image fusion techniques were employed to combine the images from different polarimetric channels. In this chapter, we develop a compressed sensing (CS)–based technique for simultaneous image formation and fusion, based on multiple measurement signals.

Recently, CS has been considered for radar imaging due to its ability to reconstruct a high-resolution image from a reduced set of measurements [2,5,17,18,23,24,31]. The scene reconstruction is posed as an inverse problem, whereby a spatial map of reflections is formed from radar measurements. Most CS-based methods for TWRI exploit sparsity in a single polarimetric channel only or assume the targets have invariant reflections at different polarizations; in other words, the interchannel correlations are not fully exploited by the image formation process.

In this chapter, we present a CS approach based on the *multiple measurement vectors* (MMV) model, where multiple measurements from several polarizations are combined in a CS framework to compute a sparse representation of the scene. Preliminary results were presented in [29]. The scene reconstruction using multiple polarization channels is formulated as finding a sparse matrix that satisfies the measurement constraints. Compared with existing CS-based methods, the proposed approach performs simultaneous image formation and fusion, and enforces the same sparsity support across all channels. Experimental results on synthetic and real data, acquired with a stepped frequency radar, are presented, which demonstrate that the proposed method improves image quality by enhancing target reflections and attenuating background clutter.

The remainder of the chapter is organized as follows. Section 7.2 reviews the existing image formation techniques for TWRI, including delay-and-sum beamforming and compressed sensing. An extension of the single-channel CS model to multiple polarizations is also presented in this section. Section 7.3 describes the proposed MMV-based image formation approach. Section 7.4 presents experimental results, which illustrate the effectiveness of the proposed method. Finally, Section 7.5 concludes the chapter.

7.2 Through-the-Wall Radar Imaging

To obtain high-resolution images that can reveal objects inside an enclosed building, a ground-based TWR with a long array aperture and large bandwidth is required. The array aperture can be physical or synthesized by moving a transceiver parallel to the front wall. The data received at all antenna locations are then collected and processed to form the image of the scene behind the wall. Here, we assume the front wall reflections are removed prior to image formation [27,28]. Before discussing compressed sensing for TWRI, we first present a brief review of image formation using the traditional delay-and-sum (DS) beamforming in the next section.

7.2.1 Delay-and-Sum Beamforming

Consider a monostatic stepped-frequency TWR system with a synthetic array aperture containing M antennas [4]. At each antenna location, the transceiver transmits and receives N monochromatic signals of frequencies:

$$f_n = f_1 + (n-1)\Delta f, \quad (n = 1,\ldots,N), \tag{7.1}$$

where
  f_1 is the initial frequency
  Δf is the frequency step size

The transceiver is moved horizontally parallel to the wall to synthesize an array aperture. Given P targets in the scene, the monochromatic signal of frequency f_n received at the mth antenna location is given by

$$z_{mn} = \sum_{p=1}^{P} \sigma_p \exp\left(-j2\pi f_n \tau_{mp}\right), \tag{7.2}$$

where
  σ_p is the complex reflectivity of the pth target
  τ_{mp} is the round-trip signal propagation delay from the mth antenna
    location to the pth target

The delay τ_{mp} is given by

$$\tau_{mp} = \frac{2\left(d_a + \sqrt{\epsilon_w}\, d_w\right)}{c}, \tag{7.3}$$

where
  d_a is the distance traveled through the air
  ϵ_w is the dielectric constant of the wall
  d_w is the distance traveled through the wall
  c is the speed of light in the air

Assume that the scene behind the wall is represented as a rectangular grid comprising N_x and N_y pixels along the crossrange and downrange directions, respectively. Let $\mathbf{x}$ denote the one-dimensional vector containing the image pixels arranged in a lexicographical ordering:

$$\mathbf{x} = \left[x_1,\ldots,x_q,\ldots,x_Q\right]^T, \tag{7.4}$$

where $Q = N_x N_y$. The complex amplitude of the qth pixel can be obtained by summing the delayed monochromatic signals received at all M antenna locations and N frequencies:

$$x_q = \frac{1}{MN} \sum_{m=1}^{M} \sum_{n=1}^{N} z_{mn} \exp\left(j\,2\pi f_n \tau_{mq}\right), \tag{7.5}$$

where τ_{mq} is the focusing delay between the mth antenna location and the qth pixel. DS beamforming utilizes all measurements to compute the complex amplitude of every pixel, thereby increasing the requirements for data acquisition and computation cost. To alleviate this problem, CS-based image formation methods have been developed for TWRI, where a reduced set of measurements is usually sufficient to recover a sparse scene. In the following section, the reconstruction of a TWRI scene from a single polarization is formulated as a single measurement vector (SMV) CS model.

7.2.2 Single-Polarization Imaging Using SMV Model

With the rapidly increasing demand on large-scale signal processing, it is not surprising to see CS emerging as one of the most important research areas in the past decade. CS has received considerable attention recently for its ability to perform data acquisition and compression simultaneously [1,9,16,21]. It can be used to reconstruct an approximation of a sparse or compressible signal from far fewer measurements than required by the sampling theorem. In TWRI, this has the advantage of reducing the number of measurement samples and data acquisition and processing time. A number of CS-based methods were proposed for TWRI in recent years [5,8,23,24,30,31]. In this section, we review briefly the CS-based approach for solving the image formation problem as an inverse problem using the single measurement vector model.

Suppose the received monochromatic signals are arranged into a column vector $\mathbf{z}$ of length MN,

$$\mathbf{z} = [z_{11}, z_{21}, \ldots, z_{mn}, \ldots, z_{MN}]^T, \tag{7.6}$$

where z_{mn} is the signal received by the mth antenna at the nth frequency, see Equation 7.2. Then, Equation 7.2 can be expressed in matrix–vector form as

$$\mathbf{z} = \Psi \mathbf{x}, \tag{7.7}$$

where $\Psi = [\psi_{ij}]$ is the so-called *sensing* or *steering* matrix. The element ψ_{ij} is given by

$$\psi_{ij} = \exp(-j2\pi f_n \tau_{mj}), \tag{7.8}$$

where
  $m = i \bmod M$
  $n = 1 + (i - m)/M$
  τ_{mj} is the round-trip propagation delay between the mth antenna location and the jth pixel

In the absence of measurement noise and clutter, the qth pixel value is ideally given by

$$x_q = \begin{cases} \sigma_p, & \text{if } q\text{th pixel includes } p\text{th target} \\ 0, & \text{otherwise} \end{cases} \tag{7.9}$$

where σ_p is the complex reflectivity of the pth target; in other words, the pixel value is nonzero only if a target exists at that pixel location. In practice, this may not be the case due to shadows and ghosts formed by multipath and multiple reflections between targets. However, in TWRI the imaged scene is usually sparsely populated, and hence the number of nonzero pixels is expected to be much smaller than the image size.

Suppose the vector $\mathbf{x}$ is K-sparse; that is, $\mathbf{x}$ contains at most K nonzero elements with $K \ll Q$. Given a linear measurement process, represented by a matrix Φ of size $R \times MN$ (where $R < Q$), the measurement vector $\mathbf{y}$ can be expressed as

$$\mathbf{y} = \Phi\mathbf{z} = \Phi\Psi\mathbf{x} = \mathbf{D}\mathbf{x}, \tag{7.10}$$

where $\mathbf{D} = \Phi\Psi$ is known as the *dictionary*. CS theory allows the reconstruction of a K-sparse vector $\mathbf{x}$ from the measurements $\mathbf{y}$ by solving the following problem:

$$\min \|\mathbf{x}\|_1 \quad \text{subject to } \mathbf{y} = \mathbf{D}\mathbf{x}, \tag{7.11}$$

where $\|.\|_1$ denotes the p-norm (with $p = 1$). Alternatively, if the measurement vector is corrupted by noise, a sparse signal $\mathbf{x}$ can be recovered by solving

$$\min \|\mathbf{x}\|_1 \quad \text{subject to } \|\mathbf{y} - \mathbf{D}\mathbf{x}\|_2 \le \epsilon, \tag{7.12}$$

where ϵ is an upper bound on the noise level (see Candes and Wakin in [1]).

The problem now is how to design a measurement matrix Φ that ensures stable recovery of a K-sparse vector $\mathbf{x}$ from a reduced set of measurements $\mathbf{y}$. A sufficient condition for stable recovery is incoherence between the matrices Φ and Ψ. For an orthogonal measurement matrix Φ, the signal $\mathbf{x}$ can be recovered almost perfectly provided the number of measurements $R \approx \mathcal{O}\left(K^2\mu(\Phi, \Psi)\right)\log(Q)$ [8], where $\mu(\Phi, \Psi)$ is the *mutual coherence* between Φ and Ψ. If Φ contains only one nonzero element in each row, then this is equivalent to selecting a subset of antennas and frequencies to perform the measurements; in this case, Φ is referred to as a *selection matrix*.

7.2.3 Multipolarization Imaging Using SMV Model

The extension of the SMV model to multiple polarimetric channels is given in this section. Assume we have L polarimetric channels. The measurement vector of each channel can be represented as

$$\mathbf{y}_i = \mathbf{D}_i \mathbf{x}_i, \quad (i = 1, \ldots, L), \tag{7.13}$$

where
$\mathbf{D}_i = \Phi_i \Psi$
$\mathbf{x}_i$ is a column vector containing the image of the ith polarimetric channel arranged in a lexicographical ordering

Here, without loss of generality, we assume all polarimetric channels have the same steering matrix Ψ. The SMV model given in Equation 7.12 can be applied to each polarization channel separately. Alternatively, the problem can be formulated as a single SMV model comprising all the channels. Let $\tilde{\mathbf{y}} = \left[\mathbf{y}_1^T, \ldots, \mathbf{y}_L^T\right]^T$ and $\tilde{\mathbf{x}} = \left[\mathbf{x}_1^T, \ldots, \mathbf{x}_L^T\right]^T$ denote the composite measurement and image vectors, respectively. The corresponding composite dictionary $\tilde{\mathbf{D}}$ is obtained by arranging the individual channel dictionaries $\mathbf{D}_i$ $(i = 1, \ldots, L)$ along the main diagonal and setting the off-diagonal elements to zero:

$$\tilde{\mathbf{D}} = \begin{bmatrix} \mathbf{D}_1 & 0 & 0 \\ 0 & \ddots & 0 \\ 0 & 0 & \mathbf{D}_L \end{bmatrix}. \tag{7.14}$$

The SMV model can now be applied to recover the composite vector $\tilde{\mathbf{x}}$ by solving

$$\min \|\tilde{\mathbf{x}}\|_1 \quad \text{subject to } \|\tilde{\mathbf{y}} - \tilde{\mathbf{D}}\tilde{\mathbf{x}}\|_2 \le \epsilon. \tag{7.15}$$

Although the extension of SMV model to multipolarization TWRI is straightforward, the drawback is that the recovered vectors $\mathbf{x}_i$ $(i = 1, \ldots, L)$ are not guaranteed to have the same sparse support. Furthermore, the SMV model of (7.15) does not exploit interchannel correlations. Finally, the complexity of the problem increases with increased size of the matrix $\tilde{\mathbf{D}}$, thereby requiring larger storage and more computation time to solve the CS problem (7.15). In the next section, we present a model that exploits the interchannel correlations using a joint sparse representation to enforce the same sparsity support on the recovered signals.

7.3 Multipolarization Imaging Using MMV Model

In this section, we present an MMV-based image formation method for multipolarization TWRI. First, a brief description of the multiple measurement vectors CS model is given in the next section. Then, the multipolarization TWRI problem is formulated as an inverse MMV problem.

7.3.1 MMV CS Model

The MMV model processes several measurement vectors simultaneously to produce a sparse matrix solution [10,12]. Consider a matrix of measurements $Y \in \mathbb{C}^{R \times L}$, comprising L measurement vectors, and a known dictionary $\mathbf{D}$ containing Q atoms. The MMV model aims to find a sparse matrix X by solving the following problem [10]:

$$\min \; \mathcal{S}_0(X) \quad \text{subject to } Y = \mathbf{D}X, \tag{7.16}$$

where $\mathcal{S}_0(X)$ denotes the sparsity rank of the matrix X, which is the number of nonzero rows in X. In other words, the aim is to find a sparse matrix solution whose columns possess the same sparsity profile. Let $\mathbf{r}_i$ denote the ith row of the matrix X. Furthermore, let us define a column vector $\mathbf{s}$ whose ith element $s_i = \|\mathbf{r}_i\|_p$, with $p \geq 2$. The sparsity rank of the matrix X is given by

$$\mathcal{S}_0(X) = \|\mathbf{s}\|_0, \tag{7.17}$$

where $\|\cdot\|_0$ denotes the zero pseudo-norm or cardinality of the vector argument.

However, minimizing $\mathcal{S}_0(X)$ is NP-hard because an exhaustive enumeration is required in terms of all possible locations of nonzero rows in X. Therefore, the zero pseudo-norm is usually replaced with the one-norm, $\|\cdot\|_1$, resulting in the following problem:

$$\min \; \mathcal{S}_1(X) = \|\mathbf{s}\|_1 \quad \text{subject to } Y = \mathbf{D}X. \tag{7.18}$$

In [10], the authors proved that minimization of $\mathcal{S}_1(X)$ is equivalent to minimization of $\mathcal{S}_0(X)$ when the sparsity rank of X is sufficiently low.

7.3.2 Joint Image Fusion and Formation Using MMV

For a stepped-frequency TWRI system, the steering matrix Ψ is related to the bandwidth of the signal and the image size, see Equation 7.8. Since the radar interrogates the same scene, all polarization channels share the same sensing matrix, that is, $\Psi_i = \Psi$ ($i = 1, \ldots, L$). Moreover, we can choose the selection matrix to be identical for all channels, $\Phi_i = \Phi \ \forall i$. This means the same dictionary can be used across all channels, $D_i = D$ ($i = 1, \ldots, L$). Substituting D for D_i in (7.13) yields the single channel measurement vector

$$y_i = Dx_i, \quad i = 1, \ldots, L. \tag{7.19}$$

Since the vectors x_i ($i = 1, \ldots, L$) represent images of the same scene, a final composite image of the scene can be easily obtained using image fusion techniques [22]. By contrast, here we propose first to combine the raw measurement vectors from different polarimetric channels, then perform image reconstruction based on the fused measurement vector. Specifically, a composite measurement vector $\widehat{y}$ is defined as a linear combination of the measurement vectors of different polarimetric channels:

$$\widehat{y} = \sum_{i=1}^{L} w_i y_i = D \sum_{i=1}^{L} w_i x_i = D\widehat{x}, \tag{7.20}$$

where w_is are positive weights satisfying $\sum_{i=1}^{L} w_i = 1$. Here we employ a criterion based on mutual information (MI) to compute the weights w_i. Mutual information is used to estimate the coherence between two measurement vectors y_i and y_j:

$$I\left(y_i, y_j\right) = H\left(y_i\right) + H\left(y_j\right) - H\left(y_i, y_j\right), \quad (i \neq j), \tag{7.21}$$

where
$H\left(y_i\right)$ is the marginal entropy
$H\left(y_i, y_j\right)$ is the joint entropy

The weights w_i ($i = 1, 2, \ldots, L$) are computed as

$$w_i = \frac{H\left(y_i\right)^2 - \sum_{j \neq i}^{L} I\left(y_j, y_i\right) + \sum_{j \neq i}^{L} I\left(y_j, y_i | \cup_{k \neq i, j} y_k\right)}{H\left(y_1, y_2, \ldots, y_L\right) H\left(y_i\right)}, \tag{7.22}$$

where $I\left(y_j, y_i | \cup_{k \neq j, i} y_k\right)$ is the conditional mutual information. The first term in the numerator of (7.22) defines the importance of the ith measurement vector compared with other measurement vectors, whereas the second and third terms are used to remove the overlapping information in the first term. The rationale for computing a weighted-average measurement vector is to have a sparse solution that represents the final output image with better signal-to-noise ratio. It is relatively simple to show that for additive i.i.d.[*] measurement noise, the weighted linear combination (7.20) reduces the noise variance in $\widehat{\mathbf{y}}$, compared with the noise variance in $\mathbf{y}_i$.

Since the radar images the same scene, it is reasonable to assume that the images of the L polarimetric channels share the same sparse support but may have different nonzero coefficients. Therefore, using an augmented measurement matrix $Y = [\mathbf{y}_1, \dots, \mathbf{y}_L, \widehat{\mathbf{y}}]$, we can simultaneously reconstruct the vectors $\mathbf{x}_i$ and the composite vector $\widehat{\mathbf{x}} = \sum_{i=1}^{L} w_i \mathbf{x}_i$, using the MMV model of Equation 7.18. However, in general, the measurements are corrupted by noise; therefore, we replace the MMV problem (7.18) with the following mixed-norm regularized least-squares problem:

$$\min \ ||X||_{1,2} \quad \text{subject to} \ ||Y - \mathbf{D}X||_F \leq \epsilon, \tag{7.23}$$

where

$||\cdot||_{1,2}$ denotes the mixed $(1,2)$-norm, which is the sum of the Euclidean norms of the rows of X

$||\cdot||_F$ is the Frobenius norm

ϵ is an upper bound on the noise level

The resulting solution matrix X contains in its first L columns the images corresponding to the individual polarimetric channels, and in the last column the image corresponding to the composite measurement vector. Consequently, this approach may be viewed as performing joint image fusion and image formation.

7.4 Experimental Results

In this section, the proposed MMV-based method for multipolarization TWRI is evaluated on both synthetic and real radar data. For comparison, DS beamforming and the SMV model are implemented and tested. The effectiveness of the proposed method is tested in terms of the number of selected measurements and the signal-to-noise ratio of the received signals. To evaluate different image formation algorithms, the target-to-clutter ratio (TCR) is used as a performance measure:

$$\text{TCR} = 10 \log \left(\frac{1/N_{\mathcal{B}} \sum_{q \in \mathcal{B}} |x_q|^2}{1/N_{\mathcal{C}} \sum_{q \in \mathcal{C}} |x_q|^2} \right), \tag{7.24}$$

---

[*] Independent and identically distributed.

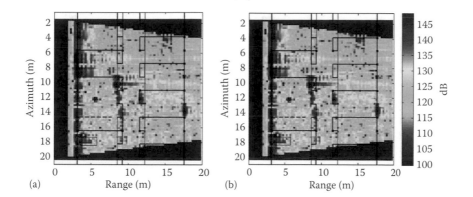

FIGURE 2.7
B-scan obtained with the matched filter tuned to point-like scatterers, VV polarization (a) and HH polarization (b).

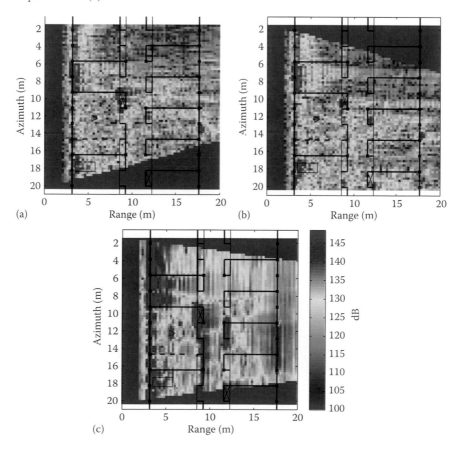

FIGURE 2.10
B-scans obtained with the *forward-looking* matched filter (a), the *backward-looking* matched filter (b), and the *linear phase* matched filter (c), VV polarization.

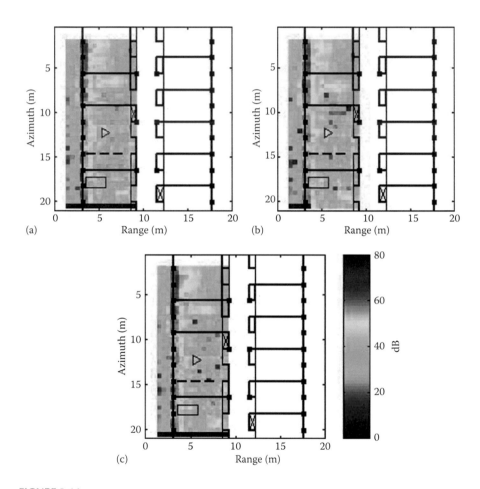

FIGURE 2.14
Equivalent B-scans obtained with OCD: the *forward-looking* part (a), the *backward-looking* part (b), and the *linear phase* part (c), VV polarization.

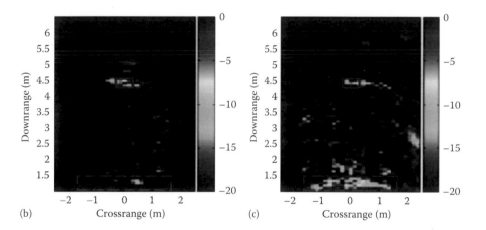

FIGURE 4.2
(b) Beamforming-based imaging result after background subtraction. (From Lagunas, E. et al., Joint wall mitigation and compressive sensing for indoor image reconstruction, *IEEE Trans. Geosci. Rem. Sens.*, 51(2), 891, February 2013; Amin, M.G. and Ahmad, F., Compressive sensing for through-the-wall radar imaging, *J. Electron. Imag.*, 22 (3), 030901, July 2013. © 2013 IEEE. With permission.) (c) Beamforming-based imaging result after subspace projection based wall mitigation.

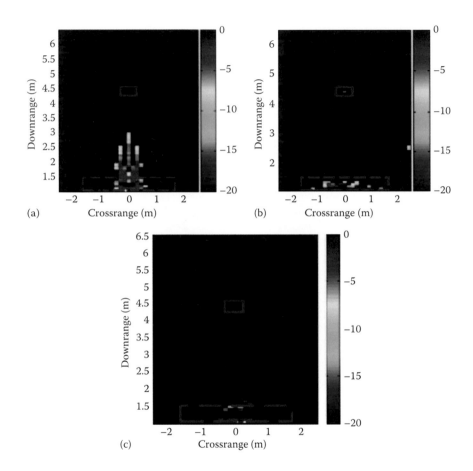

FIGURE 4.3
(a) CS-based imaging result using all of the raw data. (b) CS-based imaging result using 10.2% data volume with the same frequency set at each antenna, averaged over 100 trials. (c) CS-based imaging result using 10.2% of the data volume with different frequency sets from different antennas, averaged over 100 trials. (From Lagunas, E. et al., Joint wall mitigation and compressive sensing for indoor image reconstruction, *IEEE Trans. Geosci. Rem. Sens.*, 51(2), 891, February 2013. © 2013 IEEE. With permission.)

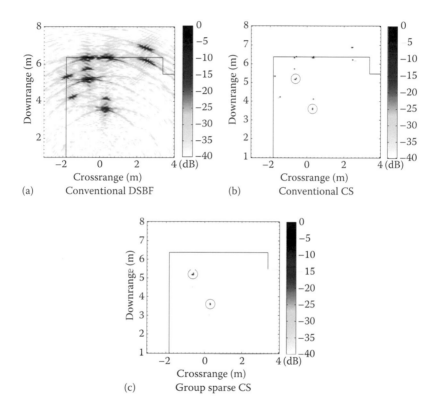

FIGURE 5.8

Reconstruction results using different algorithms for the simulated scene with two point targets. One-fourth of the array elements and one-eighths of the frequency bins were used for the formation of (b) and (c). Image (a) was created using the full set of measurements. (From Leigsnering, M. et al., Multipath exploitation in through-the-wall radar imaging using sparse reconstruction, *IEEE Trans. Aerosp. Electron. Syst.*, in press. With permission. © 2013 IEEE.)

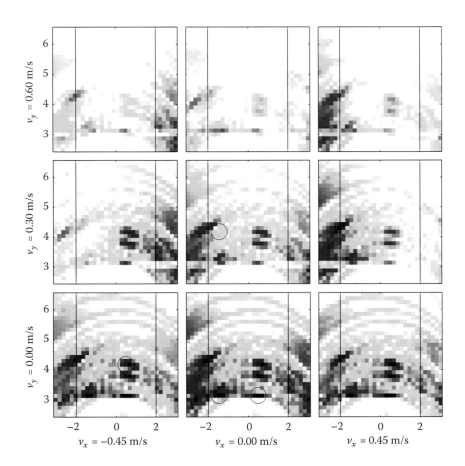

FIGURE 5.10

Delay-and-sum beamforming result using full data.

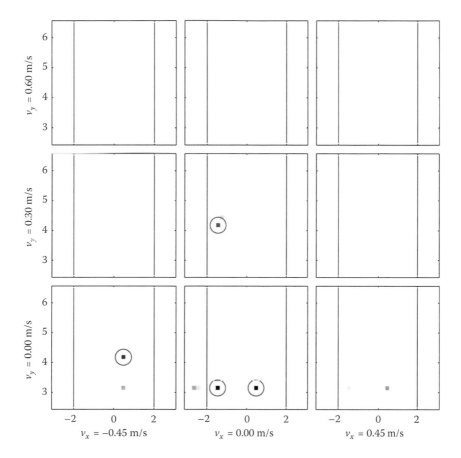

FIGURE 5.11
CS reconstruction using 7.8% of the measurements.

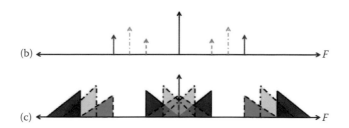

FIGURE 6.3
Frequency-domain convolution of signal and measurement kernel. (a) Frequency spectrum of original signal, (b) frequency spectrum of the measurement kernel, and (c) frequency spectrum of the time-domain product of the signal and measurement kernel.

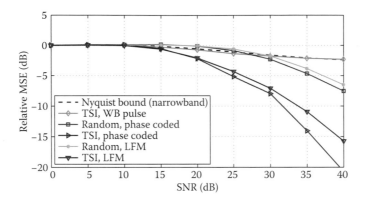

FIGURE 6.10
Reconstruction performance in free space without sequential sampling structure enforced and prior knowledge of object orientation between 20° and 50°.

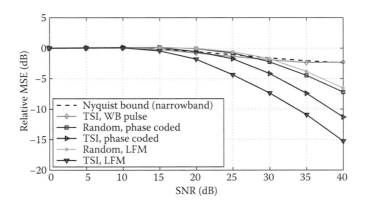

FIGURE 6.11
Reconstruction performance in free space with prior knowledge of object orientation between 20° and 50°. The sequential sampling structure has now been enforced.

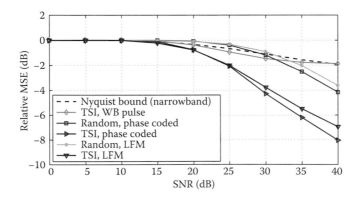

FIGURE 6.12
Reconstruction performance for an object located behind and in close proximity to a wall with uncertain thickness and permittivity. Prior pdf on orientation window is 20°–50°.

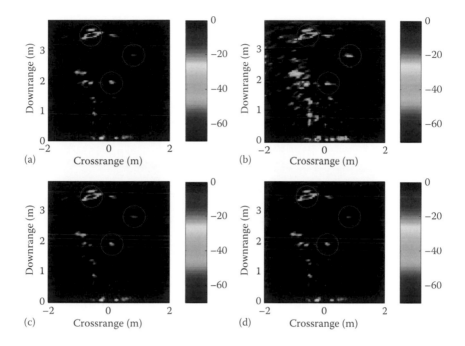

FIGURE 7.9
Reconstructed images using MMV approach with 15% measurements. (a) HH image, (b) HV image, (c) VV image, and (d) composite image.

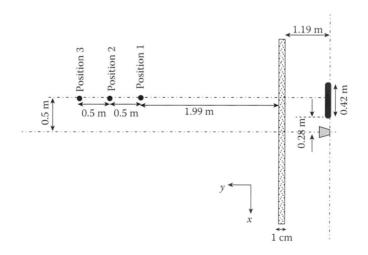

FIGURE 8.2
Scene layout for the target undergoing translational motion. (From Ahmad, F. and Amin, M., Through the wall human motion indicator using sparsity-driven change detection, *IEEE Trans. Geosci. Rem. Sens.*, 51(2), 881, February 2013. With permission. © 2010 IEEE.)

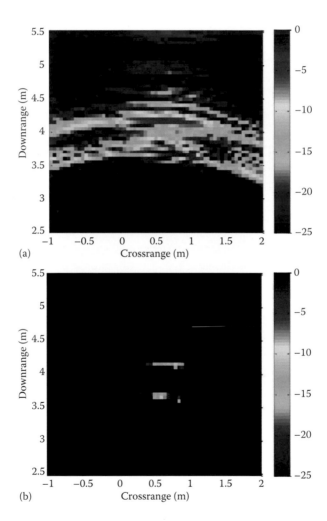

FIGURE 8.3
(a) Backprojection-based CD image using the full data set. (b) Sparsity-based CD image using 5% of the data volume, averaged over 100 trials. (From Ahmad, F. and Amin, M., Through the wall human motion indicator using sparsity-driven change detection, *IEEE Trans. Geosci. Rem. Sens.*, 51(2), 881, February 2013. With permission. © 2010 IEEE.)

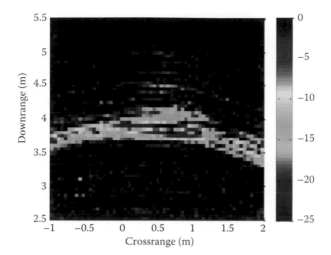

FIGURE 8.5
Backprojection-based CD image using full data volume for the target undergoing short sudden movement. (From Ahmad, F. and Amin, M., Through the wall human motion indicator using sparsity-driven change detection, *IEEE Trans. Geosci. Rem. Sens.*, 51(2), 881, February 2013. With permission. © 2010 IEEE.)

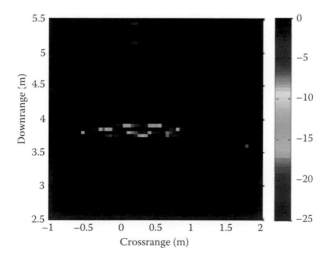

FIGURE 8.6
Sparsity-based composite images with 5% data volume for the subimage combining approach in Equation 8.25. The image is the averages of 100 reconstructions. (From Ahmad, F. and Amin, M., Through the wall human motion indicator using sparsity-driven change detection, *IEEE Trans. Geosci. Rem. Sens.*, 51(2), 881, February 2013. With permission. © 2010 IEEE.)

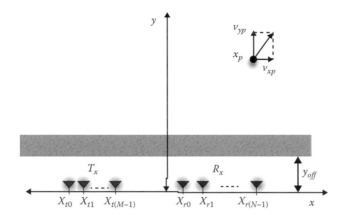

FIGURE 8.7
Geometry on transmit and receive. (From Ahmad, F. and Amin, M., Through the wall human motion indicator using sparsity driven change detection, *IEEE Trans. Geosci. Rem. Sens.*, 51(2), 881, February 2013. With permission. © 2010 IEEE.)

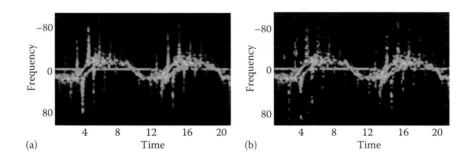

FIGURE 9.5
The STFT of critically sampled signal (a) and the OMP-based TFR from randomly undersampled signal (b).

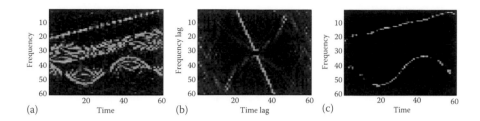

(a) Time   (b) Time lag   (c) Time

FIGURE 9.7
(a) Wigner distribution, (b) ambiguity function, and (c) resulting sparse TFR.

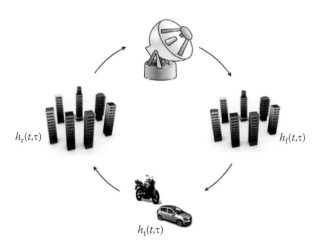

$h_r(t,\tau)$ $h_f(t,\tau)$

$h_t(t,\tau)$

FIGURE 10.1
Block diagram showing the forward, reverse transmission channels, and the targets.

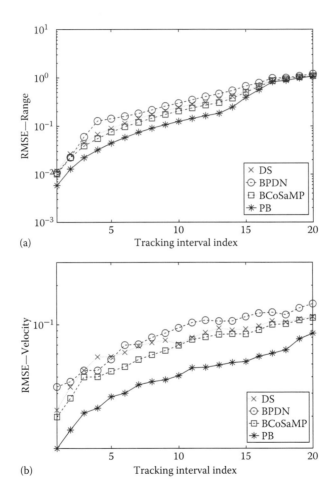

FIGURE 10.2
Root-mean-square error for various tracking algorithms. (a) RMSE in range and (b) RMSE in velocity.

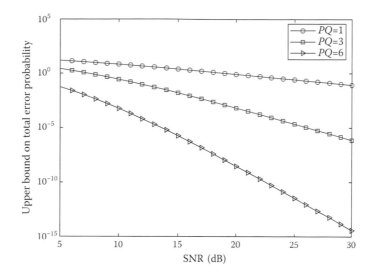

FIGURE 10.4
Upper bound on error probability for an optimal reconstruction.

FIGURE 11.4
L_p regularized LS reconstructions of a civilian vehicle from the GOTCHA data set (3-D, side and top views, $p = 1$ and $\lambda = 10$). (Reproduced with permission from Austin, C.D., Ertin, E., and Moses, R.L., Sparse signal methods for 3-D radar imaging, *IEEE J. Select. Top. Signal Process.*, 5(3), 420, 2011. © 2011 IEEE.)

FIGURE 11.8
Wide-angle multipass IFSAR reconstructions of a civilian vehicle from the GOTCHA data set (3-D, side and top views). (Reproduced with permission from Austin, C.D., Ertin, E., and Moses, R.L., Sparse signal methods for 3-D radar imaging, *IEEE J. Select. Top. Signal Process.*, 5(3), 420, 2011. © 2011 IEEE.)

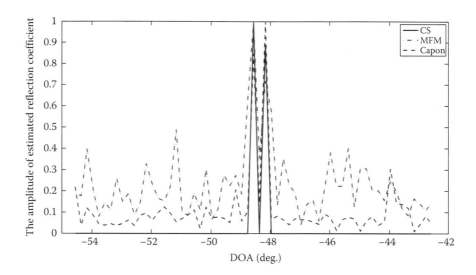

FIGURE 12.4
DOA Estimates of CS, Capon, and MEM using 16 received samples.

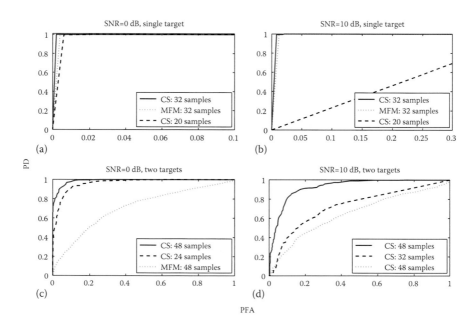

FIGURE 12.6
ROC performances of range–DOA estimation for CS, Capon, and MFM methods with $M_t = 8$, $N_r = 5$. (a) SNR $= 0$ dB, single target, (b) SNR $= -10$ dB, single target, (c) SNR $= 0$ dB, two targets, and (d) SNR $= -10$ dB, two targets.

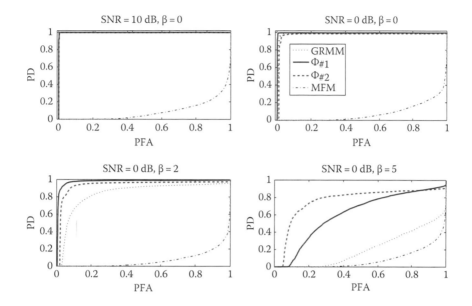

FIGURE 12.16
ROC curves for CS-MIMO radars using $\Phi_{\#1}$, $\Phi_{\#2}$ and the GRMM and for MIMO radars using the MFM ($M_t = N_r = 4$ and $\lambda = 1.5$).

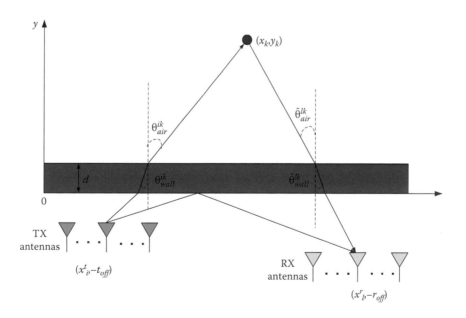

FIGURE 12.17
Geometry of through-the-wall propagation and wall reflection.

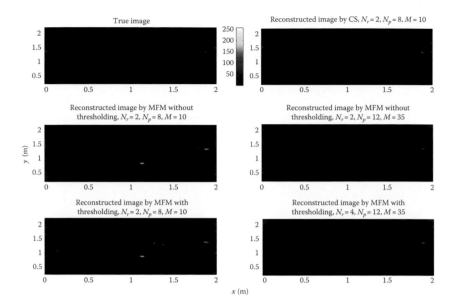

FIGURE 12.18
Reconstructed image by CS and MFM.

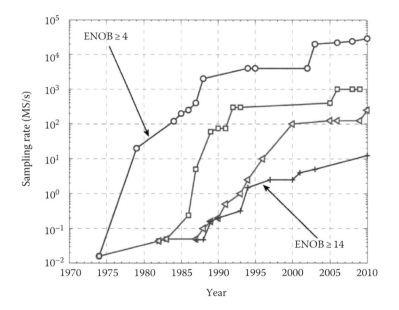

FIGURE 13.1
Trends in the progress of ADC technology since 1975 to 2010, adapted from Jonsson (2010). ENOB stands for *effective number of bits* and refers to the achievable number of quantization levels. The different colors represent the progress of ADC technology for different values of ENOB. Blue line: ENOB ≥ 4. Red line: ENOB ≥ 8. Green line: ENOB ≥ 12. Black line: ENOB ≥ 14. (From Jonsson, B.E., A servey of A/D converter performance earlution,in: 17th IEEE International Conference on Electronics, Circuits, and Systems (ICECS), Athens, Greece, 2010, pp. 766–769. With permission.)

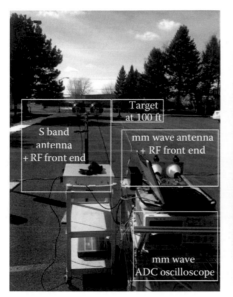

FIGURE 13.4
Photographs of radar system and target.

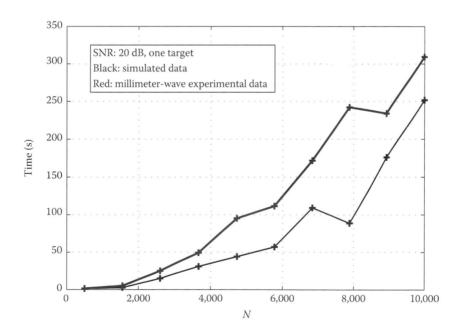

FIGURE 13.14
Computational complexity of SPGL 1 convex optimization signal recovery in terms of running time of algorithm on an Intel i7 2.8 GHz with 8 GB of RAM.

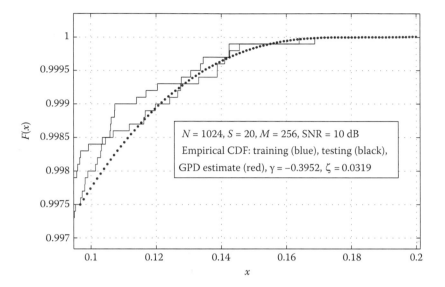

FIGURE 13.15
Plot showing application of Algorithm 13.1 for GPD-based extrapolation of the cumulative distribution of empirical residue data. Compressive recovery with $N = 1024$.

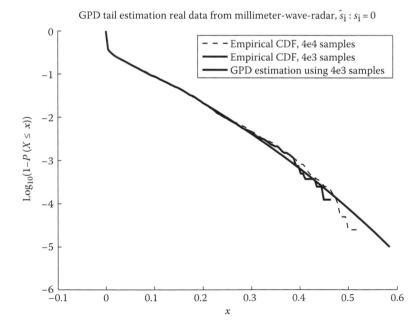

FIGURE 13.17
Tail estimation for real data. We observe that the empirical CDF corresponds closely to the GPD estimate for the tail. The target scenario involved one corner reflector target at 100 ft imaged multiple times using the millimeter wave radar described in Section 2.

where
x_q represents the qth pixel from the image
$\mathcal{B}$ and $\mathcal{C}$ denote the target and clutter regions
$N_\mathcal{B}$ and $N_\mathcal{C}$ are the number of pixels in the regions $\mathcal{B}$ and $\mathcal{C}$, respectively

7.4.1 Experimental Results Using Synthetic Data

In this section, experiments are conducted using synthetic data. The simulated TWRI system is a monostatic synthetic aperture radar consisting of 71 transceivers with an aperture of 2.0 m. The stepped-frequency signal covers a 2 GHz bandwidth, ranging from 1 to 3 GHz, with a 5 MHz frequency step. Thus, the radar system transmits and receives 28,471 (71 × 401) monochromatic signals. The imaged scene is situated behind a wall of thickness 0.15 cm and a dielectric constant $\epsilon_w = 7.5$. It covers an area of 5.0 m × 5.0 m, that is, 5.0 m wide and 5.0 m deep. The scene is partitioned into 128 × 128 pixels along the crossrange and downrange directions. Three point targets are placed behind the wall at coordinates $(-1.5, 4)$m, $(0, 1.5)$m, and $(1.5, 3)$m. Here, we assume the scattering matrix is symmetric, hence only three polarization channels are considered: HH, VV, and HV. Figure 7.1 illustrates the images reconstructed from the three polarimetric channels using DS

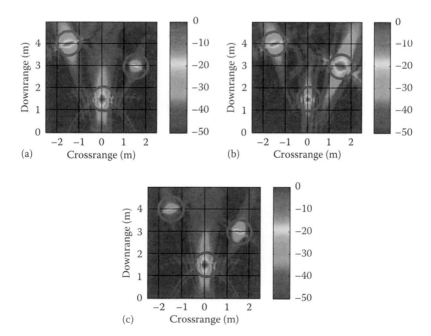

FIGURE 7.1
DS beamforming with full measurement set. The targets are simulated with different reflection coefficients at different polarization. (a) HH image, (b) HV image, and (c) VV image.

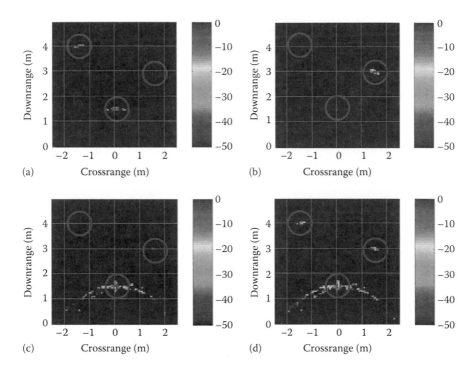

FIGURE 7.2
Images formed with 5% measurements using SMV model of Equation 7.15. (a) HH image, (b) HV image, (c) VV image, and (d) Composite image.

beamforming with the full set of measurements. As can be observed, the formed images contain many large sidelobes around the targets.

Next, CS-based algorithms are employed to reconstruct the image of the scene using 5% of the full measurement set. Figure 7.2 illustrates the images formed by the SMV model given in Equation 7.15; the image displayed in Figure 7.2d is obtained from the composite measurement vector of the three channels, given by Equation 7.20. Clearly, the images in Figure 7.2 do not share the same support, and some images fail to detect all targets. One possible reason is that the SMV model does not exploit interchannel correlations and does not impose the same sparsity profile on the reconstructed images. By contrast, Figure 7.3 depicts the images formed by the proposed MMV model given in Equation 7.23. These images share the same support, the targets are clearly visible, and the clutter is significantly reduced.

To further assess the effectiveness of the CS-based methods, the MMV and SMV models are tested on measurements corrupted by noise. Various scenarios are considered by varying the SNR and the number of measurement used to reconstruct the image of the scene. Each experiment is repeated 20 times, and the average TCR is recorded as a measure of performance.

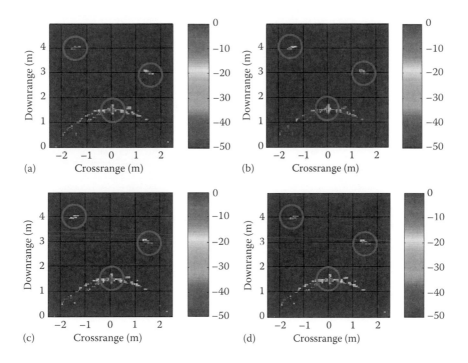

FIGURE 7.3
Images formed with 5% measurements using MMV model of Equation 7.23. (a) HH image,
(b) HV image, (c) VV image, and (d) Composite image.

The TCR is computed from the final output images obtained from the
composite measurement vector $\hat{y}$ (see Figures 7.2d and 7.3d). Figure 7.4 illustrates the average TCR as a function of the number of measurements for
different SNR values. As expected, increasing the percentage of measurements improves the quality of the reconstructed image for both SMV and
MMV; however, MMV-based method achieves consistently higher TCR than
does its SMV-based counterpart.

7.4.2 Experimental Results Using Real Data

Real radar data were collected at the Radar Imaging Lab of the Center for
Advanced Communications, Villanova University, PA, USA. An Agilent network analyzer, model ENA-5071B with an operation frequency range of
300 kHz to 8.5 GHz, was used to generate a stepped-frequency signal in
the range [0.7, 3] GHz with a frequency step of 2.875 MHz. A 57-element
array was synthesized by mounting a horn antenna, model ETS-Lindgren
3164-04, with an operational bandwidth of 0.7–6 GHz on a scanner and moving the scanner horizontally at an interspacing of 0.022 m. A room of size
7.62 × 7.62 m was padded with radio frequency–absorbing material for data

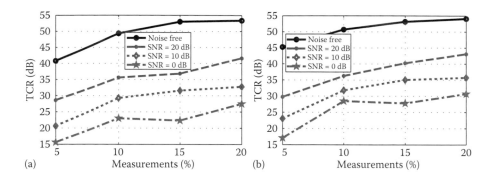

(a)   (b)

FIGURE 7.4
Average TCR as a function of number of measurements. The measurements are corrupted by additive white Gaussian noise with different SNR values. (a) SMV and (b) MMV.

collection; the interested reader is referred to [14] for more details about the room setting, the data acquisition, and the specification of the imaging system. The TWRI system was used to interrogate a scene populated with three targets placed at different locations behind a concrete wall of thickness 0.14 m and dielectric constant 7.6632. Figure 7.5a shows an image of the real scene with the three targets: a dihedral, a sphere, and a trihedral. Figure 7.5b presents a schematic diagram of the imaged scene depicting the target locations in the downrange–crossrange plane. The total number of collected measurement samples is 45,657 (801 frequencies × 57 antennas). The region of interest behind the wall is set to 4 m × 4 m, and the pixel size is

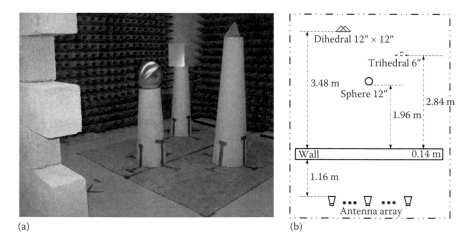

(a)   (b)

FIGURE 7.5
(a) A real scene populated with three targets: a dihedral, a sphere, and a trihedral. (b) Schematic diagram of a horizontal cross section of the imaged scene.

set to 3.125 cm × 3.125 cm, resulting in an image of size 128 × 128 pixels. The received signals are preprocessed by removing the strong wall returns, using background subtraction, and the common signal at each antenna location. If the background signal is not available, then the strong wall returns can be mitigated using wall parameter estimation techniques that do not require the full measurement set [25].

The received signals from each polarization channel are first processed using DS beamforming to form three images. Figure 7.6 depicts the reconstructed images using the full measurement set. All images contain significant amount of clutter, but the targets are visible in the images of the copolarized channels, HH and VV. Figure 7.7 presents the images obtained from 15% of the measurements, selected randomly from the full measurement set. Clearly, the level of clutter increases in all images as the number of measurements is reduced.

In the remaining experiments, the CS methods are tested on reduced measurements from real data. Figure 7.8 depicts the images generated by the SMV model using 15% of the measurements. We can observe that some targets are missing from the single polarization images, and the composite image in Figure 7.8d contains a high level of clutter. In contrast, Figure 7.9 illustrates the images produced by the proposed MMV method, where all

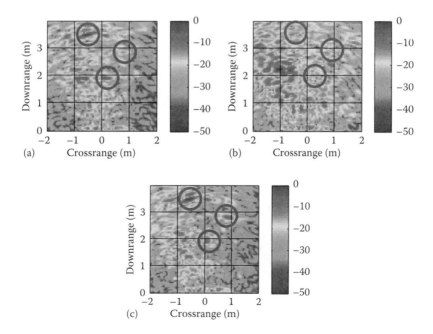

FIGURE 7.6
Reconstructed images using DS beamforming with full measurements. (a) HH image, (b) HV image, and (c) VV image.

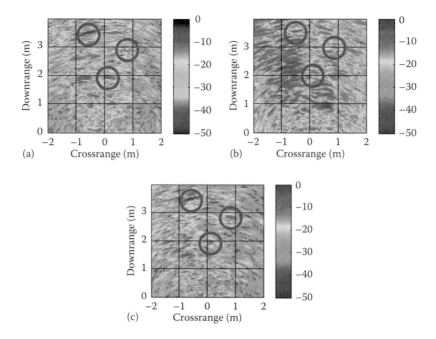

FIGURE 7.7
Reconstructed images using DS beamforming with 15% measurements. (a) HH image, (b) HV image, and (c) VV image.

images display the targets correctly and the level of clutter is much reduced compared with the amount of clutter in the SMV generated images. Table 7.1 lists the TCR of the reconstructed composite images for different measurement percentages using DS beamforming and SMV and MMV models. Again, the quality of the reconstructed image improves as the number of measurements increases, but the MMV approach always yields a higher TCR than DS beamforming and SMV model. These experimental results confirm the superiority of the proposed image formation approach.

7.5 Conclusion

This chapter presented an image formation approach for multipolarization TWRI based on the multiple measurement vectors model of compressed sensing. In the MMV model, the measurement vectors obtained from different polarimetric channels are arranged into columns of a measurement matrix, which is then used to recover a sparse matrix solution whose columns constitute the images of different polarimetric channels. This is in

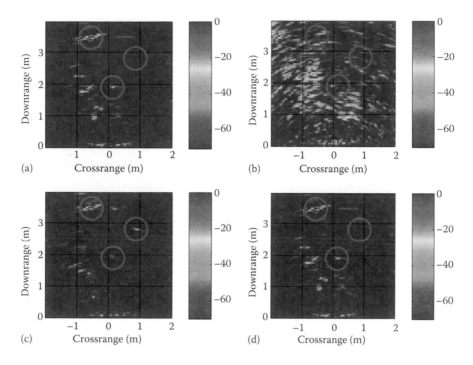

FIGURE 7.8
Reconstructed images using SMV approach with 15% measurements (a) HH image, (b) HV image, (c) VV image, and (d) Composite image.

contrast to the single measurement vector model, where each measurement vector is processed independently or, alternatively, all measurements are concatenated into a single measurement vector. Notwithstanding the differences in target reflectivity at different polarizations, the reconstructed channel images should ideally have the same sparsity support since they represent the same scene. The MMV model enforces the same support on all columns of the solution matrix by exploiting the interchannel correlations between the different measurement vectors. Furthermore, in the proposed approach, all channel measurement vectors are combined together to form a composite vector, which is used to reconstruct a fused image of the scene. Therefore, the proposed method can be viewed as a joint image formation and fusion approach. Experimental evaluation was conducted using synthetic and real data. The performance of the proposed approach was compared with that of the single measurement vector CS model and the traditional DS beamforming. Experimental results prove that the proposed approach reconstructs images with higher target-to-clutter ratio than does SMV or DS beamforming. In particular, the MMV images tend to have much less clutter and share the same support.

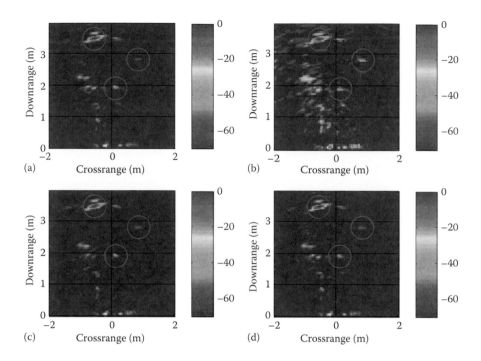

FIGURE 7.9
(See color insert.) Reconstructed images using MMV approach with 15% measurements. (a) HH image, (b) HV image, (c) VV image, and (d) composite image.

TABLE 7.1

TCR of Reconstructed Composite Images Using DS Beamforming, SMV, and MMV Models

% Measurements	5%	10%	15%
DS (dB)	5.2	6.7	7.9
SMV (dB)	39.9	41.6	47.5
MMV (dB)	44.7	46.4	50.6

References

1. Special issue on sensing, sampling, and compression. *IEEE Signal Processing Magazine*, **25**(2), March 2008.
2. E. Aguilera, M. Nannini, and A. Reigber. Multisignal compressed sensing for polarimetric SAR tomography. *IEEE Geoscience and Remote Sensing Letters*, 9(5):871–875, 2012.

3. F. Ahmad and M. G. Amin. Multi-location wideband synthetic aperture imaging for urban sensing applications. *Journal of the Franklin Institute*, **345**(6):618–639, 2008.
4. F. Ahmad, M. G. Amin, and S. A. Kassam. A beamforming approach to stepped-frequency synthetic aperture through-the-wall radar imaging. In *IEEE International Workshop on Computational Advances in Multi-Sensor Adaptive Processing*, Philadelphia, PA, pp. 24–27, 2005.
5. M. G. Amin and F. Ahmad. Compressive sensing for through-the-wall radar imaging. *Journal of Electronic Imaging*, **22**(2): Article 030901, 1–21, 2013.
6. M. G. Amin and K. Sarabandi. Special issue on remote sensing of building interior. *IEEE Transactions on Geoscience and Remote Sensing*, **47**(5):1267–1268, 2009.
7. M. G. Amin (Ed.). *Through-the-Wall Radar Imaging*. Boca Raton, FL: CRC Press, 2011.
8. E. Candes and J. Romberg. Sparsity and incoherence in compressive sampling. *Inverse Problems*, **23**:969985, 2006.
9. E. Candes, J. Romberg, and T. Tao. Robust uncertainty principles: Exact signal reconstruction from highly incomplete frequency information. *IEEE Transactions on Information Theory*, **52**:489–509, 2006.
10. J. Chen and X. Huo. Theoretical results on sparse representations of multiple-measurement vectors. *IEEE Transactions on Signal Processing*, **54**(12):4634–4643, 2006.
11. S. R. Cloude. *Polarisation: Applications in Remote Sensing*. London, U.K.: Oxford University Press, 2010.
12. S. F. Cotter, B. D. Rao, K. Engan, and K. Kreutz-Delgado. Sparse solutions to linear inverse problems with multiple measurement vectors. *IEEE Transactions on Signal Processing*, **53**(7):2477–2488, 2005.
13. C. Debes, A. M. Zoubir, and M. G. Amin. Enhanced detection using target polarization signatures in through-the-wall radar imaging. *IEEE Transactions on Geoscience and Remote Sensing*, **50**(5):1968–1979, 2012.
14. R. Dilsavor, W. Ailes, P. Rush, F. Ahmad, W. Keichel, G. Titi, and M. Amin. Experiments on wideband through the wall imaging. In *Algorithms for Synthetic Aperture Radar Imagery XII, Vol. 5808 of Proceedings of SPIE*, pp. 196–209, March 28, 2005.
15. T. Dogaru and C. Le. Through-the-wall small weapon detection based on polarimetric radar techniques. ARL-TR-5041, 2009.
16. D. Donoho. Compressed sensing. *IEEE Transactions on Information Theory*, **52**:1289–1306, 2006.
17. A. C. Gurbuz, J. H. McClellan, and W. R. Scott. A compressive sensing data acquisition and imaging method for stepped frequency gprs. *IEEE Transactions on Signal Processing*, **57**(7):2640–2650, 2009.
18. Q. Huang, L. Qu, B. Wu, and G. Fang. UWB through-wall imaging based on compressive sensing. *IEEE Transactions on Geoscience and Remote Sensing*, **48**(3):1408–1415, 2010.
19. J. S. Lee and E. Pottier. *Polarimetric Radar Imaging: From Basics to Applications*. Boca Raton, FL: CRC Press, 2009.
20. A. A. Mostafa, C. Debes, and A. M. Zoubir. Segmentation by classification for through-the-wall radar imaging using polarization signatures. *IEEE Transactions on Geoscience and Remote Sensing*, **50**(9):3425–3439, 2012.

21. H. Rauhut, K. Schnass, and P. Vandergheynst. Compressed sensing and redundant dictionaries. *IEEE Transactions on Information Theory*, **54**:2210–2219, 2008.
22. C. H. Seng, A. Bouzerdoum, M. G. Amin, and S. L. Phung. Two-stage fuzzy fusion with applications to through-the-wall radar imaging. *IEEE Geoscience and Remote Sensing Letters*, **10**(4):687–691, July 2013.
23. V. H. Tang, A. Bouzerdoum, and S. L. Phung. Enhanced through-the-wall radar imaging using Bayesian compressive sensing. In *Compressive Sensing II, Vol. 8717 of Proceedings of SPIE*, Article 87170I, pp. 1–12, 2013.
24. V. H. Tang, A. Bouzerdoum, and S. L. Phung. Two-stage through-the-wall radar image formation using compressive sensing. *Journal of Electronic Imaging*, **22**(2): Article 021006, 1–10, 2013.
25. C. Thajudeen, A. Hoorfar, and W. Zhang. Estimation of frequency-dependent parameters of unknown walls for enhanced through-the-wall imaging. In *Proceedings of the IEEE International Symposium on Antennas and Propagation (AP-S/URSI)*, Spokane, WA, pp. 3070–3073, July 2011.
26. M. Thiel and K. Sarabandi. Ultrawideband multi-static scattering analysis of human movement within buildings for the purpose of stand-off detection and localization. *IEEE Transactions on Antennas and Propagation*, **59**(4):1261–1268, 2011.
27. F. H. C. Tivive and A. Bouzerdoum. An improved svd-based wall clutter mitigation method for through-the-wall radar imaging. In *Proceedings of the 14th IEEE Workshop on Signal Processing Advances in Wireless Communications (SPAWC)*, Darmstadt, Germany, pp. 425–429, June 16–19, 2013.
28. F. H. C. Tivive, A. Bouzerdoum, and M. G. Amin. An svd-based approach for mitigating wall reflections in through-the-wall radar imaging. In *Proceedings of the IEEE Radar Conference (RadarCon 2011)*, Kansas City, MO, pp. 19–524, 2011.
29. J. Yang, A. Bouzerdoum, F. H. C. Tivive, and M. G. Amin. Multiple-measurement vector model and its application to through-the-wall radar imaging. In *Proceedings of the IEEE International Conference on Acoustics, Speech and Signal Processing (ICASSP)*, Prague, Czech Republic, pp. 2672–2675, 2011.
30. Y.-S. Yoon and M. G. Amin. Through-the-wall radar imaging using compressive sensing along temporal frequency domain. In *IEEE International Conference on Acoustics Speech and Signal Processing (ICASSP)*, Dallas, TX, pp. 2806–2809, 2010.
31. Y.-S. Yoon and M. G. Amin. Compressed sensing technique for high-resolution radar imaging. In *Signal Processing, Sensor Fusion, and Target Recognition XVII, Vol. 6968 of Proceedings of SPIE*, pp. 69681A 1–10, April 3, 2008.

8

Sparsity-Aware Human Motion Indication

Moeness G. Amin

CONTENTS

ABSTRACT This chapter considers sparsity-driven moving target indication. A radar imaging system can recognize an animate target indoors by a change in target position. The course of that change can be continuously or nonconsecutively monitored by the radar. The target location over time can be guided by a particular model or characterized by a specific motion profile. In this case, using a sequence of radar pulses, target velocity and its derivatives, including acceleration, can be obtained by estimating the corresponding Doppler frequency and higher-order motion parameters. Alternatively, the radar may illuminate the scene at particular time instants to capture the target at different positions. Detection of target position variations by subtraction of data or images at different time instants is the foundation of change detection where details of motion profile are difficult to discern and become irrelevant to motion indication. These instants can be widely separated, several pulses apart, or consecutive to detect short human movements. Whether motion modeling and character- ization is performed or change detection is applied, we show in this chapter that the targets, maintaining or migrating their range, can be localized with significantly reduced amount of data due to sparseness in space and velocity.

8.1 Introduction

One of the primary goals of through-the-wall radar imaging (TWRI) and urban sensing is the detection and localization of human targets [1–7]. Humans belong to the class of animate objects, which are characterized by the motion of the torso and limbs, breathing, and heartbeat. These features make animate objects distinguishable from other objects and clutter, and they enable target detection to proceed based on changes in the phase of the scattered radar signals over time. Fans, pendulums, and other inanimate moving targets exhibit a type of motion that is confined to oscillations, rota- tions, or vibrations and, as such, can be easily distinguished from large or small translation motions that signify human movements.

For urban sensing environments, humans living in their residences and inside their homes do not typically generate fixed prolonged Doppler fre- quencies. On the contrary, hand waving, arm movements, body swaying, and head turning define most of the cross-motor activities. In essence, changes in the backscattered signal phase due to human motion do not nec- essarily lend themselves to Doppler frequency shifts. This is because the human motion can be abrupt and highly nonstationary, producing a time- dependent phase whose rate of change may fail to translate into single or multicomponent sinusoids that can be captured by different single Doppler filters. Rather, the corresponding wide spectrum of human motions can span the entire radar frequency band. In many cases, and in lieu of Doppler fil- ters, time-frequency processing can be applied to reveal the instantaneous frequency signatures [8,9]. However, time-Doppler frequency signal rep- resentations and the corresponding micro-Doppler signatures are complex, generate high false alarm rates, and are often very difficult to detect with- out a carrier frequency of tens of GHz. High-frequency signals encounter large attenuation through exterior and interior walls. Therefore, micro- Doppler processing for indoor target surveillance may not be a viable option all the time whether the nonstationary radar returns are considered to be deterministic or stochastic [10].

Change detection (CD) can be used instead of Doppler processing, wherein human detection is accomplished by subtraction of some data frames acquired over successive probing of the scene [11–16]. Change detection mitigates the heavy clutter that is caused by strong reflections from exterior and interior walls and also removes stationary objects inside enclosed structures, thereby rendering a densely populated scene sparse [17]. With sparsity, it becomes possible to handle reductions in data collections which are reflected in using fewer antenna elements or time samples or frequency steps, depending on the choice of the transmit waveform. These reductions culminate in fast data acquisition and quick actionable intelligence.

In this chapter, we consider sparsity-based moving target indication using both change detection and Doppler information. Compressive sensing (CS) for radar imaging has been shown to be an effective tool [18–21]. Its application to urban radar was first proposed in [22], and then pursued in subsequent work [23,24]. The specific urban radar sensing goal, considered in this chapter, is to detect and localize human targets inside buildings, while simultaneously offering a sizable reduction in the data volume. CD is applied to different data frames for each range bin in the unambiguous range. It is noted that the frames can be consecutive, which is suited for targets exhibiting sudden short motions at the same range gate, or nonconsecutive, with relatively long time difference, for the case in which the target changes its range gate position. Scene reconstruction is then achieved using CS techniques and sparsity-driven imaging methods. We focus on human targets undergoing translational motion as well as sudden short movements of their limbs, heads, and/or torsos. The latter is a typical situation underlying the activities in homes, lecture halls, and auditoriums as well as other sit-down human interactions. For each type of motion, we establish an appropriate model and define the corresponding sensing matrix, which is then used to achieve sparse scene reconstructions.

When target position is characterized by few motion parameters and monitored by the radar from pulse to pulse, then it is possible to incorporate common motion profiles in the forward model. This, in turn, allows us to carry out sparsity-driven joint localization of stationary and moving targets using a reduced set of data observations. We focus on reduction in spatial as well as fast- and slow-time observations. For this case, ultra-wideband (UWB) pulsed radar platforms can be used where the compact temporal support of the UWB signal may be exploited to suppress the front wall clutter through time gating. Exterior wall suppression improves sparsity and enables the application of CS for behind-the-wall scene reconstruction.

In this chapter, we demonstrate the effectiveness of the above two localization schemes using real data collected in a laboratory environment in the Radar Imaging facility at the Center for Advanced Communications, Villanova University. Target sparse scenes are imaged with humans undergoing both translation and short sudden movements. For both types of human motions, it is shown that, compared with the conventional backprojection-based change detection methods, sparsity-driven motion target indications achieve substantial reduction in the data volume without degradation in system performance.

8.2 Change Detection

8.2.1 Backprojection-Based Change Detection

A sequential multiplexing of the transmitters with simultaneous reception at multiple receivers is assumed. The sequential transmit operation is the salient feature of three known TWRI systems; one is built by the Army Research Lab [25], the other by the Defense Research and Development Canada [26], and the third by MIT Lincoln Lab [13]. With the assumption of sequential multiplexing, a signal model can thus be developed based on single active transmitters. We note that the timing interval for each data frame is assumed to be a fraction of a second so that the moving target appears stationary during each data collection interval. Depending on the number of transmitters, the nature of multiplexing, and the employed coherent integration period for signal-to-noise ratio (SNR) enhancement, the frame may include one pulse or span several pulse repetition periods.

Let $s(t)$ be the wideband baseband signal used for interrogating the scene. For the case of a single point target, located at $\mathbf{x}_p = (x_p, y_p)$, the pulse emitted by the mth transmitter with phase center at $\mathbf{x}_{tm} = (x_{tm}, 0)$ is received at the nth receiver with phase center at $\mathbf{x}_{rn} = (x_{rn}, 0)$ in the form

$$z_{mn}(t) = a_{mn}(t) + b_{mn}(t), \quad a_{mn}(t) = \sigma_p s(t - \tau_{p,mn}) \exp(-j\omega_c \tau_{p,mn}) \quad (8.1)$$

where
  σ_p is the complex reflectivity of the target, which is assumed to be independent of frequency and aspect angle
  ω_c is the carrier frequency
  $\tau_{p,mn}$ is the propagation delay for the signal to travel between the mth transmitter, the target at $\mathbf{x}_p$, and the nth receiver
  $b_{mn}(t)$ represents the contribution of the stationary background at the nth receiver with the mth transmitter active

For through-the-wall propagation, $\tau_{p,mn}$ will comprise the components corresponding to traveling distances before, through, and after the wall. Note that the expression in (8.1) does not consider the wall attenuation and free-space path loss encountered by the radar return, which can be easily incorporated.

In its simplest form, change detection is achieved by coherent subtraction of the data corresponding to two data frames, which may be consecutive or separated by one or more data frames. This subtraction operation is performed for each range bin. CD results in the set of difference signals, given by

$$\delta z_{mn}(t) = z_{mn}^{(L+1)}(t) - z_{mn}^{(1)}(t) = a_{mn}^{(L+1)}(t) - a_{mn}^{(1)}(t) \qquad (8.2)$$

where L denotes the number of frames between the two time acquisitions. The component of the radar return from the stationary background is the same over the two time intervals and is thus removed from the difference signal. We assume that the clutter bandwidth is zero and it is confined to the zero Doppler frequency. It is noted that $L=1$ represents the case when the two acquisitions are performed over consecutive frames. Using (8.1) and (8.2), the (m,n)th difference signal can be expressed as

$$\delta z_{mn}(t) = \sigma_p s\left(t - \tau_{p,mn}^{(L+1)}\right) \exp\left(-j\omega_c \tau_{p,mn}^{(L+1)}\right) - \sigma_p s\left(t - \tau_{p,mn}^{(1)}\right) \exp\left(-j\omega_c \tau_{p,mn}^{(1)}\right)$$

$$(8.3)$$

where $\tau_{p,mn}^{(1)}$ and $\tau_{p,mn}^{(L+1)}$ are the respective two-way propagation delays for the signal to travel between the mth transmitter, the target, and the nth receiver, during the first and the second data acquisitions, respectively. In order to generate an image of the scene being interrogated, the MN difference signals corresponding to the operation of M transmitters and N receivers are processed as follows. The region of interest is divided into a finite number of grid points in x and y, where x and y represent crossrange and downrange, respectively. The composite signal corresponding to the pixel, located at $\mathbf{x}_q = (x_q, y_q)$, is obtained by summing time delayed versions of the MN difference signals

$$\delta z_q(t) = \sum_{m=0}^{M-1} \sum_{n=0}^{N-1} \delta z_{mn}\left(t + \tau_{q,mn}\right)$$

$$= \sum_{m=0}^{M-1} \sum_{n=0}^{N-1} \left(a_{mn}^{(L+1)}\left(t + \tau_{q,mn}\right) - a_{mn}^{(1)}\left(t + \tau_{q,mn}\right)\right) \qquad (8.4)$$

where $\tau_{q,mn}$ is the focusing delay applied to the (m,n)th difference signal. It is noted that additional weighting can be applied during the summation operations of (8.4) to control the sidelobe level of the system point spread function [27]. Substituting (8.3) in (8.4) yields

$$\delta z_q(t) = \sum_{m=0}^{M-1} \sum_{n=0}^{N-1} \sigma_p \left(s \left(t + \tau_{q,mn} - \tau_{p,mn}^{(L+1)} \right) \exp\left(-j\omega_c \left(\tau_{p,mn}^{(L+1)} - \tau_{q,mn} \right) \right) \right.$$

$$\left. -s \left(t + \tau_{q,mn} - \tau_{p,mn}^{(1)} \right) \exp\left(-j\omega_c \left(\tau_{p,mn}^{(1)} - \tau_{q,mn} \right) \right) \right) \tag{8.5}$$

The complex amplitude image value $I(\mathbf{x}_q)$, for the pixel at $\mathbf{x}_q$, is then obtained by applying a filter, matched to $s(t)$ to $\delta z_q(t)$ and sampling the filtered data, as per the following equation:

$$I(\mathbf{x}_q) = \delta z_q(t) * h(t) \big|_{t=0} \tag{8.6}$$

where $h(t) = s^*(-t)$ is the impulse response of the matched filter, the superscript $*$ denotes complex conjugation, and $*$ denotes the convolution operator. The process described by (8.4) through (8.6) is repeated for all pixels in the image to generate the composite image of the scene. The general case of multiple targets can be obtained by superposition of target reflections.

Merely employing part of the signal time duration in backprojection provides an image quality that is degraded in proportion to the number of missing data [29]. Since the removal of stationary background converts a populated scene into a sparse scene of moving targets, reduction in data volume should be pursued under the CS framework.

8.2.2 Sparsity-Driven Change Detection under Translational Motion

Consider the difference signal in (8.3), reproduced later for convenience, for the case where the target is undergoing translational motion. Two non-consecutive data frames with relatively long time difference are used, that is, $L \gg 1$ [28].

$$\delta z_{mn}(t) = \sigma_p s \left(t - \tau_{p,mn}^{(L+1)} \right) \exp\left(-j\omega_c \tau_{p,mn}^{(L+1)}\right) - \sigma_p s \left(t - \tau_{p,mn}^{(1)} \right) \exp\left(-j\omega_c \tau_{p,mn}^{(1)}\right) \tag{8.7}$$

In this case, the target will change its range gate position during the time elapsed between the two data acquisitions. As seen from (8.7), the moving target will present itself as two targets, one corresponding to the target position during the first time interval and the other corresponding to the target location during the second data frame. It is noted that the imaged target at the reference position corresponding to the first data frame cannot be suppressed for the coherent change detection approach, whether employing backprojection or sparsity-driven imaging. On the other hand, the noncoherent CD approach that deals with differences of image magnitudes corresponding to the two data frames allows suppression of the reference image through a zero thresholding operation [14]. However, as the noncoherent approach requires the scene reconstruction to be performed prior to change detection, it is not a feasible option for sparsity-based imaging, which relies on coherent CD to render the scene sparse. Therefore, we rewrite (8.7) as

$$\delta z_{mn}(t) = \sum_{i=1}^{2} \tilde{\sigma}_i s\left(t - \tau_{i,mn}\right) \exp(-j\omega_c \tau_{i,mn}) \tag{8.8}$$

with

$$\tilde{\sigma}_i = \begin{cases} \sigma_p & i = 1 \\ -\sigma_p & i = 2 \end{cases} \quad \text{and} \quad \tau_{i,mn} = \begin{cases} \tau_{p,mn}^{(L+1)} & i = 1 \\ \tau_{p,mn}^{(1)} & i = 2 \end{cases} \tag{8.9}$$

Assume that the scene being imaged, or the target space, is divided into a finite number of grid points in crossrange and downrange. If we sample the difference signal $\delta z_{mn}(t)$ at times $\{t_k\}_{k=0}^{K-1}$ to obtain the $K \times 1$ vector Δz_{mn} and form the concatenated $Q \times 1$ scene reflectivity vector $\tilde{\sigma}$ corresponding to the spatial sampling grid, then using the developed signal model in (8.9), we obtain the linear system of equations

$$\Delta z_{mn} = \Psi_{mn} \tilde{\sigma} \tag{8.10}$$

The qth column of Ψ_{mn} consists of the received signal corresponding to a target at grid point x_q and the kth element of the qth column can be written as

$$[\Psi_{mn}]_{k,q} = \frac{s(t_k - \tau_{q,mn}) \exp(-j\omega_c \tau_{q,mn})}{\|s_{q,mn}\|_2},$$

$$k = 0, 1, \ldots, K - 1, q = 0, 1, \ldots, Q - 1 \tag{8.11}$$

where the kth element of the vector $s_{q,mn}$ is $s(t_k - \tau_{q,mn})$, which implies that the denominator in the RHS of (8.11) is the energy in the time signal. Therefore, each column of Ψ_{mn} has unit norm. Further, note that $\tilde{\sigma}$ in (8.10) is a weighted indicator vector defining the scene reflectivity, that is, if there is a target at the qth grid point, the value of the qth element of $\tilde{\sigma}$ should be $\tilde{\sigma}_q$; otherwise, it is zero.

The change detection model described in (8.10 and 8.11) permits the scene reconstruction within the CS framework. We measure an L ($<K$) dimensional vector of elements randomly chosen from Δz_{mn}. The new measurements can be expressed as

$$\xi_{mn} = \mathbf{\Phi}_{mn}\Delta\mathbf{z}_{mn} = \mathbf{\Phi}_{mn}\mathbf{\Psi}_{mn}\tilde{\sigma} = \mathbf{A}_{mn}\tilde{\sigma} \qquad (8.12)$$

where $\mathbf{\Phi}_{mn}$ is an $\breve{L} \times K$ measurement matrix and $\mathbf{A}_{mn} = \mathbf{\Phi}_{mn}\mathbf{\Psi}_{mn}$ is an $\breve{L} \times Q$ matrix. Several types of measurement matrices have been reported in the literature [30 and the references therein]. It was shown in [20] that the measurement matrix with random ± 1 elements requires the least amount of compressive measurements for the same radar imaging performance and permits a relatively straightforward data acquisition implementation. Therefore, we use such a measurement matrix in image reconstructions.

Given ξ_{mn} for $m = 0, 1, \ldots, M-1, n = 0, 1, \ldots, N-1$, we can recover $\tilde{\sigma}$ by solving the following equation:

$$\hat{\tilde{\sigma}} = \arg\min_{\tilde{\sigma}} \|\tilde{\sigma}\|_1 \text{ subject to } \mathbf{A}\tilde{\sigma} \approx \xi \qquad (8.13)$$

where

$$\mathbf{A} = \left[\mathbf{A}_{00}^T \, \mathbf{A}_{01}^T \, \cdots \, \mathbf{A}_{(M-1)(N-1)}^T\right]^T, \quad \xi = \left[\xi_{00}^T \, \xi_{01}^T \, \cdots \, \xi_{(M-1)(N-1)}^T\right]^T \quad (8.14)$$

A stable solution of the sparse target space reconstruction problem in Equation 8.13 is guaranteed provided that the matrix $\mathbf{A}$ satisfies the restricted isometry property (RIP), which states that all subsets of r columns taken from $\mathbf{A}$ are, in fact, nearly orthogonal, r being the sparsity of the signal $\tilde{\sigma}$ [20,21,31]. In general, it is computationally difficult to check this property and, therefore, other related measures on the matrix $\mathbf{A}$, such as mutual coherence, are often used to guarantee stable recovery through l_1-minimization. Mutual coherence of the columns of $\mathbf{A}$ can be viewed as the largest off-diagonal entry of the Gram matrix $\mathbf{A}^H\mathbf{A}$, where the columns of $\mathbf{A}$ have been normalized. We note that the problem in Equation 8.13 can be solved using convex relaxation, greedy pursuit, or combinatorial algorithms [31–33]. Also, Bayesian CS techniques, which are robust to the RIP condition, can be applied to the linear model in (8.10) [34]. In this chapter, we chose CoSaMP as the reconstruction algorithm primarily because of its ability to handle complex arithmetic [33].

Equations 8.13 and 8.14 represent one strategy that can be adopted for sparsity-based change detection approach, wherein a reduced number of time samples are chosen randomly for all the transmitter–receiver pairs constituting the array apertures. These two equations can also be extended so that the reduction in data measurements includes both spatial and time samples.

8.2.3 Sparsity-Driven Change Detection under Short Sudden Movements

Assume that consecutive ($L = 1$) data frames are employed for change detection and consider a scene comprising a human target undergoing sudden short movements of the limbs, head, and/or torso. In this case, we can model the target as a cluster of P point scatterers within the same resolution cell and only a small number, say P_1, of these scatterers move during successive data acquisitions. For example, in a round-table meeting, the upper part of the human body, especially the hands, is likely to move while the legs remain stationary over successive observations. Using (8.1), the baseband received signal, corresponding to the (m, n)th transmitter–receiver pair, for the first data frame can be expressed as

$$z_{mn}^{(1)}(t) = \sum_{p=1}^{P} \sigma_p s\left(t - \tau_{p,mn}^{(1)}\right) \exp\left(-j\omega_c \tau_{p,mn}^{(1)}\right) + b_{mn}(t) \qquad (8.15)$$

where

σ_p is the complex reflectivity of the pth point scatterer

$\tau_{p,mn}^{(1)}$ is the two-way propagation delay for the signal to travel between the (m, n)th transmitter–receiver pair and the pth scatterer during the first frame

As the P scatterers are clustered within the same resolution cell, we can rewrite (8.15) as

$$z_{mn}^{(1)}(t) = \sigma_{mn}^{(1)} s\left(t - \bar{\tau}_{mn}\right) \exp\left(-j\omega_c \bar{\tau}_{mn}\right) + b_{mn}(t) \qquad (8.16)$$

where $\bar{\tau}_{mn}$ is the propagation delay from the mth transmitter to the center of the cell and back to the nth receiver, and

$$\sigma_{mn}^{(1)} = \sum_{p=1}^{P} \sigma_p \exp\left(-j\omega_c \Delta\tau_{p,mn}^{(1)}\right) \qquad (8.17)$$

is the net target reflectivity with $\Delta\tau_{p,mn}^{(1)} = \tau_{p,mn}^{(1)} - \bar{\tau}_{mn}$.

Let the first P_1 scatterers represent the portion of the body that undergoes a short movement. Then, the mth received signal corresponding to the second data frame can be expressed as

$$z_{mn}^{(2)}(t) = \sigma_{mn}^{(2)} s(t - \bar{\tau}_{mn}) \exp\left(-j\omega_c \bar{\tau}_{mn}\right) + b_{mn}(t) \qquad (8.18)$$

with the net reflectivity $\sigma_{mn}^{(2)}$ given by

$$\sigma_{mn}^{(2)} = \sum_{p=1}^{P_1} \sigma_p \exp\left(-j\omega_c \Delta\tau_{p,mn}^{(2)}\right) + \sum_{p=P_1+1}^{P} \sigma_p \exp\left(-j\omega_c \Delta\tau_{p,mn}^{(1)}\right) \qquad (8.19)$$

and the set of differential delays, $\left\{\Delta\tau_{p,mn}^{(2)} = \tau_{p,mn}^{(2)} - \bar{\tau}_{mn}\right\}_{p=1}^{P_1}$, corresponds to the new locations of the P_1 scatterers within the same resolution cell. The difference signal corresponding to these successive data measurements is given by

$$\delta z_{mn}(t) = z_{mn}^{(2)}(t) - z_{mn}^{(1)}(t) = \left(\sigma_{mn}^{(2)} - \sigma_{mn}^{(1)}\right) s(t - \bar{\tau}_{mn}) \exp(-j\omega_c\bar{\tau}_{mn})$$

$$= \delta\sigma_{mn}s(t - \bar{\tau}_{mn}) \exp(-j\omega_c\bar{\tau}_{mn}) \qquad (8.20)$$

where $\delta\sigma_{mn}$ represents the change in reflectivity between the consecutive acquisitions.

Again, working with a discretized version of (8.20), we obtain the linear system of equations:

$$\Delta z_{mn} = \Psi_{mn}\delta\sigma_{mn} \qquad (8.21)$$

where

Ψ_{mn} is defined in (8.11)

$\delta\sigma_{mn}$ is a weighted indicator vector defining the change in scene reflectivity as observed at the nth receiver with the mth transmitter active, that is, if there is a change in target reflectivity at the qth grid point, the value of the qth element of $\delta\sigma_{mn}$ will be $\sigma_{q,mn}^{(2)} - \sigma_{q,mn}^{(1)}$ and zero otherwise

For the signal model in (8.21), we observe that the change in scene reflectivity depends on the transmitter and receiver locations. As such, the aspect-independent scattering assumption is no longer applicable and the scene reflectivity changes for each transmitter–receiver pair. To address this issue, we consider composite image formation using subapertures [35]. In this scheme, the transmit and receive arrays are divided into subapertures. Assuming isotropic scattering within the angular extent of these subapertures, subimages can be obtained, which are then combined to form a single composite image of the scene. The subaperture-based scene reconstruction can be performed within the CS framework using the change detection model of (8.21).

Assume the M-element transmit and the N-element receive arrays are divided into K_1 and K_2 nonoverlapping subapertures, respectively. The choice of K_1 and K_2 is guided by the local isotropy requirement, that is, each transmit and receive subaperture should correspond to a small aspect angle data set (typically on the order of a few degrees). In the spirit of CS, a small number of *random* measurements carry enough information to completely represent the sparse signal $\delta\sigma^{(k_1,k_2)}$, which is the *image* of the scene corresponding to the k_1th transmit and the k_2th receive subapertures. Thus, we measure a random subset of $\breve{L}$ ($<K$) samples of the difference signal for the n_{k_2}th antenna of the k_2th receive subaperture when the m_{k_1}th antenna of the k_1th transmit subaperture is active. In matrix form, the new measurements can be expressed as

$$\xi^{(k_1,k_2)}_{m_{k_1} n_{k_2}} = \Phi^{(k_1,k_2)}_{m_{k_1} n_{k_2}} \Delta z^{(k_1,k_2)}_{m_{k_1} n_{k_2}} = \Phi^{(k_1,k_2)}_{m_{k_1} n_{k_2}} \Psi^{(k_1,k_2)}_{m_{k_1} n_{k_2}} \delta\sigma^{(k_1,k_2)} = A^{(k_1,k_2)}_{m_{k_1} n_{k_2}} \delta\sigma^{(k_1,k_2)}$$

(8.22)

where $\Phi^{(k_1,k_2)}_{m_{k_1} n_{k_2}}$ is an $\breve{L} \times K$ measurement matrix corresponding to the n_{k_2}th antenna position in the k_2th receive subaperture and the m_{k_1}th antenna in the k_1th transmit subaperture and the matrix $A^{(k_1,k_2)}_{m_{k_1} n_{k_2}} = \Phi^{(k_1,k_2)}_{m_{k_1} n_{k_2}} \Psi^{(k_1,k_2)}_{m_{k_1} n_{k_2}}$ is of dimension $\breve{L} \times Q$. Given $\xi^{(k_1,k_2)}_{m_{k_1} n_{k_2}}$ for $m_{k_1} = 0, 1, \ldots, \lceil M/K_1 \rceil - 1, n_{k_2} = 0, 1, \ldots, \lceil N/K_2 \rceil - 1$, we can recover $\delta\sigma^{(k_1,k_2)}$ by solving the following equation:

$$\delta\hat{\sigma}^{(k_1,k_2)} = \arg\min_{\alpha} \|\alpha\|_1 \text{ subject to } A^{(k_1,k_2)}\alpha \approx \xi^{(k_1,k_2)}$$

(8.23)

where

$$A^{(k_1,k_2)} = \left[\left(A^{(k_1,k_2)}_{00}\right)^T \left(A^{(k_1,k_2)}_{01}\right)^T \cdots \left(A^{(k_1,k_2)}_{\lceil M/K_1 \rceil -1, \lceil N/K_2 \rceil -1}\right)^T \right]^T$$

$$\xi^{(k_1,k_2)} = \left[\left(\xi^{(k_1,k_2)}_{00}\right)^T \left(\xi^{(k_1,k_2)}_{01}\right)^T \cdots \left(\xi^{(k_1,k_2)}_{\lceil M/K_1 \rceil -1, \lceil N/K_2 \rceil -1}\right)^T \right]^T$$

(8.24)

Once the subimages $\delta\hat{\sigma}^{(k_1,k_2)}$ corresponding to all K_1 transmit and K_2 receive subapertures have been reconstructed, the composite image $\delta\hat{\sigma}$ can be formed as

$$[\delta\hat{\sigma}]_q = \arg\max_{k_1,k_2} \left| [\delta\hat{\sigma}^{(k_1,k_2)}]_q \right|$$

(8.25)

where $[\delta\hat{\sigma}]_q$ and $\left[\delta\hat{\sigma}^{(k_1,k_2)}\right]_q$ denote the qth pixel of the composite image and the subimage corresponding to the k_1th transmit and k_2th receive subapertures, respectively. Alternative methods for accounting for the different scattering coefficients across the transmitter and receiver apertures include group sparsity, which can be solved using greedy algorithms, convex optimization, or Bayesian CS techniques.

It is noted that the two aforementioned models of translation and sudden change motions can be combined to describe a change detection based on a more general motion profile.

8.2.4 Experimental Results for Change Detection

A through-the-wall wideband pulsed radar system was used for real data collection in the Radar Imaging Lab at Villanova University. The system uses a 0.7 ns pulse, shown in Figure 8.1, for scene interrogation. The pulse is up-converted to 3 GHz for transmission and down-converted to baseband through in-phase and quadrature demodulation on reception. The system operational bandwidth from 1.5 to 4.5 GHz provides a range resolution of 5 cm. The peak transmit power is 25 dBm. Transmission is through a single horn antenna, model BAE-H1479, with an operational bandwidth from 1 to 12.4 GHz, which is mounted on a tripod. An eight-element line array of Vivaldi elements with an inter-element spacing of 0.06 m is used as the receiver and is placed to the right of the transmit antenna. The center-to-center separation between the transmitter and the leftmost receive antenna is 0.28 m, as shown in Figure 8.2. A 3.65 m × 2.6 m wall segment was constructed utilizing 1-cm-thick cement board on a 2-by-4 wood stud frame. The transmit antenna and the receive array were at a standoff distance of 1.19 m from the wall. The pulse repetition frequency (PRF) is 10 MHz, providing an unambiguous range of 15 m, which is roughly three times the length of the

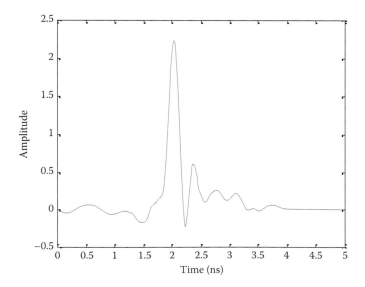

FIGURE 8.1
The wideband pulse used for imaging.

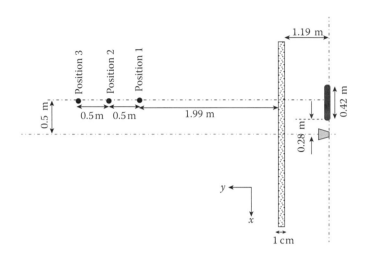

FIGURE 8.2
(See color insert.) Scene layout for the target undergoing translational motion. (From Ahmad, F. and Amin, M., Through the wall human motion indicator using sparsity-driven change detection, *IEEE Trans. Geosci. Rem. Sens.*, 51(2), 881, February 2013. With permission. © 2010 IEEE.)

room being imaged. Despite the high PRF, the system refresh rate is 100 Hz. This is because (a) equivalent time sampling is used [36], and (b) instead of simultaneous reception, the receive array elements are accessed sequentially through a multiplexer.

In order to illustrate the performance of the sparsity-driven change detection scheme under both translational and sudden short human motions, two different experiments were considered. In the first experiment, a person walked away from the wall in an empty room (the back and the side walls were covered with RF-absorbing material) along a straight-line path. The path is located 0.5 m to the right of the center of the scene, as shown in Figure 8.2. The data collection started with the target at position 1 and ended after the target reached position 3, with the target pausing at each position along the trajectory for a second. Consider the data frames corresponding to the target at positions 2 and 3. Each frame consists of 20 pulses, which are coherently integrated to improve the SNR. The imaging region (target space) is chosen to be 3 m × 3 m, centered at (0.5 m, 4 m), and divided into 61 × 61 grid points in crossrange and downrange, resulting in 3721 unknowns. The space–time response of the target space consists of 8 × 1536 space–time measurements. Figure 8.3a shows the backprojection-based CD image of the scene using all 8 × 1536 data points. In this figure and all subsequent figures in this section, we plot the image intensity with the maximum intensity value in each image normalized to 0 dB. We observe that, as the human changed its range gate position during the time elapsed between the two data

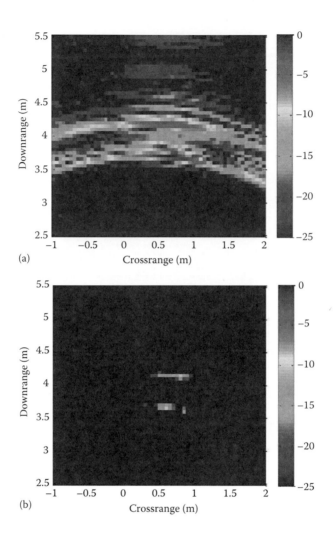

FIGURE 8.3
(See color insert.) (a) Backprojection-based CD image using the full data set. (b) Sparsity-based CD image using 5% of the data volume, averaged over 100 trials. (From Ahmad, F. and Amin, M., Through the wall human motion indicator using sparsity-driven change detection, *IEEE Trans. Geosci. Rem. Sens.*, 51(2), 881, February 2013. With permission. © 2010 IEEE.)

acquisitions, it presents itself as two targets in the image and is correctly localized at both of its positions.

For sparsity-based CD, only 5% of the 1536 time samples are randomly selected at each of the eight receive antenna locations, resulting in 8×77 space–time measured data. More specifically, the 77 time samples at each receive location were obtained as the product of the 1536 point time-domain response with a 77×1536 measurement matrix, whose elements are randomly chosen ± 1 values with a probability of $1/2$. According to CS theory, an *r*-sparse target space with Q unknowns can be recovered from $O(r \log(Q))$ measurements [37]. The human target roughly extended 0.5 m in crossrange and 0.25 m in downrange, thereby occupying 10×5 grid points. Therefore, for the data set under consideration wherein the target presents itself as two targets after change detection, the 8×77 measured data points exceed this requirement of $O(r \log(Q))$ measurements. We reconstructed the target space using sparsity-based CD with 5% data volume 100 times. For each trial, a different random measurement matrix was used to generate the reduced set of measurements, followed by sparsity-based scene reconstruction. For each of the 100 trials, we also computed the mutual coherence of the columns of the matrix **A**, which is the product of the measurement matrix $\boldsymbol{\Phi}$ with random ± 1 elemental values and the $\boldsymbol{\Psi}$ matrix defined in Equation 8.14. The average value of the mutual coherence of the columns of **A** is equal to 0.892. Figure 8.3b depicts the sparsity-based CD result, averaged over 100 trials. The higher the intensity of a grid point in this figure, the greater is the number of times that grid point was populated during the 100 reconstruction trials. We observe that, on average, the sparsity-based CD scheme detects and localizes the target accurately at both positions. Also, compared with the backprojection-based result of Figure 8.3a, the image in Figure 8.3b is less cluttered. The *cleaner* image is due to the fact that a sparse solution is enforced by the l_1 minimization in Equation 8.13.

Next, we collected data from a scene, consisting of a standing human facing the wall, located at 0.5 m crossrange and at a downrange of 3.9 m from the radar, as shown in Figure 8.4. The data were collected with the target initially looking straight at the wall and then suddenly lifting the head

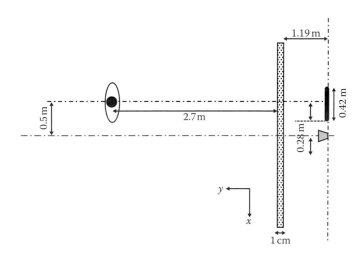

FIGURE 8.4
Scene layout for the target undergoing sudden short movement.

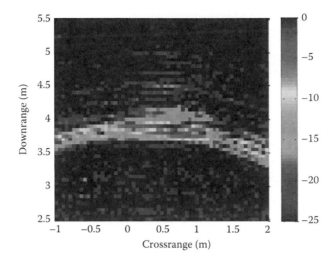

FIGURE 8.5
(See color insert.) Backprojection-based CD image using full data volume for the target under-going short sudden movement. (From Ahmad, F. and Amin, M., Through the wall human motion indicator using sparsity-driven change detection, *IEEE Trans. Geosci. Rem. Sens.*, 51(2), 881, February 2013. With permission. © 2010 IEEE.)

to look upward. As the person moved the head, there was also a slight movement of the shoulders and heaving of the chest. Two data frames of 20 pulses each corresponding to the two head positions were considered. The system parameters, the dimensions of the target space, the number of grid points, and the number of space–time measurements employed for the backprojection-based and sparsity-based reconstructions are all the same as those for the translational motion example. The target space image obtained using the backprojection-based CD with full data volume (after coherent integration) is shown in Figure 8.5. We observe that the change detection approach was able to detect the cumulative change in target reflectivity due to the head movement and associated slight outward and upward movement of the chest as the target looked upward. Compared with the translational motion image of Figure 8.3a, the sudden short movement image is more clut-tered. This is because the radar return is much weaker in this case due to a slight motion of only a small part of the body.

For the corresponding sparsity-based CD composite imaging results, we used two subapertures, each consisting of four receive antenna elements, and employed only 5% of the total data volume. That is, we used 77 time samples per antenna location within each of the subapertures. Similar to the trans-lational motion example, we performed scene reconstruction 100 times, and the averaged target space image with the subimages combined in accordance with (8.25) is provided in Figure 8.6. We observe that the sparsity-based

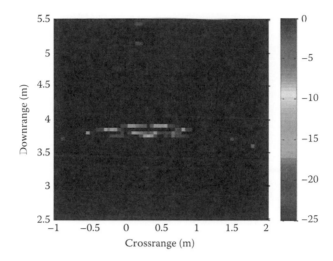

FIGURE 8.6

(See color insert.) Sparsity-based composite images with 5% data volume for the subimage combining approach in Equation 8.25. The image is the averages of 100 reconstructions. (From Ahmad, F. and Amin, M., Through the wall human motion indicator using sparsity-driven change detection, *IEEE Trans. Geosci. Rem. Sens.*, 51(2), 881, February 2013. With permission. © 2010 IEEE.)

CD approach successfully detects and localizes the target undergoing short movement using much reduced data volume.

It is noted that the work presented here only considered the sparsity of the target space and did not make any further assumptions about the support of the sparse solution during the reconstruction process. As humans are extended targets, they appear as clusters in the through-the-wall images. As such, the corresponding sparse solution support has an underlying block structure [38,39]. Future efforts will focus on exploiting this structured sparsity to further reduce the number of compressive measurements required for stable recovery.

8.3 Sparsity for Target Localization and Motion Parameter Estimation

Unlike change detection approach, the radar in this second part of the chapter is assumed to be continuously monitoring the scene. Motion, if it occurs, is characterized by the target velocity. The objective is then to locate the stationary and moving targets inside the room at any given time.

8.3.1 UWB Signal Model

Again, consider an M-element transmit array and an N-element receive array, both located along the x-axis at a standoff distance y_{off} from a homogeneous wall, as shown in Figure 8.7. Note that although the arrays are assumed to be parallel to the front wall for notational simplicity, this is not a requirement. Let $\mathbf{x}_{tm} = (x_{tm}, 0)$ and $\mathbf{x}_{rn} = (x_{rn}, 0)$ be the respective phase centers of the mth transmitter and the nth receiver. Let the transmit signal be expressed as

$$s(t) = a(t)\, \exp(j\omega_c t) \tag{8.26}$$

where
  $a(t)$ is the UWB baseband signal
  ω_c is the carrier frequency, and let T_r be the pulse repetition interval

Consider a coherent processing interval of $\breve{K}$ pulses per transmitter and a single point target moving slowly away from the origin with constant horizontal and vertical velocity components (v_{xp}, v_{yp}) as depicted in Figure 8.7. Let the target position be $\mathbf{x}_p = (x_p, y_p)$ at time $t = 0$. Assume that the timing interval for sequencing through the transmitters is short enough so that the target appears stationary during each data collection interval of length MT_r. This implies that the target position corresponding to the kth pulse is given by

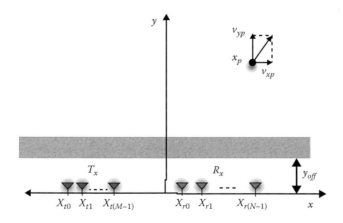

FIGURE 8.7
(See color insert.) Geometry on transmit and receive. (From Ahmad, F. and Amin, M., Through the wall human motion indicator using sparsity driven change detection, *IEEE Trans. Geosci. Rem. Sens.*, 51(2), 881, February 2013. With permission. © 2010 IEEE.)

$$\mathbf{x}_p(k) = \left(x_p + v_{xp}\breve{k}MT_r, y_p + v_{yp}\breve{k}MT_r \right) \qquad (8.27)$$

The baseband target return measured by the nth receiver corresponding to the kth pulse emitted by the mth transmitter is given by

$$\overset{p}{\underset{mn\,k}{\breve{z}}}(t) = \sigma_p a \left(t - \breve{k}MT_r - mT_r - \tau_{p,mn}(\breve{k}) \right) \exp\left(-j\omega_c \tau_{p,mn}(\breve{k}) \right) \qquad (8.28)$$

where

σ_p is the complex reflectivity of the target
$\tau_{p,mn}(\breve{k})$ is the propagation delay for the $\breve{k}$th pulse to travel from the mth transmitter to the target at $\mathbf{x}_p(k)$, and back to the nth receiver

For through-the-wall propagation, $\tau_{p,mn}(\breve{k})$ comprises the components corresponding to traveling distances before, through, and after the wall. In the presence of P point targets, the received signal component corresponding to the targets will be a superposition of the individual target returns in (8.28) with $p = 0, 1, \ldots, P-1$. Interactions between the targets and multipath returns are ignored in this model. Note that any stationary targets behind the wall are included in this model and would correspond to the motion parameter pair $(v_{xp}, v_{yp}) = (0, 0)$. Further, note that the slowly moving targets are assumed to remain within the same range cell over the coherent processing interval.

On the other hand, as the wall is a specular reflector, the baseband wall return received at the nth receiver corresponding to the kth pulse emitted by the mth transmitter can be expressed as

$$\underset{mn\,k}{\breve{z}^{wall}}(t) = \sigma_{wall} a(t - \breve{k}MT_r - mT_r - \tau_{wall,mn}) \exp(-j\omega_c \tau_{wall,mn}) + \underset{mn\,k}{\breve{B}^{wall}}(t)$$
$$(8.29)$$

where

σ_{wall} is the complex wall reflectivity
$\tau_{wall,mn}$ is the propagation delay from the mth transmitter to the wall and back to the nth receiver
$\underset{mn\,k}{\breve{B}^{wall}}(t)$ represents the wall reverberations of decaying amplitudes resulting from multiple reflections within the wall (see Figure 8.8)

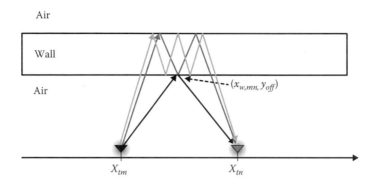

FIGURE 8.8
Wall reverberations.

The propagation delay $\tau_{wall,mn}$ is given by [38]

$$\tau_{wall,mn} = \frac{\sqrt{(x_{tm} - x_{w,mn})^2 + y_{off}^2} + \sqrt{(x_{rn} - x_{w,mn})^2 + y_{off}^2}}{c} \tag{8.30}$$

where c is the speed of light in free space, and

$$x_{w,mn} = \frac{x_{tm} + x_{rn}}{2} \tag{8.31}$$

is the point of reflection on the wall corresponding to the mth transmitter and the nth receiver, as shown in Figure 8.8. Note that, as the wall is stationary, the delay $\tau_{wall,mn}$ does not vary from one pulse to the next. Therefore, the expression in (8.29) assumes the same value for $k = 0, 1, \ldots, K - 1$.

Combining (8.28) and (8.29), the total baseband signal received by the nth receiver, corresponding to the kth pulse with the mth transmitter active, is given by

$$z'_{mnk}(t) = z^{wall}_{mnk}(t) + \sum_{p=0}^{P-1} z^{p}_{mnk}(t) \tag{8.32}$$

The returns from behind-the-wall targets, both moving and stationary, are, in general, much weaker than the front wall reflections, resulting in a low signal-to-clutter ratio (SCR). Because of the UWB nature of the transmit signal, it is natural to remedy this situation by gating out the wall return in the time domain, thereby providing access to the sparse behind-the-wall scene of a few stationary and moving targets of interest. Therefore, the time-gated received signal contains only contributions from the P targets behind the wall as well as any residuals of the wall not removed or fully mitigated by gating. In this chapter, we assume that wall clutter is effectively suppressed by gating. Therefore, using (8.32), we obtain

$$z_{\underset{mn\,k}{\smile}}(t) = \sum_{p=0}^{P-1} z^p_{\underset{mn\,k}{\smile}}(t) \tag{8.33}$$

8.3.2 Backprojection-Based Stationary and Moving Target Localization

Radar images are typically formed using the well-known backprojection algorithm. However, the presence of moving targets in the observed scene presents a problem for conventional backprojection. Unlike stationary targets, the moving targets are defocused and get smeared across the image. This makes it very difficult to detect and localize moving targets in the backprojected image [39]. Often, the approach to handling moving targets involves Doppler discrimination [40] in order to form a focused image of the moving targets. The simplest version is implemented as a fast Fourier transform (FFT) along the slow-time dimension of the raw data cube, followed by backprojection applied to the fast time versus spatial data per Doppler bin, as illustrated in Figure 8.9.

Consider the signal $\left\{ z_{\underset{mn\,k}{\smile}}(t) \right\}_{k=0}^{\breve{K}-1}$ received by the nth receiver with the mth transmitter active over the coherent processing interval, where $z_{\underset{mn\,k}{\smile}}(t)$ is given by (8.33). With the application of the Fourier transform to $\left\{ z_{\underset{mn\,k}{\smile}}(t) \right\}_{k=0}^{\breve{K}-1}$ along slow time, we let the resulting signal corresponding to the lth Doppler frequency bin be denoted by $z^l_{mn}(t)$. In order to generate the range versus crossrange image corresponding to the lth Doppler bin, the signal $z^l_{mn}(t)$ corresponding to all M transmitters and all N receivers is processed as follows.

The region of interest is divided into a finite number of pixels, say Q. The composite signal corresponding to the qth pixel, located at $\mathbf{x}_q = (x_q, y_q)$, can be obtained by applying focusing delays and then summing the results:

$$z^l_q(t) = \sum_{m=0}^{M-1} \sum_{n=0}^{N-1} z^l_{mn}(t + \tau_{q,mn}). \tag{8.34}$$

Note that the focusing delay $\tau_{q,mn}$ corresponds to the two-way signal propagation time between the mth transmitter, the qth pixel, and the nth receiver. If the target is indeed present at the qth pixel location, that is, $\mathbf{x}_q = \mathbf{x}_p$, then

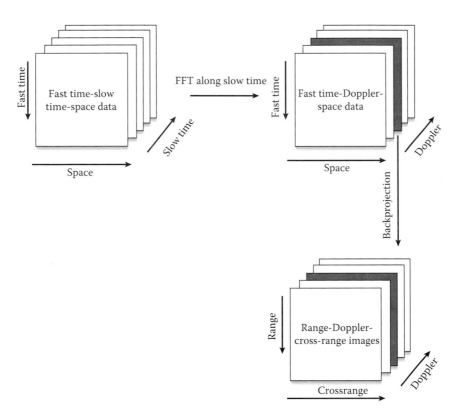

FIGURE 8.9
Flowchart of conventional backprojection-based imaging scheme.

the focusing delays align all the signals corresponding to the different transmitters and receivers, resulting in the signals being coherently combined. On
the other hand, if there is no target at the qth pixel, the same focusing time
delays will cause the various signals to stagger in time, thereby producing a
reduced combined output.

The complex amplitude image value corresponding to the qth pixel is
obtained from (8.34) as

$$I^l(\mathbf{x}_q) = z_q'^l(t)\Big|_{t=0} \tag{8.35}$$

where $z_q^l(t)$ is the output of the matched filter applied to the corresponding
Doppler bin. The process described by (8.34) and (8.35) is performed for all
Q pixels to generate the complex image of the scene corresponding to the lth
Doppler bin.

Note that it is possible to obtain the spatial and motion parameters
of targets in the field of view even if some of the spatial fast time–slow

time measurements are missing. However, merely employing some of the transmit and receive elements, a few pulses, and/or part of the signal time duration, in the backprojection-based scene reconstruction scheme degrades the image quality. CS framework is, therefore, more suitable for pursuing reduction in data volume.

8.3.3 CS-Based Stationary and Moving Target Localization

In this section, we develop the linear signal model with sensing matrices for application of CS and present the sparse reconstruction scheme for joint localization of stationary and moving targets inside enclosed structures.

8.3.3.1 Linear Model Formulation

With the observed scene divided into Q pixels in crossrange and downrange, consider N_{v_x} and N_{v_y} discrete values of the expected horizontal and vertical velocities, respectively. Therefore, an image with Q pixels in crossrange and downrange is associated with each considered horizontal and vertical velocity pair, resulting in a four-dimensional target space. Note that the considered velocities contain the (0, 0) velocity pair to include stationary targets:

Sampling the received signal $z_{mnk}(t)$ at times $\{t_k\}_{k=0}^{K-1}$, we obtain a $K \times 1$ vector z_{mnk}. For the lth velocity pair (v_{xl}, v_{yl}), we vectorize the corresponding crossrange versus downrange image into a $Q \times 1$ scene reflectivity vector $\sigma(v_{xl}, v_{yl})$. The vector $\sigma(v_{xl}, v_{yl})$ is a weighted indicator vector defining the scene reflectivity corresponding to the lth considered velocity pair, that is, if there is a target at the spatial grid point (x, y) with motion parameters (v_{xl}, v_{yl}), then the value of the corresponding element of $\sigma(v_{xl}, v_{yl})$ should be nonzero; otherwise, it is zero.

Using the developed signal model in (8.28) and (8.33), we obtain the linear system of equations

$$z_{mnk} = \Psi_{mnk}(v_{xl}, v_{yl})\sigma(v_{xl}, v_{yl}), \quad l = 0, 1, \ldots, (N_{v_x}N_{v_y} - 1) \tag{8.36}$$

where the matrix $\Psi_{mnk}(v_{xl}, v_{yl})$ is of dimension $K \times Q$. The qth column of $\Psi_{mnk}(v_{xl}, v_{yl})$ consists of the received signal corresponding to a target at pixel x_q with motion parameters (v_{xl}, v_{yl}) and the kth element of the qth column can be written as

$$\left[\Psi_{mnk}(v_{xl}, v_{yl})\right]_{k,q} = a\left(t_k - \breve{k}MT_r - mT_r - \tau_{q,mn}(\breve{k})\right)\exp\left(-j\omega_c\tau_{q,mn}(\breve{k})\right) \tag{8.37}$$

where $\tau_{q,mn}(\breve{k})$ is the two-way signal traveling time, corresponding to (v_{xl}, v_{yl}), from the mth transmitter to the qth spatial grid point and back to the nth receiver for the $\breve{k}$th pulse.

Stacking the received signal samples corresponding to $\breve{K}$ pulses from all MN transmitting and receiving element pairs, we obtain the $KMN\breve{K} \times 1$ measurement vector $\mathbf{z}$ as

$$\mathbf{z} = \mathbf{\Psi}(v_{xl}, v_{yl})\mathbf{\sigma}(v_{xl}, v_{yl}), \quad l = 0, 1, \ldots, (N_{v_x} N_{v_y} - 1) \tag{8.38}$$

where

$$\mathbf{\Psi}(v_{xl}, v_{yl}) = \left[\mathbf{\Psi}_{000}^T(v_{xl}, v_{yl}), \ldots, \mathbf{\Psi}_{(M-1)(N-1)(\breve{K}-1)}^T(v_{xl}, v_{yl}) \right]^T \tag{8.39}$$

Finally, forming the $KMN\breve{K} \times QN_{v_x}N_{v_y}$ matrix $\mathbf{\Psi}$ as

$$\mathbf{\Psi} = \left[\mathbf{\Psi}(v_{x0}, v_{y0}), \ldots, \mathbf{\Psi}(v_{x(N_{v_x} N_{v_y} - 1)}, v_{y(N_{v_x} N_{v_y} - 1)}) \right] \tag{8.40}$$

we obtain the linear matrix equation

$$\mathbf{z} = \mathbf{\Psi}\mathbf{\sigma} \tag{8.41}$$

with $\mathbf{\sigma}$ being the concatenation of target reflectivity vectors corresponding to every possible considered velocity combination.

8.3.3.2 CS Data Acquisition and Scene Reconstruction

The model described earlier permits the scene reconstruction within the CS framework. We measure an $\breve{L} < KMN\breve{K}$ dimensional vector of elements randomly chosen from $\mathbf{z}$. The reduced set of measurements can be expressed as

$$\tilde{\mathbf{z}} = \mathbf{\Phi}\mathbf{\Psi}\mathbf{\sigma} \tag{8.42}$$

where $\mathbf{\Phi}$ is a $\breve{L} \times KMN\breve{K}$ measurement matrix. For measurement reduction simultaneously along the spatial, slow time and fast time dimensions, the specific structure of the matrix $\mathbf{\Phi}$ is given by

$$\mathbf{\Phi} = \left(\mathbf{\Phi}_1 \otimes \mathbf{I}_{J\breve{K}_1 N_1} \right) \cdot \left(\mathbf{\Phi}_2 \otimes \mathbf{I}_{J\breve{K}_1 M} \right) \cdot \left(\mathbf{\Phi}_3 \otimes \mathbf{I}_{JMN} \right) \cdot \mathrm{diag}\left(\mathbf{\Phi}_4^{(0)}, \ldots, \mathbf{\Phi}_4^{(MN\breve{K}-1)} \right) \tag{8.43}$$

where

"$\otimes$" denotes the Kronecker product

$\mathbf{I}_{(\cdot)}$ is an identity matrix with the subscript indicating its dimensions

$M_1, N_1, \breve{K}_1$, and J denote the reduced number of transmit elements, receive elements, pulses, and fast time samples, respectively, with the total number of reduced measurements $\breve{L} = JM_1N_1\breve{K}_1$

The matrix $\mathbf{\Phi}_1$ is an $M_1 \times M$ matrix, $\mathbf{\Phi}_2$ is an $N_1 \times N$ matrix, $\mathbf{\Phi}_3$ is a $\breve{K}_1 \times \breve{K}$ matrix, and $\mathbf{\Phi}_4^{(i)}$, $i = 0, 1, \ldots, MN\breve{K} - 1$ is a $J \times K$ matrix for determining the reduced number of transmitting elements, receiving elements, pulses, and fast time samples, respectively. Each of the three matrices $\mathbf{\Phi}_1, \mathbf{\Phi}_2$, and $\mathbf{\Phi}_3$ consists of randomly selected rows of an identity matrix. These choices of reduced matrix dimensions amount to a selection of subsets of existing available degrees of freedom offered by the fully deployed imaging system. Any other matrix structure does not yield to any hardware simplicity or saving in acquisition time. On the other hand, three different choices are available for compressive acquisition of each pulse in fast time. That is, the matrix $\mathbf{\Phi}_4^{(i)}$, $i = 0, 1, \ldots, MN\breve{K} - 1$ can be (1) a Gaussian random matrix with entries drawn from N(0,1), (2) a random matrix with entries equal to ± 1 with probability $1/2$, or (3) a matrix consisting of randomly selected rows of an identity matrix. The three options provide tradeoff between the imaging performance and ease of hardware implementation, as discussed in [20]. A possible receiver hardware implementation for the first two types of random matrices is depicted in Figure 8.10 using the i_nth receive element with the i_mth transmit element active, where $i_n \in \{0, 1, \ldots, N - 1\}$ and $i_m \in \{0, 1, \ldots, M - 1\}$ are the indices of the randomly selected reduced set of receivers and

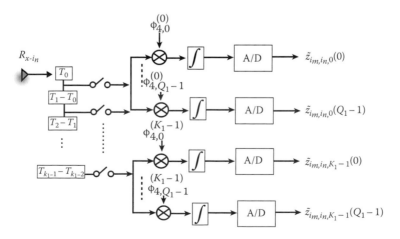

FIGURE 8.10
Receiver implementation for compressive ultrawideband through-the-wall radar.

transmitters. The reduced number of pulses is assumed to be transmitted in a random fashion, with the receiver synchronized to the transmitter. Further assume $T_0, T_1, \ldots, T_{\breve{K}_1 - 1}$ to be the time period between consecutive transmissions in the sequence of randomly selected $\breve{K}_1$ pulses, as shown in Figure 8.10. The switch implements the time gating for wall return removal and the time-gated signals are input to the subsequent random fast time measurement system. Each pulse is first multiplied by each column of the fast time measurement matrix and the product is then integrated. Both of these operations can be implemented by microwave mixers and low-pass filters. The subsequent sampling operation in the system can, thus, furnish *clean* data without the wall returns at a reduced volume.

Given the reduced measurement vector $\tilde{z}$ in (8.42), we can recover σ by solving the following equation:

$$\hat{\sigma} = \arg\min_{\sigma} \|\sigma\|_1 \text{ subject to } \boldsymbol{\Phi}\boldsymbol{\Psi}\tilde{\sigma} \approx \tilde{z} \tag{8.44}$$

The problem in (8.44) can be solved using convex relaxation, greedy pursuit, or combinatorial algorithms. In this work, we use orthogonal matching pursuit (OMP) for the CS-based reconstruction. We note that the reconstructed vector can be rearranged into $N_{v_x} N_{v_y}$ matrices, each corresponding to the Q spatial pixels, in order to depict the estimated target reflectivity for different vertical and horizontal velocity combinations. The stationary targets will be localized for the $(0, 0)$ velocity pair.

Further note that if the wall clutter is not totally mitigated and the wall residuals are comparable in strength to the target returns, then the image will contain artifacts resulting from reconstruction of the wall residuals. Also, OMP may require relatively more iterations to recover the targets in this case.

8.3.4 Experimental Results

The scene tested is as follows. The center-to-center separation between the transmitter and the leftmost receive antenna is 0.3 m. A 3.65 m × 2.6 m wall segment was constructed utilizing 1-cm-thick cement board on a 2-by-4 wood stud frame. The transmit antenna and the receive array were at a standoff distance of 1.19 m from the wall. The system refresh rate of the radar is 100 Hz.

The origin of the coordinate system was chosen to be at the center of the receive array. The scene behind the wall consisted of one stationary target and one moving target. A metal sphere of 0.3 m diameter, placed on a 1 m high Styrofoam pedestal, was used as the stationary target. The pedestal was located 1.25 m behind the wall, centered at (0.49 m, 2.45 m). A person walked toward the front wall at a speed of 0.7 m/s approximately along a straight-line path, which is located 0.2 m to the right of the transmitter. The back and the right side walls in the region behind the front wall were covered with RF absorbing material, whereas the 8-in.-thick concrete side wall on the left and the floor were uncovered. A coherent processing interval of 15 pulses was selected.

The image region is chosen to be 4 m $\times$ 6 m, centered at $(-0.31$ m, 3 m), and divided into 41 $\times$ 36 pixels in crossrange and downrange. As the human moves directly toward the radar, we only consider varying vertical velocity from -1.4 to 0 m/s, with a step size of 0.7 m/s, resulting in three velocity pixels. For the CS-based reconstruction, the random measurement matrices $\Phi_4^{(i)}$ are chosen to be the same for each pulse.

After removal of the front wall return from the received signals through time gating, the space-slow time-fast time data include 8 $\times$ 15 $\times$ 2048 measurements. The full time-gated data were used for scene reconstruction with the backprojection-based approach. The resulting images are depicted in Figure 8.11a and b for the 0 Hz and 14 Hz Doppler bins, respectively. Clearly, in the absence of the wall, the algorithm has successfully detected and localized both stationary and moving targets.

Figure 8.12 provides the corresponding result of OMP reconstruction, averaged over 50 trials. In each trial, we used all eight receivers, randomly selected five pulses (33.3% of 15) and chose 400 Gaussian random measurements (19.5% of 2048) in fast time, which amounts to using 6.5% of the total data volume. The number of OMP iterations was set to 4. Figure 8.12a through c are the respective images corresponding to the 0, -0.7, and -1.4 m/s velocities. It is apparent that with the wall removed, both the stationary and moving targets have been correctly localized even with the reduced set of measurements.

8.4 Conclusions

This chapter considered sparsity-driven localization of moving targets using two different approaches. One approach is based on change detection and the other is based on modeling of motion profiles and estimating the corresponding parameters. Removal of stationary background via change detection converts populated scenes to sparse scenes, whereby CS and sparse reconstruction become most applicable. Both translational motion and short sudden movements were considered and appropriate measurement models were developed. We only discussed sparsity from a point target perspective. As humans are extended targets, they appear as clusters in the through-the-wall images. As such, one could exploit the underlying block structure in rendering an enhanced imaging solution.

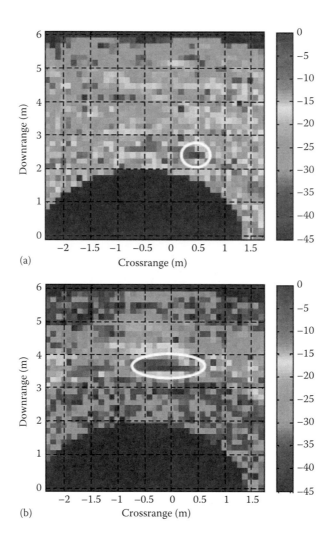

FIGURE 8.11
(a) Backprojection result after time gating for the stationary target. (b) Backprojection result after time gating for the moving target.

When a human exercises a continuous translation motion, the target becomes sparse in both space and velocity. For this case, we presented the proper linear model with the associated sensing matrix. The sparsity-based reconstruction is performed for the downrange–crossrange velocity space. Results based on real data experiments demonstrated that joint localization of stationary and moving can be achieved via sparse regularization using a reduced set of measurements without degradations in system performance.

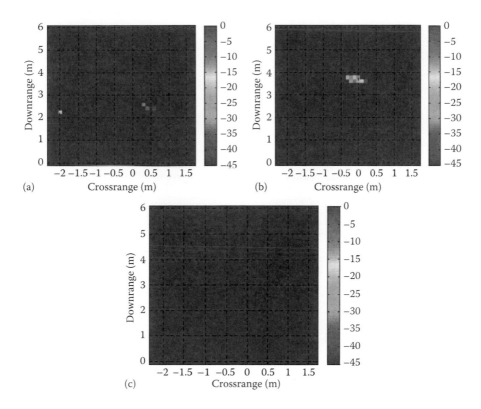

FIGURE 8.12
(a) CS reconstructed imaging result σ(0, 0). (b) CS reconstructed imaging result σ(0, −0.7). (c) CS reconstructed imaging result σ(0, −1.4).

References

1. M.G. Amin (Ed.), *Through-the-Wall Radar Imaging*, CRC Press, Boca Raton, FL, 2011.
2. M. Amin and K. Sarabandi (Eds.), Special issue on remote sensing of building interior, *IEEE Transactions on Geoscience and Remote Sensing*, 47 (5), 1270–1420, 2009.
3. A. Martone, K. Ranney, and R. Innocenti, Automatic through the wall detection of moving targets using low-frequency ultra-wideband radar, in *Proceedings of the IEEE International Radar Conference*, Washington DC, pp. 39–43, May 2010.
4. S.S. Ram and H. Ling, Through-wall tracking of human movers using joint Doppler and array processing, *IEEE Geoscience and Remote Sensing Letters*, 5 (3), 537–541, 2008.

5. M. Amin (Ed.), Special issue on advances in indoor radar imaging, *Journal of Franklin Institute*, 345 (6), 556–722, 2008.
6. T. Dogaru and C. Le, Validation of Xpatch computer models for human body radar signature, U.S. ARL technical report ARL-TR-4403, March 2008.
7. C.P. Lai and R.M. Narayanan, Through-wall imaging and characterization of human activity using ultrawideband (UWB) random noise radar, in *Proceedings of the SPIE-Sensors and C3I Technologies for Homeland Security and Homeland Defense*, May 2005, Vol. 5778, pp. 186–195.
8. S.S. Ram, Y. Li, A. Lin, and H. Ling, Doppler-based detection and tracking of humans in indoor environments, *Journal of Franklin Institute*, 345 (6), 679–699, 2008.
9. I. Orovic, S. Stankovic, and M. Amin, A new approach for classification of human gait based on time-frequency feature representations, *Signal Processing*, 91 (6), 1448–1456, 2011.
10. M.G. Amin, Time-frequency spectrum analysis and estimation for nonstationary random processes, in *Time-Frequency Signal Analysis: Methods and Applications*, B. Boashash (Ed.), Longman-Cheshire, New York, 1992.
11. A.R. Hunt, Use of a frequency-hopping radar for imaging and motion detection through walls, *IEEE Transactions on Geoscience and Remote Sensing*, 47 (5), 1402–1408, 2009.
12. N. Maaref, P. Millot, C. Pichot, and O. Picon, A study of UWB FM-CW radar for the detection of human beings in motion inside a building, *IEEE Transactions on Geoscience and Remote Sensing*, 47 (5), 1297–1300, 2009.
13. T.S. Ralston, G.L. Charvat, and J.E. Peabody, Real-time through-wall imaging using an ultrawideband multiple-input multiple-output (MIMO) phased array radar system, in *Proceedings of the IEEE International Symposium on Phased Array Systems and Technology*, Boston, MA, October 2010, pp. 551–558.
14. M.G. Amin and F. Ahmad, Change detection analysis of humans moving behind walls, *IEEE Transactions on Aerospace and Electronic Systems*, 49 (3), 1869–1896, July 2013.
15. F. Soldovieri, R. Solimene, and R. Pierri, A simple strategy to detect changes in through the wall imaging, *Progress in Electromagnetics Research M*, 7, 1–13, 2009.
16. J. Moulton, S.A. Kassam, F. Ahmad, M.G. Amin, and K. Yemelyanov, Target and change detection in synthetic aperture radar sensing of urban structures, in *Proceedings of the IEEE Radar Conference*, Rome, Italy, May 2008.
17. M.G. Amin, F. Ahmad, and W. Zhang, A compressive sensing approach to moving target indication for urban sensing, in *Proceedings of the IEEE Radar Conference*, Kansas City, MO, May 2011.
18. R. Baraniuk and P. Steeghs, Compressive radar imaging, in *Proceedings of the IEEE Radar Conference*, Waltham, MA, April 2007, pp. 128–133.
19. M. Herman and T. Strohmer, High-resolution radar via compressive sensing, *IEEE Transactions on Signal Processing*, 57 (6), 2275–2284, 2009.
20. A. Gurbuz, J. McClellan, and W. Scott Jr., Compressive sensing for subsurface imaging using ground penetrating radar, *Signal Processing*, 89 (10), 1959–1972, 2009.

21. L.C. Potter, E. Ertin, T. Parker, and M. Cetin, Sparsity and compressed sensing in radar imaging, *Proceedings of the IEEE*, 98 (6), 1006–1020, 2010.
22. Y. Yoon and M.G. Amin, Compressed sensing technique for high-resolution radar imaging, in *Proceedings of SPIE*, 6968, 69681A–69681A-I0, 2008.
23. Q. Huang, L. Qu, B. Wu, and G. Fang, UWB through-wall imaging based on compressive sensing, *IEEE Transactions on Geoscience and Remote Sensing*, 48 (3), 1408–1415, 2010.
24. M. Leigsnering, C. Debes, and A.M. Zoubir, Compressive sensing in through-the-wall radar imaging, in *Proceedings of the IEEE International Conference on Acoustics, Speech, and Signal Processing*, Prague, Czech Republic, May 2011, pp. 4008–4011.
25. K. Ranney et al., Recent MTI experiments using ARL's synchronous impulse reconstruction (SIRE) Radar, in *Proceedings of the SPIE-Radar Sensor Technology XII*, April 2008, Vol. 6947, pp. 694708-1–694708-9.
26. P. Sevigny et al., Concept of operation and preliminary experimental results of the DRDC through-wall SAR system, *Proceedings of the SPIE-Radar Sensor Technology XIV*, April 2010, Vol. 7669, pp. 766907-1–766907-11.
27. F. Ahmad, Y. Zhang, and M.G. Amin, Three-dimensional wideband beamforming for imaging through a single wall, *IEEE Geoscience and Remote Sensing Letters*, 5 (2), 176–179, April 2008.
28. F. Ahmad and M. Amin, Through the wall human motion indicator using sparsity driven change detection, *IEEE Transactions on Geoscience and Remote Sensing*, 51 (2), 881–890, 2013.
29. L. He, S.A. Kassam, F. Ahmad, and M.G. Amin, Sparse multi-frequency waveform design for wideband imaging, in *Principles of Waveform Diversity and Design*, M. Wicks, E. Mokole, S. Blunt, R. Schneible, and V. Amuso (Eds.), SciTech Publishing, Raleigh, NC, 2010, pp. 922–938.
30. X.X. Zhu and R. Bamler, Tomographic SAR inversion by L_1-norm regularization—The compressive sensing approach, *IEEE Transactions on Geoscience and Remote Sensing*, 48 (10), 3839–3846, October 2010.
31. E. Candes, J.Romberg, and T. Tao, Stable signal recovery from incomplete and inaccurate measurements, *Communications in Pure and Applied Mathematics*, 59, 1207–1223, 2006.
32. R. Tibshirani, Regression shrinkage and selection via the LASSO, *Journal of the Royal Statistical Society: Series B*, 58, 267–288, 1996.
33. D. Needell and J.A. Tropp, CoSaMP: Iterative signal recovery from incomplete and inaccurate samples, *Applied and Computational Harmonic Analysis*, 26 (3), 301–321, May 2009.
34. S. Ji, Y. Xue, and L. Carin, Bayesian compressive sensing, *IEEE Transactions on Signal Processing*, 56 (6), 2346–2356, 2008.
35. M. Cetin and R.L. Moses, SAR imaging from partial-aperture data with frequency-band omissions, *Proceedings of SPIE*, 5808, 32–43, 2005.
36. Y. Yang and A. Fathy, Development and implementation of a real-time see-through-wall radar system based on FPGA, *IEEE Transactions on Geoscience and Remote Sensing*, 47 (5), 1270–1280, 2009.
37. M.T. Alonso, P. Loìpez-Dekker, and J.J. Mallorquí, A novel strategy for radar imaging based on compressive sensing, *IEEE Transactions on Geoscience and Remote Sensing*, 48 (12), 4285–4295, December 2010.

38. F. Ahmad, M.G. Amin, and J. Qian, Through-the-wall moving target detection and localization using sparse regularization, *Proceedings of SPIE*, 8365, 83650R, 2012.
39. M. Ferrara, J. Jackson, and M. Stuff, Three-dimensional sparse aperture moving-target imaging, *Proceedings of SPIE*, 6970, 697006, 2008.
40. M. Skolnik (Ed.), *Introduction to Radar Systems*, 3rd edn., McGraw Hill, New York, 2001.

9

Time–Frequency Analysis of Micro-Doppler Signals Based on Compressive Sensing

Ljubiša Stanković, Srdjan Stanković, Irena Orović, and Yimin D. Zhang

CONTENTS

ABSTRACT Signals returned from an object containing fast-moving parts have Doppler frequency modulations, called micro-Doppler effects, which produce time-varying spectral contents. Motion analysis, which involves parameter estimation of the target micro-Doppler signatures, is important in urban sensing. One of the goals in urban radar is the classification of different types of human gait. It is equally important to separate multiple micro-Doppler components induced by rigid bodies. Time–frequency representation has been shown to be a powerful tool in the analysis of such signals. However, the information rendered by time–frequency analysis can be impacted, if not compromised or distorted, when dealing with data from random sampling or significantly reduced sampling frequency. These changes in the sampling patterns can be attributed to changes in the pulse repetition periods to avoid range or Doppler ambiguities or can be a result of deliberately discarding samples highly contaminated by disturbances. Nevertheless, the problems can be observed within the concept of compressive sensing theory and practice. After a short review of the basic compressive sensing methods, various approaches for the analysis and separation of fast-varying signal components are presented. Several methods are considered. First, we discuss direct applications of the compressive sensing algorithms to the ambiguity domain for the purpose of achieving high-resolution time–frequency representations. Next, in dealing with missing samples, we consider sparse signal reconstructions when operating on both the data and the local autocorrelation function. The latter can result from bilinear products or from performing higher-order estimations. The presented methods are illustrated using simulated and real data, demonstrating effectiveness of the compressive sensing based time–frequency approaches in the analysis of radar signals with micro-Doppler effects.

9.1 Introduction

During the last two decades, different techniques have been developed for radar signal analysis and processing in order to provide an efficient target description and identification [1]. Generally, in radar data analysis, we deal with micro-Doppler (m-D) and rigid body components. The m-D effect appears in the radar imaging when a target has one or more fast moving parts [2–8]. This effect may complicate the radar signal analysis or decrease the readability of radar images. The frequency content of the m-D signal changes over time in a wide range. Therefore, the m-D may obscure the rigid body and make it difficult to detect. On the other hand, the m-D signatures, at the same time, carry useful information about the features of moving parts (type, velocity, size, etc.) [9]. It is easier to estimate these features if the m-D effect is separated from the rigid body part of the radar image. Thus, m-D extraction has attracted significant attention [8–11]. Having in mind their nature and time-varying spectral characteristics, most of the m-D data belong to the group of highly nonstationary signals that require joint time–frequency (TF) analysis. The most efficient processing of these types of radar signals is performed in the TF domain [12–15] within the coherent integration time (CIT). The obtained TF representation (TFR) is then used to make decision whether a component belongs to the rigid body or to the fast moving target point. The TF analysis is also efficiently combined with robust processing approaches [18,19], such as the L-statistics, aiming to separate the rigid body and m-D effects.

In light of the recent developments in signal acquisition and reconstruction, radar data analysis may benefit from compressive sensing (CS) theory. The CS allows reconstruction of the entire signal from its small randomly chosen set of measurements [20–38]. This is especially important for reducing acquisition resources when the underlying signal has a high sampling rate, as in the case of radar applications. Decimated sampling may enable a single radar to handle multiple antenna sets for multistatic operation with a low cost. In certain situations, missing samples or observations occur as a consequence of separating desired signal components from undesired ones [9,10,20–31]. This is the case when one seeks to discard samples corrupted by strong disturbances or to remove the m-D components in order to reveal a rigid body. These missing samples could be recovered using the CS principles. If ignored, the missing samples produce unwanted side-effects presented as a spectral noise, which may cause a masking effect of signal components. An important issue determining the applicability of the CS signal reconstruction is related to the signal sparsity. More specifically, we need to identify a suitable domain of signal sparsity and a corresponding linear transformation that maps the signal from the acquisition to the sparsity domain. As a logical choice, we consider the use of joint variable domains, such as the TF and ambiguity domains, instead of the commonly used Fourier transform (FT) domain. Note that the radar signals containing both rigid body and m-D components, generally, are not narrowband and, as such, cannot be cast as sparse over the frequency variable.

In this chapter, the CS applications in TF analysis are observed in two directions. The first one is related to the linear TF approaches such as the short-time Fourier transform (STFT), which is combined with the L-statistics. Stationary and nonstationary signal components that highly overlap in time and frequency render their separation difficult using conventional methods involving windowing or filtering. The L-statistics is applied to remove the m-D components from the STFT, resulting in CS stationary data in the TF domain. As such, the CS reconstruction approaches can be used to recover narrowband signals corresponding to rigid body signal components. The second approach is related to the quadratic distributions, where the CS problem is observed through the linear relationship between the Wigner distribution (WD), the ambiguity function (AF), and the instantaneous autocorrelation function (IAF) [39,40]. Reduced signal observations can be acquired in the time domain or the ambiguity domain [26,27]. Missing samples in the time domain certainly affect the WD, but this influence can be tackled by the signal adaptive kernels. At the same time, it is possible to select a small set of ambiguity domain measurements and to exploit sparse reconstruction to obtain cross-term-free highly localized TFRs.

This chapter is organized as follows. The basic properties of radar data and the commonly used models for radar signals are presented in Section 9.2. Section 9.3 discusses the basis of the TF analysis in radars (m-D and rigid body signals). Section 9.4 introduces the concept of signal sparsity and the basic perspective of CS approach in the TF domain. Two cases are discussed: missing samples due to the lower sampling rate and missing samples due to the removal of unwanted signals components. Section 9.5 provides a detailed analysis of missing samples effects in discrete Fourier transform (DFT) (i.e., one window of STFT) as well as in the case of quadratic WD. The applications of CS signal reconstruction are developed and presented in details in Section 9.5. Therein, we emphasize two important approaches: one is related to the sparse TFR using ambiguity domain observations, whereas the other is related to the application of linear transforms and CS in signal components separation.

9.2 Background

In monostatic operations, the transmitted signal of a coherent radar system is reflected from the target and returned back to the radar. The returned signal contains important information about the moving target, which can be revealed using different signal analysis methods. For instance, the carrier frequency of the returned signal will be shifted compared to the transmitted frequency. This effect is known as the Doppler frequency shift, which provides information about the target velocity. Beside the target's Doppler frequency, we may also deal with vibrating and rotating target parts (rotating antennas, rotors, etc.), producing the m-D (sidebands about the Doppler frequency). Therefore, the m-D signatures bring additional important information about the nature and status of the target. Generally speaking, when a target produces a certain kind of nonuniform motions, the radar backscattering will contain frequency modulations reflected in the form of specific signatures. Therefore, as it is shown in the sequel, the mathematical model of the m-D effect can be derived by introducing nonuniform motions to conventional Doppler analysis.

9.2.1 Time-Varying Micro-Doppler Signatures

Consider a pulse-Doppler radar that transmits signals in a form of coherent series of N linear frequency modulated (LFM), or chirp, waveforms [1,13]. The received signal, reflected from a target, is delayed with respect to the transmitted signal by $t_d = 2d(t)/c$, where $d(t)$ is the target distance from the radar and c is the speed of light. This signal is demodulated to the baseband, with possible distance compensation and other preprocessing operations (such as pulse compression). In order to analyze the effect of cross-range nonstationarities, we only consider the Doppler part in the received signal of a point target, in the continuous dwell time, as it is usually done in the radar literature [13], expressed as

$$s(t) = \sigma e^{j2d(t)\omega_0/c}, \tag{9.1}$$

where
   σ is the reflection coefficient of the target
   ω_0 is the radar operating frequency

The repetition time of the LFM pulses is denoted by T_r. The CIT is $T_c = NT_r$, where N is the number of chirp pulses. The received signal, for a system of point scatterers, can be modeled as a sum of individual point scatterer responses [13]. In the inverse synthetic aperture radar (ISAR) case, the aim is to obtain a high-resolution image of a target based on the change in viewing angle of the target with respect to the fixed radar. The common ISAR imaging models assume that all point scatterers share the same angular motion after translational motion compensation. The Doppler part of the received signal, corresponding to the K rigid body points, can be written as [13,14]

$$s(t) = \sum_{i=1}^{K} \sigma_{Bi} e^{j2[R_B(t)+x_{Bi}\cos(\theta_B(t))+y_{Bi}\sin(\theta_B(t))]\omega_0/c}. \tag{9.2}$$

The target's translation and angular motion are denoted by $R_B(t)$ and $\theta_B(t)$, respectively. The initial locations of the K points, in the coordinate system whose origin is in the center of the target rotation, are (x_{Bi}, y_{Bi}). Subscript B denotes the rigid body parameters. For each point, we have used approximation $d_i(t) = \sqrt{(R + x_i)^2 + y_i^2} \cong R + x_i$ to obtain $d_{Bi}(t) \cong R_B(t) + x_{Bi}\cos(\theta_B(t)) + y_{Bi}\sin(\theta_B(t))$ [14].
   For the rigid body points $|\theta_B(t)| \ll 1$ holds during the CIT, following in $\cos\theta_B(t) \approx 1$ and $\sin\theta_B(t) \approx \theta_B(t) = \omega_B t$, where ω_B is the effective body rotation rate. Motion compensation techniques [15] can be employed to remove the effect of the translational motion, that is, the factor $R_B(t) + x_{Bi}$, thus we get $d_{Bi}(t) \sim y_{Bi}\omega_B t$.
   The previous approximations cannot be applied for fast rotating (moving) points since their angular position $\theta_R(t)$ can significantly change during the CIT. Subscript R will be used for fast moving points. Assume that there are P fast rotating points, which rotate around their central points (x_{R0i}, y_{R0i}) with radii A_{Ri}. The coordinates of these points are described by $x_p = x_{R0i} + A_{Ri}\sin(\theta_{Ri}(t))$ and $y_p = y_{R0i} + A_{Ri}\cos(\theta_{Ri}(t))$. Thus, the resulting

coordinate changes of these scatterers are $x'_p = x_p \cos \theta_B(t) + y_p \sin \theta_B(t)$ and $y'_p = -x_p \sin \theta_B(t) + y_p \cos \theta_B(t)$. Assume that the rotation speed of the ith fast rotating point is ω_{Ri}, with $\theta_{Ri}(t) = \omega_{Ri}t$. With the previous approximations for the rigid body values, after compensation for $R_B(t)$ and x_{R0i}, we have

$$d_i(t) \cong y_{R0i}\omega_B t + A_{Ri} \sin(\omega_{Ri}t). \tag{9.3}$$

The received signal, including both the rigid body points and P fast rotating m-D points, can be written as

$$s(t) = \sum_{i=1}^{K} \sigma_{Bi} e^{j2y_{Bi}\omega_B t\omega_0/c} + \sum_{i=1}^{P} \sigma_{Ri} e^{j2[y_{R0i}\omega_B t + A_{Ri}\sin(\omega_{Ri}t)]\omega_0/c}. \tag{9.4}$$

A similar form of the received signal is obtained in the case of vibrating points. These kinds of signal may be considered as sparse in the inverse Radon transform domain [16,17]. Any arbitrary motion can be easily described within the previous framework by using $x_p = x_{R0i} + x_{arb}(t)$ and $y_p = y_{R0i} + y_{arb}(t)$, where $x_{arb}(t)$ and $y_{arb}(t)$ denote arbitrary movement with respect to the central point.

Since we only consider the Doppler part of the received signal, the analysis of the radar signal reduces to one-dimensional signal (9.4) and its FT analysis. If we calculate the FT of the signal corresponding to one point of the rigid body in (9.4), we obtain

$$S_{Bi}(\Omega) = FT\left\{\sigma_{Bi} e^{j2y_{Bi}\omega_B t\omega_0/c}\right\} = 2\pi\sigma_{Bi}\delta\left(\Omega - \frac{2\omega_0}{c}\omega_B y_{Bi}\right),$$

which is a delta function at the position proportional to the cross-range coordinate y_{Bi}. The delta pulse position also depends on the carrier frequency and the rotation speed ω_B.

The Doppler part of the radar signal that corresponds to an arbitrary moving point is a frequency modulated (FM) signal with the instantaneous frequency (IF)

$$\Omega_{Ri}(t) = \frac{2\omega_0}{c}\left[y_{R0i}\omega_B + \frac{d(x_{arb}(t))}{dt}\right], \tag{9.5}$$

with assumptions as in the derivation of (9.3) and (9.4). Note that exact relation (9.2) can be used in simulations, instead of the compact form (9.4), which is appropriate for the qualitative analysis.

9.2.2 Human Gait Modeling

Over several years, human gait analysis has been receiving significant interest in numerous applications, such as surveillance, identification, and security applications. As in other radar applications, the human gait analysis requires transmitting signal/waveform and receiving the radar returns to estimate motion parameters using the Doppler effect. The human motion includes various components of the body, especially arms and legs, which produce high-frequency modulations that appear as m-D sidebands about the Doppler frequency corresponding to the translation motion of the torso. The m-D analyses are usually performed for the purpose of motion classification. For example, in [41–44] human walking is classified into one of the following categories: free arm-motion, partial arm-motion, or no arm-motion, as these three categories are considered important in law enforcement and homeland security operations.

Because different human body parts move with different velocities, the returned radar signals have different phase changes due to the variations in range. The signal returned from the swinging arms may include frequency modulation that produces the sidebands around the torso Doppler. The received Doppler can be modeled as in (9.1), $s(t) = \sigma e^{j\phi(t)}$, where again σ is the reflectivity of the chosen reflecting point and $\phi(t)$ is the time-varying phase. For an oscillating/vibrating object, $\phi(t) = \frac{2\omega_0 D_v}{c} \sin(\omega_v t)$, where parameter D_v represents the amplitude of vibration, or the maximum deviation from the center of the motion. The corresponding induced m-D frequency is the derivative of the phase and is given by

$$\Omega(t) = \frac{d\phi(t)}{dt} = \frac{2\omega_0}{c} D_v \omega_v \cos(\omega_v t). \tag{9.6}$$

Hence, in this case, m-D represents the sinusoidal function of time at the frequency ω_v.

9.3 Time–Frequency Analysis of the m-D and Rigid Body Signals

The analysis of radar signals in the frequency domain using the FT can provide the information about the presence of m-D components since they appear as a deviation around the central frequency. Nevertheless, the Fourier domain analysis does not provide the information about the temporal behavior of m-D spectral components or the rate of the rotation/vibration processes. Thus, the time-varying spectral components should be analyzed using the TFR. A simplest way to localize the signal behavior in shorter intervals, within the CIT, is to apply a window function to the standard FT. The resulting STFT is defined as

$$STFT(m,k) = \sum_{i=0}^{N-1} s(i)w(m-i)e^{-j2\pi ik/N}, \tag{9.7}$$

where $w(m)$ is a window function used to truncate the considered signal. The squared absolute value of the STFT is called the spectrogram. The window width is M, that is, $w(m) \neq 0$ for $-M/2 \leq m \leq M/2-1$. In our applications, the window is zero padded up to the total signal duration N, the same number of samples as the original signal, so that we have the same frequency grid in the STFT as in the FT. Then, we can easily reconstruct the FT, without interpolation, with the concentration close or equal to the concentration of the original FT. By using a lag window $w(i)$ in the STFT, the concentration in frequency is reduced as compared with the original FT. The concentration could be restored to the original one by summing all the low concentrated STFT (complex) values over m. The reconstruction formula, for the case when the signal is not zero padded, is

$$\sum_{m=0}^{N-1} STFT(m,k) = \sum_{i=0}^{N-1} s(i) \left[\sum_{m=0}^{N-1} w(m-i) \right] e^{-j2\pi ik/N} = S_w(k). \tag{9.8}$$

In the case when the STFT is calculated for each time instant m (time step is one), the resulting window $\sum_{m=0}^{N-1} w(m-i)$ is constant for any window $w(m)$. It means that, during the CIT interval, we have the normalized equivalent rectangular window. This means that we will be able to reconstruct the FT with a concentration as in the original FT by using low concentrated STFTs calculated with narrow windows. In this way, we can restore the high-concentrated radar image, although we used low-concentrated STFT in the analysis. The analysis is not restricted to the step one case in the STFT calculation. The same resulting window would be obtained for a step equal to a half of the window width ($M_w/2$) and a Hanning, Hamming, triangular, or rectangular window. The same is valid for steps equal to $M/4, M/8, \ldots$.

The presented mechanism of restoring the original concentration of the FT, in the conjunction with the knowledge of the TF patterns behavior of fast moving and rigid scattering points, leads to an algorithm for the m-D free, highly concentrated, radar images [6,14]. The rigid body and the fast-moving points behave differently in the TFR of the returned radar signal, within the CIT. The rigid body signal is almost constant in time (stationary), while the fast varying m-D part of the signal is highly nonstationary. This part of the signal changes its position in the frequency domain.

For illustration, assume that the signal is returned from one-point rigid body scatterer and one-point fast rotating (m-D) scatterer. We analyze two

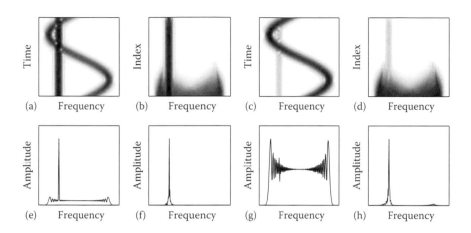

FIGURE 9.1
Simulated radar signals corresponding to a rigid body reflector with $\sigma_B = 1$ and a rotating reflector with reflection coefficient $\sigma_R = 0.8$ (a–d) and $\sigma_R = 15$ (e–f). (a, e) Absolute value of the STFT, (b, f) sorted STFT values, (c, g) the original FTs, (d, h) the reconstructed FTs.

cases with different strengths of the m-D reflection. In the first case, the reflection coefficient of the rigid body is $\sigma_B = 1$, while the reflection coefficient of the fast-moving scatterer is $\sigma_R = 0.8$. The STFT representation of the resulting signal is shown in Figure 9.1a. The second case is with a strong m-D, $\sigma_R = 15$ and the same σ_B as in the previous case. The STFT representation of this signal is shown in Figure 9.1e. In both cases, the rigid body part has a constant Doppler frequency for all t within the CIT, while the fast rotating part has time-dependent frequency. If we perform sorting over the time axis, as in Figure 9.1b and f, we will not change the result of the summation in (9.8), since it is a commutative operation. By summing the STFT values over time, from either of these two plots, presented in Figure 9.1a and b or Figure 9.1e and f, we will get the original FT of the corresponding signal Figure 9.1c and g. Note that any value of σ_R from (and including the case without m-D) $\sigma_R = 0$ up to $\sigma_R \gg \sigma_B$ will not significantly change the pattern. The results presented in Figure 9.1d and h will be explained within the next section.

9.3.1 Missing STFT Samples due to the m-D Removal

The basic idea of separating rigid body and fast rotating parts is in the sorting of STFT values of the returned radar signal along the time axis within the CIT. Since the rigid body return is stationary, the sorting procedure will not significantly change the distribution of its values. However, the fast varying m-D part of the signal is highly nonstationary, occupying different frequency bins at different time instants. Its existence is short in time for each frequency, but it spans a wide range of frequencies. Thus, after sorting the STFT along the time axis, the m-D part of the signal has strong values at a wide frequency range, but for a few samples only. By removing several strongest values of the sorted STFT, for each frequency, we eliminate most or all of the m-D components. Summing the rest of the STFT values over time results in the rigid body radar spectrum.

Consider a set of M (or $M - M_w$ if the signal is not zero padded) elements of the STFT, for a given frequency k,

$$S_k(m) = \{STFT(m,k), m = 0, 1, \ldots, M - 1\}.$$

After sorting $S_k(m)$ along the time, for a given frequency k, we obtain a new ordered set of elements $O_k(m) \in S_k(m)$ such that their absolute values satisfy $|O_k(0)| \leq |O_k(1)| \leq \cdots \leq |O_k(M - 1)|$. Because the addition is a commutative operation, using the entire data set yields

$$\sum_{m=0}^{M-1} STFT(m,k) = \sum_{m=0}^{M-1} O_k(m) = S(k).$$

By discarding $M_Q = M - M_A$ highest values of O_k for each k, we produce an estimate of $S(k)$, denoted by $S_L(k)$, as

$$S_L(k) = \sum_{m=0}^{M_A-1} O_k(m), \qquad (9.9)$$

where $M_A = \text{int}[M(1 - Q/100)]$, with $\text{int}[\cdot]$ denoting the integer part, and Q is the percent of discarded values.

To illustrate this procedure, we eliminated 40% of the strongest values of the STFT from the previous example. In this way, we completely eliminate the m-D components from the TFR. We are left with the 60% lowest STFT values, which correspond only to the rigid body. The FTs, reconstructed from these values only, are shown in Figure 9.1d and h, respectively, for the cases of a weak and a strong m-D. The FT of the rigid body is successfully reconstructed in both cases by summing the remained 60% of sorted STFT samples. Note that the result is not significantly influenced by the value of σ_R since the points corresponding to the m-D signature are removed, meaning that their values are not so important.

In the data analysis, this approach, based on elimination of a part of data, before analyzing the rest of the data, is known as the L-statistics [18].

Since we have eliminated some of the TFR values, we will analyze the influence of the incomplete sum in (9.9). This is the same theory as the

L-statistics applied to the noisy or noise-free data [18]. Assume that only points in $m \in D_k$ are used in summation:

$$S_L(k) = \sum_{m \in D_k} STFT(m,k), \tag{9.10}$$

where, for each k, D_k is a subset of $\{0,1,2,\ldots,M-1\}$ with M_A elements. Within the framework of the previous analysis, it means that there is a highly concentrated component $S(k)$ surrounded by several low-concentrated values $\sum_{m \notin D_k} STFT(m,k)$. Note that the amplitude of $STFT(m,k)$ is M times lower than the amplitude $S(k)$ since $S(k)$ is obtained as a sum of M values of the STFT. In general, by removing, say, M_Q values in m, we will obtain one very highly concentrated pulse, as in $S(k)$, and M_Q values of low-concentrated components of $STFT(m,k)$, being spread around the peak of $S(k)$ and summed up by different random phases. Only the peak value is summed in phase. Consider:

1. Case for $k = k_0$ corresponding to the position of the rigid body point: At this frequency, all terms in the sum are the same and equal to $W(0)$. Thus, the value of $S_L(k)$ does not depend on the positions of the removed samples. Its value is $S_L(k_0) = (M - M_Q)W(0)$.

2. Case for $k = l + k_0$, where $l \neq 0$: The removed terms are of the form $x_l(m) = W(l)e^{j2\pi ml/M}$. They assume values from the set $\phi_l = \left\{ W(l)e^{j2\pi ml/M},\ m = 0,1,2,\ldots,M-1 \right\}$, with equal probability, for a given l. The statistical mean of these values is $E\{x_l(m)\} = 0$ for $l \neq 0$, resulting in $E\{S_L(l+k_0)\} = 0$.

The resulting statistical mean for any k is

$$E\left\{S_L(k)\right\} = (M - M_Q)\,W(0)\delta\left(k - k_0\right).$$

The higher-order statistical analysis of this process could be performed in detail, but it is out of the scope of this chapter.

Example: In the analysis of the rigid body with uncompensated acceleration, we should first compensate the remaining acceleration. This is not possible in the original signal since the m-D signatures prevent us from doing so. However, the application of the proposed method for the m-D removal can solve this problem. We use the first-order local polynomial Fourier transform (LPFT), defined as [14]

$$LPFT(t, \Omega) = STFT\left\{s(\tau)e^{-j\alpha\tau^2}\right\} = \int\limits_{-\infty}^{\infty} s(\tau)w(\tau - t)e^{-j(\Omega\tau + \alpha\tau^2)}d\tau, \tag{9.11}$$

which is the STFT with an additional term $\exp(-j\alpha\tau^2)$ used to compensate the LFM of the rigid body part of the signal. The parameter α is not known in advance, but assumed to take values from a set $\Lambda = [-\alpha_{max}, \alpha_{max}]$, where α_{max} is the chirp rate corresponding to the maximal expected acceleration (positive or negative). In this example, we use $\Lambda = [-2, 2]$ with a step size of 0.25. Now, $\hat{\alpha}$ can be estimated as the value from the set Λ for which we obtain the highest concentration of the reconstructed rigid body (compensated FT) based on the LPFT and the L-statistics with, for example, $Q = 50\%$. The reconstructed FT, by using 50% of the lower LPFT values, is denoted by $S_{L,\alpha}(k)$. Its concentration is measured using the l_1-norm form [45], expressed as

$$H(\alpha) = \sum_{k=0}^{M-1} |S_{L,\alpha}(k)| = \|S_{L,\alpha}(k)\|_1 . \tag{9.12}$$

The LPFT, calculated with the estimated optimal value of $\hat{\alpha} = 1.25$, which results from $H(\alpha)$, is shown in Figure 9.2e. The LFM is compensated by $\hat{\alpha}$ in (9.11), resulting in almost constant frequencies in the TFR. In this way, we successfully reconstruct the rigid body and remove the m-D part, as presented in Figure 9.2h. The procedure is not very sensitive to $\hat{\alpha}$, and good results are obtained with neighboring values $\hat{\alpha} = 1.0$ and $\hat{\alpha} = 1.5$. Note that it would be impossible to estimate the chirp rate $\hat{\alpha}$ from the original signal without employing the proposed algorithm for m-D removal.

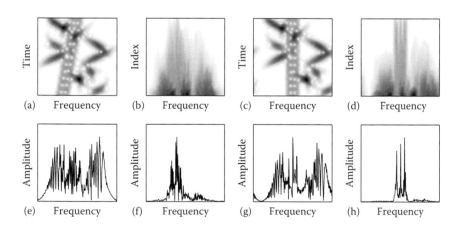

FIGURE 9.2
Accelerating rigid body with a complex form of the m-D. (a) TFR of the signal without motion compensation. (b) Sorted TFR of the original signal. (c) TFR of the signal after motion compensation. (d) Sorted TFR of acceleration compensated signal. (e) Original FT of the analyzed signal. (f) Reconstructed FT of the accelerating rigid body without motion compensation. (g) The FT of the original signal with motion compensation. (h) Reconstructed FT of the accelerating rigid body with motion compensation.

Here, we have demonstrated how a simple removal of the corrupted samples may produce a satisfactory result in the rigid body analysis. This was possible due to the fact that the form of the rigid body signal is simple, consisting of several nonzero values in the Fourier domain (without or with motion compensation). For these kinds of signals, we say that they are sparse in the TF domain. Sparse signals may be fully recovered with a reduced number of available observations using the CS theory that will be presented in the next section.

9.4 Sparse Compressive Sensed Signals and Time–Frequency Analysis

A large number of signals that appear in real applications (array signal processing, indoor and ISAR/SAR imaging, communications, remote sensing, and biomedical and multimedia applications) are sparse in their own or a certain transform domain. The signal sparsity refers to the property that the signals have concise representations when expressed in a proper basis. In general, a signal that is K-sparse in a specific domain can be completely characterized by M measurements ($M > K$), although the total number of samples required by the Shanon–Nyquist theorem is far above M. Furthermore, the entire signal can be recovered from a small incomplete set of samples by applying different optimization algorithms (e.g., convex optimization and greedy algorithms). This concept is known as CS. CS has led to the development of new data acquisition devices, which require significantly fewer resources than the devices that follow Shannon–Nyquist sampling theorem. The ability to recover missing samples opened up a wide area of applications in signal processing, where we are faced with the "wrong" or "unwanted" samples, which could be declared as the missing ones. In the sequel, we discuss two possible scenarios of missing samples: (1) those that are due to the reduced sampling rate and (2) those that are due to the removal of "unwanted components," such as clutter and noise. In both scenarios, the missing samples produce spectral noise, which significantly degrades signal components and the signal sparsity in the frequency and the TF domain. The analysis of missing sample effects is provided in this section.

9.4.1 Missing Samples due to Reduced Sampling Rate

When dealing with a signal that, according to the sampling theorem, requires a high sampling rate, the acquisition of samples demands large data storage and transmission capacities. In these circumstances, it would be very feasible to explore the possibility of sampling at far lower rates and, afterward, reconstruct the rest of the signal for the purpose of analysis and representation.

To achieve this goal, we usually need to sample randomly and identify the domain of signal sparsity.

The mathematical foundation of CS lies in the fact that it is possible to reconstruct a sparse signal exactly from an underdetermined linear system of equations and that can be done in a computationally efficient manner via convex programming. A discrete signal,

$$\mathbf{x} = [x(0), \ x(1), \ \ldots, \ x(N-1)]^T,$$

is sparse if the number of components in the transform domain $\mathbf{X}$, defined by linear transformation matrix $\mathbf{\Psi}$,

$$\mathbf{x} = \mathbf{\Psi X},$$

satisfies $\|\mathbf{X}\|_0 = K \ll N$, where N is the number of samples in the acquisition domain. Due to the sparsity of $\mathbf{X}$ (very reduced set of its nonzero values to $K \ll N$), one could recover its values by solving the following optimization problem:

$$\min \|\mathbf{X}\|_0 \quad \text{s.t.} \quad \mathbf{y} = \mathbf{AX}, \tag{9.13}$$

where $\mathbf{y}$ is a vector of observations or measurements from $\mathbf{x}$, while the combined matrix, $\mathbf{A} = \mathbf{\Phi \Psi}$, is referred to as the representation dictionary or CS matrix, where $\mathbf{\Phi}$ models the measurement process.

However, solving (9.13) with the l_0-norm is practically not feasible. It also means that any small nonzero value in the sparse signal, caused by extremely small noise in the analyzed signal, will be counted as a nonzero transform value [20,21,45]. Instead, we consider its convex relaxation using the l_1 norm:

$$\min \|\mathbf{X}\|_1 \quad \text{s.t.} \quad \mathbf{y} = \mathbf{AX}, \tag{9.14}$$

which can be solved efficiently via linear or quadratic programming techniques. Under certain conditions on the matrix $\mathbf{A}$ and the sparsity of $\mathbf{X}$, both (9.13) and (9.14) have the same unique solution.

Consequently, the important issue is to choose the suitable domain of signal sparsity, which is related by a linear transform to the domain of data acquisition. The domain of sparsity is also related to the signal characteristics and varies for different classes of signals. For instance, when narrowband signals are considered, one of the commonly used domains is the FT domain. However, the signals that are not typically narrowband cannot be cast as sparse over the frequency variable. Wideband nonstationary signals, such as FM signals, are often sparse in the TF domain, but not in the FT domain. In this case, the sparsity depends on the choice of the TFR, which can be crucial for signal reconstruction efficiency.

The CS concept based on the TF sparsity of nonstationary signals should provide the IF and m-D signature estimation as in the case of the original dataset. In certain circumstances, the signal phase nonlinearity cannot be easily estimated and the simplest solution could be to use the STFT. Hence, starting from the assumption that we have an incomplete set of data in the time domain, the idea of CS is to perform sparse signal reconstructions over overlapping intervals defined by the different window positions to provide TF signal representations. As one of the simplest CS reconstruction solutions, the orthogonal matching pursuit (OMP) can be employed [47]. The OMP is a greedy algorithm that provides an approximate solution for the minimization problem defined in (9.14). In each iteration, it searches for the maximum correlation between the measurements and the dictionary matrix. Thus, through the iterations it selects a certain number of dictionary matrix columns, where this number is defined by the given number of iterations. The least-square optimization is performed afterward in the subspace spanned by all previously selected columns.

It is interesting to note that the IF estimate from the first iteration of OMP is the same as that provided by the location of highest peak value in spectrogram [48]. A higher number of iterations becomes necessary if any improvement over spectrogram is to be achieved. Consider the case of a chirp signal represented by 45% of randomly chosen measurements. The initial STFT, calculated according to (9.7), uses M samples around each n. The OMP reconstructed STFT in 15 iterations, denoted by $STFT_R(n,k)$, is shown in Figure 9.3. Note that the OMP does not always provide a good reconstruction, especially when the number of measurements is small. The results can be improved by using the quadratic distribution such as the S-method, defined as

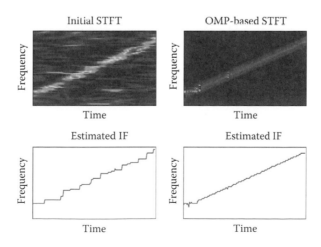

FIGURE 9.3
TFR and the IF estimation using OMP with the STFT.

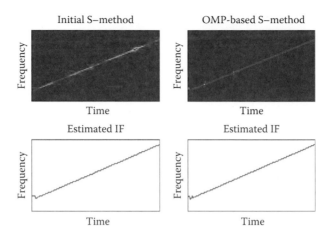

FIGURE 9.4
TFR and the IF estimation using OMP with the S-method.

$$SM(n,k) = \sum_{i=-L}^{L} STFT_R(n,k+i)STFT_R^*(n,k-i).$$

In that case, the OMP-based S-method (i.e., S-method applied to the OMP-based STFT recovery) will provide almost the same IF estimation results as the S-method of the standard STFT without missing data (Figure 9.4) since the S-method can provide an ideal chirp representation just as in the case of the WD, as far as L is sufficiently large for a complete summation over the auto-term in the STFT. Here, just a few samples will produce the result as in the WD case.

A similar approach can be applied to radar data corresponding to human gait signatures [48]. The OMP is employed to recover the m-D signature under missing samples. In order to verify the approach, the data are uniformly sampled at the critical Nyquist rate, followed by data sample removal. The motion signature using spectrogram on full data depicts detailed torso and limbs movements, as shown in Figure 9.5a. Under missing samples, the spectrogram performance suffers from very noisy and cluttered motion TF signature. The OMP-based result successfully reconstructs the human motion signatures from random samples (Figure 9.5b).

9.4.2 Analysis of Missing Samples in the FT (STFT) Domain

The missing samples, whether they are a product of random sampling strategy or the result of the L-statistics, generate undesirable side lobes in the spectral analysis of nonuniformly sampled data sequences. The question is how many missing samples we can afford so as to keep the noise

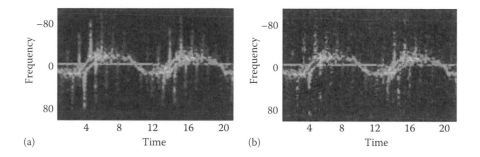

FIGURE 9.5
(See color insert.) The STFT of critically sampled signal (a) and the OMP-based TFR from randomly undersampled signal (b).

influence small to the signal reconstruction performance. Here, the analysis of the influence of missing samples on the transform domain values is performed from a noise perspective. As a fundamental starting point, we will first observe the case of the spectral analysis using the FT, which is at the core of the STFT (which is the FT of a windowed signal), as well as many other TFRs.

It is shown that the missing samples, when using the FT, could be described by a new type of noise that deteriorates signal representation. Increasing the number of missing samples increases the corresponding noise level, lowering the signal's sparsity. Consequently, signal component detection becomes more difficult. In order to deal with this issue, we need the expressions that relate the number of missing samples to the statistics of the spectral noise. This relation is also crucial for the analysis of the initial steps (transforms) in the sparse signal reconstruction algorithms.

Consider a set of N signal values, $s(1)$, $s(2),\ldots,$ $s(N)$, that belong to the sparse signal with K sinusoidal frequency components defined by the amplitudes A_i and frequencies k_i, $i=1,\ldots,K$. The DFT of this signal is defined as

$$S(k) = \sum_{n=0}^{N-1} \sum_{i=1}^{K} A_i \exp(-j2\pi(k - k_i)n/N).$$

Now, we might observe the following values:

$$\mathbf{x} = \{x(n), n = 0, 1, \ldots, N - 1\}$$

$$= \left\{ A_i \exp(-j2\pi(k - k_i)n/N), \ n = 0, 1, \ldots, N - 1 \right\},$$

where we assume that $\sum_{i=0}^{N-1} x(i) = 0$. Furthermore, consider a set of $M \leq N$ available samples from $\mathbf{x}$ corresponding to the CS signal $\mathbf{y} = \{y(1), y(2), \ldots, y(M)\}$. Hence, we have a number of $M_Q = N - M$ samples unavailable or they are just discarded in the case of applying the L-statistics to a noisy signal. Initially, we observe a simple case where only a single component is present: $K = 1, A_1 = 1, k = k_1$. The DFT over the available set of samples can be written as follows:

$$F(k) = \sum_{n=1}^{M} y(n) = \sum_{n=1}^{N} \{x(n) - \varepsilon(n)\},$$

where at the positions of missing samples the noise can be modeled by

$$\varepsilon(n) = x(n) = \exp\left(j2\pi(k - k_1)n/N\right),$$

while $\varepsilon(n) = 0$ otherwise. Therefore, the disturbance $\varepsilon(n)$ actually contains a set of $M_Q = N - M$ missing signal values. Obviously, $E\{F_{k=k1}\} = M$, while $E\{F_{k\neq k1}\} = 0$ holds. The noise variance in F can be calculated as follows:

$$var(F) = E\left\{[y(1) + \cdots + y(M)] \cdot [y(1) + \cdots + y(M)]^*\right\} = M\frac{N-M}{N-1},$$

where $(\cdot)^*$ denotes complex conjugate, and the following equalities have been used:

$$E\left\{x(i)x^*(i)\right\} = E\left\{y(i)y^*(i)\right\} = 1,$$

$$E\left\{x(i)x^*(j)\right\} = E\left\{y(i)y^*(j)\right\} = -\frac{1}{N-1}, \quad i \neq j.$$

Next, we provide the ratio between the DFT value at the signal and nonsignal positions. Starting from the assumption that the absolute values of $F_{k\neq k1}$ are Rayleigh distributed, we have

$$R_{95} = \left|\frac{F_{k\neq k1}}{F_{k=k1}}\right| < \frac{\sqrt{6}\sigma}{M} = \sqrt{\frac{3(N-M)}{M(N-1)}}$$

with probability 0.95, where $\sigma^2 = var(F)$. Hence, a random variable $|F_{k\neq k1}|$ is above $\sqrt{6}\sigma$ with probability 0.05. Depending on the value of M, the noise along DFT samples $F_{k\neq k1}$ will have the amplitude level $|F_{k=k1}|\sqrt{3(N-M)/M(N-1)}$ with probability 0.05. This analysis can be used to define M such that the noise is below the signal components. In this way, we can ensure that we are able to detect signal components either in the CS initial transform or in the L-estimate transform. The DFT of a multi-component signal for different numbers of available samples M is shown in Figure 9.6, where $N = 512$.

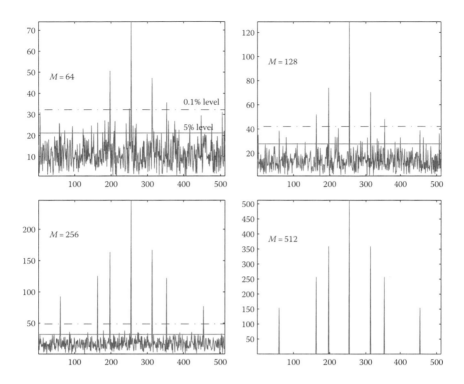

FIGURE 9.6
Illustration of the DFT of multicomponent signal for different M. Threshold R_{95} is plotted by a horizontal solid line.

In the case of multicomponent signal, the variable $x(n)$ is given in the form:

$$x(n) = \sum_{i=1}^{K} A_i \exp(-j2\pi(k - k_i)n/N),$$

for a certain n. The mean value of M available randomly positioned samples $\mathbf{y} = \{y(1), \ldots, y(M)\} \subset \{x(1), \ldots, x(N)\}$ is given by $E\{F\} = \sum_{i=1}^{K} MA_i \delta(k - k_i)$, whereas the variance of DFT values at the nonsignal and signal positions can be calculated according to

$$\sigma_\varepsilon^2 = \mathrm{var}(F_{k \neq ki}) = M\frac{N - M}{N - 1} \sum_{j=1}^{K} A_j^2, \text{ and}$$

$$\sigma_s^2 = \mathrm{var}(F_{k=ki}) = M\frac{N - M}{N - 1} \sum_{j=1, j \neq i}^{K} A_j^2.$$

As already mentioned, the DFT values at the nonsignal positions are Rayleigh distributed. In the sequel, they will be referred to as the noise-alone values. Using the Rayleigh distribution, we can now define the probability that the noise-alone DFT value is below a certain value T:

$$P(T) = 1 - \int\limits_{T}^{\infty} \frac{2z}{\sigma_\varepsilon^2} \exp\left(-z^2/\sigma_\varepsilon^2\right) dz = 1 - \exp\left(-T^2/\sigma_\varepsilon^2\right).$$

Consequently, the probability that all noise-alone DFT values are below T is $P(T)^{(N-K)}$. When estimating signal components, the error appears when the DFT signal value is in the range T and $T+dT$, while at least one noise-alone value is above T. By assuming that the DFT values of the ith signal component is equal to MA_i, we can define an approximative form of error probability (probability of wrong detection of the ith signal component) as follows:

$$P_e = 1 - \left(1 - \exp\left(-\frac{M^2 A_i^2}{\sigma_\varepsilon^2}\right)\right)^{N-K}.$$

A better approximation, in some cases, is obtained by using $M^2 A_i^2 - \sigma_s^2$ instead of $M^2 A_i^2$.

An analysis of the relationship between the sampling and aliasing has also been considered in the literature [33]. It is based on the Fourier random sampling problem that was studied before introducing the CS paradigm and perspective. It was proven that, when dealing with random sampling intervals, the aliases are not periodical replicas of the original spectrum as in the uniform sampling case. The random distribution of sampling intervals produces aliases in the form of a noise floor, whose power is proportional to the signal spectrum power. It follows that the aliasing noise power is proportional to the spectrum occupancy of the original signal. If the original spectrum is sparse, then the aliasing noise power will not dominate, allowing this signal to be successfully reconstructed. This theory of discrete random sampling is extended from the standard continuous time random sampling, which is usually difficult to implement in the practice. Therefore, the discrete random sampling provides the possibility of designing a more feasible sampling hardware solution, which can be integrated into existing standard analog–digital converters [33]. This hardware can be used for efficient sampling and reconstruction of spectrally sparse signals at an average sampling rate significantly below the Nyquist one.

9.4.3 Effects of Missing Samples to Bilinear Time–Frequency Distributions

In the sequel, we analyze the effects of missing samples to the quadratic TF distributions, such as the widely used WD [40]. The difference to the STFT is that, instead of the signal in time domain, we deal with the IAF, whose FT with respect to the time lag becomes the WD. The missing data samples yield missing entries in the IAF and, consequently, cause artifacts, which, similar to additive noise, spread over the entire WD, thereby degrading the TFR. Consider a discrete-time signal, $x(n)$, $n = 1, \ldots, N$, which comprises a single or multiple components of FM signals. Assume that M samples from $x(n)$ are observed ($M \leq N$). Denote $y(n)$ as the observation vector with $M_Q = N - M$ missing samples, which are randomly and uniformly distributed over time. Consequently, the missing data can be expressed using the missing samples mask $\Delta(n)$ as

$$x_m(n) = x(n) \cdot \Delta(n) = \sum_{i=1}^{M_Q} x(n_i)\delta(n - n_i), \quad n_i \notin \mathbf{S}, \quad (9.15)$$

where $\mathbf{S}$ is the set of observed time instants with a cardinality of card$\{\mathbf{S}\} = M$. Thus, the observed data can be expressed as

$$y(n) = x(n) - x_m(n) = x(n) - \sum_{i=1}^{M_Q} x(n_i)\delta(n - n_i). \quad (9.16)$$

The IAF of $x(n)$ is defined as

$$C_{xx}(n, m) = x(n + m)\, x^*(n - m), \quad (9.17)$$

where m is the time lag. Denote $R(n) = 1(n) - \Delta(n)$ as the observation mask, where $1(n) = 1$, $\forall n$, is an all-one function. Then, the IAF of $y(n)$ is calculated as

$$C_{yy}(n, m) = y(n + m)\, y^*(n - m) = C_{xx}(n, m)C_{RR}(n, m), \quad (9.18)$$

where $C_{xx}(n, m)$ and $C_{RR}(n, m)$ are, respectively, the IAFs of $x(n)$ and $R(n)$. In particular, $C_{RR}(n, m)$ can be expressed as

$$C_{RR}(n, m) = C_{11}(n, m) + C_{\Delta\Delta}(n, m) - C_{1\Delta}(n, m) - C_{\Delta1}(n, m), \quad (9.19)$$

where $C_{11}(n, m)$ and $C_{\Delta\Delta}(n, m)$ are the IAF of $1(n)$ and $\Delta(n)$, respectively, and $C_{1\Delta}(n, m)$ and $C_{\Delta1}(n, m)$ are two IAF cross-terms between $\Delta(n)$ and $1(n)$. The difference in the mask IAF due to the missing data samples can be expressed as

$$C_D(n, m) = C_{11}(n, m) - C_{RR}(n, m) \quad (9.20)$$

$$= C_{1\Delta}(n, m) + C_{\Delta1}(n, m) - C_{\Delta\Delta}(n, m). \quad (9.21)$$

In reality, the IAF is affected by the window size due to zero padding. The length of the rectangular window along the m dimension depends on n and is expressed as

$$Q_m(n) = N - |N + 1 - 2n|, \quad n = 1, \ldots, N. \tag{9.22}$$

By taking this into account, an N-sample function $1(n)$ would have $N^2/2$ nonzero entries in $C_{11}(n, m)$ if N is even, or $(N^2 + 1)/2$ entries if N is odd. Without loss of generality, we consider an even value of N here. In this case, the number of unit-value entries of $C_D(n, m)$, in the presence of M_Q missing data samples, can be well approximated as $\widetilde{M}_Q \approx M_Q N - M_Q^2/2$. Because the missing data sample positions are randomly and uniformly distributed in time, the positions of the missing IAF entries can also be considered randomly and uniformly distributed over n and m. Then, for a specific n, the number of missing entries in $C_{yy}(n, m)$ is

$$K_m(n) = \frac{\widetilde{M}_Q}{N^2/2} Q_m(n) \approx \frac{2M_Q N - M_Q^2}{N^2} Q_m(n). \tag{9.23}$$

The FT of the IAF $C_{yy}(n, m)$ with respect to lag coordinate m is the WD:

$$W_{yy}(n, k) = FT_m \left[C_{yy}(n, m) \right] = \sum_{m \in \mathbf{S}_m(n)} C_{xx}(n, m) e^{-j4\pi km}, \tag{9.24}$$

where $\mathbf{S}_m(n)$ is the set of nonzero m entries for a specific n with a cardinality of card$\{\mathbf{S}_m(n)\} = L_m(n) = Q_m(n) - K_m(n)$, and the value of m is bounded by $-[Q_m(n) - 1]/2$ and $[Q_m(n) - 1]/2$. Using the uniform distribution of the missing entries in n and m, it is straightforward to verify that $E[W_{yy}(n, k)] = \xi W_{xx}(n, k)$, where $\xi = L_m(n)/Q_m(n) = M^2/N^2$. That is, $W_{yy}(n, k)$ is an unbiased estimator of $W_{xx}(n, k)$, up to a scaling factor ξ, for every n and k. Write $W_{yy}(n, k)$ as

$$W_{yy}(n, k) = W_{xx}(n, k) - W_D(n, k), \tag{9.25}$$

where $W_D(n, k)$ denotes the artifacts in the WD due to the missing data samples. Then, from this discussion, we obtain $E[W_D(n, k)] = (1 - \xi)W_{xx}(n, k)$. Because $W_{xx}(n, k)$ is deterministic, we have

$$var\left[W_{yy}(n, k)\right] = var\left[W_D(n, k)\right] = \sum_{m \notin \mathcal{S}_m(n)} |C_{xx}(n, m)|^2 - \frac{K_m(n)}{Q_m^2(n)} |W_{xx}(n, k)|^2. \tag{9.26}$$

We can conclude that the missing data samples yield spreading artifacts that are randomly distributed over the entire TF domain, and the overall variance increases as the number of missing data samples increases. Similar analysis holds for the ambiguity function, which is the dual domain representation of the WD [40]. Note that $var[W_{yy}(n,k)]$ varies with n in a zero-padded case, and a large variance is observed when n is around $N/2$ in which a large number of samples are missing.

9.5 CS Reconstructions in Time–Frequency Domain

This section covers different scenarios and possibilities for nonstationary signal reconstructions in the joint variable domains (e.g., TF and AF domains). According to the analysis presented in the previous section, we consider two interesting cases. The first one is related to the CS problem formulated using bilinear transform domains such as AF, IAF, and WD, aiming at reconstruction of sparse TFR. The second one explores the possibility of using the combination of DFT and STFT to provide the total reconstruction of sparse components from the generally nonsparse and nonstationary signals (appearing in radars).

9.5.1 Signal Reconstruction from Bilinear Transforms

In the previous section, we have shown that the effects of missing samples reflect as additive noise in the WD domain. The noise spreads over the entire TF plane. On the other side, it is known that most of the interferences that generally appear in the WD could be mitigated in its counterpart domain known as the AF domain. Therefore, in the sequel, we first formulate the AF-based CS problem and show that this approach can be efficiently used to deal with both the cross-terms and strong impulse noise. Also, it is possible to apply CS reconstruction to recover IAF samples, which leads to high-quality TFR.

9.5.1.1 Ambiguity Domain-Based CS

The CS applications in the AF and TF domains are initiated by Flandrin and Borgnat [25]. The WD is a two-dimensional (2D) DFT of the AF:

$$AF(\theta, m) = \frac{1}{N}\sum_{k=1}^{N}\sum_{n=1}^{N}WD(n,k)e^{-j2\pi(n\theta-km)/N}. \tag{9.27}$$

In the classical TF analysis, a lowpass kernel is multiplied to the AF to attenuate the cross-terms. Instead of using the classical reduced-interference kernels with lowpass characteristics [45], a few samples of the AF around the origin are selected to constitute reduced observations [25], thus forming a "CS AF." Generally speaking, the sparsity assumption is defined as follows: the $N \times N$ representation of signal with K components should have at most $K \times N$ nonzero points. Since we usually deal with multicomponent signals, attenuating or removing the cross-terms necessitates discarding ambiguity domain points that are far from the origin. Thus, the AF domain becomes the preferred observation domain: The AF values away from the origin are considered unavailable, whereas a small region around the origin, where the auto-terms are located, is considered available AF values. The WD is assumed to be the transformation domain where the signal is sparse. In this case, sparse signal reconstruction techniques (ℓ_1 norm reconstruction algorithms) will provide high-resolution TFR without cross-terms, thereby allowing accurate IF estimations [46]. Hence, the desired TFR (in matrix notation **TFR** with elements $TFR(n, k)$) is obtained as a result of applying the standard CS minimization procedure defined in the form of linear programming:

$$\min \sum_{k=1}^{N}\sum_{n=1}^{N} |TFR(n,k)| \quad \text{s.t.} \quad \frac{1}{N}\sum_{k=1}^{N}\sum_{n=1}^{N} TFR(n,k)e^{-j2\pi(n\theta-km)/N} = y_c,$$

for M selected points in (θ, m). In a matrix form, this relation can be rewritten as

$$\min \|\mathbf{TFR}\|_1 \quad \text{s.t.} \quad \mathbf{y}_c = \mathbf{A} \cdot \mathbf{TFR},$$

where $\mathbf{y}_c$ is the vector containing the available AF values and $\mathbf{A}$ is the corresponding CS matrix obtained after discarding the elements from 2D DFT matrix related to the missing AF values.

Example: Consider the multicomponent signal, which consists of a chirp and a sinusoidal frequency modulated component:

$$s(n) = \exp(j(16/5\cos(3/2\pi n) + 6\cos(\pi n) + 12\pi n))$$

$$+ \exp(-j(5\pi n^2 + 20\pi n)) + v(n),$$

where $v(n)$ is Gaussian noise. In order to provide faster computations, we consider the TFR of a small size 60×60 (3600 points). The central region of size 7×7 (1.4% of the total number of points) in the ambiguity domain is considered as available AF values. The resulting number of nonzero points in the sparse representation is approximately 130 (estimated from different experiments), which is a small percentage of the total number of points in the TF domain. The original WD and AF are shown in Figure 9.7a and b, respectively. The resulting sparse representation is illustrated in Figure 9.7c. It can be observed that the CS approach not only provides the signal power localization but also improves TF signal resolution, while mitigating cross-terms.

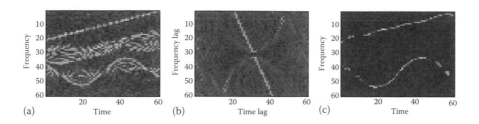

FIGURE 9.7
(See color insert.) (a) Wigner distribution, (b) ambiguity function, and (c) resulting sparse TFR.

9.5.1.2 IAF Reconstruction Yielding Sparse TFR

Similar to the treatment of TF cross-terms in the ambiguity domain, it is possible to mitigate the artifacts due to missing samples by applying kernels in the IAF domain. As such, a less cluttered TF distribution (TFD) can be constructed with the suppression of both missing data artifacts and undesired cross-terms. The best kernel in this case is the one that only keeps the signal signature, whereas the other regions are filtered out. One of such choices is the adaptive optimal kernel (AOK) [49], which is known to provide signal-adaptive filtering capability in the AF domain. The TFD reconstruction is based on the FT that relates the IAF and the WD. For illustration purposes, we use a two-component FM signal, where the instantaneous phase laws of the two polynomial phase components are respectively expressed as

$$\phi_1(n) = 0.05n + 0.05n^2/N + 0.1n^3/N^2,$$

$$\phi_2(n) = 0.15n + 0.05n^2/N + 0.1n^3/N^2, \tag{9.28}$$

for $t = 1, \ldots, N$, where $N = 128$, and 50% (or 64 samples) of the data are missing. The two polynomial phase signal components have the same power. The artifacts in the WD spread evenly over the frequency axis, but have stronger presence in the central portion of the time axis (Figure 9.8a). The TFD obtained from the AOK is shown in Figure 9.8b. It is evident that this kernel type substantially mitigates the missing data artifacts.

Denote $\mathbf{x_c}(n) = [C_{xx}(n, m), \forall m]$ as a vector that contains all IAF entries along the m dimension corresponding to time n, and $\mathbf{X_c}(n)$ as an FT vector collecting all the TFD entries for the same time n. Note that $\mathbf{x_c}(n)$ has missing entries due to missing data and may be smoothed as a result of applying kernel. Then, these two vectors are related by the inverse DFT (IDFT), expressed as

$$\mathbf{x_c}(n) = \mathbf{\Psi} \mathbf{X_c}(n), \quad \forall n, \tag{9.29}$$

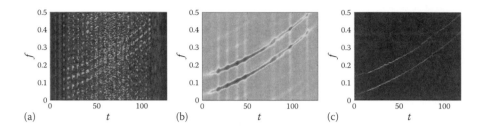

FIGURE 9.8
TFDs obtained from the WD (a), the AOK (b), and the OMP (c).

where $\boldsymbol{\Psi}$ is a matrix performing the IDFT. By removing the IAF entries with zero or negligible values, we can construct a vector $\mathbf{y}_c(n)$, which becomes

$$\mathbf{y}_c(n) = \mathbf{A}\mathbf{X}_c(n), \tag{9.30}$$

where $\mathbf{A}$ is the resulting CS matrix after removing the corresponding rows from IFT matrix $\boldsymbol{\Psi}$. Because the signals are sparsely represented in the TF domain, $\mathbf{X}_c(n)$ can be reconstructed through sparse signal recovery techniques:

$$\min \|\mathbf{X}_c(n)\|_1 \quad \text{s.t.} \quad \mathbf{y}_c(n) = \mathbf{A}\mathbf{X}_c(n). \tag{9.31}$$

This process should be repeated for all n. For instance, as in the previous examples, we can apply the OMP for each time instant n. We depict the TFD results obtained from the observed two-component FM signal with missing samples by exploiting sparse signal reconstructions. Figure 9.8c shows the reconstructed TFDs using the OMP algorithm from the IAF after applying the AOK, with two iterations used for each time instant. In this case, the reconstructed result shows a TFD with very little clutter.

9.5.1.3 Robust Ambiguity Domain-Based CS in the Presence of Impulse Noise

Now we consider the capability of dealing with ambiguity domain observations in the presence of noise. The starting assumption is that the ambiguity domain measurements are seriously affected by a significant amount of impulsive noise. The underlying problem is accurate reconstruction of sparse TFR when impulse noise is encountered in the ambiguity domain (as domain of observations). In these circumstances, the standard CS reconstruction techniques would no longer be able to provide desirable results. Therefore, we may discard the noise by applying the robust statistics. In that way, we are able to map the noisy ambiguity domain observations directly to the noise-free sparse TFR. This approach significantly eliminates the influence of noisy pulses as well as the influence of cross-terms without applying a kernel function. The observation vector is obtained as a set of measurements from the AF:

$$\mathbf{y} = \{AF(\theta, m) \mid (\theta, m) \in \Omega\}, \tag{9.32}$$

where, in order to avoid the cross-terms, the measurements are again taken from a narrow ambiguity region Ω around the origin. Furthermore, the transform domain matrix $\mathbf{\Psi}$ is obtained as a 2D DFT.

Problem arises, however, in the presence of noise, especially when the underlying ambiguity domain measurements are corrupted by impulsive noise. That is, the linear measurements are severely degraded, with original information masked by large noise amplitudes spreading across the measurements. The corrupted samples will cause standard reconstruction algorithms to fail in their attempts to recover an accurate sparse TFR. One solution could be to define a robust measurement procedure, which is not based on linear projections in order to avoid the impulsive noise. In our case, we have a predefined measurement defined by the ambiguity domain mask around the origin. Thus, we propose a solution that includes robust statistics into CS reconstruction technique in the ambiguity domain. Namely, we seek to provide a noise-free version of the initial transform domain vector or, in other words, robust initial transformation to the TF domain. One efficient robust approach is based on the concept of L-estimation and α-trimmed filter form. The L-estimators are defined as linear combinations of order statistics and can be used even for the mixed noise type. The L-statistics approach involves sorting out data samples and then removing the highest values.

The minimization problem can be formulated as follows:

$$TFR = \arg\min_{\theta} \|\mathbf{X_L}\|_1 \quad \text{s.t.} \quad \mathbf{y} - \mathbf{A}\mathbf{X_L} = 0 \mid_{(0,m)\in\Omega}, \tag{9.33}$$

where $\mathbf{A}$ is CS matrix corresponding to the ambiguity region Ω, while the initial transform is

$$\mathbf{X_{L0}} = \sum_{i=1}^{M_a} L_i(\theta, m), \quad M_a = N(1 - 2\alpha) + 4\alpha. \tag{9.34}$$

The sorted vector of AF samples multiplied by the corresponding 2D DFT basis functions is denoted by $L_i(\theta, m)$. It reads

$$L_i(\theta, m) = \text{sort}\left\{AF(\theta, m) \exp\left(-j2\pi mk/N\right) \exp\left(-j2\pi\theta l/N\right)\right\},$$

where $(\theta, m) \in \Omega$ and $\text{card}\{\Omega\} = N$. Note that the sorting operation is performed in nondecreasing order with respect to the amplitudes. For the purpose of noise elimination, we discard $2\alpha(M - 2)$ of the highest elements in $L(\theta, m)$, while the mean is calculated over the rest of the values. Therefore, the proposed modified approach can be observed as the L-statistics based ℓ_1-norm minimization.

In order to illustrate the advantages of the proposed method, let us observe a set of noisy measurements in the ambiguity domain. We are dealing with a nonstationary multicomponent signal where, in addition to noise, strong cross-terms between each component pairs can obscure the individual component power distribution. The signal consists of two cosine frequency modulated components:

$$s(n) = \exp\big(j(16/5\cos(3/2\pi n) + 6\cos(\pi n) + 6\pi n)\big)$$
$$+ \exp\big(j(16/5\cos(3/2\pi n - 3\pi/2) + 6\cos(\pi n - 3\pi/2) - 14\pi n)\big).$$

The original signal AF and WD are shown in Figure 9.9a and b, respectively. The 10% of the measurements in the AF domain are corrupted by strong noisy peaks. The results of standard ℓ_1-based reconstruction applied directly to the noisy measurements are shown in Figure 9.9c, and we might observe that it is seriously corrupted by noise. However, the L-statistics-based minimization provides results that are both cross-term-free and noise-free, as shown in Figure 9.9b.

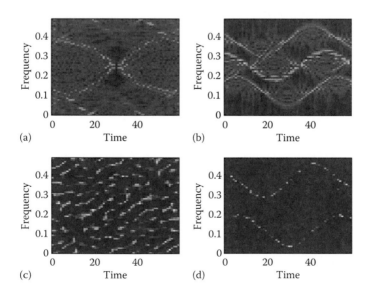

FIGURE 9.9
(a) Standard ambiguity domain representation of original signal; (b) standard WD of original signal; (c) sparse TF reconstruction results from the set of noisy measurements in the ambiguity domain; and (d) sparse TF reconstruction results obtained using the proposed method.

9.5.2 Compressive Sensing Reconstruction from Linear Transforms

In this part, we deal with a class of problems where the missing samples or observations are not due to Nyquist sampling relaxation, but rather occur as a consequence of attempting to separate desired from undesired nonstationary signal components. Separation in the TF domain becomes difficult if the respective TF signatures reside over common TF regions or encounter several TF intersection points. Namely, it is often the case that significant signal components overlap in TF plane, which further means that in the AF these components will be considered a single one. Ambiguity domain analysis [50], including the common CS-based one, cannot deal with these kinds of signals. This case makes TF masking difficult and renders TF synthesis methods ineffective. The CS approach to the underlying problem is based on the premise that missing samples are undesired samples, which are removed from consideration due to interference contribution.

The main goal here is to use the CS approach to recover narrowband signals when contaminated with highly nonstationary signals. In this case, the desired sparse representation is achieved using the Fourier basis. The samples in the TF domain are selected to favor the sinusoidal signals, which are both locally and globally sparse. The local behavior of both stationary and nonstationary signals is revealed by taking the STFT. The TF regions corresponding to the nonstationary signals over all windows are identified and removed from consideration and, therefore, cast as missing observations. For a successful removal of these regions, the L-statistics-based analysis is used. A simplified example is presented in Figure 9.10, where the m-D is induced by four rotating parts reflecting continuously and several

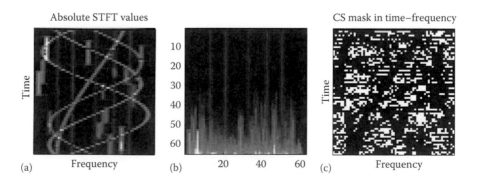

FIGURE 9.10
TFR of the radar signal, at a given range, corresponding to five rigid body points, four fast rotating reflectors, and several points reflecting during a short time interval, producing flashes: (a) the spectrogram within the coherent integration time; (b) the sorted spectrogram values, along time, for each frequency (cross-range); and (c) the matrix showing available (in white) and omitted (in black) values in the TF plane, after the L-statistics approach with 60% of omitted values is used. The Hanning window is used in TF analysis.

rotating flashing parts. The sparse (rigid body) signals intersect with the nonstationary components at various TF points (Figure 9.10). We remove a large number of overlapping points or intervals, and retain only those TF observations belonging to the narrowband signals [51].

The analyzed problem can be stated as follows. Let us consider a composite signal,

$$x(n) = x_{sp}(n) + x_{ns}(n), \tag{9.35}$$

where

$x_{sp}(n)$ is the stationary and sparse signal part
$x_{ns}(n)$ is the highly nonstationary part

This kind of signal composition is inspired by the radar signal returning from moving targets, where a rigid body and m-D components can be both present. The DFT of this signal can, therefore, be defined as

$$X(k) = X_{sp}(k) + X_{ns}(k), \tag{9.36}$$

$$X_{sp}(k) \neq 0 \text{ for } k \in \{k_{01}, k_{02}, \ldots, k_{0K}\}, \quad K \ll N,$$

where N is the total number of signal samples. Thus, $X_{sp}(k)$ is stationary and sparse in k, whereas $X_{ns}(k)$ is nonstationary and nonsparse, and may assume nonzero values even for all frequencies k. In the aforementioned signal model, the sparse and nonstationary part of signal may significantly overlap in frequency. Further, certain frequency components in $X_{ns}(k)$ could be much stronger than their counterparts in $X_{sp}(k)$. The nonstationary components at some frequencies $k_j \neq k_{0i}, i = 1, 2, \ldots, K$ could also be much stronger than the total signal components at the desired frequencies, that is, $|X_{ns}(k_j)| \gg |X(k_{0i})|$.

Classical spectrum analysis tools that deal with signals separation would prove ineffective in handling signals adhering to the model in (9.36), even if we assume that the frequencies of desired signals $k_{01}, k_{02}, \ldots, k_{0K}$ are known. For instance, consider an ideal notch filter (its inverse form) applied to a priori known desired frequency (which is not the case in practice). When disturbance appears at the same frequency as the desired component, it cannot be filtered out even with the use of an ideal notch filter. In this case, the signal amplitude obtained at the filter output will have an incorrect value, as will be illustrated in the examples. On the other hand, we also assume that the sparse part of the signal $x_{sp}(n)$ and the highly nonstationary part $x_{ns}(n)$ overlap in a significant number of time samples, that is, for any (or all) n, $x(n) \neq x_{sp}(n)$ may hold. This renders time-domain signal separation difficult, if not impossible, even in a small number of instants. It prevents the application of the CS algorithms in the time domain. Finally, since we actually have the full set of data available, we may consider the application of l_1 norm-based robust processing, which will again fail to extract the sparse part due to time and frequency overlapping with sporadically stronger nonstationary part.

The highly nonstationary signal components invite TF signal representation:

$$\rho_x(n,k) = \rho_{sp}(n,k) + \rho_{ns}(n,k), \tag{9.37}$$

where $\rho_x(n,k)$ is a linear TF representation (quadratic representations give rise to cross-terms). We deal with the general case where $\rho_{sp}(n,k)$ and $\rho_{ns}(n,k)$ overlap in a significant number of points (n,k) in the TF plane. The significant number of disturbance values in the overlapping regions can be much stronger than their sparse signal counterparts. The reconstruction of the desired sparse signal is considered impossible unless we discard the overlapping values. An effective way to proceed is to apply the CS methods tailored to the TF problem formulation. Note that satisfactory results will be obtained as long as the number of available data is larger than sparsity of signal.

In the TF analysis, we use the simplest representation based on the STFT. It is evident from the aforementioned problem description that the main challenge lies in properly selecting the signal observations in the STFT domain, avoiding all disturbances, and in establishing a linear relationship between TF and Fourier domain. Here, the L-statistics is adapted to properly select TF regions. It involves sorting out data samples along the time axis for a given frequency and then removing some of them. If we have only the nonstationary component at specific frequency in TF representation, it is clear that eliminating the highest values along that frequency would actually eliminate the undesired nonstationary interference. For the frequency lines with contributions from both the nonstationary interference and the desired sparse sinusoidal components, the highest values correspond to common or overlapping regions. Removing most of the highest values along the frequency line will remove the interference contribution as well. Another possible case is when the nonstationary and sparse signals are of the same order of amplitude, but the opposite phases produce low values at the intersection points. In this case, the solution is to remove some of the lowest values, in addition to the highest ones. As such, we avoid interference contamination by keeping the middle part of sorted values. The L-statistics is applied to $\rho_x(n,k)$, separately for each frequency $k = k_i$:

$$L\rho_x(n,k_i) = O_{k_i}(p) \quad \text{for } p \in [\alpha N, \beta N],$$

where $O_{k_i}(p) = \text{sort}_n \{\rho_x(n,k_i)\}$, $n, p \in \{0, \dots, N-1\}$, and L denotes the L-estimation operator, while the parameters α and β are defined as $0 \leq \alpha < \beta \leq 1$. It means that for each k, instead of the original N points, $n = 0, \dots, N-1$, we are left with a certain arbitrary set of time intervals where

transformation values $N_1(k_i), N_2(k_i), \ldots, N_B(k_i)$, are used in further calculation. These intervals will be referred to as the frequency-dependent arbitrary
positioned time intervals. The signal values that appear within these intervals can be considered available CS measurements, containing the stationary
and sparse part of the signal only. Namely, after the L-statistics, we have
$L\rho_{ns}(n,k) \ll L\rho_{sp}(n,k)$ for $n \in [N_1(k), N_2(k), \ldots, N_B(k)]$, where the L-statistics
outputs $L\rho_x(n,k)$ belonging to the nonstationary and sparse components are
denoted as $L\rho_{ns}(n,k)$ and $L\rho_{sp}(n,k)$, respectively. Note that $L\rho_{ns}(n,k)$ is negligible, while $L\rho_{sp}(n,k)$ represents the available sparse signal TF values. For
example, using the L-statistics with $Q=70\%$ ($Q=100(1-\beta+\alpha)\%$) discarded values, for a given k, implies that the total duration of intervals
$N_1(k), N_2(k), \ldots, N_B(k)$, where the sparse signal TF values are available, for a
given frequency, is just 30% of the original observations.

9.5.2.1 Reconstruction Based on the CS Methods

Consider a discrete-time signal $x(n)$ of length N and its DFT $X(k)$. When a
rectangular window of width M is used, the STFT, defined in (9.7), can be
written in the matrix form as

$$\mathbf{STFT}_M(n) = \mathbf{W}_M \mathbf{x}(n), \tag{9.38}$$

where $\mathbf{STFT}_M(n)$ and $\mathbf{x}(n)$ are vectors:

$$\mathbf{STFT}_M(n) = [STFT(n,0), \ldots, STFT(n, M-1)]^T,$$
$$\mathbf{x}(n) = [x(n), x(n+1), \ldots, x(n+M-1)]^T, \tag{9.39}$$

and $\mathbf{W}_M$ is the $M \times M$ DFT matrix with coefficients $W(m,k) = \exp(-j2\pi km/M)$.
Considering nonoverlapping contiguous data segments, the next STFT will
be calculated at instant $n + M$ as follows $\mathbf{STFT}_M(n+M) = \mathbf{W}_M \mathbf{x}(n+M)$.
Assuming that N/M is an integer, the last STFT at instant $n + N - M$ is
expressed as $\mathbf{STFT}_{N-M}(n+N-M) = \mathbf{W}_M \mathbf{x}(n+N-M)$. Combining all STFT
vectors in a single equation, we obtain

$$\begin{bmatrix} \mathbf{STFT}_M(0) \\ \mathbf{STFT}_M(M) \\ \vdots \\ \mathbf{STFT}_M(N-M) \end{bmatrix} = \begin{bmatrix} \mathbf{W}_M & \mathbf{0}_M & \cdots & \mathbf{0}_M \\ \mathbf{0}_M & \mathbf{W}_M & \cdots & \mathbf{0}_M \\ \vdots & \vdots & \ddots & \vdots \\ \mathbf{0}_M & \mathbf{0}_M & \cdots & \mathbf{W}_M \end{bmatrix} \mathbf{x},$$

$$\mathbf{STFT} = \mathbf{W}_{M,N}\, \mathbf{x}. \tag{9.40}$$

In order to avoid confusion with notations, we emphasize again that $STFT(n,k)$ represents a scalar STFT value at a given time n and frequency k, while the boldface notation $\mathbf{STFT}_M(n)$ with one argument and one index represents the vector of STFT values (at M frequencies for a given instant n). Finally, boldface notation $\mathbf{STFT}$ without arguments denotes vector of STFT values for all frequencies k and all instants n. The vector $\mathbf{x}$ is the signal vector since

$$\mathbf{x} = \left[\mathbf{x}^T(0), \mathbf{x}^T(M), \ldots, \mathbf{x}^T(N-M)\right]^T = [x(0),\ x(1), \ldots, x(N-1)]^T.$$

Expressing the aforementioned vector in the Fourier domain, $\mathbf{x} = \mathbf{W}_N^{-1}\mathbf{X}$, where $\mathbf{W}_N^{-1}$ denotes the IDFT matrix of dimension $N \times N$ and $\mathbf{X}$ is the DFT vector, we have the relation between the STFT and DFT values as follows:

$$\mathbf{STFT} = \mathbf{\Psi}\mathbf{X}. \tag{9.41}$$

The transformation matrix is in this case defined as $\mathbf{\Psi} = \mathbf{W}_{M,N}\mathbf{W}_N^{-1}$, and it maps the global frequency information in $\mathbf{X}$ into local frequency information in $\mathbf{STFT}$. By removing a set of TF points using L-statistics, as discussed in Section 9.2, only a few elements in the observation vector $\mathbf{STFT}$ remain. This is accomplished as follows. For each frequency k, a vector of $\mathbf{STFT}$ in time is formed as

$$\mathbf{S}_k(n) = \left\{STFT(n,k),\ n = 0, \ldots, N-1\right\}. \tag{9.42}$$

After sorting the elements of $\mathbf{S}_k(n)$, we obtain the new ordered set of elements:

$$O_k(n) \in \mathbf{S}_k(n),$$

such that

$$|O_k(0)| \leq \cdots \leq |O_k(N-1)|. \tag{9.43}$$

A percentage Q of the high- and low-value elements are removed from consideration. As previously explained, these values capture most of the interference TF samples. The remaining STFT values belong to the sinusoidal desired signal.

Denote the vector of available STFT values by $\mathbf{STFT}_{CS}$. The corresponding CS matrix $\mathbf{A}$, relating the sparse DFT vector $\mathbf{X}$ to $\mathbf{STFT}_{CS}$, is formed by omitting the rows in $\mathbf{\psi} = \mathbf{W}_{M,N}\mathbf{W}_N^{-1}$ corresponding to the removed STFT values. Each row corresponds to one time and frequency point (n,k). We maintain that the reduced observations, the sparse DFT domain, and the linear relationship provide the necessary ground for a CS problem. The goal is to reconstruct the original sparse stationary signal since it produces the best concentrated DFT $X(k)$. Therefore, the corresponding minimization problem can be defined as follows:

$$\min \|\mathbf{X}\|_1 = \min \sum_{k=0}^{N-1} |X(k)| \quad \text{s.t.} \quad \mathbf{STFT}_{CS} = \mathbf{AX}. \tag{9.44}$$

The amount of discarded samples allowing recovery is not very restrictive for performance, as long as the corrupted samples are discarded and sufficient information about the desired signal remains.

A special case of this kind of analysis, with $M = 1$ when $\mathbf{STFT} = \mathbf{x}$, is presented in [38] for the separation of the stationary sparse part of the signal from a strong impulsive noise disturbance.

Norm ℓ_1 is used in minimization of (9.44) as a simpler form for the realization than the ℓ_0 form that would count the number of nonzero components. The norms ℓ_p with $0 \leq p \leq 1$ are used for the optimization of the TFR parameters in [45].

Example: We present a very simple case illustrating how the TF localization of the signal and disturbance can make signal recovery possible. Consider a sparse signal $x(n) = x_{sp}(n) + x_{ns}(n) = [-2.2500 + j2, -5.2374 + j2.1768, -3 - j1.25, 4.2374 + j1.1768, -1.75, 1.2374 - j0.1768, 5 + j2.25, 3.7626 + j1.8232]^T$ with most of its samples heavily corrupted by a strong disturbance (corresponding to a simplified two-point rigid body in the radar image). If we apply a direct DFT calculation or any L-statistics form in the time domain, we will not get a result that can be used in the CS recovery methods. In the time domain, there are only two noncorrupted samples, which are not sufficient for the CS reconstruction. These results are shown in Figure 9.11a and b. However, if we use the STFT calculation with $M = 2$ and then remove the STFT outlier values (there are three of them, $STFT_2(0,0)$, $STFT_2(6,0)$, $STFT_2(4,1)$, Figure 9.11a), we get a CS formulation in the TF domain as

$$\begin{bmatrix} STFT_2(0,1) \\ STFT_2(2,0) \\ STFT_2(2,1) \\ STFT_2(4,0) \\ STFT_2(6,1) \end{bmatrix} = \mathbf{A} \begin{bmatrix} X(0) \\ X(1) \\ X(2) \\ \vdots \\ X(7) \end{bmatrix}.$$

These kinds of problems may be solved using the CS methods. For example, if we perform the initial reconstruction by setting these three STFT values to 0 (Figure 9.11c), then we get a good initial result. We can easily conclude that the sparse part of the signal consists of $K = 2$ components at frequencies $k_{01} = 3$ and $k_{02} = 5$, Figure 9.11c. A simple

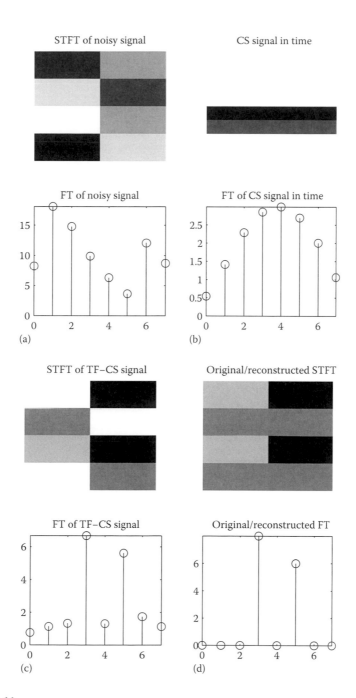

FIGURE 9.11
Illustration of the CS calculation based on the STFT. (a) Results for the noisy signal, (b) CS signal in time, (c) TF-CS signal, and (d) reconstruction results.

recovery may be done by varying the values of the STFT coefficients, being set to zero, in order to minimize $\|\mathbf{X}\|_1$. Minimal $\|\mathbf{X}\|_1$ is obtained with $STFT(0,0) = 0.5126 + j0.1768i$, $STFT(6,0) = -1.2374 + j0.0732$ and $STFT(4,1) = -1.2374 - j0.4268$. Thus, the recovered DFT of the sparse part of signal $x(n)$ is $X_{sp}(k) = [0, 0, 0, 8, 0, 6, 0, 0]$ (Figure 9.11d). It corresponds to the exact original sparse part of signal $x_{sp}(n) = \exp(j2\pi 3n/8) + 0.75 \exp(j2\pi 5n/8)$. Another way to get this simple result is to set all $X(k)$ for $k \neq 3$ and $k \neq 5$ to zero and solve the system of two unknowns $X(3)$ and $X(5)$ using five remaining STFT values.

9.5.3 Windows with Overlapping

In general, for a signal $x(n)$ of duration N we may use an arbitrary window $w(m)$ of an arbitrary duration. The time step M_s in the STFT calculation may also be arbitrary. The STFT in terms of the signal's DFT reads

$$STFT(n,k) = \sum_{m=0}^{M-1} w(m) \left[\frac{1}{N} \sum_{p=0}^{N-1} X(p) e^{j2\pi p(m+n)/N} \right] e^{-j2\pi mk/M}$$

$$= \frac{1}{N} \sum_{p=0}^{N-1} X(p) e^{j2\pi pn/N} W(k - pM/N) = \psi_n \mathbf{X}, \qquad (9.45)$$

where $W(k) = \sum_{m=0}^{N-1} w(m) e^{-j2\pi mk/N}$ and the matrix Ψ_n coefficients are defined by $\Psi_n(k,p) = \frac{1}{N} W(k - pM/N) e^{j2\pi pn/N}$.

A matrix notation of the STFT, with a step M_s in time, is $\mathbf{STFT} = \mathbf{\Psi X}$. Note that each STFT is actually a weighted linear combination of signal samples. These combinations are linearly independent only in the nonoverlapping STFT case. In the case of overlapping STFT, they become dependent because the same signal samples are used to calculate several STFT values. For instance, using a step $M/4$ instead of M, we get $4N$ STFT values, where the same samples are involved in four (overlapping) different STFTs. This further means that if just one sample is corrupted by a disturbance, several TF values will be removed in a row.

In order to eliminate the nonstationary part of the STFT, containing $x_{ns}(n)$, we remove $Q\%$ of the STFT values, for each frequency. Then we proceed to solve the CS problem similar to (9.44) with corresponding $\mathbf{A}$ and $\mathbf{STFT}_{CS}$.

If we want to keep the frequency grid in the STFT as in the original DFT of the signal, then the windows should be zero padded up to N, with $M = N$ in (9.45). The zero padding will provide that for each signal component with a constant frequency k_{0i} on the DFT grid, there is a frequency-direction line in the TF plane where its STFT values have the same and constant phase for all considered instants. Calculating the STFT in this way will increase the number of equations. However, an efficient CS solution may be obtained using the OMP methods with the initial recovered signal values obtained by setting the missing STFT values to zero [6,32].

Example: Consider a signal that consists of four stationary sinusoids:

$$x(n) = e^{j256\pi n/N} + 1.2e^{-j256\pi n/N+j\pi/8} + 0.7e^{j512\pi n/N+j\pi/4} + e^{-j512\pi n/N-j\pi/3},$$

with short duration pulses and strong transient signals as disturbances

$$x_{ch}(n) = \sum_{i=1}^{Ic} A_i e^{j\omega_i n} e^{-(n-n_i)^2/d_i^2}.$$

This case can be presented within the CS framework in the TF domain. The STFT of the original signal is shown in Figure 9.12a. After the L-statistics approach is performed on the sorted STFT values with 60% of the largest values and 10% of the lowest values being removed ($Q = 70\%$), the result is presented in Figure 9.12b, which amounts to the CS form of the STFT. The omitted STFT values are indicated in black (Figure 9.12c). The reconstruction is performed based on the CS values of the STFT given in Figure 9.12d.

The DFT of the sparse part of signal is obtained (Figure 9.13c) with both the amplitude and the phase preserved. The DFT of original signal is presented in Figure 9.13a. The DFT obtained by using the ℓ_2 norm, in lieu of the ℓ_1 norm when solving (9.44), is shown in Figure 9.13b.

In the sequel, we use the same example to demonstrate a low sensitivity of the proposed algorithm to the choice of Q. Namely, a very wide range of Q can be successfully used in the applications. Thus, the CS algorithms provide efficient signal reconstruction as far as most of nonstationary disturbances are removed. However, if we omit more than 85%–90% of values, we do not have enough information for signal reconstruction. The influence of various values of Q has been measured by the MSE between the desired and reconstructed Fourier transform. The results are shown in Figure 9.14 (for the proposed ℓ_1-based CS and for the ℓ_2 reconstruction). Observe that the MSE decreases as the amount of removed disturbances increases, being almost negligible between $Q = 50\%$ and $Q = 80\%$. Namely, in this range of Q the disturbances are negligible, while we still have enough useful information for the CS-based reconstruction. Further increase of Q ($Q \geq 80\%$) increases the MSE due to the lack of signal information required for the efficient CS algorithm application. Thus, we may conclude that a good choice will be the highest value of Q for which the CS reconstruction can still be used.

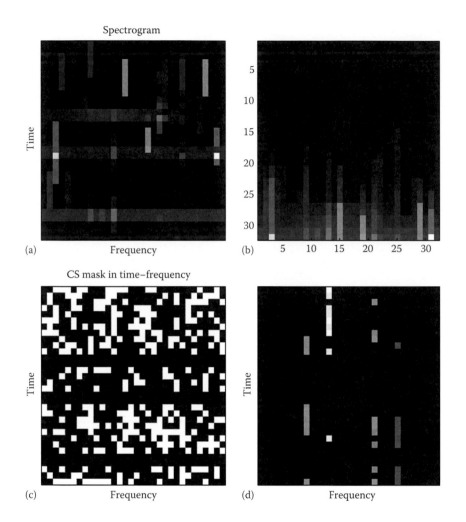

FIGURE 9.12
(a) The STFT of the composite signal with impulse disturbances in time domain. (b) Its sorted values. (c) The CS mask corresponding to the L-statistics-based STFT values (d) The STFT values that remain after applying the CS mask on the absolute values of the STFT.

The result with ℓ_2 norm is obtained as a least-squared solution of the previous minimization problem (9.44b). The ℓ_2 norm on the dataset that remains after the L-statistics is applied. It can be written in the form

$$\mathbf{STFT}_{CS} = \mathbf{AX} \text{ or } \mathbf{A}^H\mathbf{STFT}_{CS} = \mathbf{A}^H\mathbf{AX}$$

with

$$\mathbf{X} = \left(\mathbf{A}^H\mathbf{A}\right)^{-1}\mathbf{A}^H\mathbf{STFT}_{CS},$$

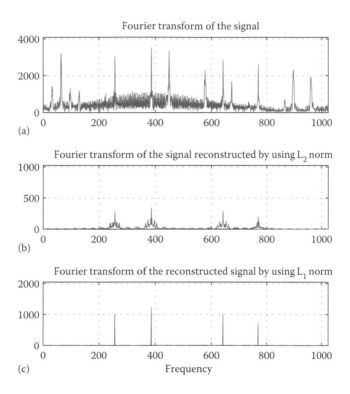

FIGURE 9.13
The DFT of the original signal (a); The signal reconstructed by using the L2 norm, on the set of data after L-statistics is applied in TF domain (b); and the reconstructed FTm by using the proposed method with the CS values of the STFT (c).

where ^{H} denotes Hermitian transpose. The results obtained with ℓ_2 norm could be used as an initial representation in the OMP algorithm. The results from [6] could also be used as the initial representation in this sense.

It is clear that the MSE is significantly decreased when using the CS (Figure 9.13). In this simulation example, a simple a posteriori check of the accuracy was possible since the exact signal values are known. The exact recovery of a sparse signal is considered using the restricted isometry property (RIP) [52]. The sufficient condition for recovery is defined by using the squared ℓ_2 norms (energies) of the sparse signal and the available measurements. In general, the exact recovery depends on the signal and samples (measurements) positions. For example, a single discrete sinusoid can be exactly recovered using just a few of its samples, with a high probability. However, the exact recovery is not guaranteed, even for a very large number of available samples. For example, with 512 out of 1024 samples (taking every other sample), we will not be able even to detect a high-frequency discrete sinusoid. A simple stochastic analysis of the results can be performed similarly as in [32].

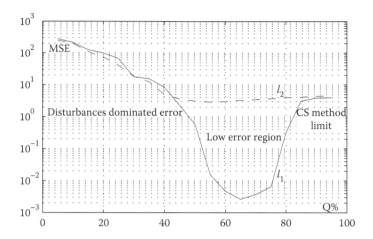

FIGURE 9.14
MSE between the desired and reconstructed FT obtained using the proposed ℓ_1-based approach (solid line) and ℓ_2-based approach (thick line), for different values of parameter Q. The MSE value normalized by the number of samples N is presented.

Example: In general, if a disturbance component covers all considered time interval, and if it is not quasi-stationary, then an appropriate window function should be used to localize the disturbance in frequency. The case with a Hanning localization window, in the overlapping STFT calculation, is considered in this example. The data and the reconstruction results are presented in Figure 9.10. The sparse signal is given in the form

$$x_{sp}(n) = e^{j640\pi n/N} + 0.8e^{-j\pi/8} + 1.2e^{-j768\pi n/N+j\pi/4}$$

$$+ 1.2e^{j1152\pi n/N+j\pi/4} + e^{-j120\pi n/N+j\pi/4}$$

with $N = 2048$. The disturbance consists of four sinusoidally modulated signals and 22 shorter duration signals of the form $A_i \exp(jw_i n/N + j\phi_i) \exp(-[(n - n_{0i})/d_i]^p)$. Different amplitudes are assumed for 22 components all in the range between $A_i = 1$ and $A_i = 5.5$, with durations (defined by d_i). Some of the disturbance terms appear at the same frequency w_i as the signal (stationary) components. Additive noise with standard deviation $\sigma = 0.25$ is present as well. The STFT is calculated by using Hanning window of the width $M = 64$. A Hanning window, zero padded up to the full signal length, could also be used to provide a fine frequency grid in the TF analysis [6]. The STFTs are calculated with the step $M_s = 32$, that is, with a half of the overlapping window. The STFT absolute value is presented in Figure 9.10a. The values are sorted and the L-statistics is performed (Figure 9.10b) with

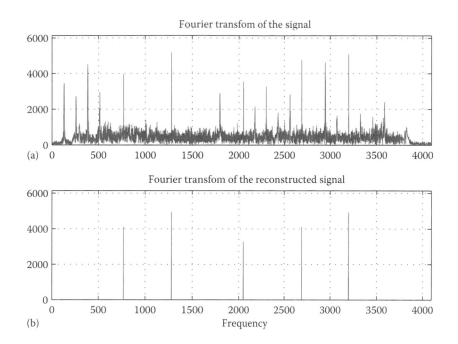

FIGURE 9.15
The DFT of the original composite signal (a) and the reconstructed Fourier transform by using the proposed method with the CS values of the STFT (b).

$Q = 70\%$ discarded samples. The CS mask is presented in Figure 9.10c, while the result of applying the CS method to the STFT (with CS mask) is shown in Figure 9.15.

9.6 Conclusion

Various sparsity-driven approaches for the analysis and separation of fast varying micro-Doppler components are presented. The analysis is based on the fact that the rigid body signals are instantaneously narrowband, that is, sparse in frequency, while the micro-Doppler signals are wideband. The direct applications of the CS algorithms to the ambiguity and local auto-correlation function domains are considered in order to achieve accurate time–frequency signature estimation of Doppler and micro-Doppler signal components. The effectiveness of the compressive sensing-based time–frequency approaches in the analysis of radar signals with micro-Doppler effects is illustrated using simulated and real data.

References

1. V. C. Chen, *The Micro-Doppler Effect in Radar*, Artech House, Norwood, MA, 2011.
2. V. C. Chen, F. Li, S.-S. Ho, and H. Wechsler, Analysis of micro-Doppler signatures, *IEE Proc. Radar Sonar Navig.*, 150(4), 271–276, Aug. 2003.
3. F. Totir and E. Radoi, Superresolution algorithms for spatial extended scattering centers, *Dig. Signal Process.*, 19(5), 780–792, Sept. 2009.
4. M. Martorella, Novel approach for ISAR image cross-range scaling, *IEEE Trans. Aerosp. Electron. Syst.*, 44(1), 281–294, Jan. 2008.
5. T. Thayaparan, L. Stanković, M. Daković, and V. Popović, Micro-Doppler parameter estimation from a fraction of the period, *IET Signal Process.*, 4(3), 201–212, Jan. 2010.
6. L. Stankovic, T. Thayaparan, M. Dakovic, and V. Popovic-Bugarin, Micro-Doppler removal in the radar imaging analysis, *IEEE Trans. Aerosp. Electron. Syst.*, 49(2), 1234–1250, Apr. 2013.
7. M. Martorella and F. Berizzi, Time windowing for highly focused ISAR image reconstruction, *IEEE Trans. Aerosp. Electron. Syst.*, 41(3), 992–1007, July 2005.
8. T. Sparr and B. Krane, Micro-Doppler analysis of vibrating targets in SAR, *IEE Proc. Radar Sonar Navig.*, 150(4), 277–283, Aug. 2003.
9. B. Lyonnet, C. Ioana, and M. G. Amin, Human gait classification using microDoppler time–frequency signal representations, in *Proc. IEEE Radar Conf.*, Washington, DC, pp. 915–919, May 2010.
10. T. Thayaparan, S. Abrol, and E. Riseborough, Micro-Doppler feature extraction of experimental helicopter data using wavelet and time–frequency analysis, in *Proc. Int. Conf. Radar Systems*, Toulouse, France, October 19–22, 2004.
11. L. Stanković, T. Thayaparan, and I. Djurović, Separation of target rigid body and micro-Doppler effects in ISAR imaging, *IEEE Trans. Aerosp. Electron. Syst.*, 41(4), 1496–1506, Oct. 2006.
12. S. L. Marple, Special time–frequency analysis of helicopter Doppler radar data, in *Time–Frequency Signal Analysis and Processing*, ed. B. Boashash, Elsevier, New York, 2004.
13. V. C. Chen and H. Ling, *Time–Frequency Transforms for Radar Imaging and Signal Analysis*, Artech House, Boston, MA, 2002.
14. L. Stankovic, M. Dakovic, and T. Thayaparan, *Time–Frequency Signal Analysis with Applications*, Artech House, Boston, MA, 2013.
15. Y. Wang, H. Ling, and V. C. Chen, ISAR motion compensation via adaptive joint time–frequency techniques, *IEEE Trans. Aerosp. Electron. Syst.*, 38(2), 670–677, 1998.
16. L. Stankovic, T. Thayaparan, and I. Djurovic, Separation of target rigid body and micro-Doppler effects in ISAR imaging, *IEEE Trans. Aerosp. Electron. Syst.*, 42(7), 1496–1506, Oct. 2006.
17. M. Dakovic and L. Stankovic, Estimation of sinusoidally modulated signal parameters based on the inverse Radon transform, in *Proc. ISPA 2013*, Trieste, Italy, Sept. 2013.
18. I. Djurović, L. Stanković, and J. F. Bohme, Robust L-estimation based forms of signal transforms and time–frequency representations, *IEEE Trans. Signal Proc.*, 51(7), 1753–1761, July 2003.

19. P. J. Huber, *Robust Statistics*, John Wiley, New York 1981.
20. E. Candès, J. Romberg, and T. Tao, Robust uncertainty principles: Exact signal reconstruction from highly incomplete frequency information, *IEEE Trans. Inform. Theory*, 52(2), 489–509, 2006.
21. D. Donoho, Compressed sensing, *IEEE Trans. Inform. Theory*, 52(4), 1289–1306, 2006.
22. D. Angelosante, G.B. Giannakis, and E. Grossi, Compressed sensign of time-varying signals, in *Proc. Int. Conf. Digital Signal Processing*, pp. 1–8, 2009.
23. S. Stankovic, I. Orovic, and E. Sejdic, *Multimedia Signals and Systems*, Springer, New York, 2012.
24. R. Baraniuk, Compressive sensing, *IEEE Signal Proc. Mag.*, 24(4), 118–121, 2007.
25. P. Flandrin and P. Borgnat, Time–frequency energy distributions meet compressed sensing, *IEEE Trans. Signal Proc.*, 58(6), 2974–2982, 2010.
26. S. Stankovic, I. Orovic, and M. Amin, Compressed sensing based robust time–frequency representation for signals in heavy-tailed noise, in *Proc. ISSPA*, Montreal, Quebec, Canada, July 2012.
27. I. Orovic, S. Stankovic, and M. Amin, Compressive sensing for sparse time–frequency representation of nonstationary signals in the presence of impulsive noise, in *Proc. SPIE Defense, Security and Sensing*, Baltimore, MD, April 2013.
28. F. Ahmad and M. G. Amin, Through-the-wall human motion indication using sparsity-driven change detection, *IEEE Trans. Geosci. Remote Sens.*, 51(2), 881–890, 2013.
29. Y. Yoon and M. G. Amin, Compressed sensing technique for high-resolution radar imaging, in *Proc. SPIE*, 6968, 6968A1–6968A10, 2008.
30. E. Sejdic, A. Cam, L.F. Chaparro, C.M. Steele, and T. Chau, Compressive sampling of swallowing accelerometry signals using TF dictionaries based on modulated discrete prolate spheroidal sequences, *EURASIP J. Adv. Signal Proc.*, 101, 2012. doi:10.1186/1687-6180-2012-101.
31. L. Stankovic, I. Orovic, S. Stankovic, and M. Amin, Compressive sensing based separation of nonstationary and stationary signals overlapping in time–frequency, *IEEE Trans. Signal Proc.*, 61(18), 4562–4572, 2013.
32. L. Stankovic, S. Stankovic, and M. Amin, Missing samples analysis in signals for applications to L-estimation and compressive eensing, *Signal Process.*, 94, 401–408, 2014.
33. C. Luo and J. H. McClellan, Discrete random sampling theory, in *Proc. IEEE ICASSP*, Vancouver, British Columbia, Canada, pp. 5430–5434, May 2013.
34. S. Stankovic, I. Orovic, and M. Amin, L-statistics based modification of reconstruction algorithms for compressive sensing in the presence of impulse noise, *Signal Process.*, 93(11), 2927–2931, Nov. 2013.
35. S. Stankovic and I. Orovic, An ideal OMP based complex-time distribution, in *Proc. Mediterranean Conf. Embedded Computing*, Budva, Montenegro, pp. 109–112, June 2013.
36. A. Tarczynski and N. Allay, Spectral analysis of randomly sampled signals: suppression of alising and sampler jitter, *IEEE Trans. Signal Proc.*, 52(12), 3324–3334, Dec. 2004.
37. P. Babu and P. Stoica, Spectral analysis of nonuniformly sampled data – A review, *Dig. Signal Proc.*, 20, 359–378, 2010.

38. L. Stankovic, S. Stankovic, I. Orovic, and M. Amin, Robust time–frequency analysis based on the L-estimation and compressive sensing, *IEEE Signal Proc. Lett.*, 20(5), 499–502, 2013.
39. Y. D. Zhang and M. G. Amin, Compressive sensing in nonstationary array processing using bilinear transforms, in *Proc. IEEE Sensor Array and Multichannel Signal Processing Workshop*, Hoboken, NJ, June 2012.
40. Y. D. Zhang, M. G. Amin, and B. Himed, Reduced interference time–frequency representations and sparse reconstruction of undersampled data, in *Proc. European Signal Proc. Conf.*, Marrakech, Morocco, Sept. 2013.
41. B. G. Mobasseri and M. G. Amin, A time–frequency classifier for human gait recognition, in *Proc. SPIE Symposium on Defense, Security, and Sensing*, Orlando, FL, April 2009.
42. S. S. Ram, C. Christianson, Y. Kim, and H. Ling, Simulation and analysis of human microDopplers in through-wall environments, *IEEE Trans. Geosci. Remote Sensing*, 48, 2015–2023, April 2010.
43. I. Orovic, S. Stankovic, and M. Amin, A new approach for classification of human gait based on time–frequency feature representations, *Signal Process.*, 91(6), 1448–1456, 2011.
44. F. Tivive, A. Bouzerdoum, and M. G. Amin, A human gait classification method based on radar Doppler spectrograms, *EURASIP J. Adv. Signal Process.*, 2010, Article ID 389716, 2010.
45. L. Stanković, A measure of some time–frequency distributions concentration, *Signal Process.*, 81(3), 621–631, March 2001.
46. S. Stankovic, I. Orovic, and C. Ioana, Effects of Cauchy integral formula discretization on the precision of IF estimation: Unified approach to complex-lag distribution and its L-form, *IEEE Signal Proc. Lett.*, 16(4), 307–310, 2009.
47. J. A. Tropp and A. C. Gilbert, Signal recovery from random measurements via orthogonal matching pursuit, *IEEE Trans. Inform. Theory*, 53(12), 4655–4666, 2007.
48. B. Jokanovic, M. Amin, and S. Stankovic, Instantaneous frequency and time–frequency signature estimation using compressive sensing, in *Proc. SPIE Defense, Security and Sensing*, Baltimore, MD, April 2013.
49. D. L. Jones and R. G. Baraniuk, An adaptive optimal-kernel time–frequency representation, *IEEE Trans. Signal Proc.*, 43(10), 2361–2371, Oct. 1995.
50. M. G. Amin, A. Belouchrani, and Y. Zhang, The spatial ambiguity function and its applications, *IEEE Signal Proc. Lett.*, 7(6), 138–140, 2000.
51. J. Lerga, V. Sucic, and B. Boashash, An efficient algorithm for instantaneous frequency estimation of nonstationary multicomponent signals in low SNR, *EURASIP J. Adv. Signal Process.*, ASP/725189, Jan. 2011.
52. E. J. Candes and T. Tao, Near-optimal signal recovery from random projections: Universal encoding strategies? *IEEE Trans. Inform. Theory*, 52(12), 5406–5425, Dec. 2006.

10

Urban Target Tracking Using Sparse Representations*

Phani Chavali and Arye Nehorai

CONTENTS

ABSTRACT In this chapter, we propose a novel sparsity-based algorithm for multiple-target tracking in a time-varying multipath environment. A sparse measurement model for the received signal is obtained by considering a finite-dimensional representation of the time-varying system function that characterizes the transmission channel. This sparse measurement model allows us to exploit the joint delay-Doppler diversity offered by the environment. We reformulate the problem of multiple-target tracking as a block support recovery problem and derive an upper bound on the overall error probability of wrongly identifying the support of the sparse signal. Using this bound, we prove that spread-spectrum waveforms are ideal candidates for signaling. In addition, under spread-spectrum signaling, the dictionary of the sparse measurement model exhibits a special structure. We exploit this

* This work was supported by the Department of Defense under the AFOSR Grant FA9550-11-1-0210, the ONR Grant N000141310050.

structure to develop a computationally inexpensive support recovery algorithm by projecting the received signal on to the row space of the dictionary. Numerical simulations show that tracking using proposed algorithm for support recovery performs better when compared with tracking using other sparse reconstruction algorithms. The proposed algorithm takes significantly less time when compared with the time taken by other methods.

10.1 Introduction

The problem of tracking multiple targets moving in dense urban environments is a complex, yet important, subject. Urban environments provide two major challenges for the operation of conventional radar systems (Krolik et al., 2006): complete obscuration of the target due to tall buildings and the presence of multiple scatterers in the environment. Radar systems operating in these environments therefore suffer severe performance degradation compared with the conventional radar systems designed for the line-of-sight (LOS) propagation environments. In order to overcome the problem of obscuration, radars generally maintain a steep grazing angle. This, however, dramatically decreases the coverage area (Krolik et al., 2006). The degradation due to the presence of scatterers can be compensated by treating the signals arriving on the non-LOS paths as interference, and thus mitigating it (Krach and Weigel, 2009; Rigling, 2008). Recently, however, there has been a growing interest in exploiting the multipath nature of the environment to obtain a better performance (Chakraborty et al., 2010; Chavali and Nehorai, 2010). The presence of multiple scatterers in the environment introduces a delay spread in the urban transmission channel and the relative motion between the moving targets and the scatterers introduces time variations in the channel, which manifests as Doppler spread. Thus, urban environments with multiple scatterers are characterized by a time-varying multipath channel. The delay and the Doppler spread provide additional diversity for the problem of target tracking, and by employing methods that exploit this diversity, the performance of the tracking system can be significantly improved.

The traditional method for tracking a single target in the absence of multipath and clutter is to pass the received signal through a bank of filters, each matched to a particular delay and Doppler (Levanon and Mozeson, 2004). The peaks of the matched filter output will then correspond to the delay and Doppler of the targets, which are usually considered measurements. Sometimes, beamforming is used in addition to matched filtering to obtain the azimuth and the elevation angle information. The measurements thus obtained are then used to obtain an estimate of the unknown target state using Bayesian filters such as the extended Kalman filter or the particle filter (Arulampalam et al., 2002). When there are multiple targets, each measurement is assigned to a target or is identified as a false alarm, before state estimation is performed. This step is called data association. State estimation is then performed either in a global fashion, where one filter is used to estimate the joint state vector of all the targets, or on an individual basis, where several filters operating in a parallel fashion are used. Several practical methods are developed to obtain good suboptimal solutions, like the probabilistic data association filter (Bar-Shalom and Tse, 1975), joint probabilistic data association filter (Fortman et al., 1983), multiple hypothesis tracking filter (Reid, 1979), probabilistic MHT (PMHT) (Gauvrit et al., 1997), and, more recently, Monte-Carlo methods such as Vermaak et al. (2005), Särkkä et al. (2004), Chavali and Nehorai (2013), and random finite set methods (Mahler, 2007). Similar ideas have been extended for target tracking in multipath environments (Algeier et al., 2008), where measurements arriving on various paths are identified as either valid measurements or spurious measurements depending on whether they are target generated or not, respectively.

Sparse modeling has drawn a lot of attention in the radar community (Herman and Strohmer, 2009) recently. In the context of radar, sparse models are used to reconstruct the target scene by representing each target as a grid point in the delay-Doppler plane. Since the number of targets is in general less than the number of grid points, the target scene can be represented as a sparse signal and it can be reconstructed by finding the support* of the sparse signal. Sparsity-based models process the received signal directly without the need to use a matched filter. In addition to sparsity-based estimation, there is also a growing interest in developing algorithms for sparsity-based tracking. In sparsity-based tracking, in addition to the nonzero coefficients, the support of the sparse signal changes dynamically with time. This structure is used to improve reconstruction accuracy further. Recent work on least-square CS residual (Vaswani, 2010), KF-CS residual (Vaswani, 2008), and the modified CS residual (Vaswani and Lu, 2010) studied the problems of the time-varying sparsity patterns using fewer measurements than the measurements used by the traditional sparse models. The reconstruction performance of these methods is significantly faster too.

In this chapter (see also Chavali and Nehorai, 2012), a finite-dimensional measurement model is developed for the tracking problem by modeling multipath propagation channel as a time-varying linear system. This measurement model captures both the delay and the Doppler spread that is introduced into the system. The dimensionality of the linear representation has a physical significance, and it represents the additional diversity that the environment offers. Next, sparse modeling is used to transform the problem of multiple-target tracking into a problem of support recovery of a block-sparse signal. Note that we address tracking as an estimation problem at

---

* Support of a vector x is defined as the set of indices where the vector is nonzero, that is, $\text{supp}(x) = \{i \mid x_i \neq 0\}$.

each time, but vary the model at each time. Further, a computationally inexpensive algorithm for the support recovery of the sparse signal is introduced in this chapter. This method works by projecting the received signal vector onto the row space of the dictionary.[*] The dictionary corresponding to the block-sparse model consists of delayed and Doppler shifted versions of the transmitted waveforms. Under spread-spectrum signaling, we prove that the dictionary of the sparse model exhibits a special structure, which enables efficient support recovery. We refer to this algorithm as projection-based (PB) support recovery. We use PB support recovery algorithm for target tracking and demonstrate using numerical simulations that target tracking using PB algorithm takes significantly less time compared with the time taken by the standard sparse signal reconstruction-based tracking methods, while giving good performance.

Further, an upper bound on the overall error probability of wrongly identifying the support of the sparse target scene is derived. Using this bound, it will be shown that spread-spectrum waveforms are ideal candidates for signaling, as they have good time-bandwidth properties, which are essential to obtain the full diversity provided by the delay and the Doppler spread.

The rest of the chapter is organized as follows. We describe the signal model, the state model, and the measurement model in Section 10.2. The time-varying multipath channel is modeled as a linear and time-varying system, and a representation of the received signal by sampling the time-varying system response is developed. In Section 10.3, we develop a block-sparse measurement model by exploiting the sparsity present in the target scene. We follow an approach similar to Sen and Nehorai (2011) and sample the delay-Doppler plane to develop a block-sparse measurement model (Eldar and Kuppinger, 2010). In Section 10.4.1, we briefly describe the existing sparse signal support reconstruction methods. In Section 10.4.2, to demonstrate the advantage of the multipath modeling and the spread-spectrum signaling, we derive an upper bound on the overall error probability for an optimal support recovery algorithm. In Section 10.4.3, we describe a PB support recovery algorithm. We provide several numerical results in Section 10.5 and draw conclusions in Section 10.6.

10.2 System Model

10.2.1 Multipath Environment Model

Consider a monostatic pulse radar system operating in an urban environment as shown in Figure 10.1. Assume that the radar antenna is

---

[*] The dictionary of a sparse model corresponds to the overcomplete set of basis functions obtained by discretizing the variable of interest.

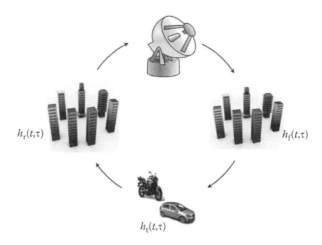

FIGURE 10.1
(See color insert.) Block diagram showing the forward, reverse transmission channels, and the targets.

omnidirectional. This assumption ensures that the antenna transmits equal power in all the directions and is capable of receiving the signal from all the directions. The radar transmits a known electromagnetic signal to obtain an image of the target. The transmitted signal propagates through a forward transmission channel before it reaches the target. The target responds to the transmitted waveform and backscatters a modified waveform into the environment. The modified waveform propagates back to the radar receiver through a reverse transmission channel. The received signal is then processed to obtain the necessary information about the target. This is a standard model for radar systems operating in a multipath environment (Bell, 1993; Jin et al., 2010).

In the absence of the time-varying multipath effect due to the surrounding environment, the forward transmission channel and the reverse transmission channel do not have any effect on the backscattered waveform, except possibly for a propagation loss. In such cases, the target is the object that causes the uncertainty in the received waveform. The electromagnetic signal that is reflected off the target is modified as a function of the target's relative position, velocity, and its physical characteristics, referred to as the radar cross section (RCS) (Peebles, 1998). If the target is assumed to be a point target, the signal reflected off it can be expressed as

$$r(t) = \alpha s(t - \tau_d)e^{j2\pi v_d t}, \qquad (10.1)$$

where
  $s(t)$ is the transmitted waveform
  $r(t)$ is the backscattered waveform from the target

α is the unknown RCS

τ_d is the round trip delay

v_d is the Doppler shift in the frequency due to the relative motion between the radar and the target

For simplicity, assume initially that there is only one target present in the region of interest. The extension to the case of multiple targets will be made in Section. 10.2.4. Without the loss of generality, consider that the radar is located at $(0,0)$. With this assumption,

$$\tau_d = \frac{2}{c}\sqrt{x^2 + y^2}, \tag{10.2}$$

$$v_d = \frac{2f_c}{c}\frac{x\dot{x} + y\dot{y}}{\sqrt{x^2 + y^2}}, \tag{10.3}$$

where

(x, y) and $(\dot{x}, \dot{y})$ denote the position and velocity of the target, respectively, in the two-dimensional plane

f_c is the carrier frequency

c is the speed of propagation

In the presence of time-varying multipath channel, the uncertainty in the received waveform is due to the joint effect of the forward transmission channel, the target itself, and the reverse transmission channel. Since the propagation of electromagnetic waves obeys superposition principle, it is reasonable to model the forward transmission channel and the reverse transmission channel as linear time-varying systems (Bell, 1993). Denoting the time-varying system responses of the forward transmission channel and the reverse transmission channel using $h_f(t, \tau)$ and $h_r(t, \tau)$, respectively, the received signal can be expressed as

$$y(t) = h_r(t, \tau) * r(t) * h_f(t, \tau) + w(t),$$

$$= \int h(t, \tau)r(t - \tau)d\tau + w(t), \tag{10.4}$$

where

$y(t)$ is the received waveform in the presence of time-varying multipath channel

$h(t, \tau)$ is the response of the overall multipath channel at delay τ and time t given as $h(t, \tau) = h_r(t, \tau) * h_f(t, \tau)$

$w(t)$ is circularly symmetric, complex additive white Gaussian noise

Considering Fourier transform of $h(t, \tau)$, the signal $y(t)$ can be expressed as

$$y(t) = \int \int H(f, \tau)r(t - \tau)e^{j2\pi ft}d\tau df + w(t), \tag{10.5}$$

where $H(f, \tau)$ is the Fourier transform $h(t, \tau)$ at time τ. It can be assumed that $h(t, \tau)$ is a wide-sense stationary Gaussian random process in the variable t and the responses due to different scatterers at delays τ_1 and τ_2 to be uncorrelated. This model incorporating the wide-sense stationarity and the uncorrelated scattering is widely used for characterizing time-varying systems responses (Bello, 1963). Under this assumption, the power spectral density of $H(f, \tau)$, which denotes the average power output of the time-varying multipath channel as a function of time and frequency, is given by the Wiener–Khinchin theorem (Proakis, 2001):

$$\psi(f, \tau) = \mathrm{E}\big[|H(f, \tau)|^2\big]. \tag{10.6}$$

The range of values of delay, τ, and the frequencies (two-sided), f, over which $\psi(f, \tau)$ is nonzero are defined as the delay spread (T_d) and the Doppler spread (B_d) of the channel, respectively. The inverse of the delay spread is defined as the coherence bandwidth (B_c), and the inverse of the Doppler spread is defined as the coherence time (T_c) of the channel. Coherence time and coherence bandwidth denote the range of time scales and frequencies over which the variations caused due to the channel are constant. In other words, for two frequencies f_1, f_2 such that $|f_1 - f_2| \le B_c$, and two time instants τ_1 and τ_2 such that $|\tau_1 - \tau_2| \le T_c$, the power spectral density satisfies $\psi(f_1, \tau_1) = \psi(f_2, \tau_2)$ (Proakis, 2001). A finite-dimensional representation of Equation 10.5 is obtained by sampling the variables τ and f at a resolution $\Delta\tau$ and Δf such that $\tau \times f \in [0, T_d] \times [-B_d/2, B_d/2]$. Specifically, each delay-Doppler grid of size $\Delta\tau \times \Delta f$ consists of all the paths whose delay and Doppler shifts fall within that grid cell.

$$y(t) = \sum_{q=-Q/2}^{Q/2-1} \sum_{p=0}^{P-1} H(p, q) r(t - p\Delta\tau) e^{j2\pi q \Delta f t} + w(t), \tag{10.7}$$

where $Q = \lceil B_d/\Delta f \rceil$ and $P = \lceil T_d/\Delta\tau \rceil$. In a rich scattering like urban environment, each of the grid points is populated by at least one path. Further, if the sampling resolutions are chosen such that $\Delta\tau > T_c$ and $\Delta f > B_c$, the coefficients $H(p, q)$ are independent of each other (Sayeed and Aazhang, 1999). Thus, the vector h defined as $h = \mathrm{vec}(H)$ is a multivariate Gaussian random variable with independent entries. h will be referred to as the channel state vector and H as the channel state matrix from now on. For simplicity, in this chapter, assume that the processing interval and the signal energy are chosen so that the radar can estimate the channel state vector in each tracking interval without any error. This assumption is equivalent to considering that the channel state information is available at the receiver.

It is evident from the representation given in Equation 10.7 that the received signal is a linear combination of independent time shifted (multipath) and frequency shifted (Doppler) versions of the back scattered signal from the target. Hence, the representation in Equation 10.7 provides two kinds of additional diversity: delay diversity and the Doppler diversity. Similar representations are used in problems related to communication in fading wireless channels (Sayeed and Aazhang, 1999).

10.2.2 Signal Model

In this section, the signal model is described. The radar transmits a coherent pulse train of L pulses, with a pulse repetition interval of t_p seconds in each tracking interval. Pulse train signaling enables the radar receiver to estimate both the range and the Doppler without any ambiguity (Levanon and Mozeson, 2004). The corresponding transmitted signal is given as

$$s(t) = \sqrt{E} \sum_{l=0}^{L-1} a_l(t - lt_p), \quad 0 \le t \le t_i, \tag{10.8}$$

where
   $a_l(t)$ is the unit energy transmitted waveform in the lth pulse
   E is the signal energy per pulse

In each pulse, a spread-spectrum waveform (Proakis, 2001) is transmitted. Apart from a few works (Dobrosavljevic and Dukic, 1996; Wang et al., 1997), there has not been much research in the literature that use spread-spectrum waveforms in the radar context. We demonstrate in subsequent sections that spread-spectrum waveforms are useful for exploiting the full delay-Doppler diversity offered by the environment and enforce a unique structure on the dictionary of the sparse model. The spread-spectrum waveform $a(t)$ takes the form

$$a(t) = \sum_{g=0}^{G-1} a_g v(t - gt_c), \quad \text{such that } \sum_{g=0}^{G-1} a_g a_{g-k} \approx 0, \quad k \ne g, \tag{10.9}$$

where
   t_c is the chip duration
   $v(t)$ is a rectangular waveform of duration t_c
   G is the number of chips in each pulse
   $\{a_g, g = 0, \ldots, G - 1\}$ is the spreading code corresponding to the spread-spectrum waveform $a(t)$

The bandwidth corresponding to this waveform is given by $B = 1/t_c$. Several code sequences are used for spread-spectrum signaling like the pseudorandom code sequences, Gold code sequences, etc. There is a vast literature available on the design and properties of such code sequences (Gold, 1967; Sarwate and Pursley, 1980).

10.2.3 State-Space Model

Consider K targets moving in a two-dimensional plane and denote the position and velocity of the kth target as (x^k, y^k) and $(\dot{x}^k, \dot{y}^k)$, respectively. The target state at time i is then represented by the vector $\theta_i^k = [x_i^k, y_i^k, \dot{x}_i^k, \dot{y}_i^k]^T$, and the equation describing the evolution of the target state across time is given as

$$\theta_{i+1}^k = F^k \theta_i^k + v_i^k, \tag{10.10}$$

where F^k is the state transition matrix for the kth target. Assuming that all the targets move along linear trajectories with constant velocity, the state transition matrix $F^k, k = 1, \ldots, K$ is given as

$$F^k = \begin{bmatrix} 1 & 0 & t_i & 0 \\ 0 & 1 & 0 & t_i \\ 0 & 0 & 1 & 0 \\ 0 & 0 & 0 & 1 \end{bmatrix}, \tag{10.11}$$

where t_i is the system sampling time, which corresponds to the time interval after which the processing is done. This time, t_i, is referred to as the tracking interval and is indexed using the variable i. The error in the state model v^k is assumed to be Gaussian distributed with zero mean and a covariance matrix given by Bar-Shalom et al. (2001)

$$\Sigma_v^k = \epsilon^k \begin{bmatrix} \frac{1}{3}t_i^3 & 0 & \frac{1}{2}t_i^2 & 0 \\ 0 & \frac{1}{3}t_i^3 & 0 & \frac{1}{2}t_i^2 \\ \frac{1}{2}t_i^2 & 0 & t_i & 0 \\ 0 & \frac{1}{2}t_i^2 & 0 & t_i \end{bmatrix}, \tag{10.12}$$

where ϵ^k is the intensity of the noise process for the kth target. By concatenating the state vectors of all the targets, an overall target state transition equation is obtained as

$$\theta_{i+1} = F\theta_i + v_i, \tag{10.13}$$

where

θ_i is the $4K \times 1$ joint target state defined as $\theta_i = \left[(\theta_i^1)^T, \ldots, (\theta_i^K)^T\right]^T$

F is the $4K \times 4K$ matrix representing the overall state transition matrix, which is defined as $F = \text{blkdiag}\{F^1, \ldots, F^K\}$

v_i is the $4K \times 1$ overall additive noise given as $v_i = \left[(v_i^1)^T, \ldots, (v_i^K)^T\right]^T$ with noise covariance matrix $\Sigma_v = \text{blkdiag}\{\Sigma_v^1, \ldots, \Sigma_v^K\}$

10.2.4 Measurement Model

Using Equations 10.1 and 10.8 in Equation 10.7, the received signal due to a single target can be expressed as

$$
\tilde{y}(t) = \sqrt{E} \sum_{q=-Q/2}^{Q/2-1} \sum_{p=0}^{P-1} \tilde{\alpha} H(p,q)
$$

$$
\times \sum_{l=0}^{L-1} a_l(t - lt_p - p\Delta\tau - \tau_d) e^{j2\pi v_d(t-p\Delta\tau)} e^{j2\pi q\Delta ft} + w(t),
$$

$$
\approx \sqrt{E} \sum_{q=-Q/2}^{Q/2-1} \sum_{p=0}^{P-1} \tilde{\alpha} H(p,q) \sum_{l=0}^{L-1} a_l(t - lt_p - p\Delta\tau - \tau_d) e^{j2\pi v_d lt_p} e^{j2\pi q\Delta flt_p} + w(t).
$$

$$(10.14)$$

The term $e^{j2\pi v_d t}$ is approximated as $e^{j2\pi v_d lt_p}$ (constant) and $e^{j2\pi q\Delta ft}$ is approximated as $e^{j2\pi q\Delta flt_p}$ within each $a(t)$ since it is a narrowband pulse. Also, the term $e^{-j2\pi v_d p\Delta\tau}$ is ignored as the variations in this term are negligible when compared with the variations in other terms. The superscript $\sim$ is used to emphasize that the signal and the scattering coefficient correspond to a single target. When there are multiple targets, the received signal due to all the targets is expressed as

$$
y(t) = \sum_{k=1}^{K} \sqrt{E} \sum_{q=-Q/2}^{Q/2-1} \sum_{p=0}^{P-1} \tilde{\alpha}^k H(p,q)
$$

$$
\times \sum_{l=0}^{L-1} a_l(t - lt_p - p\Delta\tau - \tau_d^k) e^{j2\pi v_d^k lt_p} e^{j2\pi q\Delta flt_p} + w(t), \qquad (10.15)
$$

where $\tau_d^k, v_d^k, \tilde{\alpha}^k$ represent the delay, Doppler, and the RCS of the kth target, respectively. The distance between the targets is smaller compared with the distance between the targets the radar and hence channel state matrix is assumed to remain the same for all the targets.

10.3 Sparse Modeling

In this section, we develop a sparse measurement model for Equation 10.15. Consider the received signal in Equation 10.15 and discretize the delay-Doppler plane into grid points such that the delay and Doppler corresponding to each target falls within one specific grid point that is,

$$\tau_{\mathrm{d}}^k = m_1^k \Delta\tau,$$
$$v_{\mathrm{d}}^k = m_2^k \Delta v,$$ (10.16)

for $k = 1, 2, \ldots, K$, where m_1^k and m_2^k represent the indices of the discretized delay and Doppler of the kth target. If the delay-Doppler plane is discretized into M_1 points along the delay dimension and M_2 points along the Doppler dimension, corresponding to a total of $M = M_1 M_2$ grid points in the region of interest, the received signal for the kth target can be expressed as

$$\tilde{y}^k(t) = \sqrt{E} \sum_{q=-Q/2}^{Q/2-1} \sum_{p=0}^{P-1} \tilde{\alpha}^k H(p, q)$$

$$\times \sum_{l=0}^{L-1} a_l(t - lt_{\mathrm{p}} - p\Delta\tau - m_1^k \Delta\tau) e^{j2\pi(m_2^k \Delta f + q\Delta f)lt_{\mathrm{p}}} + w(t).$$ (10.17)

The received signal is now sampled at a rate $f_{\mathrm{s}} = B$ so that one sample from each chip is collected. Consider N samples around a reference point* in each pulse. The sampling resolution of the delay-Doppler grid is chosen to commensurate with the sampling of the signal, that is, $\Delta\tau = 1/f_{\mathrm{s}}$ and $\Delta f = 1/t_{\mathrm{i}}$. The corresponding discrete-time signal is then given by

$$\tilde{y}^k(nt_{\mathrm{s}}) = \sqrt{E} \sum_{q=-Q/2}^{Q/2-1} \sum_{p=0}^{P-1} \tilde{\alpha}^k H(p, q)$$

$$\times \sum_{l=0}^{L-1} a_l(nt_{\mathrm{s}} - lt_{\mathrm{p}} - pt_{\mathrm{s}} - m_1^k t_{\mathrm{s}}) e^{j2\pi(m_2^k + q)l\frac{t_{\mathrm{p}}}{t_{\mathrm{i}}}} + w(t).$$ (10.18)

Expressing Equation 10.18 in a matrix form,

$$\tilde{y}^k = \sqrt{E} \sum_{q=-Q/2}^{Q/2-1} \sum_{p=0}^{P-1} \tilde{\alpha}^k H(p, q) \underbrace{\left(F(q, m_2^k) \otimes J(p, m_1^k) \right)}_{\Phi(p, q, m_1^k, m_2^k)} s + w,$$ (10.19)

where

$\tilde{y}^k$ is the $LN \times 1$ received signal vector corresponding to the kth target

$F(q, m_2^k)$ is the $L \times L$ Doppler modulation matrix defined as $\mathrm{diag}\{1, e^{\frac{j2\pi(q + m_2^k)}{L}},$

$\ldots, e^{\frac{j2\pi(q + m_2^k)(L-1)}{L}}\}$

---

* The choice of the reference point can be arbitrary. In this chapter, it is chosen to be the predicted state of the first target.

$J(p, m_1^k)$ is the $N \times G$ time shift matrix defined as $\begin{bmatrix} \mathbf{0}^T_{(m_1^k+p) \times G} & I_G \end{bmatrix}$

$\mathbf{0}^T_{(N-G-m_1^k-p) \times G} \Big]^T$

s is the $LG \times 1$ column vector obtained by stacking the spreading code a, L times as $[a^T, a^T, \ldots, a^T]^T$

w is the $LN \times 1$ complex additive white Gaussian noise with zero mean and covariance matrix $\sigma_w^2 I_{LN}$

In obtaining Equation 10.19, it is assumed that all the samples of the received waveform $\tilde{y}^k(t)$ fall within the sampling window of size N. By further simplifying Equation 10.19,

$$\tilde{y}^k = \tilde{\alpha}^k \sqrt{E} \Phi_k h + w, \tag{10.20}$$

where

Φ_k is the $LN \times PQ$ defined as $\Phi_k = [\phi(1,1,m_1^k,m_2^k), \ldots, \phi(p,q,m_1^k,m_2^k), \ldots, \phi(P,Q,m_1^k,m_2^k)]$

h is the $PQ \times 1$ defined as $\text{vec}(H)$, where H is a $P \times Q$ matrix with elements $[H]_{pq} = H(p,q)$ and $h \sim \mathbb{CN}(0, \Sigma_h)$

When there are K targets, the received signal due to all the targets can be expressed as

$$\tilde{y}(t) = \sqrt{E} \sum_{k=1}^{K} \sum_{q=-Q/2}^{Q/2-1} \sum_{p=0}^{P-1} \tilde{\alpha}^k H(p,q)$$

$$\times \sum_{l=0}^{L-1} a_l(t - lt_p - p\Delta\tau - m_1^k \Delta\tau) e^{j2\pi(m_2^k \Delta f + q\Delta f)lt_p} + w(t). \tag{10.21}$$

As described earlier, m_1^k and m_2^k represent the indices of the discretized delay and Doppler of the kth target, and $M = M_1 M_2$ is the total number of grid points in the region of interest. In general, the number of targets K is much smaller than the number of grid points M, that is, $(K \ll M)$. Therefore, Equation 10.21 can be simplified by changing the outer summation over the targets to a summation over the grid points. In order to change the summation, we collect all grid points where the targets are located in a set $\mathcal{K}$ and define

$$\alpha^m = \begin{cases} \tilde{\alpha}^k & \text{if } m \in \mathcal{K} \\ 0 & \text{if } m \notin \mathcal{K}. \end{cases}$$

With this definition and with $m_1 = (m - 1) \mod M_1$, $m_2 = \lfloor (m - 1)/M_1 \rfloor$, and each of the K targets located at one grid point m, we can rewrite Equation 10.21 as a sparse measurement model

$$y(t) = \sqrt{E} \sum_{m=1}^{M} \sum_{q=-Q/2}^{Q/2-1} \sum_{p=0}^{P-1} \alpha^m H(p, q)$$

$$\times \sum_{l=0}^{L-1} a_l(t - lt_p - p\Delta\tau - m_1\Delta\tau)e^{j2\pi(q+m_2)\Delta f l l_p} + w(t), \qquad (10.22)$$

As before, expressing the received signal in a vector form gives

$$y = \sum_{m=1}^{M} \alpha^m \sqrt{E} \Phi_m h + w, \qquad (10.23)$$

By further simplifying Equation 10.23

$$y = \sqrt{E} \Phi \zeta + w, \qquad (10.24)$$

where

Φ is the $LN \times MPQ$ dictionary of block-sparse model defined as $\Phi = [\Phi_1 \ldots \Phi_m, \ldots, \Phi_M]$

ζ is the $MPQ \times 1$ block-sparse vector (Eldar and Kuppinger, 2010) defined as $\zeta = [\alpha^1 h^T, \ldots, \alpha^M h^T]^T = \alpha \otimes h$

It should be noted here that we did not combine the channel vector into the dictionary matrix of the model, but instead we incorporated it into the unknown vector ζ. The motivation behind such a modeling is to enforce a special structure on the dictionary matrix. Observe that the dictionary matrix consists of time–frequency shifted versions of the transmitted signal in its columns. The dictionary matrix satisfies a special structural property, which will be stated as a theorem. But before that, a notation D_κ^ξ is introduced.

Let $D_1, D_2 \ldots D_U$ denote diagonal matrices each of order V and let $D = \text{diag}\{D_1, D_2 \ldots D_U\}$ be a block diagonal matrix. Denote by D_κ^ξ, a matrix obtained by moving the diagonal entries of all the matrices $D_1, D_2 \ldots D_U$ in D to ξth subdiagonal below or above their respective principal diagonals, depending on whether ξ is positive or negative. Further, all the diagonal matrices in the block diagonal matrix D are moved to the κth subdiagonal below or above the principal diagonal of D, depending on whether κ is positive or negative. Thus, the superscript denotes the offset of the elements of $D_1, D_2 \ldots D_U$ from their respective principal diagonals and the subscript denotes the offset of the matrices $D_1, D_2 \ldots D_U$ from the principal diagonal of D. An example for the notation is provided in the example.

Example: Let

$$
D = \begin{bmatrix}
d_{11} & 0 & 0 & 0 & 0 & 0 \\
0 & d_{12} & 0 & 0 & 0 & 0 \\
0 & 0 & d_{13} & 0 & 0 & 0 \\
0 & 0 & 0 & d_{21} & 0 & 0 \\
0 & 0 & 0 & 0 & d_{22} & 0 \\
0 & 0 & 0 & 0 & 0 & d_{23}
\end{bmatrix} = \begin{bmatrix} D_1 & 0 \\ 0 & D_2 \end{bmatrix},
$$

then the matrix D_1^1 is derived as follows. Since the subscript is 1, the matrices D_1 and D_2 in the block diagonal matrix D are moved to the first subdiagonal below the principal diagonal. Further, the diagonal elements of D_1 are moved to the first subdiagonal giving

$$
D_1^1 = \begin{bmatrix}
0 & 0 & 0 & 0 & 0 & 0 \\
0 & 0 & 0 & 0 & 0 & 0 \\
0 & 0 & 0 & 0 & 0 & 0 \\
0 & 0 & 0 & 0 & 0 & 0 \\
d_{11} & 0 & 0 & 0 & 0 & 0 \\
0 & d_{12} & 0 & 0 & 0 & 0
\end{bmatrix} = \begin{bmatrix} 0 & 0 \\ \tilde{D}_1^1 & 0 \end{bmatrix}.
$$

Similarly,

$$
D_{-1}^{-1} = \begin{bmatrix}
0 & 0 & 0 & 0 & d_{11} & 0 \\
0 & 0 & 0 & 0 & 0 & d_{12} \\
0 & 0 & 0 & 0 & 0 & 0 \\
0 & 0 & 0 & 0 & 0 & 0 \\
0 & 0 & 0 & 0 & 0 & 0 \\
0 & 0 & 0 & 0 & 0 & 0
\end{bmatrix} = \begin{bmatrix} 0 & \tilde{D}_1^{-1} \\ 0 & 0 \end{bmatrix},
$$

with

$$
\tilde{D}_1^1 = \begin{bmatrix}
0 & 0 & 0 \\
d_{11} & 0 & 0 \\
0 & d_{12} & 0
\end{bmatrix} \quad \text{and} \quad \tilde{D}_1^{-1} = \begin{bmatrix}
0 & d_{11} & 0 \\
0 & 0 & d_{12} \\
0 & 0 & 0
\end{bmatrix}.
$$

Theorem 10.1 *If $L \geq (M_2 + Q)/2$ and $\sum_{g=0}^{G-1} a_g a_{g-k} = 0, k \neq g$, then the dictionary $\boldsymbol{\Phi}$ satisfies*

$$\boldsymbol{\Phi}_m^H \boldsymbol{\Phi}_n = D_{\kappa_{M_1}(n,m),}^{\xi_{M_1}(n,m)} \tag{10.25}$$

where $\xi_{M_1}(n,m) = \lfloor (n-1)/M_1 \rfloor - \lfloor (m-1)/M_1 \rfloor$, $\kappa_{M_1}(n,m) = \{(n-1) \bmod M_1\} - \{(m-1) \bmod M_1\}$, $D = \text{blkdiag}(\underbrace{LI_P, \ldots, LI_P}_{Q \text{ times}})$.

Proof: From the definition of $\boldsymbol{\Phi}$,

$$\left[\boldsymbol{\Phi}_m^H \boldsymbol{\Phi}_n \right]_{ij} = a^H \sum_{l=0}^{L-1} e^{\frac{j2\pi l}{L} \xi_{M_1}(n,m) + \xi_P(j,i)} \left(J\Big(\kappa_{M_1}(n,m), \kappa_P(j,i) \Big) a \right). \tag{10.26}$$

1. First, consider the case $m = n$, that is, within the same block. For this case, $\xi_{M_1}(n,m) = 0$ and $\kappa_{M_1}(n,m) = 0$. Therefore,

$$\left[\boldsymbol{\Phi}_m^H \boldsymbol{\Phi}_m \right]_{ij} = \sum_{l=0}^{L-1} e^{\frac{j2\pi l \xi_P(j,i)}{L}} \sum_{g=0}^{G-1} a_g^* a_{g - \kappa_P(j,i)}. \tag{10.27}$$

When $\xi_P(j,i) \neq 0$, that is, $\lfloor (j-1)/P \rfloor - \lfloor (i-1)/P \rfloor \neq 0$, corresponding to all the $P \times P$ submatrices that are not along the principal diagonal in the $PQ \times PQ$ matrix $\boldsymbol{\Phi}_m^H \boldsymbol{\Phi}_m$, we have

$$\left[\boldsymbol{\Phi}_m^H \boldsymbol{\Phi}_m \right]_{ij} = 0. \tag{10.28}$$

When $\xi_P(j,i) = 0$, that is, $\lfloor (j-1)/P \rfloor - \lfloor (i-1)/P \rfloor = 0$, corresponding to all the $P \times P$ submatrices along the principal diagonal of $\boldsymbol{\Phi}_m^H \boldsymbol{\Phi}_m$, we again consider two possibilities. First, $\kappa_P(j,i) = 0$. In this case,

$$\left[\boldsymbol{\Phi}_m^H \boldsymbol{\Phi}_m^H \right]_{ij} = L \sum_{g=0}^{G-1} a_g^* a_g = L. \tag{10.29}$$

Since $\xi_P(j,i) = 0$, this corresponds to the case when $i = j$, or equivalently all the diagonal entries of the matrix $\boldsymbol{\Phi}_m^H \boldsymbol{\Phi}_m$ are L. When $\kappa_P(j,i) \neq 0$ or equivalently, $i \neq j$,

$$\left[\boldsymbol{\Phi}_m^H \boldsymbol{\Phi}_m \right]_{ij} = L \sum_{g=0}^{G-1} a_g a_{g-(j-i)}^* \approx 0. \tag{10.30}$$

Hence, for $m = n$, $\boldsymbol{\Phi}_m^H \boldsymbol{\Phi}_n = LI$.

2. Next, consider the case when $m \neq n$. In this case, $\xi_{M_1}(n,m) + \xi_P(j,i) = 0$ only when $\xi_{M_1}(n,m) = -\xi_P(j,i)$ or equivalently $\xi_{M_1}(n,m) = \xi_P(i,j)$. This corresponds to all the $P \times P$ submatrices along the $\xi_{M_1}(n,m)$th diagonal. For all other i,j, since $L \geq (M_2 + Q)/2$ and $\xi_{M_1}(n,m) + \xi_P(j,i) \neq 0$, we have $\left[\Phi_m^H \Phi_n\right]_{ij} = 0$. The constraint on L essentially eliminates the possibility of an ambiguity in the Doppler estimation. Within each submatrix along the $\xi_{M_1}(n,m)$th diagonal, $\kappa_{M_1}(n,m) + \kappa_P(j,i) = 0$, when $\kappa_{M_1}(n,m) = \kappa_P(i,j)$. These correspond to the indices along the $\kappa_{M_1}(n,m)$th diagonal within each submatrix. For all these indices, $\left[\Phi_m^H \Phi_n\right]_{ij} = L$. When $\kappa_{M_1}(n,m) + \kappa_P(j,i) \neq 0$, $\left[\Phi_m^H \Phi_n\right]_{ij} = 0$. Hence the structure.

10.4 Sparsity-Based Multiple-Target Tracking

The problem of target tracking is to estimate the target state that comprises the positions and velocities of all the targets in each tracking interval i. Mathematically, given a measurement model describing the relationship between the observations and the unknown state, and a state model describing the state evolution, tracking finds an estimate of the state at current time using all the measurements up to current time. For a detailed description of this topic, the readers can refer to Bar-Shalom et al. (2011). Tracking is solved in two stages—prediction and update—using Bayesian filtering algorithms such as Kalman or particle filters (Kay, 1993; Arulampalam et al., 2002). In the prediction stage, a predicted value of the current state at time i, say $\tilde{\theta}_i$ is obtained. In the update stage, the measurement at time i is used to update the predicted value of the target state to obtain the estimate of the target state, $\hat{\theta}_i$. This process is repeated in each tracking interval.

In a sparsity-based tracking procedure, we first compute the predicted state $\tilde{\theta}_i^k = [\tilde{x}_i^k, \tilde{y}_i^k, \tilde{\dot{x}}_i^k, \tilde{\dot{y}}_i^k]^T$ of each target, followed by computation of the state estimate of each target. First, we find the support of the block-sparse signal* ζ using the measurement at time i given by Equation 10.24. From the support of the block-sparse signal, we compute the estimates of the delay and Doppler of each target using Equation 10.16. Second, we find the target state estimates for each target using the following equations:

---

* The support of a block-sparse vector is defined as the set of blocks which have nonzero norm, that is, bsupp(a) as bsupp(a) = $\{l \mid \|a[l]\|_{\ell_2} \neq 0\}$, where $a[l]$ denotes the lth block of elements of the vector a.

$$(\hat{x}_i^k, \hat{y}_i^k) = \frac{c\hat{\tau}_{d,i}^k}{2}\tilde{u}_i^k,$$

$$(\hat{\dot{x}}_i^k, \hat{\dot{y}}_i^k) = \frac{c\hat{v}_{d,i}^k}{2f_c(\tilde{u}_i^k)^T\tilde{\tilde{u}}_i^k}\tilde{\tilde{u}}_i^k,$$

(10.31)

where

$\hat{\tau}_{d,i}^k$ and $\hat{v}_{d,i}^k$ are the estimates of delay and Doppler of *k*th target at time *i*, respectively

$\tilde{u}_i^k$ and $\tilde{\tilde{u}}_i^k$ are the unit vectors in the direction of the position and velocity of *k*th target at time *i*, respectively, given by

$$\tilde{u}_i^k = \left(\frac{\tilde{x}_i^k}{\sqrt{\left(\tilde{x}_i^k\right)^2 + \left(\tilde{y}_i^k\right)^2}}, \frac{\tilde{y}_i^k}{\sqrt{\left(\tilde{x}_i^k\right)^2 + \left(\tilde{y}_i^k\right)^2}} \right),$$

$$\tilde{\tilde{u}}_i^k = \left(\frac{\tilde{\dot{x}}_i^k}{\sqrt{\left(\tilde{\dot{x}}_i^k\right)^2 + \left(\tilde{\dot{y}}_i^k\right)^2}}, \frac{\tilde{\dot{y}}_i^k}{\sqrt{\left(\tilde{\dot{x}}_i^k\right)^2 + \left(\tilde{\dot{y}}_i^k\right)^2}} \right).$$

(10.32)

Using this procedure, the problem of multiple-target tracking boils down to the problem of estimating the support of the block-sparse signal ζ. The support of ζ can be found by reconstructing the sparse signal ζ using one of the following standard sparse signal reconstruction techniques, followed by thresholding.

10.4.1 Standard Sparse Signal Reconstruction Techniques

In its most canonical form, a sparse model consists of an $N \times 1$ measurement vector of the form $y = \Phi x + w$, where Φ is an $N \times M$ dictionary of basis vectors and x is an $M \times 1$ sparse vector, with sparsity level K, $K \ll M$, and w is the additive noise. Several algorithms have been proposed in the literature to reconstruct the sparse signal x, all of which can be broadly classified into three main categories. In the absence of noise, the signal x can be recovered by solving

$$\mathrm{P}_0: \quad \arg\min_x \|x\|_{\ell_0} \text{ subject to } y = \Phi x.$$

Unfortunately, P_0 is a nonconvex optimization problem and it is NP hard to solve (Natarajan, 1995). A convex relaxation to P_0 is obtained by replacing the ℓ_0 norm with the ℓ_1 norm or ℓ_p norm in general. Solutions thus obtained fall under the category of convex relaxation. Basis pursuit (BP) (Chen et al., 1998) solves the problem

$$\text{BP}: \quad \arg\min_{x} \|x\|_{\ell_1} \text{ subject to } y = \Phi x.$$

In the presence of noise, a variation of basis pursuit called basis pursuit denoising (BPDN) (Candés et al., 2006; Chen et al., 1998) is used. BPDN solves

$$\text{BPDN}: \quad \arg\min_{x} \|x\|_{\ell_1} \text{ subject to } \|y - \Phi x\|_{\ell_2} \leq \epsilon,$$

where $\epsilon > 0$ is the tuning parameter chosen based on the noise level. Dantzig selector (DS) (Candés and Tao, 2007), another convex relaxation optimization introduced recently, solves the optimization problem

$$\text{DS}: \quad \arg\min_{x} \|x\|_{\ell_1} \text{ subject to } \|\Phi^{H}(y - \Phi x)\|_{\ell_\infty} \leq \mu,$$

where $\mu > 0$ is the tuning parameter. When the standard deviation of the additive noise is known, and the columns of the dictionary are normalized, the parameters ϵ and μ are chosen to be $\sqrt{2 \log N}\sigma$ (Chen et al., 1998). A second category of reconstruction methods perform a greedy iterative search for the solutions of P_0 by making locally optimal choices. Orthogonal matching pursuit (OMP) (Tropp, 2004; Tropp and Gilbert, 2007) and compressive sampling matching pursuit (CoSaMP) (Needell and Tropp, 2009) fall under this category. The algorithm terminates when the squared error of the estimate between the consecutive iterations is below a predetermined threshold. The third category of solutions enforces a sparsity based prior on the signal x and recover the signal x by solving for the MAP estimate. Most common methods in this category are Bayesian compressive sensing (BCS) (Ji et al., 2008) and sparse Bayesian learning (SBL) (Wipf and Rao, 2004). The choice of the sparsity enforcing priors is problem dependent and there are no universal priors that guarantee good performance for all the models.

Using these standard sparse signal reconstruction algorithms to reconstruct the target scene in a multipath is often computationally expensive as they involve high dimensional vectors and matrices. Hence, these reconstruction algorithms are not well suited for target tracking, which requires real time processing. In Section. 10.4.3, a PB support recovery algorithm that exploits the structural property of the dictionary is described. PB support recovery algorithm is a simple single-step procedure that does not require convex optimization or an iterative greedy search.

10.4.2 Effect of Multipath Environment

In this section, the effect of the time-varying multipath channel and the signaling on the performance of the support recovery algorithm is analyzed. As shown in Section 10.4, multiple-target tracking problem is equivalent to a support recovery problem. In particular, we are interested in estimating the set $\mathcal{K}$ of indices that correspond to the nonzero blocks of the block sparse vector $\tilde{\zeta}$ in Equation 10.24. Since there are K targets, there are $S = \binom{M}{K}$ possible locations where the blocks of nonzero entries can be placed in $\tilde{\zeta}$. Let $\mathcal{S}$ denote the set containing all the possible locations of the nonzero blocks. An error in the estimate of the delay or the Doppler of a target occurs whenever the index corresponding a true target location is associated with a different index. To study the performance of such a scheme, the total error probability of a multiple hypothesis testing problem is used as a performance metric (Tang and Nehorai, 2010). In such a scheme, each hypothesis corresponds to a candidate support. We label these hypothesis as $\mathcal{H}_0, \ldots, \mathcal{H}_S$, then

$$
\begin{cases}
\mathcal{H}_0 : & \text{bsupp}(\zeta) = \mathcal{S}_0, \\
\vdots & \\
\mathcal{H}_S : & \text{bsupp}(\zeta) = \mathcal{S}_S
\end{cases}
\tag{10.33}
$$

where the set $\mathcal{S}_m$ denotes the mth potential target location set. Evaluating the total error probability for an arbitrary reconstruction method is, in general, not feasible. Hence, we use an upper bound on total error probability for an optimal maximum likelihood–based decision criterion. If all the hypothesis are assumed to be equally likely, the optimal decision rule is given as (Trees, 2001)

$$
m^* = \arg \max_m p(y | \mathcal{H}_m).
\tag{10.34}
$$

The following theorem upper bounds the error probability of wrongly identifying the support.

Theorem 10.2 *For the sparse signal model defined in (10.23), the total error probability, P_K^M, is upper bounded by*

$$
P_K^M \leq \frac{1}{2S} \sum_{m=1}^{S} \sum_{\substack{n=1 \\ n \neq m}}^{S} \prod_{\rho=1}^{PQ} \frac{1}{1 + \lambda_\rho(m, n)},
\tag{10.35}
$$

where $\lambda_1(m, n) \cdots \lambda_{KPQ}(m, n)$ are the eigenvalues of the $KPQ \times KPQ$ matrix $\Lambda(m, n)$ defined as $\Lambda(m, n) = SNR/2L\Sigma_{\tilde{\zeta}}\Delta\tilde{\Phi}_{m,n}^{H}\Delta\tilde{\Phi}_{m,n}$, $SNR = EL/\sigma^2$ is the signal-to-noise ratio, σ is the standard deviation of the additive white Gaussian noise w, $\Sigma_{\tilde{\zeta}} = diag\{\alpha^1, \ldots, \alpha^K\} \otimes \Sigma_h$, $\Delta\tilde{\Phi}_{m,n} = \tilde{\Phi}_n - \tilde{\Phi}_m$, and $\tilde{\Phi}_m$ is obtained by concatenating the K blocks that correspond to the nonzero locations of the mth hypothesis.

Proof: For the system model described by Equation 10.23, first consider a binary case where there are only two supports, indexed by m and n, that is, the block support of ζ is either $\mathcal{S}_m$ or $\mathcal{S}_n$. Denote the probability of wrongly associating the support $\mathcal{S}_m$ to $\mathcal{S}_n$ as $P_c(m \to n|m)$. This probability is evaluated for an optimal maximum likelihood–based hypothesis test. For the binary case, given that $\mathrm{bsupp}(\zeta) = \mathcal{S}_m$, Equation 10.23 can be written as

$$y = \sqrt{E}\tilde{\Phi}_m\tilde{\zeta} + w, \tag{10.36}$$

where $\tilde{\Phi}_m$ (not to be confused with Φ_m used earlier) is obtained by concatenating K blocks each of size $NL \times PQ$ that correspond to the nonzero block locations in the block sparse vector ζ, and $\tilde{\zeta}$ is obtained by removing the zeros in the block-sparse vector ζ.

Using the optimal decision rule given by Equation 10.34, an error in wrongly associating support is given as

$$P_c(m \to n|m) = \Pr\{P(y|\mathrm{bsupp}(\zeta) = \mathcal{S}_n) > P(y|\mathrm{bsupp}(\zeta) = \mathcal{S}_m)|m\}. \tag{10.37}$$

The subscript c emphasizes that the error probability is conditional on $\tilde{\zeta}$. Under the hypothesis $\mathcal{H}_m : \mathrm{bsupp}(\zeta) = \mathcal{S}_m$, $y \sim \mathbb{CN}(\sqrt{E}\tilde{\Phi}_m\tilde{\zeta}, \Sigma)$. Thus, the conditional error probability given $\tilde{\zeta}$ is given as

$$
\begin{aligned}
P_c(m \to n|m) = \Pr\{ & e^{-(y-\sqrt{E}\tilde{\Phi}_n\tilde{\zeta})^H\Sigma^{-1}(y-\sqrt{E}\tilde{\Phi}_n\tilde{\zeta})} \\
& \geq e^{-(y-\sqrt{E}\tilde{\Phi}_m\tilde{\zeta})^H\Sigma^{-1}(y-\sqrt{E}\tilde{\Phi}_m\tilde{\zeta})}|m\}, \\
= \Pr\{ & \sqrt{E}y^H\Sigma^{-1}\Delta\tilde{\Phi}_{m,n}\tilde{\zeta} + \sqrt{E}\tilde{\zeta}^H\Delta\tilde{\Phi}_{m,n}^H\Sigma^{-1}y \\
& \geq E\tilde{\zeta}^H\tilde{\Phi}_n^H\Sigma^{-1}\tilde{\Phi}_n\tilde{\zeta} - E\tilde{\zeta}^H\tilde{\Phi}_m^H\Sigma^{-1}\tilde{\Phi}_m\tilde{\zeta}|m\}.
\end{aligned}
$$

However, $y^H\Sigma^{-1}\Delta\tilde{\Phi}_{m,n}\tilde{\zeta} + \tilde{\zeta}^H\Delta\tilde{\Phi}_{m,n}^H\Sigma^{-1}y$ follows a complex normal distribution with mean $\sqrt{E}\tilde{\zeta}^H\tilde{\Phi}_m^H\Sigma^{-1}\Delta\tilde{\Phi}_{m,n}\tilde{\zeta} + \sqrt{E}\tilde{\zeta}^H\Delta\tilde{\Phi}_{m,n}^H\Sigma^{-1}\tilde{\Phi}_m\tilde{\zeta}$ and variance–covariance matrix $\tilde{\zeta}^H\Delta\tilde{\Phi}_{m,n}^H\Sigma^{-1}\Delta\tilde{\Phi}_{m,n}\tilde{\zeta}$. Hence,

$$
\begin{aligned}
P_c(m \to n|m) = \Pr\{ & \sqrt{E}(y^H\Sigma^{-1}\Delta\tilde{\Phi}_{m,n}\tilde{\zeta} + \tilde{\zeta}^H\Delta\tilde{\Phi}_{m,n}^H\Sigma^{-1}y) \\
& \geq E(\tilde{\zeta}^H\tilde{\Phi}_n^H\Sigma^{-1}\tilde{\Phi}_n\tilde{\zeta} - \tilde{\zeta}^H\tilde{\Phi}_m^H\Sigma^{-1}\tilde{\Phi}_m\tilde{\zeta})|m\}, \\
= Q\left(\sqrt{E\tilde{\zeta}^H\Delta\tilde{\Phi}_{m,n}^H\Sigma^{-1}\Delta\tilde{\Phi}_{m,n}\tilde{\zeta}} \right), & \tag{10.38}
\end{aligned}
$$

where $Q(\gamma)$ is the complementary error function defined as $Q(\gamma) = \int_{x=\gamma}^{\infty} 1/\sqrt{2\pi} e^{-\frac{x^2}{2}} dx$. Using the bound $Q(\gamma) \leq 1/2 e^{-\gamma^2/2}$ (Wozencraft and Jacobs, 1965),

$$P_c(m \to n|m) < \frac{1}{2} e^{-\frac{E\tilde{\zeta}^H \Delta \tilde{\Phi}_{m,n}^H \Sigma^{-1} \Delta \tilde{\Phi}_{m,n} \tilde{\zeta}}{2}}. \tag{10.39}$$

The unconditional probability of error is obtained by averaging $P_c(m \to n|m)$ over the distribution of $\tilde{\zeta}$. Since $\tilde{\zeta} \sim \mathbb{CN}_{KPQ}(0, \Sigma_{\zeta})$,

$$P(m \to n|m) < \frac{1}{2} \int e^{-\frac{E\tilde{\zeta}^H \Delta \tilde{\Phi}_{m,n}^H \Sigma^{-1} \Delta \tilde{\Phi}_{m,n} \tilde{\zeta}}{2}} \frac{1}{(\pi)^{KPQ} |\Sigma_{\zeta}|} e^{-\tilde{\zeta}^H \Sigma_{\zeta}^{-1} \tilde{\zeta}} d\tilde{\zeta},$$

$$= \frac{1}{2} \frac{2|\Sigma_{\zeta}^{-1}|}{\left| E\Delta \tilde{\Phi}_{m,n}^H \Sigma^{-1} \Delta \tilde{\Phi}_{m,n} + 2\Sigma_{\zeta}^{-1} \right|}$$

$$\times \int \frac{\left| E\Delta \tilde{\Phi}_{m,n}^H \Sigma^{-1} \Delta \tilde{\Phi}_{m,n} + 2\Sigma_{\zeta}^{-1} \right|}{2(\pi)^{KPQ}} e^{-\tilde{\zeta}^H \left(\frac{E\Delta \tilde{\Phi}_{m,n}^H \Sigma^{-1} \Delta \tilde{\Phi}_{m,n} + 2\Sigma_{\zeta}^{-1}}{2} \right) \tilde{\zeta}} d\tilde{\zeta},$$

$$= \frac{1}{2} \frac{1}{\left| E\Sigma_{\zeta} \Delta \tilde{\Phi}_{m,n}^H \Sigma^{-1} \Delta \tilde{\Phi}_{m,n}/2 + I \right|}. \tag{10.40}$$

The total probability of error is now obtained by using the union bound:

$$P_{\text{err}} = \sum_{m=1}^{S} \Pr\{\text{err}|m\} P(m),$$

$$= \frac{1}{S} \sum_{m=1}^{S} \Pr\left\{ \bigcup_{n \neq m} P(m \to n|m) \right\},$$

$$< \frac{1}{S} \sum_{m=1}^{S} \sum_{\substack{n=1 \\ n \neq m}}^{S} P(m \to n|m),$$

$$< \frac{1}{2S} \sum_{m=1}^{S} \sum_{\substack{n=1 \\ n \neq m}}^{S} \frac{1}{\left| \frac{E\Sigma_{\zeta} \Delta \tilde{\Phi}_{m,n}^H \Sigma^{-1} \Delta \tilde{\Phi}_{m,n}}{2} + I \right|}. \tag{10.41}$$

Denote $\Lambda(m,n) = \frac{E\Sigma_{\zeta} \Delta \tilde{\Phi}_{m,n}^H \Sigma^{-1} \Delta \tilde{\Phi}_{m,n}}{2}$ and let $\lambda_1(m,n), \lambda_2(m,n), \ldots, \lambda_{KPQ}(m,n)$ be the eigenvalues of $\Lambda(m,n)$. Then,

$$P_{\mathrm{err}} < \frac{1}{2S} \sum_{\substack{m=1}}^{S} \sum_{\substack{n=1 \\ n\neq m}}^{S} \prod_{\rho=1}^{KPQ} \left(\frac{1}{1 + \lambda_\rho(m,n)} \right). \tag{10.42}$$

Several comments are in order here. First, note that the performance metric, P_K^M, can be difficult to compute due to exponentially large number of terms in the summation. Nevertheless, it gives various insights on how a support recovery algorithm depends on the signaling parameters and the channel. P_K^M, varies inversely with the product $PQ = T_d B_d TB$. For a given signaling scheme, that is, for a given TB, P_K^M decreases as the product $T_d B_d$ increases. This performance advantage is due to the joint delay-Doppler diversity that is inherently present in the measurement model due to a time-varying multipath channel. By using a model that captures this joint diversity, significant performance improvement is achieved. For a given channel, that is, for a given $T_d B_d$, P_K^M decreases as the time-bandwidth product, TB, of the signaling increases. To maximize the product PQ, and to exploit the full diversity offered by the model, a signaling scheme that maximizes the time-bandwidth product should be used. For nonspread-spectrum signals, the signaling duration and the bandwidth cannot be controlled independently. As a result, by increasing the signaling duration, the diversity offered due to the Doppler spread is lost and by increasing the signaling bandwidth, the diversity offered due to the delay spread is lost. For spread-spectrum signals, signaling duration and the bandwidth can be controlled independently. In particular, the time-bandwidth product will be proportional to G, which can be made arbitrarily large in principle. Hence, spread-spectrum waveforms are optimal for sparsity-based tracking in multipath scenarios.

Second, the eigenvalues λ_ρ depend on the dictionary, Φ, which in turn depend on the spread-spectrum sequence chosen for the transmission. This leads to a signal optimization problem. The optimal sequence to be transmitted in each tracking interval can be found by solving the following optimization problem:

$$a^* = \arg\min_{a} \sum_{\substack{m=1}}^{S} \sum_{\substack{n=1 \\ n\neq m}}^{S} \left(\frac{1}{1 + \lambda_{\min}(m,n,a)} \right)^{KPQ}, \tag{10.43}$$

where $\lambda_{\min}(a)$ corresponds to the minimum eigenvalue of the matrix Λ defined earlier. However, this design problem is not considered in this chapter.

Finally, note also that the upper bound given in Theorem 10.2 may not be a tight for all SNR values. Due to the large number of terms in the summation, the bound P_K^M can be greater than one for some values of SNR, in which case the bound becomes trivial.

10.4.3 PB Support Recovery Algorithm

In this section, a PB support recovery algorithm is discussed. The PB algorithm is a single-step approach for finding the support of the block-sparse vector. First, construct the dictionary $\boldsymbol{\Phi}$ for the sparse model by discretizing the delay-Doppler plane, as described in Section 10.3. Next, project the received vector onto the row space of the dictionary and coherently combine the energy in each block of the projection vector. The projection z can be expressed as

$$z = (I_M \otimes h^H)\boldsymbol{\Phi}^H y. \tag{10.44}$$

The block support of ζ can be then estimated by finding the indices that correspond to the maximum absolute values of the vector z. The following discussion demonstrates why the algorithm works. For a noise-free case,

$$z = (I_M \otimes h^H)\boldsymbol{\Phi}^H y = \sqrt{E}(I_M \otimes h^H)\boldsymbol{\Phi}^H \boldsymbol{\Phi}\zeta. \tag{10.45}$$

Rewriting ζ as $\zeta = \alpha \otimes h$, and expressing the rth element of z as

$$z_r = \sqrt{E}h^H \sum_{k \in \mathcal{K}} \alpha^k \boldsymbol{\Phi}_r^H \boldsymbol{\Phi}_k h. \tag{10.46}$$

The summation over other terms is zero as $\alpha^m = 0$ for $m \notin \mathcal{K}$.

For $r \in \mathcal{K}$, (10.46) simplifies as

$$
\begin{aligned}
z_r &= \alpha^r \sqrt{E}h^H \boldsymbol{\Phi}_r^H \boldsymbol{\Phi}_r h + \sqrt{E}h^H \sum_{k \in \mathcal{K}-\{r\}} \alpha^k \boldsymbol{\Phi}_r^H \boldsymbol{\Phi}_k h, \\
&= \alpha^r \sqrt{E}Lh^H h + \sqrt{E} \sum_{k \in \mathcal{K}-\{r\}} \alpha^k h^H D_\kappa^\xi h.
\end{aligned}
\tag{10.47}
$$

The result from Theorem 10.1 is used to express $\boldsymbol{\Phi}_r^H \boldsymbol{\Phi}_k$ as $\boldsymbol{\Phi}_r^H \boldsymbol{\Phi}_k = D_\kappa^\xi$, where ξ is denoted as $\xi_{M_1}(k,r)$ and κ is denoted as $\kappa_{M_1}(k,r)$ for simplicity. By expanding D_κ^ξ, using the notation described in Section 10.3,

$$h^H D_\kappa^\xi h = L \sum_{i=1}^{Q-\kappa} \sum_{j=1}^{P-\xi} h_{i+\kappa,j+\xi}^* h_{ij}. \tag{10.48}$$

Since h_{ij} are independent Gaussian random variables, they are uncorrelated. Therefore, by the law of large numbers, the summation in (10.48) is approximately zero. Hence

$$z_r = \alpha^r \sqrt{EL} h^H h \quad \text{for } r \in \mathcal{K}. \tag{10.49}$$

For $r \notin \mathcal{K}$, (10.46) simplifies as

$$z_r = \sqrt{E} h^H \sum_{k \in \mathcal{K}} \alpha^k \Phi_r^H \Phi_k h,$$

$$= \sqrt{EL} \sum_{k \in \mathcal{K}} \alpha^k \sum_{i=1}^{Q+1-\kappa} \sum_{j=1}^{P+1-\xi} h_{i+\kappa, j+\xi}^* h_{ij},$$

$$\approx 0. \tag{10.50}$$

Thus, in the absence of noise, the indices corresponding to the support of the block-sparse vector ζ are exactly the same as the indices that correspond to nonzero elements in the vector z. When there is noise in the system, z will no longer be sparse and the locations of K maximum absolute values of z determine the support of the block-sparse vector. Note that we are implicitly assuming that the number of targets K is known a priori. After finding the support of the block-sparse vector ζ, the estimates of the delay and the Doppler for each target will be computed using Equation 10.16, and the exact position and velocity for each target will be computed using Equation 10.31. In order to associate each nonzero index to a specific target, Equation 10.24 is solved for ζ using a regular least-squares approach. After ζ is known, the estimates the target RCS, α^k for $k = 1, \ldots, K$ are computed. Using the RCS values, each target is assigned to one of the K absolute maximum values. As long as the noise is bounded such that the vector z does not have spikes at the indices that do not correspond to the exact target locations, the proposed algorithm gives a perfect target scene reconstruction. This process is repeated in each tracking interval i to obtain the target state. The pseudo-code for the overall tracking method is given in Algorithm 10.1.

Algorithm 10.1 PB support recovery algorithm for target tracking in multipath environment

---

1: **for** $i = 1 : N_{TI}$ **do**
2:     Compute $\tilde{\theta}_i = F\hat{\theta}_{i-1}$.
3:     Construct the dictionary Φ_i using $\tilde{\theta}_i$.
4:     Construct a $M \times 1$ vector $z_i = (I_M \otimes h_i^H) \Phi_i^H y_i$.
5:     Find support $\{s_k \mid |z_{i,s_k}| > |z_{i,r}|, r \in \{1, \ldots, M\}, r \neq s_k, k = 1, \ldots, K\}$.
6:     Using s_k, compute $\hat{\tau}_{k,i,d}, \hat{v}_{k,i,d}, k = 1, \ldots, K$.
7:     Compute the estimate of the target state using Equations 10.31 and 10.32.
8: **end for**

Tracking using the proposed PB support recovery is computationally less expensive when compared with the computational complexity of other tracking methods. This computational advantage is achieved because the PB support recovery does not need a convex optimization or an iterative greedy search. Tracking using the proposed PB support recovery performs better compared with the performance of the other tracking methods. This performance advantage is achieved because the projection vector z is obtained by coherently combining the energy from all the paths. Therefore, the algorithm exploits the joint delay-Doppler diversity that is inherently present in the problem.

10.5 Numerical Results

Several numerical examples are shown in this section to demonstrate the performance and computational advantage of proposed tracking method. These examples also show the effect of the time-varying multipath modeling on the performance of the support recovery-based tracking methods.

The simulated scenario had two crossing targets ($K = 2$) moving in the region of interest. The initial position of the first target was $(1200, 900)$m, and it was moving with a constant velocity of $(18, 24)$m/s; the initial position of the second target was at $(900, 1559)$m, and it was moving with a constant velocity of $(15, -26)$m/s. The initial parameters were estimated before the tracking process. Such an estimation can be performed using several methods, including the maximum likelihood–based estimation (Altes, 1979), beamforming-based estimation (Xu et al., 2008), Bayesian estimation (Min et al., 2010), and sparsity-based estimation. Interested readers can refer to these works for more information. Both targets were moving along linear trajectories, and hence their state transition equations were described by Equation 10.10 and the covariance matrices of the modeling error are given by Equation 10.12 with $\epsilon^1 = \epsilon^2 = 4$. The tracking interval length (t_i) was chosen to be 0.5 s. The RCS of the first target, $\tilde{\alpha}^1$, was 1, while that of the second target, $\tilde{\alpha}^2$, was 1.4.

The carrier frequency, f_c, of the transmitted waveforms was 1 GHz and the bandwidth, B, of each pulse was 100 MHz. We used $L = 4$ pulses in each tracking interval. In each pulse, a spread waveform generated from a pseudorandom noise sequence of length, $G = 16$ was used. Since there was no signal optimization or pulse-to-pulse diversity employed, the same signal was used over all the 4 pulses and the 20 tracking intervals. In each tracking interval, the entries of the matrix H were generated independently from a complex Gaussian distribution. The entries were scaled later such that variance of the coefficients corresponding to different delays decayed exponentially. PQ was chosen to be 6. A window of 300 samples ($N = 300$)

per pulse around the predicted state of the first target in each tracking interval was considered. Such a large window size is required to accommodate the received signal vector of all the targets, when they are moving away from each other over time. The entries of w were drawn independently from a complex Gaussian distribution and were scaled to obtain the required SNR defined as

$$\text{SNR} = \frac{EL}{\sigma_w^2}. \tag{10.51}$$

The performance of the proposed PB tracking method was compared with other sparsity-based tracking methods that employed BPDN, DS, and BCoSaMP algorithms for support recovery. The delay-Doppler plane was divided into $M_1 = 5$ and $M_2 = 5$ grid points for each target, with the grid size $\Delta\tau = 1/f_s$ and $\Delta f = 1/t_i$. This corresponded to a position and velocity estimate resolution of 1.5 m and 0.3 m/s, respectively in the x and the y directions. Support recovery using BPDN and DS were done by solving the optimization problems labeled BPDN and DS in Section 10.4.1, followed by thresholding to obtain the sparse vector. Then the locations corresponding to the largest K values is chosen to be the support (note that we are exploiting the known signal sparsity level in this step). MATLAB's® CVX package (Grant and Boyd, 2009) was used to solve the convex optimization problem and the tuning parameters were chosen based on the noise variance, which was assumed to be known. Both these algorithms do not use the additional information about the block sparsity present in the problem. A BCoSaMP algorithm similar to the one used in Sen and Nehorai (2011) and (Eldar and Kuppinger, 2010) was used to solve the support recovery problem. The BCoSaMP algorithm used the additional information about the block sparsity in the problem.

All the simulations were averaged over 50 Monte-Carlo iterations, and we used an SNR of 25 dB.

In Figure 10.2, the cumulative root-mean-square error (RMSE) in the range and velocity estimates are plotted, and in Figure 10.3, the actual and estimated trajectories for both the targets using different methods are plotted.

It can be seen that for all the sparsity-based methods the RMSEs in the range and the velocity estimates were under 2 m and 1 m/s, respectively. The RMSE in the velocity estimates using PB support recovery was slightly better when compared with other methods. Further, it should also be emphasized that the methods based on DS and BPDN assume that the noise variance is known for selecting the tuning parameter. The time taken by each of these algorithms is tabulated in Table 10.1.

In the second set of simulations, the performance improvement obtained due to the joint delay-Doppler diversity in the model was demonstrated.

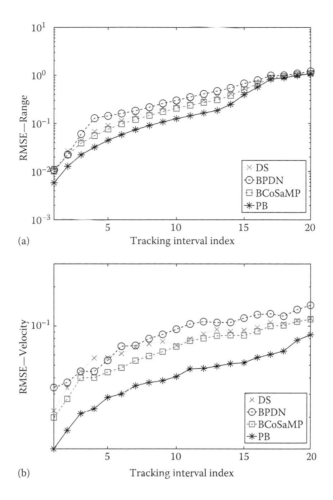

FIGURE 10.2
(See color insert.) Root-mean-square error for various tracking algorithms. (a) RMSE in range and (b) RMSE in velocity.

In Figure 10.4, the upper bound on the error probability is shown against the SNR for various values of P and Q. It can be seen that as the product PQ increased, the bound on the error probability decreased significantly. This is due to additional degrees of freedom provided by the joint delay and Doppler diversity present in the urban environment. Also, for $PQ = 1$, the bound was trivial and was greater than one for the range of SNR considered. To get a useful bound, the SNR has to be increased. In Figure 10.5, the actual RMSE in the range and the velocity estimates obtained using the PB tracking for various values of P and Q. It can be seen that when the product $PQ = 1$,

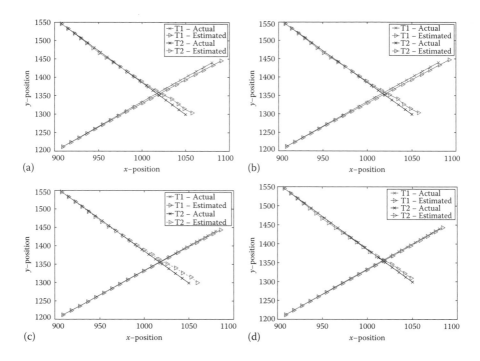

FIGURE 10.3
Actual vs estimated trajectories using various tracking algorithms. (a) Dantzig selector-based tracking, (b) BPDN-based tracking, (c) BCoSaMP-based tracking, and (d) projection-based tracking.

TABLE 10.1

Average CPU Time (in) for Various Sparsity-Based Tracking Algorithms

	DS	BPDN	BCoSaMP	PB
5 × 5 grid	8.56	7.04	2.78	2.16
9 × 9 grid	94.2	19.42	10.31	6.20

RMSE increased and was unbounded. For the same SNR, the performance in the presence of time-varying multipath channel model was significantly better.

10.6 Conclusions

In this chapter, we considered the problem of multiple-target tracking in a time-varying multipath channel. We developed a sparse model by

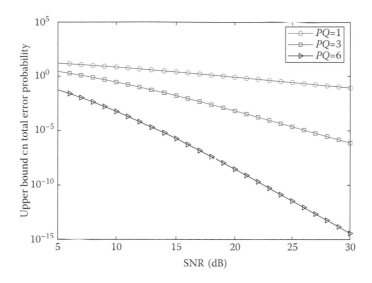

FIGURE 10.4
(See color insert.) Upper bound on error probability for an optimal reconstruction.

considering a finite-dimensional representation of the time-varying system function characterizing the channel. After discretizing the delay-Doppler plane, the target scene is represented as a sparse signal, and the tracking problem is transformed into a support recovery problem.

We then derived an upper bound on the error probability for an optimal maximum likelihood–based support recovery algorithm. Using this bound, we showed that the performance of the support recovery depends inversely on the time-bandwidth product of the signaling; hence, spread-spectrum waveforms are optimal in the sense that signaling duration and bandwidth can be controlled independently. With spread-spectrum signaling, the dictionary of the block-sparse model exhibits a special structure. We exploited this special structure to propose a new PB support recovery algorithm. As the PB support recovery algorithm did not involve an iterative search or the need to solve a convex optimization problem, target tracking using this algorithm was computationally less intensive. Further, in the scenarios where the additive noise is bounded, PB algorithm gives exact target scene reconstruction and the corresponding tracking performance was better when compared with the performance obtained using other standard sparse signal reconstruction algorithms. Numerical simulations demonstrated that the proposed tracking algorithm takes less time when compared with the time taken by the standard sparse signal reconstruction-based tracking, and it produced lower mean-squared error (MSE) in the position and velocity estimates of the targets when compared with the MSE obtained using other methods.

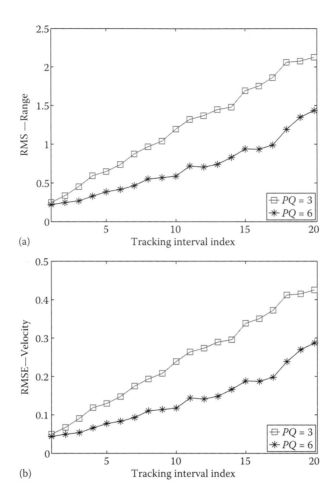

FIGURE 10.5
Root-mean-square error for various values of PQ. (a) RMSE in range and (b) RMSE in velocity.

References

V. Algeier, B. Demissie, W. Koch, and R. Thoma, State space initiation for blind mobile terminal position tracking, *EURASIP Journal on Advances in Signal Processing*, 2008, January 2008. doi: 10.1155/2008/394219, Article ID:394219. http://asp.eurasipjournals.com/content/pdf/1687-6180-2008-394219.pdf.

R. A. Altes, Target position estimation in radar and sonar, and generalized ambiguity analysis for maximum likelihood parameter estimation, *Proceedings of IEEE*, 67 (6), 920–930, June 1979.

M. S. Arulampalam, S. Maskell, N. J. Gordon, and T. Clapp, A tutorial on particle filters for online nonlinear/non-Gaussian Bayesian tracking, *IEEE Transactions on Signal Processing*, 50 (2), 174–188, February 2002.

Y. Bar-Shalom, X. -R. Li, and T. Kirubarajan, *Estimation with Applications to Tracking and Navigation*. New York: Wiley, 2001.

Y. Bar-Shalom and E. Tse, Tracking in a cluttered environment with probabilistic data association, *Automatica*, 11 (5), 451–460, 1975.

Y. Bar-Shalom, P. Willet, and X. Tian, *Tracking and Data Fusion: A Handbook of Algorithms*, 3rd edn. Storrs, CT: YBS Publishing, 2011.

M. R. Bell, Information theory and radar waveform design, *IEEE Transactions on Information Theory*, 39 (5), 1578–1597, September 1993.

P. Bello, Characterization of randomly time-variant linear channels, *IEEE Transactions on Communications*, 11 (4), 360–393, December 1963.

E. Candés, J. Romberg, and T. Tao, Stable signal recovery from incomplete and inaccurate information, *Communications of Pure and Applied Mathematics*, 59, 1207–1233, August 2006.

E. J. Candés and T. Tao, The Dantzig selector: Statistical estimation when p is much larger than n, *Annals of Statistics*, 35 (6), 2313–2351, 2007.

B. Chakraborty, Y. Li, J. J. Zhang, T. Trueblood, A. Papandreou-Suppappola, and D. Morrell, Multipath exploitation with adaptive waveform design for target tracking in urban terrain, in *International Conference on Acoustics, Speech, and Signal Processing*. Dallas, TX, March 14–19, 2010, pp. 3894–3897.

P. Chavali and A. Nehorai, Cognitive radar for target tracking in multipath scenarios, in *Proceedings of the International Waveform Diversity and Design (WDD) Conference*, Niagara Falls, Canada, August 2010, pp. 110–114.

P. Chavali and A. Nehorai, A low-complexity multi-target tracking algorithm in urban environments using sparse modeling, *Signal Processing*, 92 (9), 2199–2213, September 2012.

P. Chavali and A. Nehorai, Concurrent particle filtering and data association using game theory for tracking multiple maneuvering targets, *IEEE Transactions on Signal Processing*, 61 (20), 4934–4948, 2013.

S. Chen, D. Donoho, and M. Saunders, Atomic decomposition by basis pursuit, *SIAM Journal on Scientific Computing*, 20 (1), 33–61, 1998.

Z. S. Dobrosavljevic and M. L. Dukic, A method of spread spectrum radar polyphase code design by nonlinear programming, *European Transactions on Telecommunications*, 7 (3), 239–242, May 1996.

Y. C. Eldar and P. Kuppinger, Block sparse signals: Uncertainty relations and efficient recovery, *IEEE Transactions on Signal Processing*, 58 (6), 3042–3054, June 2010.

T. Fortman, Y. Bar-Shalom, and M. Scheffe, Sonar tracking of multiple targets using joint probabilistic data association, *IEEE Journal of Oceanic Engineering*, 8, 173–184, 1983.

H. Gauvrit, J. -P. L. Cadre, and C. Jauffret, A formulation of multitarget tracking as an incomplete data problem, *IEEE Transactions on Aerospace and Electronic Systems.*, 33, 1242–1257, October 1997.

R. Gold, Optimal binary sequences for spread spectrum multiplexing, *IEEE Transactions on Information Theory*, 13, 619–621, 1967.

M. Grant and S. Boyd, CVX: Matlab software for disciplined convex programming. Stanford University. http://stanford.edu/boyd/cvx, web page and software, June 2009.

M. A. Herman and T. Strohmer, High-resolution radar via compressive sensing, *IEEE Transactions on Signal Processing*, 57 (6), 2275–2284, June 2009.

S. Ji, Y. Xue, and L. Carin, Bayesian compressive sensing, *IEEE Transactions on Signal Processing*, 56, 2346–2356, June 2008.

Y. Jin, J. M. Moura, and N. O'Donoughue, Time reversal in multiple-input multiple-output radar, *IEEE Journal of Selected Topics in Signal Processing*, 4 (1), 210–225, February 2010.

S. M. Kay, *Fundamentals of Statistical Signal Processing, Estimation Theory*, Vol. 1. Englewood Cliffs, NJ: Prentice-Hall, 1993.

B. Krach and R. Weigel, Markovian channel modeling for multipath mitigation in navigation receivers, in *European Conference on Antennas and Propagation*, Berlin, Germany, March 2009, pp. 1441–1445.

J. L. Krolik, J. Farrell, and A. Steinhardt, Exploiting multipath propagation for GMTI in urban environments, in *IEEE Conference on Radar*, Verona, NY, IEEE, April 24–27, 2006, pp. 65–68.

N. Levanon and E. Mozeson, *Radar Signals*. New York: Wiley, 2004.

R. P. S. Mahler, *Statistical Multisource-Multitarget Information Fusion*. Norwood, MA: Artech House, Inc., 2007.

J. Min, R. Niu, and R. S. Blum, Bayesian target location and velocity estimation for MIMO radar, *IET Radar, Sonar and Navigation*, 60972152, 1–10, 2010.

B. K. Natarajan, Sparse approximate solutions to linear systems, *SIAM Journal on Computing*, 24 (2), 227–234, 1995.

D. Needell and J. A. Tropp, CoSaMP: Iterative signal recovery from incomplete and inaccurate samples, *Applied and Computational Harmonic Analysis*, 26, 301–321, May 2009.

P. Z. Peebles, *Radar Principles*. New York: Wiley, 1998.

J. G. Proakis, *Digital Communications*, 4th edn. New York: McGraw-Hill, 2001.

D. B. Reid, An algorithm for tracking multiple targets, *IEEE Transactions on Automatic Control*, AC-24, 843–854, 1979.

B. D. Rigling, Urban RF multipath mitigation, *IET Radar, Sonar and Navigation*, 2 (6), 419–425, December 2008.

S. Särkkä, A. Vehtari, and J. Lampinen, Rao-Blackwellized Monte-Carlo data association for multiple target tracking, in *Proceedings of the Seventh International Conference on Information Fusion*, Stockholm, Sweden, June 2004, pp. 583–590.

D. Sarwate and M. Pursley, Crosscorrelation properties of pseudorandom and related sequences, *Proceedings of the IEEE*, 68 (5), 593–619, May 1980.

A. Sayeed and B. Aazhang, Joint multipath-Doppler diversity in mobile wireless communications, *IEEE Transactions on Communications*, 47 (1), 123–132, January 1999.

S. Sen and A. Nehorai, Sparsity-based multi-target tracking using OFDM radar, *IEEE Transactions on Signal Processing*, 59 (4), 1902–1906, April 2011.

G. Tang and A. Nehorai, Performance analysis for sparse support recovery, *IEEE Transactions on Information Theory*, 56, 1383–1399, March 2010.

H. L. V. Trees, *Detection, Estimation and Modulation Theory*, Vol. 1. New York: Wiley, 2001.

J. A. Tropp, Greed is good: Algorithmic results for sparse approximation, *IEEE Transactions on Information Theory*, 50, 2231–2242, October 2004.

J. A. Tropp and A. C. Gilbert, Signal recovery from random measurements via orthogonal matching pursuit, *IEEE Transactions on Information Theory*, 53, 4655–4666, December 2007.

N. Vaswani, Kalman filtered compressed sensing, in *Proceedings of the 15th IEEE International Conference on Image Processing*, San Diego, CA, 2008, pp. 893–896.

N. Vaswani, Ls-cs-residual (ls-cs): Compressive sensing on least squares residual, *IEEE Transactions on Signal Processing*, 58 (8), 4108–4120, 2010.

N. Vaswani and W. Lu, Modified-cs: Modifying compressive sensing for problems with partially known support, *IEEE Transactions on Signal Processing*, 58 (9), 4595–4607, 2010.

J. Vermaak, S. Godsill, and P. Perez, Monte-Carlo filtering for multi-target tracking and data association, *IEEE Transactions on Aerospace and Electronic Systems*, 41 (1), 309–332, January 2005.

Y. Wang, X. Li, and Y. Wang, Novel spread-spectrum radar waveform, *Proceedings of SPIE, Radar Sensor Technologies*, 3066, 186–193, June 1997.

D. P. Wipf and B. Rao, Sparse Bayesian learning for basis selection, *IEEE Transactions on Signal Processing*, 52, 2153–2164, August 2004.

J. M. Wozencraft and I. M. Jacobs, *Principles of Communication Engineering*, 1st edn. London, U.K.: Wiley, 1965.

L. Xu, J. Li, and P. Stoica, Target detection and parameter estimation for MIMO radar systems, *IEEE Transactions on Aerospace and Electronic Systems*, 44 (3), 927–939, July 2008.

11

Three-Dimensional Imaging of Vehicles from Sparse Apertures in Urban Environment

Emre Ertin

CONTENTS

ABSTRACT   Three-dimensional synthetic aperture radar (SAR) imaging of
vehicles in urban setting is made possible by new data collection capabilities, in which airborne radar systems interrogate a large scene persistently
and over a large range of aspect angles. Wide-angle 3-D reconstructions of
vehicles can be useful in applications such as automatic target recognition
(ATR) and fingerprinting. The backscatter data collected by the airborne
platform at each pulse can be interpreted as 1-D lines of the 3-D Fourier
transform of the scene, and the aggregation of radar returns over the flight
path defines a conical manifold of data in the scenes 3-D Fourier domain.
Generating high-resolution 3-D images using traditional Fourier processing
methods requires that radar data be collected over a densely sampled set
of points in both azimuth and elevation angles. This method of imaging
requires very large collection times and storage requirements and may be
prohibitively costly in practice. There is thus motivation to consider more
sparsely sampled data collection strategies, where only a small fraction of
the data required to perform traditional high-resolution imaging is collected.

In this chapter, we review several techniques that have been proposed for 3-D reconstruction data collected from sparsely apertures. Particular emphasis is given to sparsity-regularized least-square approaches to wide-angle 3-D radar reconstruction for arbitrary, sparse apertures. We provide a comprehensive set of comparative results using data from the GOTCHA data collection campaign.

11.1 Introduction

In this chapter, we consider the problem of three-dimensional target reconstruction from radar data obtained from wide-angle sparse synthetic apertures. The problem of three-dimensional reconstruction of vehicles in urban setting is motivated by an increasingly difficult class of surveillance and security challenges, including vehicle detection and activity monitoring in urban scenes. These reconstructions are enabled by new data collection capabilities, in which airborne SAR systems are able to interrogate a scene, such as a city, persistently and over a large range of aspect angles. Additional information provided by wide-aspect 3-D reconstructions can be useful in applications such as ATR and fingerprinting.

We consider an airborne radar sensor that transmits wideband pulsed waveforms along a flight path and records the backscatter response from the scene. The returned echoes can be interpreted as 1-D line segments of the 3-D Fourier transform of the scene, and the aggregation of radar returns over the flight path defines a conical manifold of data in the scenes 3-D Fourier domain. Generating high-resolution 3-D images using traditional Fourier processing methods requires that radar data be collected over a densely sampled set of points in both azimuth and elevation angles. This method of imaging requires very large collection times and data storage and may be prohibitively costly in practice since the aircraft has to make a large number of passes to sample along the elevation dimension. There is thus motivation to consider more sparsely sampled data collection strategies, where only a small fraction of the data required to perform traditional high-resolution imaging is collected. However, when Fourier imaging is applied to sparsely sampled apertures, reconstruction quality suffers in either resolution or high sidelobes, or both.

In this chapter, we survey several of the techniques [1–11] that have been recently proposed for 3-D reconstruction data collected from sparsely apertures, where the radar sensor collects information from few passes at slightly different elevation angles. These techniques rely on some basic properties of scattering physics and exploit signal sparsity (in the reconstruction domain) of radar scenes. Man-made target scenes are dominated by a small number of dominant isolated scattering centers; dominant returns result from objects such as corner or plate reflectors made from electromagnetic conductive material of different dielectric properties. Therefore, the inversion problem can be regularized using sparsity constraints in the image domain. We will review both 3-D versions of L_p norm regularized least-square (LS) approaches, popularized by compressive sensing field, as well as wide-angle extension of multibaseline interferometric synthetic aperture radar (IFSAR) approaches that rely on spectral-estimation methods in the height dimension.

Wide-angle imaging problems differ from traditional techniques designed for *narrow*-angle collection geometries, where isotropic scattering assumption is well approximated. For wide-angle scenes, radar scattering is typically anisotropic over wide angles and violates the isotropic point scattering assumption of traditional radar imaging. Anisotropic scattering over wide angles is addressed in each case by using noncoherent subaperture imaging, where scattering is assumed to be isotropic over narrow-angle subapertures.

In the following, we first review the system model for wide-angle SAR collections and discuss the shortcomings of the traditional Fourier imaging methods when applied to sparse circular synthetic aperture (CSAR) collections. Throughout the chapter, we use AFRL GOTCHA CSAR data set [12] for comparative empirical results. The GOTCHA CSAR data set is fully polarimetric and consists of eight complete circular passes, with each pass being at a slightly different elevation angle. The radar used in the GOTCHA data collection has a center frequency of 9.6 GHz and a bandwidth of 640 MHz. Next, we present a direct 3-D imaging approach based on sparsity-regularized inversion of the CSAR measurement operator with empirical results and discuss computational complexity of the direct approach. Then, we review multibaseline IFSAR techniques for this problem and consider interpolated array methods and discrete Fourier transform (DFT) peak detection methods, and present empirical results in each case. We conclude with a discussion of practical considerations such as data registration and autofocus for obtaining accurate reconstructions.

11.2 System Model

In this section, we present the system model for circular synthetic aperture radar (CSAR) data collections and briefly review traditional Fourier imaging with CSAR data. We assume that the SAR system collects coherent backscatter measurements $g_{i,p}(f_k)$ on circular apertures parameterized with azimuth angles $\{\phi_i\}$ covering $[0, 2\pi]$ and at a set of elevation angles

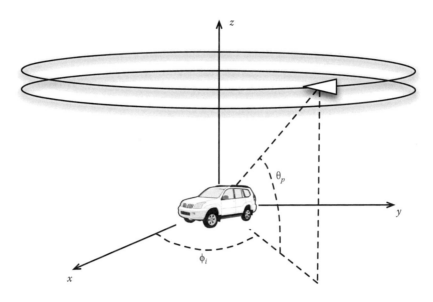

FIGURE 11.1
Multipass circular SAR data collection geometry.

$\{\theta_p\}_{p=1}^{P}$ collected over different passes of the aircraft. Figure 11.1 shows the multipass CSAR collection geometry. The radar transmits a wideband signal with bandwidth B centered at a center frequency f_c. Such a signal could be a frequency modulation chirp signal or a stepped-frequency signal, but other wideband signals can also be used. We also assume that the transmitter is sufficiently far away from the scene so that wavefront curvature is negligible, and we use a plane wave model for reconstruction; this assumption is valid, for example, when the extent of the scene being imaged is much smaller than the standoff distance from the scene to the radar.

The backscatter measurements are taken at discrete set of frequencies given by the set $\{f_k\}$. The imaging problem is to estimate the three-dimensional reflectivity function of the spotlighted scene $f(x, y, z)$ from the set of radar returns $\{g_{i,p}(f_k)\}$ collected by the radar. The three-dimensional version of the projection-slice theorem shows that the backscatter measurements represent samples of the 3-D spatial Fourier transform of the reflectivity function [13]. Specifically, the 3-D Fourier transform $G(k_x, k_y, k_z)$ of the reflectivity function $g(x, y, z)$ is given by

$$G(k_x, k_y, k_z) = \int g(x, y, z) e^{-j(k_x x + k_y y + k_z z)} \, \mathbf{dr} \qquad (11.1)$$

where $\mathbf{r} = (x, y, z)$ is a vector of spatial coordinates. Then, the radar measurements $\{g_{i,p}(f_k)\}$ correspond to the samples of $G(k_x, k_y, k_z)$ on a two-dimensional conical manifold at points $(k_x^{i,p,k}, k_y^{i,p,k}, k_z^{i,p,k})$:

$$k_x^{i,p,k} = \frac{4\pi f_k}{c} \cos(\theta_p) \cos(\phi_i)$$

$$k_y^{i,p,k} = \frac{4\pi f_k}{c} \cos(\theta_p) \sin(\phi_i)$$

$$k_z^{i,p,k} = \frac{4\pi f_k}{c} \sin(\theta_p)$$

The inverse Fourier transform of the data calculated on each conical manifold indexed by the elevation cut p results in a wide-angle volumetric image $I_p(x, y, z)$. The sum of all the volumetric images $I_p(x, y, z)$ results in the final coherent wide-angle volumetric image $I(x, y, z)$:

$$I(x, y, z) = \sum_p I_p(x, y, z) \tag{11.2}$$

We note that any 2-D slice from $I_p(x, y, z)$ contains all the information from the single pass, at elevation angle θ_p. Therefore all 2-D slices can be regenerated from the ground plane image $I_p(x, y, 0)$ using

$$I_p(x, y, z) = \mathcal{F}_{(x,y)}^{-1} \left[\mathcal{F}_{(x,y)} \left[I_j(x, y, 0) \right] e^{-j\sqrt{k_x^2 + k_y^2}\, \tan(\theta_p)z} \right] \tag{11.3}$$

Consequently, a 2-D ground plane image $I_p(x, y, 0)$ from each elevation θ_p is sufficient to construct the coherent volumetric image $I(x, y, z)$. However, the set of wide-angle images $I_p(x, y, 0)$ is not an efficient data representation for the radar returns because it requires a high spatial sampling rate to prevent aliasing of the circular bandpass signature given by

$$\delta_x < \frac{c}{4 \cos(\theta)(f_c + B/2)} \tag{11.4}$$

resulting in a Nyquist sampling rate of 1 cm for an X-band radar. In addition, 360° imagery is matched to an isotropic point scatterer that persists over the entire circular aperture. To minimize the storage of CSAR data, while providing image products matched to scatterers with limited persistence, we adopt image sequences $\{I_{p,m}(x, y, 0)\}_m$, where each image is the output of a filter matched to a limited-persistence reflector over the azimuth angles in window $\mathcal{W}_m(\phi)$. Specifically, the mth subaperture image is constructed using

$$I_{p,m}(x,y,0) = \mathcal{F}_{(x,y)}^{-1}\left[F\left(k_x,k_y,\sqrt{k_x^2+k_y^2}\tan(\theta_p)\right)\mathcal{W}_m\left(\tan^{-1}\frac{k_x}{k_y}\right)\right] \quad (11.5)$$

where the azimuthal window function $\mathcal{W}_m(\phi)$ is defined as

$$\mathcal{W}_m(\phi) = \begin{cases} W\left(\dfrac{\phi-\phi_m}{\Delta}\right), & -\Delta/2 < \phi < \Delta/2 \\ 0, & \text{otherwise} \end{cases} \quad (11.6)$$

Here, ϕ_m is the center azimuth angle for the mth window and Δ describes the hypothesized persistence width. The window function $W(\cdot)$ is an invertible tapered window used for crossrange sidelobe reduction. We also note that unlike the full 360° image, each image can be modulated to baseband and sampled at a lower resolution without causing aliasing. Each baseband image $I_{p,m}^B(x,y,z)$ is calculated as

$$I_{p,m}^B(x,y,0) = I_{p,m}(x,y,0)e^{-j(k_x^0(m)x+k_y^0(m)y)} \quad (11.7)$$

where the center frequency $(k_x^0(m),k_y^0(m))$ is determined by the center aperture ϕ_m, mean elevation angle $\bar{\theta}$, and center frequency f_c:

$$k_x^0(m) = \frac{4\pi f_c}{c}\cos(\bar{\theta})\cos(\phi_m),$$

$$k_y^0(m) = \frac{4\pi f_c}{c}\cos(\bar{\theta})\sin(\phi_m)$$

For small azimuth windows Δ, the Nyquist sampling rate for each image $I_{p,m}^B(x,y,0)$ is dictated by the radar bandwidth and results in a much smaller storage requirement for CSAR data. We note that the reduction in sampling requirement is a result of subaperture imaging and not baseband processing. Subaperture imaging limits the 2-D spectral extent to a patch whose size is proportional to the radar bandwidth. In contrast, full 360° CSAR image occupies an annulus at Fourier domain with radius dictated by the center frequency, necessitating much higher sampling frequency.

We note that the center frequency $(k_x^0(m),k_y^0(m))$ used in baseband modulation is independent of the elevation angle θ_p. The use of common center frequency preserves the relative phase information between the elevation cuts. The relative phase information between the images $I_{p,m}(x,y,0)$ corresponding passes is key to resolving the height dimension and producing three-dimensional imagery as described in Sections 11.4 and 11.5.

The image sequence $\left\{I_{p,m}^{B}(x,y,0)\right\}_{m}$ can be enhanced by deconvolving the subaperture point spread function (PSF) [14] and can be visualized in many different ways. One possibility is to use the generalized likelihood ratio test (GLRT) imaging proposed by Moses and Potter [15]. The GLRT image $I_G(x,y,z)$ can be obtained by taking a maximum over the subaperture imagery:

$$I_G(x,y,z) = \max_{m} \left| \sum_{p} I_{p,m}^{B}(x,y,z) \right| \tag{11.8}$$

11.3 Case Study for 3-D SAR: AFRL GOTCHA Volumetric SAR Data Set

To provide empirical results for the 3-D imaging techniques presented in this chapter, we use the AFRL GOTCHA CSAR data set [12], which is fully polarimetric and consists of eight complete circular (360°) passes, with each pass being at a different elevation angle, θ_p; the radar used in the GOTCHA data collection has a center frequency of 9.6 GHz, giving a center wavelength of $\lambda_c = 0.031$ and a bandwidth of 640 MHz. Figure 11.2 shows the eight passes in a global coordinate system, where the z dimension is height as measured from the ground plane in meters. This figure demonstrates the change in elevation angle across each pass.

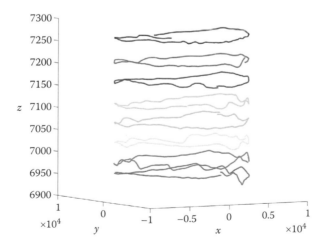

FIGURE 11.2
Eight passes from GOTCHA CSAR collection. (Unit dimension is meter.)

Defining the height dimension with respect to the slant plane coordinate system as z_S, and height in a global ground plane coordinate system as z_G, if data are uniformly sampled in z_S, resolution ρ in the height dimension of the respective coordinate systems is given by

$$\rho_S = \frac{\lambda_c}{2\theta_{ext}} \text{ m} \quad \rho_G = \frac{\lambda_c \cos(\theta)}{2\theta_{ext}} \text{ m} \tag{11.9}$$

and spatial aliasing in the height dimension occurs at

$$\text{Alias}_S = \frac{\lambda_c}{2\Delta\theta} \text{ m} \quad \text{Alias}_G = \frac{\lambda_c \cos(\theta)}{2\Delta\theta} \text{ m} \tag{11.10}$$

where θ_{ext} is the extent of the aperture in elevation angle. For the GOTCHA data set, images formed using all eight ideal passes, $\Delta\theta = 0.18°$, $\theta_{ext} = 1.29°$, a slant plane elevation of $\theta = 45°$, have resolution in the height dimension of $\rho_S = 0.69$ m, and $\rho_G = 0.49$ m, and aliasing in the height dimension of $\text{Alias}_S = 4.97$ m, and $\text{Alias}_G = 3.51$ m. We note that since the actual flight paths vary in elevation and are not uniformly sampled in height, the PSF of the SAR imager will not, in general, have a sinc-like structure. As a result, sidelobes of the PSF function have non-negligible magnitude limiting the resolution beyond the limits derived under the uniform sampling assumption. Figure 11.3 shows a traditional volumetric SAR image of a Ford Taurus station wagon. The side of the car is parallel to the y axis, with the front of the car being at the most negative y value. The images show a profile of the car with the front portion of the car sloping up to the rear domed-shape section of the car. Nonlinear flight path imaging artifacts result in a PSF with strong sidelobes in slant plane height direction. This manifests itself as artifacts both above and below the car, degrading image quality.

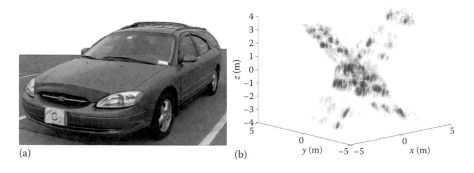

(a)   (b)

FIGURE 11.3
Photo of a Taurus station wagon and 3-D backprojection image formed using eight passes of GOTCHA data. (a) Photo of a Taurus station wagon. (b) 3-D GLRT SAR image. (Reproduced with permission from Ertin, E. et al., *IET Radar Sonar Navig.*, 4(3), 471, 2010. © IET 2010.)

11.4 Direct Approach to Sparsity-Regularized 3-D Construction

The direct approach to 3-D imaging presented in this section is applicable to arbitrary data collection scenarios including the sparse collections that is of interest to urban scenarios. This approach [1,5] assumes that the number of 3-D locations in which nonzero backscattering occurs is sparse in the 3-D reconstruction space and applies sparse reconstruction techniques to regularize the resulting inverse problem. The reconstruction problem is posed as an L_p regularized LS problem, where the regularizing term encourages sparse solutions. This L_p regularized LS imaging algorithm tries to maximize to fit an image domain scattering model to the measured k-space data using a regularization term that penalizes the number of nonzero voxels. The algorithm makes the assumption that the complex magnitude response of each scattering center is approximately constant over the aspect angles and across the radar frequency bandwidth. In contrast to the algorithms presented in Sections 11.5.1 and 11.5.2, which apply to collections consisting of approximately parallel apertures, the direct approach presented in this section does not make any a priori assumptions on the collection geometry. Let $C = \{(x_n, y_n, z_n)\}_{n=1}^{N}$ be the set of volumetric N grid points in image reconstruction space. Typically these locations are chosen on a uniform grid to facilitate fast computations. The $M \times N$ data measurement matrix is given by

$$\Phi = \left[e^{-j(k_{x,m}x_n + k_{y,m}y_n + k_{z,m}z_n)} \right] \tag{11.11}$$

where
    m indexes M measured k-space frequencies down rows
    n indexes the N coordinates in the volumetric grid C across columns

Under the assumption that scattering center amplitude is constant over the aspect angle extent and radar bandwidth considered, the measured (subaperture) data from the scattering center model can be written in matrix form as

$$y = \Phi x + n \tag{11.12}$$

where x is the N-dimensional vectorized 3-D image representing the scene in the view of the radar. The scene x is assumed to have nonzero complex reflectivity in the jth row if a scattering center is present at the location (x_j, y_j, z_j) and is zero in row j otherwise; the spatial vector x maps to the 3-D image, $I(x_j, y_j, z_j)$, through $I(x_j, y_j, z_j) = x_j$. Column i of Φ corresponds to the response of a unit amplitude scatterer placed at location (x_i, y_i, z_i). The noise vector n is an M dimensional i.i.d. circular complex Gaussian noise vector with zero mean and variance σ_n^2, and the measurement vector y is an M-dimensional vector of noisy k-space measurements. The reconstructed image, x, is the solution to the sparse optimization problem [16,17]

$$x^* = \arg\min_x \left\{ \|y - \Phi x\|_2^2 + \lambda \|x\|_p^p \right\} \tag{11.13}$$

where the L_p norm is denoted by $L \cdot L_p$, $0 < p \leq 1$, and λ is a sparsity penalty weighting parameter. Many algorithms exist for solving (11.13) or the constrained version of this problem when $p = 1$ (e.g., [18–21]), or in the more general case, when $0 < p \leq 1$ (e.g., [17,22]). We note that the model in (11.1) assumes the scatterers are anisotropic for each polarization, the image must be formed for each narrow-angle subaperture and polarization, and then combine noncoherently using (11.8). Alternative approaches for joint reconstruction of multiple images proposed in [6] may also be applied to simultaneously reconstruct all polarizations for each subaperture.

We form 3-D reconstructions of two spotlighted areas of the CSAR GOTCHA scene centered on a Toyota Camry parked stationary. For the L_p regularized LS reconstructions, 5° subapertures with no overlap were used, for a total of 72 subaperture images that are combined by (11.8). Reconstructed L_p regularized LS image voxels are spaced at 0.1 m in all three dimensions. The dimensions of the reconstructed tophat and Camry images in (x, y, z) dimensions are $[-2, 2) \times [-2, 2) \times [-2, 2)$ and $[5, 5) \times [-5, 5) \times [-5, 5)$ (all in meters), respectively. For the results shown in Figure 11.4, we chose $p = 1$ and $\lambda = 10$ experimentally to generate images that produce qualitatively good reconstructions. For each figure, only voxels whose energy is within the top 40 dB of image maximum are shown. To highlight vehicle structure, images are displayed using a smoothing interpolation with Gaussian kernel standard deviation of $\sigma = 0.1$ m. The apparent artifacts below the front of the car and to the side of the car in the 3-D view are scattering from an adjacent vehicle that is not completely removed by the spotlighting process.

FIGURE 11.4
(See color insert.) L_p regularized LS reconstructions of a civilian vehicle from the GOTCHA data set (3-D, side and top views, $p = 1$ and $\lambda = 10$). (Reproduced with permission from Austin, C.D., Ertin, E., and Moses, R.L., Sparse signal methods for 3-D radar imaging, *IEEE J. Select. Top. Signal Process.*, 5(3), 420, 2011. © 2011 IEEE.)

11.4.1 Algorithmic and Computational Considerations

In general, sparsity-regularized direct inversion of the 3-D problem requires considerable memory and computational resources. For our simulation results, we use the iterative majorization-maximization algorithm due to [17] to solve (11.13), which is applicable to the case $0 < p \leq 1$. In each iteration of the algorithm, a majorizing function $J(x, x^{(k)})$ that lower-bounds the original cost function is optimized, and the sequence of optimum solutions converges to the solution of the original optimization problem. The majorizing cost function $J(x, x^{(k)})$ is defined as

$$J\left(x, x^{(k)}\right) = \|y - \Phi x\|_2^2 + \lambda \sum_{i=1}^{N} h\left(x_i, x_i^{(k)}\right) \tag{11.14}$$

where
  superscript (k) is the iteration number
  subscript i is the component index of the vector x

and

$$h\left(x_i, x_i^{(k)}\right) = |x_i^{(k)}|^p + \text{Re}\left\{ px_i^{(k)} |x_i^{(k)}|^{p-2} \left(x_i - x_i^{(k)}\right)\right\} + \frac{1}{2}p|x_i^{(k)}|^{p-2}|x_i - x_i^{(k)}|^2 \tag{11.15}$$

It was shown in [17] that the sequence of solutions

$$x^{k+1} = \arg \min J(x, x^{(k)}) \tag{11.16}$$

$$= \left[\Phi^H \Phi + \frac{\lambda}{2} D(x^{(k)}) \right]^{-1} \Phi^H y \tag{11.17}$$

where $D(x^{(k)}) = \text{diag}\{p|x_i^{(k)}|^{p-2}\}$ converges to the solution of the L_p regularized LS problem. For the image reconstruction problems considered here, conjugate gradient (CG) algorithm [16] provides a computationally efficient method to perform the matrix inversion in (11.17). This results in an algorithm with two nested loops: The outer loop iterates on the solution $x^{(k)}$, and the inner loop is the CG loop that computes the matrix inversion.

Proper termination criteria for both loops have to be implemented to achieve a proper solution. Typically, the outer loop is terminated when the relative change in the original objective function is small between iterations, and the inner CG loop is terminated, when the relative magnitude of the residual becomes small.

In practice, the inner loop terminates after very few iterations for a Fourier operator, as in the case here. In our experience, this type of algorithm terminates faster than a split Bregman iteration approach [21] for the imaging problems considered here. Let $\Delta_x, \Delta_y, \Delta_z$ be the voxel spacings in the uniform rectilinear grid C. Then, the coordinates in the C consist of all permutations of (x, y, z) coordinates from the partitioned axes; and the set C defines a uniform 3-D grid on the scene. If, in addition, we assume that the k-space samples lie on a uniform 3-D frequency grid centered at the origin, the operation $\mathbf{\Phi}x$ can be implemented using the computationally efficient 3-D fast Fourier transform (FFT) operation. In general, the measured k-space samples are not on a uniform grid, and therefore the FFT cannot be used directly. A potential solution would be to first interpolate the k-space samples and then apply the standard FFT operation to the interpolated samples. A computationally efficient, alternative approach is to use Type 2 nonuniform FFTs (NUFFT)s as the operator $\mathbf{\Phi}$ to process data directly on the nonuniform k-space grid, as given in [23,24]. Nonuniform FFT algorithms perform an interpolation step for each evaluation of $\mathbf{\Phi}$; whereas, in interpolate then FFT approach, the interpolation occurs only once. As a result for an iterative algorithm where repeated evaluation of the $\mathbf{\Phi}$ is required, the latter approach might be preferred. For apertures scaled proportionally to the bandwidth to obtain uniform sampling in spatial domain, simple nearest neighbor interpolation is adequate.

For an iterative algorithm, like the one utilized here, only the data vector y as well as the current iterate of x and a gradient with the same dimension as x needs to be stored. For example, to reconstruct a scene of size of a single vehicle, $N = 256^3 \approx 1.7 \times 10^7$ voxels are required. Therefore, for reconstructing a vehicle, a minimum of three vectors of size 1.7×10^7 should be stored at double precision with complex valued variables. For algorithms that utilize a CG approach for matrix inversion, it is necessary to store a conjugate vector of dimension N as well. Whereas in a Newton–Raphson approach, it is necessary to store a Hessian of dimension $N \times N$. During each iteration of the recovery algorithm, the operator $\mathbf{\Phi}$ and its adjoint has to be computed. These operations can become very computationally expensive when the problem size grows and may result in a computationally intractable algorithm, unless a fast operator such as the FFT is employed. Specifically, since $\mathbf{\Phi}$ is an $M \times N$ matrix, direct multiplication of $\mathbf{\Phi}x$ requires MN multiplies and additions per evaluation. For the examples considered in this section, the average value of these M values is 10^5, requiring $MN \approx 10^{12}$ operations. In contrast after an initial interpolation step, an FFT implementation of $\mathbf{\Phi}x$ requires $\mathcal{O}(D^3 \log(D^3))$ operations, where D is the maximum number of samples across the image dimensions. For the imaging example, with $D = 256$, FFT implementation results in computational savings on the order of 10^3.

11.5 Multiple Elevation IFSAR

In this section, we consider a parametric method for high-resolution 3-D reconstruction for multipass circular SAR. Multibaseline generalizations of IFSAR have been considered for linear collection geometries in [25,26], here we consider parametric spectral estimation techniques for 3-D target reconstruction in circular SAR systems.

The input to the multipass IFSAR algorithm is a set of baseband modulated ground plane images $\left\{ I^B_{p,m}(x, y, 0) \right\}_p$ at a given subaperture centered ϕ_m of data collected at elevation pass θ_p. For notational simplicity, we drop common indices and denote the image sequence as $\{I_p(x, y)\}$ and consider without loss of generality $\phi_m = 0$. We consider a finite number of scattering centers at each resolution cell (x, y) and reparametrize the scene reflectivity $g(x, y, z)$ as

$$g_q(x, y) - g(x, y, h_q(x, y)) \tag{11.18}$$

where $g_q(x, y)$ denotes the reflectivity of the qth scattering center at location $(x, y, h_q(x, y))$. In general, the number of scatterers per resolution cell varies spatially and needs to be estimated from the data. Then, the ground plane image for elevation θ_i can be written as

$$I_p(x_l, y_l) = s(x, y) * \sum_{q=1}^{Q(x_l, y_l)} g(x, y) e^{-jk^0_x \tan(\theta_p) h_q(x, y)} e^{-jk^0_x x} \tag{11.19}$$

where

$s(x, y)$ is the inverse Fourier transform of the 2-D windowing function used in imaging,

$k^0_x = (4\pi f_c / c) \cos(\bar{\theta})$ is the center frequency used in baseband modulation,

$Q(x_l, y_l)$ is the number of scatterers in the resolution cell (x_l, y_l)

The ground locations $(x, y, h_q(x, y))$ and the image coordinates (x_l, y_l) are related through layover

$$x_l = x + \tan(\theta_p) h_q(x, y) \quad y_l = x_l \tag{11.20}$$

We assume that the difference between the elevation angles for the different passes is small enough so that for each elevation pass the scattering center $(x, y, h_q(x, y))$ falls to the same resolution cell (x_l, y_l). Without loss of generality, we consider $P + 1$ circular passes at elevation angles $\theta_p = \bar{\theta} + p\Delta\theta$ for $p = -P/2, \ldots, P/2$. Then the baseband images from each pass can be modeled as

$$I_p(x_l, y_l) = \sum_{q=1}^{Q(x_l, y_l)} \tilde{g}_q(x_l, y_l) e^{-jk_x^0 \tan(\theta_p) h_q(x_l, y_l)} \tag{11.21}$$

Using the approximation

$$\tan(\theta_p) \approx \tan(\bar{\theta}) + \frac{1}{\cos^2(\bar{\theta})} p\Delta\theta \tag{11.22}$$

we obtain the sum of complex exponential model

$$I_p(x_l, y_l) = \sum_{q=1}^{Q(x_l, y_l)} \tilde{g}_q(x_l, y_l) e^{-jk_q(x_l, y_l)p} \tag{11.23}$$

where the complex constant $e^{-j\tan(\bar{\theta})k_x^0 h_q(x_l, y_l)}$ is absorbed into the reflectivity $\tilde{g}_q(x_l, y_l)$ and the frequency factor k_q is given by

$$k_q(x_l, y_l) = \frac{4\pi f}{c \cos(\bar{\theta})} \Delta\theta h_q(x_l, y_l) \tag{11.24}$$

The frequency estimates $\hat{k}_q$ are then transformed into height estimates $\hat{h}_q$ using

$$\hat{h}_q = \hat{k}_q \frac{c \cos(\bar{\theta})}{4\pi f_c \Delta(\theta)} \tag{11.25}$$

Each estimated scattering location $(x_l, y_l, h_p(x_l, y_l))$ is then mapped to image coordinates by inverting Equation 11.20 at the mean elevation angle $\bar{\theta}$:

$$x = x_l - \tan(\bar{\theta}) h_q(x_l, y_l) y = y_l \tag{11.26}$$

Estimation of parameters of complex exponentials in noise is a fundamental problem in spectral estimation and array signal processing [27]. If the number of distinct complex exponentials is known, several high-resolution methods can be used to estimate the frequencies. Model order selection for sum of complex exponential model has been studied widely in the literature [28,29].

In general, the elevation spacing of the CSAR flight paths is not equispaced. As an example, for the GOTCHA CSAR data collection, experiment conducted by the Air Force Research Laboratory (AFRL) [12] features eight complete circular passes collected at nominal 45° elevation angle. Each pass has a planned (ideal) separation of $\Delta\theta = 0.18°$ in elevation. Actual flight paths differ from the planned paths, with elevation samples at $44.27°, 44.18°, 44.1°, 44.01°, 43.92°, 43.53°, 43.01°, 43.06°$. In addition, in each pass, the elevation varied as the aircraft circled the scene. Figure 11.5 shows the variation in elevation angle over 10° azimuth window.

The harmonic retrieval problem of multiple complex exponential terms from a short, nonuniformly sampled data set is nontrivial; common techniques such as MUSIC do not readily apply since there is only one snapshot from which to form a covariance matrix estimate.

In Sections 11.5.1 and 11.5.2, we review two methods of spectral estimation that are applicable to CSAR collections with approximately parallel

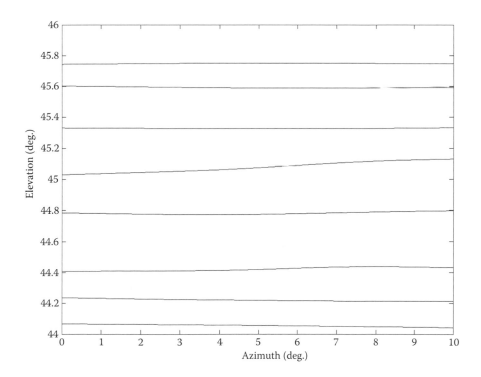

FIGURE 11.5
Variation in elevation angle over a 10° azimuth window. (Reproduced with permission from Ertin, E. et al., *IET Radar Sonar Navig.*, 4(3), 471, 2010. © IET 2010.)

flight paths. The first technique is based on interpolation of the CSAR data to a uniformly spaced vertical grid using sparsity regularized LS techniques and then use traditional spectral estimation methods such as ESPRIT [30] to the resulting oversampled covariance matrix. The second technique is based on the observation that a majority of the pixels contain only a single scattering center in the height dimension. This single scatterer per resolution voxel approximation to the system model enables computationally efficient maximum-likelihood estimation using nonuniform DFT of the data.

11.5.1 Sparsity-Regularized Interpolation Approach to m-IFSAR

One approach to solve the spectral estimation problem from nonuniformly sampled data is to first interpolate to uniform grid and then apply classical technical techniques that rely on uniformly spaced samples. In particular, algorithms known as *interpolated array methods* estimate the output of a uniform virtual array by interpolating the outputs of the actual array [31,32]. The simplest method of interpolation is to use linear interpolation to a regular grid. As we will see in the simulation example, simple linear interpolation leads to degraded performance. Here, we outline a new interpolated array method based on sparsity-regularized reconstruction of single-pulse image $I_\phi(r,h)$ obtained by coherently processing returns for multiple elevations $\{\theta_p\}$ at a single azimuth angle ϕ. The range (r) and height (h) are measured with respect to slant plane coordinates.

The relationship between the single-pulse image $I_\phi(r,h)$ and the projection of the scene reflectivity function $g_\phi(r,h)$ on the azimuth plane ϕ is given by

$$I_\phi(r,h) = \mathbf{\Phi}_\phi g_\phi(r,h) + n(r,h) \tag{11.27}$$

where
   $\mathbf{\Phi}_\phi$ is the convolution matrix of the system PSF corresponding to the elevation spacing at azimuth angle ϕ
   $n(r,h)$ represents noise and modeling errors

The deconvolution problem aims to reconstruct the scene reflectivity function $g_\phi(r,h)$ from the measured single Pulse image $I_\phi(r,h)$, given the knowledge of the convolution kernel $\mathbf{\Phi}_\phi$. The convolution kernel $\mathbf{\Phi}_\phi$ acts like a low-pass filter and does not have a bounded inverse; therefore, in the absence of any constraints on $g_\phi(r,h)$, the deconvolution problem is ill-posed [33]. Here, we again consider the majorization-maximization method reviewed in Section 11.4, which enforces sparsity in the reconstruction process. Specifically, we obtain an enhanced single-pulse image through minimization of

$$\hat{g}_\phi(r,h) = \arg\min_g \left\{ \|I_\phi - \Phi_\phi g\|_2^2 + \lambda \|g\|_p \right\} \tag{11.28}$$

The first term in the minimization captures the consistency of the reconstructions with the observed data through Equation 11.27; the second term in the minimization favors sparse solutions for g for $p \le 1$. The real-valued scalar parameter λ controls the relative weight of the two factors and is determined based on the expected signal-to-clutter ratio [14]. The unconstrained optimization problem in (11.28) can be efficiently solved using the iterative algorithm reviewed in Section 11.4.1 to reveal the location and amplitude and results in the location and amplitude of major scattering centers in the subaperture image. We note that a wide variety of methods have been proposed for the solution of (11.28) in the literature [34,35].

Fourier inversion of the enhanced image $\hat{g}_\phi(r,h)$ results in phase histories from a virtual array with equal spacing. Once interpolated histories are found, the sum of complex exponential model given in (11.23) is applicable and therefore spectral estimation methods can be employed to detect and resolve scatterers in the height dimension at each pixel (x_l, y_l) in the slant plane. Here, we employ a simple model order selection method based on thresholding the eigenvalues of the sample covariance matrix for the vector $\{l_p(x_l, y_l)\}_p$ to estimate the model order $Q(x_l, y_l)$. Using this model order, we then use the ESPRIT [30] method to estimate the frequencies $k_q(x_l, y_l)$ from the signal eigenvectors of the sample covariance matrix.

In the following, we illustrate the sparsity-regularized interpolated array method to multipass circular SAR data using the AFRL GOTCHA CSAR data set. Here we divided the data on 36 nonoverlapping windows of width $\Delta = 10°$ centered at $\phi_m \in \{0°, 10°, \ldots, 350°\}$ and used the entire 640 MHz bandwidth centered at 9.6 GHz for the single VV polarization. For each subaperture window, we created a virtual array of 32 uniformly sampled passes covering the same elevation range achieved by the SAR sensor in that subaperture. Using the sparsity-regularized interpolation method, we interpolated the phase history data collected at nonlinear flight paths to the data collected at virtual array geometry as shown in Figure 11.6. In constructing the single-pulse images, we used the prior knowledge about the target dimensions to restrict the height and range support of the single-pulse image to 5 m. For each of the virtual 32 passes and for each of the subapertures, ground plane images are constructed using classic backprojection. Next, we applied the ESPRIT-based parametric spectral estimation method to all pixels whose amplitude is within the 20 dB of image maximum to construct three-dimensional points representing observed strong scattering mechanisms. The 3-D point clouds from each subaperture window are rotated and overlayed to a common reference frame.

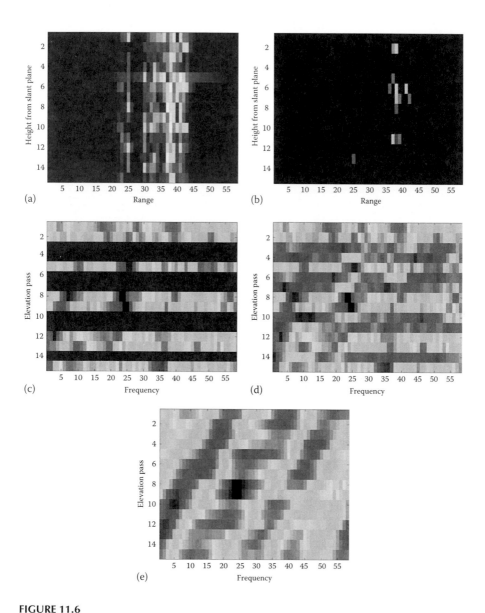

FIGURE 11.6
Single-pulse image from eight nonuniform spaced elevation passes: (a) original, (b) sparsity-regularized enhancement and the corresponding phase histories, (c) original, (d) linearly interpolated, and (e) sparsity-regularized interpolation for $\lambda = 0.1$ and $p = 1.0$. (Reproduced with permission from Ertin, E. et al., *IET Radar Sonar Navig.*, 4(3), 471, 2010. © IET 2010.)

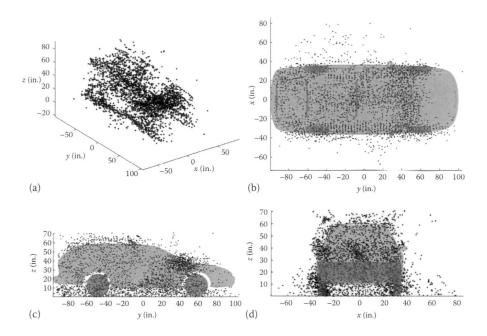

FIGURE 11.7

Three-dimensional reconstruction of a Taurus station wagon using multipass circular SAR data. (a) 3-D perspective, (b) top view, (c) side view, and (d) front view. (Reproduced with permission from Ertin, E. et al., *IET Radar Sonar Navig.*, 4(3), 471, 2010. © IET 2010.)

Figure 11.7 shows the resulting reconstruction overlayed with the CAD model of the station wagon. We observe that point cloud encompasses the CAD model and strong returns from the ground plane-side panel (double-bounce mechanism) and the curved surfaces (single-bounce mechanism) are clearly visible. We note that the car was parked on smooth asphalt surface with no closeby cars or buildings that would have introduced multipath reflections.

11.5.2 DFT Peak Detection Approach for m-IFSAR

For many vehicles, the horizontal imaging geometry ensures that only a few scatterers are present in each image pixel (x_l, y_l). Specifically, in a recent study using CSAR simulated X-band data of vehicles [3], the estimated model order was $Q = 1$ in a large majority of cases. Thus, the complex exponential signal model considered in (11.23) is sparse, with typically only 1 scattering center in the height dimension. This presents a computational advantage, because for the single-exponential case, a maximum likelihood estimator of its frequency in white measurement noise is given by the peak of the Fourier transform of the data, and this Fourier transform is easy to compute even in the case of nonuniform samples. We can thus estimate for each pixel (x_l, y_l) the location $k_q(x_l, y_l)$ as the peak of the Fourier transform of the nonuniformly sampled values for that pixel and calculate the height using (11.26). The complex amplitude of the Fourier transform at the peak provides an estimate of the amplitude of the scattering center.

We note that using this multipass IFSAR approach the original 3-D problem has been converted to a set of 2-D and 1-D processing steps. First, 2-D images are formed for each azimuth subaperture and each elevation angle. Then, 1-D processing is applied to each $P \times 1$ vector obtained by stacking the set of P elevation images and selecting the P values at a pixel location of interest. The processing reduction is afforded by the particular structure of the sparse measurement geometry provided by CSAR collections.

In general, for the multipass IFSAR approach, reconstructed scattering centers are not constrained to lay on a grid in the height dimension. To have a volumetric image defined on a uniform grid, data should be first interpolated to a grid with uniform voxel spacing in each dimension. Empirically we have found that a Gaussian kernel with standard deviation of grid spacing provides visually appealing results when used for interpolation.

Figure 11.8 shows the results of the multipass IFSAR approach applied to the CSAR GOTCHA data for Toyota Camry, previously used to illustrate the performance of the direct sparsity-regularized approach shown in Figure 11.4. The top 20 dB points are shown. The results are shown for VV polarization. Scattering is assumed to be above the ground plane in calculations; so, unlike in the L_p regularized LS reconstruction, there are no nonzero voxels below the vehicle. As in the L_p regularized LS reconstruction, a set of 72 subaperture image sets were formed, each with 5° azimuth extent, and the image-domain subaperture reconstructions were combined using (11.8).

The multipass IFSAR reconstructions appear to be more filled than the L_p regularized LS reconstructions. This is due to relatively lower downrange and crossrange resolution of the Fourier imaging techniques employed. The detriment of spatial spread is application specific. For visualization, this spreading may be desirable, favoring the multipass IFSAR reconstructions.

FIGURE 11.8
(See color insert.) Wide-angle multipass IFSAR reconstructions of a civilian vehicle from the GOTCHA data set (3-D, side and top views). (Reproduced with permission from Austin, C.D., Ertin, E., and Moses, R.L., Sparse signal methods for 3-D radar imaging, *IEEE J. Select. Top. Signal Process.*, 5(3), 420, 2011. © 2011 IEEE.)

In other applications, such as ATR, fine 3 D resolution features may be desirable, and L_p regularized LS reconstructions may be superior.

11.6 Practical Considerations: Autofocus and Registration

In this chapter, we have reviewed several methods for 3-D target reconstruction using sparsity assumption on the target support. All these methods can provide high-resolution volumetric imagery from multipass circular SAR collections, provided that data from the multiple passes are properly registered and phase coherent. While many system errors have decreased with improved collection platform reference systems and more accurate clocks, remaining system errors may still degrade imagery. Many system errors, such as platform position errors, manifest as phase errors, be it pulse to pulse or pass to pass. An oscillator scale factor error may induce different spatial translations of 2-D imagery. Amplifier thermal effects or aspect-dependent antenna gain may adversely affect gain across passes. The strong requirement for phase coherence in SAR dictates these errors be very small. For instance, Jakowatz et al. [36] suggests no more than $\pi/4$ radians of phase error, equivalently small relative range error, be tolerated for acceptable autofocus of pulse-to-pulse errors. Autofocus methods aim to estimate these errors from the radar data and compensate them. While several effective methods exist for within-pass collections, such as Phase Gradient Autofocus [36], other methods are needed for 3-D autofocus in situations when wide-aspect persistent scatterers are not available for prominent point autofocus [37]. Recently, interpass 3-D autofocus methods have been proposed [38–40]. These autofocus methods can be formulated as a set of linear filters in the spatial frequency domain applied to each pass p

$$H_p(k_x, k_y) = e^{c_p + j\theta_p + j2\pi(x_p k_x + y_p k_y)} \tag{11.29}$$

where
  c_p is a constant gain term
  θ_p is a constant phase term
  (x_p, y_p) is a linear phase term in the frequency domain or equivalently a
    (x_p, y_p) translation in the image domain in accordance with the Fourier
    shift theorem

The image domain filter is expressed as $h_p(x, y) = \mathcal{F}^{-1}\{H_p(k_x, k_y)\}$. The application of the registration filters can be performed in spatial domain as

$$\bar{I}_p(x, y; \Theta) = I_p(x, y) * h_p(x, y, \Theta) \tag{11.30}$$

The autofocus problem may be posed as a joint optimization problem. Considering P elevation passes, the registration filter of Equation 11.29 specifies a parameter vector of

$$\Theta = \begin{bmatrix} c_1 & \cdots & c_{P-1} & \theta_1 & \cdots & \theta_{P-1} & (x_1, y_1) & \cdots & (x_{P-1}, y_{P-1}) \end{bmatrix}^T \quad (11.31)$$

where the registration parameters for one of the P passes is held constant or fixed to some ground calibration feature to avoid ambiguity. The autofocus problem is to find Θ^* that jointly minimizes some appropriate objective function $C(\{\bar{I}_p(x, y, \Theta)\})$ that evaluates coherency of the multipass data:

$$\Theta^* = \arg\min_{\Theta} C\left(\{\bar{I}_p(x, y, \Theta)\}\right) \quad (11.32)$$

Interpass 3-D autofocus methods in the literature [38–40] differ in their optimization criteria. Kragh [39] considers Rényi entropy over the normalized voxel energy of the volumetric images as the optimization metric for 3-D autofocus

$$S_\alpha(g) = \frac{1}{1-\alpha} \log \sum_{n=1}^{N} q_n^\alpha \quad (11.33)$$

where $\alpha > 0$ is the entropy order parameter. To compute entropy over a volumetric image, let $g = (g_1, \ldots, g_N)$ be a vector of image samples from $\bar{I}(x, y, z; \Theta)$ obtained from focused ground plane imagery using (11.2), and let $q_n = |g_n|^2 / \sum_n |g_n|^2$ be the normalized voxel energy. Evaluating the Rényi entropy for $\alpha = 2$ gives the quadratic entropy criterion of the image used for autofocus:

$$C\left(\bar{I}_p(x, y; \Theta)\right) = 2\log \sum_{n=1}^{N} |g_n|^4 + 2\log \sum_{n=1}^{N} |g_n|^2 \quad (11.34)$$

Elkin [38] considers the LS optimization criteria as sum over the image using the model order assumption of one for the height dimension to obtain a single scatterer at height $z^*(x, y)$ per image pixel (x, y):

$$C\left(\{\bar{I}_p(x, y; \Theta)\}\right) = \sum_{x,y} \sum_{p} |I_p * h_p(x, y, \Theta)|^2$$

$$- \frac{1}{P} \left| \sum_{p} (I_p * h_p(\Theta))_{(x,y)} e^{-jk_x^0 \tan(\theta_p) z^*(x,y)} \right|^2 \quad (11.35)$$

Boss et al. [40] proposed a 3-D autofocus method based on maximizing coherence factor computed over the dominant scatterers in an image scene given by

$$C\left(\{\bar{I}_p(x, y; \Theta)\}\right) = \sum_{x,y} \frac{\left|\sum_p (I_p * h_p(\Theta))_{(x,y)} e^{-jk_x^0 \tan(\theta_p)z^*(x,y)}\right|^2}{P \sum_p \left|I_p * h_p(x, y, \Theta)\right|^2} \tag{11.36}$$

We note that for a typical registration problem the total energy of the registered images is constant; therefore, the LS criterion simplifies to

$$\Theta^* = \arg\max_{\Theta} \sum_{x,y} \left|\sum_p \left(I_p * h_p(\Theta)\right)_{(x,y)} e^{-jk_x^0 \tan(\theta_p)z^*(x,y)}\right|^2 \tag{11.37}$$

which is equivalent to maximizing the DFT amplitudes of the registered images (or equivalently summing up the numerator of the coherence factor). As a result, LS registration emphasizes coherence of higher amplitude pixels, which could be preferable in certain applications, whereas coherence factor metric provides the flexibility to target a set of pixels (e.g., pixels on target), maximizing coherence across those pixels with equal weight.

References

1. E. Ertin, L. C. Potter, and R. L. Moses, Enhanced imaging over complete circular apertures, in *Fortieth Asilomar Conference on Signals, Systems and Computers (ACSSC 06)*, October 29–November 1, 2006, pp. 1580–1584.
2. E. Ertin, C. D. Austin, S. Sharma, R. L. Moses, and L. C. Potter, GOTCHA experience report: Three-dimensional SAR imaging with complete circular apertures, in *Algorithms for Synthetic Aperture Radar Imagery XIV*, E. G. Zelnio and F. D. Garber, Eds., Orlando, FL, April 9–13, 2007, SPIE Defense and Security Symposium.
3. E. Ertin, R. L. Moses, and L. C. Potter, Interferometric methods for 3-D target reconstruction with multi-pass circular SAR, *IET Radar, Sonar and Navigation*, 4 (3), 464–473, 2010.
4. C. D. Austin and R. L. Moses, Wide-angle sparse 3D synthetic aperture radar imaging for nonlinear flight paths, in *IEEE National Aerospace and Electronics Conference (NAECON) 2008*, July 16–18, 2008, pp. 330–336 .

5. C. D. Austin, E. Ertin, and R. L. Moses, Sparse multipass 3D SAR imaging: Applications to the GOTCHA data set, in *Algorithms for Synthetic Aperture Radar Imagery XVI*, E. G. Zelnio and F. D. Garber, Eds., Orlando, FL, April 13–17, 2009, SPIE Defense and Security Symposium.

6. N. Ramakrishnan, E. Ertin, and R. Moses, Enhancement of coupled multichannel images using sparsity constraints, *IEEE Transactions on Image Processing*, 19 (8), 2115–2126, August 2010.

7. C. D. Austin, E. Ertin, and R. L. Moses, Sparse signal methods for 3-D radar imaging, *IEEE Journal of Selected Topics in Signal Processing*, 5 (3), 408–423, 2011.

8. K. E. Dungan and L. C. Potter, 3-D imaging of vehicles using wide aperture radar, *IEEE Transactions on Aerospace and Electronic Systems*, 47 (1), 187–199, 2011.

9. A. Budillon, A. Evangelista, and G. Schirinzi, Three-dimensional SAR focusing from multipass signals using compressive sampling, *IEEE Transactions on Geoscience and Remote Sensing*, 49 (1), 488–499, 2011.

10. X. Zhu and R. Bamler, Super-resolution power and robustness of compressive sensing for spectral estimation with application to spaceborne tomographic SAR, *IEEE Transactions on Geoscience and Remote Sensing*, 50 (1), 247–258, January 2012.

11. X. Zhu and R. Bamler, Demonstration of super-resolution for tomographic SAR imaging in urban environment, *IEEE Transactions on Geoscience and Remote Sensing*, 50 (8), 3150–3157, August 2012.

12. C. H. Casteel, L. A. Gorham, M. J. Minardi, S. Scarborough, and K. D. Naidu, A challenge problem for 2D/3D imaging of targets from a volumetric data set in an urban environment, in *Algorithms for Synthetic Aperture Radar Imagery (Proc. SPIE Vol. 6568)*, E. G. Zelnio and F. D. Garber, Eds., Orlando, FL, April 9–13, 2007. SPIE Defense and Security Symposium.

13. C. V. Jakowatz and P. A. Thompson, A new look at spotlight mode synthetic aperture radar as tomography: Imaging 3D targets, *IEEE Transactions on Image Processing*, 4 (5), 699–703, May 1995.

14. R. Moses, L. Potter, and M. Çetin, Wide angle SAR imaging, in *Algorithms for Synthetic Aperture Radar Imagery XI (Proc. SPIE Vol. 5427)*, E. G. Zelnio, Ed., Orlando, FL, April 2004. SPIE Defense and Security Symposium.

15. R. L. Moses and L. C. Potter, Noncoherent 2D and 3D SAR reconstruction from wide-angle measurements, in *13th Annual Adaptive Sensor Array Processing Workshop*, MIT Lincoln Laboratory, Lexington, MA, June 2005.

16. M. Cetin and W. C. Karl, Feature enhanced synthetic aperture radar image formation based on nonquadratic regularization, *IEEE Transactions on Image Processing*, 10, 623–631, April 2001.

17. T. Kragh and A. Kharbouch, Monotonic iterative algorithms for SAR image restoration, in *IEEE 2006 International Conference on Image Processing*, Atlanta, GA, October 2006, pp. 645–648.

18. M. Figueiredo, R. Nowak, and S. Wright, Gradient projection for sparse reconstruction: Application to compressed sensing and other inverse problems, *IEEE Journal of Selected Topics in Signal Processing*, 1 (4), 586–597, December 2007.

19. A. Beck and M. Teboulle, A fast iterative shrinkage-thresholding algorithm for linear inverse problems, *SIAM Journal of Imaging Sciences*, 2 (1), 183–202, 2009.

20. I. Daubechies, M. Defrise, and C. D. Mol, An iterative thresholding algorithm for linear inverse problems with a sparsity constraint, *Communications on Pure and Applied Mathematics*, 57 (11), 1413–1467, 2004.

21. T. Goldstein and S. Osher, The split Bregman method for L1-regularized problems, *SIAM Journal on Imaging Sciences*, 2 (2), 323–343, 2009.
22. R. Saab, R. Chartrand, and O. Yilmaz, Stable sparse approximations via nonconvex optimization, in *33rd International Conference on Acoustics, Speech, and Signal Processing (ICASSP)*, March 30–April 4, 2008, Las Vegas, NV, 2008.
23. L. Greengard and J. Y. Lee, Accelerating the nonuniform fast Fourier transform, *SIAM Review*, 43 (3), 443–454, 2004.
24. J. Fessler and B. Sutton, Nonuniform fast Fourier transforms using min-max interpolation, *IEEE Transactions on Signal Processing*, 51 (2), 560–574, February 2003.
25. S. Xiao and D. C. Munson, Spotlight-mode SAR imaging of a three-dimensional scene using spectral estimation techniques, in *Proceedings of the IEEE International Geoscience and Remote Sensing Symposium (IGARSS 98)*, vol. 2, Seattle, WA, 1998, pp. 624–644.
26. F. Gini and F. Lombardini, Multibaseline cross-track SAR interferometry: A signal processing perspective, *IEEE AES Magazine*, 20 (8), 71–93, August 2005.
27. P. Stoica and R. Moses, *Spectral Estimation of Signals*, Prentice Hall, Upper Saddle River, NJ, 2005.
28. M. Wax and T. Kailath, Detection of signals by information theoretic criteria, *IEEE Transactions on Acoustics, Speech and Signal Processing*, 33, 387–392, April 1985.
29. D. N. Lawley, Tests of significance of the latent roots of the covariance and correlation matrices, *Biometrica*, 43, 128–136, 1956.
30. R. Roy and T. Kailath, ESPRIT—Estimation of signal parameters via rotational invariance techniques, *IEEE Transactions on Acoustics, Speech and Signal Processing*, 37 (7), 984–995, 1989.
31. B. Friedlander, The root-MUSIC algorithm for direction finding in interpolated arrays, *Signal Processing*, 30 (1), 15–19, January 1993.
32. F. Bordoni, F. Lombardini, F. Gini, and A. Jacabson, Multibaseline cross-track SAR interferometry using interpolated arrays, *IEEE Transactions on Aerospace and Electronic Systems*, 41 (4), 1472–1481, October 2005.
33. H. Stark, Ed., *Image Recovery: Theory and Application*, Academic Press, Orlando, FL, 1987.
34. S. Wright, R. Nowak, and M. Figueiredo, Sparse reconstruction by separable approximation, *IEEE International Conference on Acoustics, Speech and Signal Processing (ICASSP 2008)*, Las Vegas, NV, pp. 3373–3376, March 2008.
35. K. Herrity, R. Raich, and A. Hero, Blind deconvolution for sparse molecular imaging, *IEEE International Conference on Acoustics, Speech and Signal Processing (ICASSP 2008)*, Las Vegas, NV, pp. 545–548, March 2008.
36. C. V. Jakowatz Jr., D. E. Wahl, P. H. Eichel, D. C. Ghiglia, and P. A. Thompson, *Spotlight-Mode Synthetic Aperture Radar: A Signal Processing Approach*, Kluwer Academic Publishers, Boston, MA, 1996.
37. M. Ferrara, J. A. Jackson, and C. Austin, Enhancement of multi-pass 3D circular SAR images using sparse reconstruction techniques, in *Algorithms for Synthetic Aperture Radar Imagery XVI*, E. G. Zelnio and F. D. Garber, Eds., Orlando, FL, April 13–17, 2009, SPIE Defense and Security Symposium.
38. F. L. Elkin, Autofocus for 3D imaging, in *Algorithms for Synthetic Aperture Radar Imagery (Proc. SPIE Vol. 6970)*, E. G. Zelnio and F. D. Garber, Eds., Orlando, FL, April 2008.

39. T. J. Kragh, Minimum-entropy autofocus for three-dimensional SAR imaging, in *Algorithms for Synthetic Aperture Radar Imagery XVI (Proc. SPIE Vol. 7337)*, E. G. Zelnio and F. D. Garber, Eds., Orlando, FL, April 2009.

40. N. Boss, E. Ertin, and R. Moses, Autofocus for 3d imaging with multipass SAR feature extraction algorithm for 3D scene modeling and visualization using monostatic SAR, in *Algorithms for Synthetic Aperture Radar Imagery XVII (Proc. SPIE Vol. 7699)*, E. G. Zelnio and F. D. Garber, Eds., Orlando, FL, April 2010.

12

Compressive Sensing for MIMO Urban Radar

Yao Yu, Athina Petropulu, and Rabinder N. Madan

CONTENTS

ABSTRACT Benefiting from the developments in MIMO communication systems, MIMO radars have received considerable attention in recent years. Unlike phased-array radars, in which all transmit antennas emit a scaled version of the same waveform, a MIMO radar system transmits multiple independent or correlated waveforms from its antennas. This enables improved target angle and Doppler resolution, lower minimum detectable velocity, and lower probability of intercept. These advantages render MIMO radars a highly desirable technology for military and civilian applications, such as homeland defense, medicine, and marine science.

This chapter presents a class of MIMO radars that use compressive sensing (CS), namely, CS-MIMO radars, which were initially proposed in [9–12]. CS-MIMO radars capitalize on the sparsity of the target returns in the target space, to achieve the high resolution of MIMO radars but with significantly fewer measurements, or significantly improved performance for the same number of measurements. A reduction of the volume of required data translates into shorter acquisition time, and in bandwidth and power savings.

Both military and civilian applications are increasingly interested in networked radars, in which the antennas are placed on sensors and communicate their findings to a fusion center via a wireless link. Reliable surveillance, however, requires collection, communication, and fusion of vast amounts of data from a range of sensors, thus requiring high-communication overhead in terms of bandwidth and transmission power. CS-MIMO radars appear as good candidates for networked radars as they substantially reduce the amount of data measured and transmitted by each sensor through the network.

12.1 Introduction

A MIMO radar system consists of multiple transmit and receive antennas, and is advantageous in two different configurations, namely, widely separated antennas [1–3] and collocated antennas [4–6]. In the widely separated scenario, the transmit antennas are located far apart from each other relative to their distances to the target. The MIMO radar system transmits independent probing signals from its antennas that follow independent paths and thus each target return carries independent information about the target. Joint processing of the target returns results in diversity gain, which enables the MIMO radar to achieve improved target parameters estimation. In the collocated scenario, transmit and receive antennas are located close to each other relative to the target so that all antennas view the same aspect of the target. In this configuration, the phase differences induced by transmit and receive antennas can be exploited to form a long virtual array, with the number of elements equal to the product of the numbers of transmit and receive antennas (nodes). As a result, the MIMO radars enable superior resolution in terms of direction of arrival (DOA) estimation and parameter identification.

As in MIMO radars, in CS-MIMO radars, the transmit antennas transmit uncorrelated waveforms. Each receive antenna employs compressive sampling to obtain a small number of samples of the target returns and subsequently forwards the samples to a fusion center. For a small number of targets which are sparsely located in the search space, based on the samples forwarded, the fusion center can obtain target information by formulating and solving a sparse signal estimation problem.

CS-MIMO radars achieve the superior resolution of MIMO radars with far fewer samples. This advantage implies shorter acquisition time, and also bandwidth and power savings during the communication phase between the receive antennas and the fusion center. These savings make CS-MIMO radars good candidates for surveillance in power-limited, infrastructureless scenarios, for example, when the antennas are placed on battery-operated mobile nodes.

12.1.1 Outline of the Chapter

This chapter considers colocated MIMO radars. Section 12.2 presents the problem formulation, the assumptions, and the solution under ideal conditions. Section 12.3 discusses some issues that would be critical in a practical setting, such as computational complexity, sensitivity to clutter, and frequency/phase synchronization. Section 12.4 describes some optimal design techniques for CS-MIMO radars, which enable further performance improvement. Finally, Section 12.5 presents the application of CS-MIMO technique to through-the-wall radar.

12.2 Colocated CS-MIMO Radars

In the following, we consider the extraction of target information, that is, DOA, range, and speed, for CS-MIMO radars with colocated antennas.

12.2.1 Problem Formulation and Solution

Let us consider a MIMO radar system consisting of M_t transmit (TX) and N_r receive (RX) antennas, which are uniformly distributed on a disk (see Figure 12.1). The radius of the disk is small as compared with the distance between the antennas and the targets. In the far field of the antennas, there are K slowly moving targets. Without loss of generality, the antennas and the targets are all on the same plane. Let (r_i^t, α_i^t) and (r_i^r, α_i^r) denote the locations, in polar coordinates, of the ith TX and RX antenna, respectively. All nodes are time and phase synchronized so that the received signal at all RX antennas can be coherently combined to extract target information. For simplicity, we assume that the environment is clutter-free. The presence of clutter will be discussed in Section 12.3.3.

Since the inter-antenna distances are much smaller than the antenna-target distances, the targets will be treated as point targets, and will be represented by their center of gravity. The kth target is at azimuth angle θ_k and moves with constant radial speed v_k. Its range equals $d_k(t) = d_k(0) - v_k t$, where $d_k(0)$ is the distance between this target and the origin at time zero.

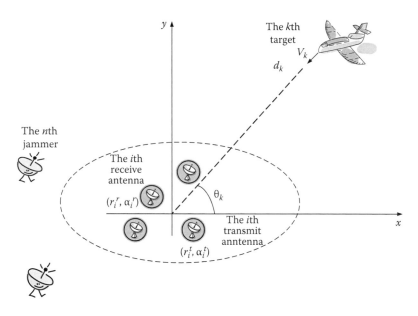

FIGURE 12.1
Colocated MIMO radars.

Under the far-field assumption, that is, $d_k(t) \gg r_i^{t/r}$, the distance between the ith TX/RX antenna and the kth target d_{ik}^t / d_{ik}^r can be approximated as

$$d_{ik}^{t/r}(t) \approx d_k(t) - \eta_i^{t/r}(\theta_k) = d_k(0) - v_k t - \eta_i^{t/r}(\theta_k) \qquad (12.1)$$

where $\eta_i^{t/r}(\theta_k) = r_i^{t/r} \cos(\theta_k - \alpha_i^{t/r})$.

Each TX antenna emits a pulse of duration T_p and pulse repetition interval (PRI) T. Let $x_i(t)e^{j2\pi ft}$ denote the continuous-time waveform during the pulse duration, where $x_i(t)$ is a narrowband signal and f is the carrier frequency.

Due to reflections by the targets, the demodulated baseband signal received by the lth antenna during the mth pulse equals

$$z_{lm}(t) = \sum_{k=1}^{K} \sum_{i=1}^{M_t} \beta_k x_i \left(t - (m-1)T - \frac{d_{ik}^t(t) + d_{lk}^r(t)}{c} \right) e^{-j2\pi f \frac{d_{ik}^t(t) + d_{lk}^r(t)}{c}} + n_{lm}(t)$$

$$\approx \sum_{k=1}^{K} \sum_{i=1}^{M_t} \beta_k x_i \left(t - (m-1)T - \frac{2d_k(0)}{c} \right) e^{-j2\pi f \frac{d_{ik}^t(t) + d_{lk}^r(t)}{c}} + n_{lm}(t)$$

$$l = 1, \ldots, M_r, \ m = 1, \ldots, N_p \qquad (12.2)$$

where $\{\beta_k, k=1,\ldots,K\}$ are the target reflectivities and $n_{lm}(t)$ represents interference, which is assumed to be independent of the transmit waveforms. The interference could be due to thermal noise or jamming.

Due to the point target assumption, all RX antennas see the same aspect of each target, that is, the β_ks are the same for all TX/RX antenna pairs. In the aforementioned equation, the time delay due to the kth target does not change with time over the pulse duration, and is the same for all antennas. This approximation is enabled by the assumption of slowly moving targets and colocated antennas.

The lth RX antenna compressively samples the return signal to obtain M samples per pulse (refer to Figure 12.2 for a schematic of the receiver). Let L denote the number of T_s-spaced samples (fast time) of the transmitted waveforms within one pulse, that is, $T_p = LT_s$. The CS receiver, shown in Figure 12.2, premultiplies the received signal in the continuous time domain by a set of functions $\Phi_l^i(t)$, $i = 1,\ldots,M$; then the result is integrated and sampled to produce M samples. In Figure 12.2, the functions $\Phi_l^i(t), i = 1,\ldots,M$ are typically formed by converting Gaussian or Bernoulli discrete-time sequences into continuous-time signals.

The samples collected in the aforementioned fashion by the lth antenna, during the mth pulse, can be expressed in vector form as

$$\mathbf{r}_{lm} \triangleq \Phi_l[z_{lm}((m-1)T + 0T_s),\ldots,z_{lm}((m-1)T + (L+\tilde{L}-1)T_s)]^T$$

$$= \sum_{k=1}^{K} \beta_k e^{j2\pi p_{lmk}} \Phi_l \mathbf{D}(f_k)\mathbf{C}_{\tau_k}\mathbf{X}\mathbf{v}(\theta_k) + \Phi_l \mathbf{n}_{lm} \qquad (12.3)$$

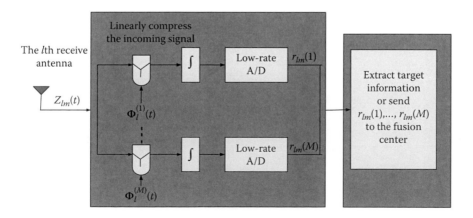

FIGURE 12.2
The compressive receiver. (From Yu, Y. et al., MIMO radar using compressive sampling, *IEEE J. Select. Top. Signal Process.*, 4(1), 146, February 2010. With permission. © 2010 IEEE.)

where

$\mathbf{\Phi}_l \in \mathcal{C}^{M \times (L+\tilde{L})}$ is a matrix whose ith row consists of $(L + \tilde{L})$ T_s-spaced samples of the function $\Phi_l^i(t)$

$p_{lmk} = -2d_k(0)f/c + \eta_l^r(\theta_k)f/c + f_k(m-1)T$, where $f_k = 2v_kf/c$ is the Doppler shift induced by the kth target

$\mathbf{X} \in \mathcal{C}^{L \times M_t}$ contains the waveforms of the M_t TX antennas as its columns

$\mathbf{v}(\theta_k) = [e^{j\frac{2\pi f}{c}\eta_1^t(\theta_k)}, \ldots, e^{j\frac{2\pi f}{c}\eta_{M_t}^t(\theta_k)}]^T \in \mathcal{C}^{M_t}$ is the transmit steering vector

$\tau_k = \lfloor 2d_k(0)/cT_s \rfloor$ and $\mathbf{C}_{\tau_k} = [\mathbf{0}_{L \times \tau_k}, \mathbf{I}_L, \mathbf{0}_{L \times (\tilde{L}-\tau_k)}]^T \in \mathcal{C}^{(L+\tilde{L}) \times L}$. Here, we assume that the target returns fall completely within the sampling window of length $(L+\tilde{L})T_s$, and that T_s is small enough so that the rounding error in the delay is small, that is, $x_i(t - \tau_k) \approx x_i(t - \lfloor 2d_k(0)/cT_s \rfloor)$

$\mathbf{D}(f_k) = \text{diag}\{[e^{j2\pi f_k 0 T_s}, \ldots, e^{j2\pi f_k(L+\tilde{L}-1)T_s}]\} \in \mathcal{C}^{(L+\tilde{L}) \times (L+\tilde{L})}$. For slowly moving targets, the Doppler shift within a pulse can be ignored and so $\mathbf{D}(f_k)$ is reduced to an identity matrix

$\mathbf{n}_{lm} \in \mathcal{C}^{(L+\tilde{L}) \times 1}$ is the interference at the lth receiver during the mth pulse, which includes a jammer's signal and thermal noise

Let us discretize the angle, speed, and range space on a fine grid, that is, respectively, $[\tilde{a}_1, \ldots, \tilde{a}_{N_a}]$, $[\tilde{b}_1, \ldots, \tilde{b}_{N_b}]$, and $[\tilde{c}_1, \ldots, \tilde{c}_{N_c}]$, as shown in Figure 12.3. Let the grid points be arranged first angle-wise, then range-wise, and finally speed-wise to yield the grid points $(a_n, b_n, c_n), n = 1, \ldots, N_a N_b N_c$. Through this ordering, the grid point $(\tilde{a}_{n_a}, \tilde{b}_{n_b}, \tilde{c}_{n_c})$ is mapped to point (a_n, b_n, c_n), where $n = (n_b - 1)n_a n_c + (n_c - 1)n_a + n_a$. We assume that the discretization step is small enough so that each target falls on some angle–speed–range grid point. Then, (12.3) can be rewritten as

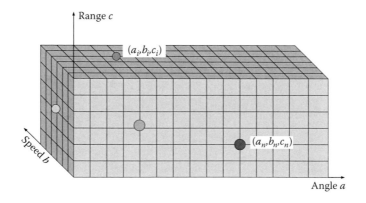

FIGURE 12.3
Sparse target representation in the search space.

$$\mathbf{r}_{lm} = \mathbf{\Phi}_l \left(\sum_{n=1}^{N} s_n e^{j2\pi q_{lmn}} \mathbf{D}\left(\frac{2b_n f}{c}\right) \mathbf{C}_{\lfloor \frac{2c_n}{cT_s} \rfloor} \mathbf{X}\mathbf{v}(a_n) + \mathbf{n}_{lm} \right) \tag{12.4}$$

where

$$s_n = \begin{cases} \text{reflection coefficient of the target,} & \text{if there is a target at } (a_n, b_n, c_n) \\ 0, & \text{if there is no target at } (a_n, b_n, c_n) \end{cases}, \tag{12.5}$$

$$N = N_a N_b N_c,$$

and

$$q_{lmn} = \frac{-2c_n f}{c} + \frac{\eta_l^r(a_n)f}{c} + \frac{2b_n f(m-1)T}{c}. \tag{12.6}$$

We can rewrite the aforementioned equation in a more compact form as

$$\mathbf{r}_{lm} = \mathbf{\Theta}_{lm}\mathbf{s} + \mathbf{\Phi}_l \mathbf{n}_{lm} \tag{12.7}$$

where

$$\mathbf{s} = [s_1, \dots, s_N]^T, \tag{12.8}$$

and

$$\mathbf{\Theta}_{lm} = \mathbf{\Phi}_l \underbrace{[e^{j2\pi q_{lm1}} \mathbf{D}(2b_1 f/c)\mathbf{C}_{\lfloor \frac{2c_1}{cT_s} \rfloor} \mathbf{X}\mathbf{v}(a_1), \dots, e^{j2\pi q_{lmN}} \mathbf{D}(2b_N f/c)\mathbf{C}_{\lfloor \frac{2c_N}{cT_s} \rfloor} \mathbf{X}\mathbf{v}(a_N)]}_{\mathbf{\Psi}_{lm}}$$

$$\in \mathcal{C}^{M \times N}. \tag{12.9}$$

If the number of targets is small, then **s** will contain mostly zero elements. The location of the nonzero elements will correspond to target angle speed and range.

The receive antennas forward their samples to the fusion center. Combining the samples from N_p pulses forwarded by each of N_r receive antennas, the fusion center can formulate the equation

$$\mathbf{r} \triangleq [\mathbf{r}_{11}^T, \dots, \mathbf{r}_{1N_p}^T, \dots, \mathbf{r}_{N_r N_p}^T]^T$$

$$= \mathbf{\Phi}\mathbf{\Psi}\mathbf{s} + \mathbf{n} \tag{12.10}$$

where

$$\boldsymbol{\Psi} = [(\boldsymbol{\Psi}_{11})^T, \ldots, (\boldsymbol{\Psi}_{1N_p})^T, \ldots, (\boldsymbol{\Psi}_{N_rN_p})^T]^T \in \mathcal{C}^{(L+\tilde{L})N_rN_p \times N}$$

$$\boldsymbol{\Phi} = \mathrm{diag}\{[\boldsymbol{\Phi}_1, \ldots, \boldsymbol{\Phi}_1, \boldsymbol{\Phi}_2, \ldots, \boldsymbol{\Phi}_2, \ldots, \boldsymbol{\Phi}_{N_r}, \ldots, \boldsymbol{\Phi}_{N_r}]\} \in \mathcal{C}^{N_rN_pM \times (L+\tilde{L})N_rN_p}$$

$$\tag{12.11}$$

and

$$\mathbf{n} = [(\boldsymbol{\Phi}_1\mathbf{n}_{11})^T, \ldots, (\boldsymbol{\Phi}_1\mathbf{n}_{1N_p})^T, \ldots, (\boldsymbol{\Phi}_{N_r}\mathbf{n}_{N_rN_p})^T]^T$$

$$\in \mathcal{C}^{N_rN_pM \times 1} \tag{12.12}$$

Recovering **s** from **r** in (12.10) is a sparse signal estimation problem, with $\boldsymbol{\Theta}$ playing the role of the sensing matrix. Assuming that only K elements of the vector **s** are nonzero, with K being much smaller than the dimension of **s**, and under certain conditions on $\boldsymbol{\Theta}$, **s** can be recovered. For a general sensing matrix, and in the noise-free case, exact recovery is possible if in every set of columns of the sensing matrix with cardinality less than K the columns are orthonormal. This condition is described mathematically by the uniform uncertain principle (UUP) [15–18].

Based on the predefined measurement matrices, $\boldsymbol{\Phi}_l$, $l = 1, \ldots, N_r$, the discretization of the angle–speed–range space, knowledge of the antenna locations, and the waveform matrix **X**, the fusion center can construct $\boldsymbol{\Theta}$, and subsequently obtain an estimate of **s**, that is, $\hat{\mathbf{s}}$ by applying any of the existing CS algorithms, that is, matching pursuit algorithms [19–22] and convex relaxation (basis pursuit) algorithms [23–25]. For example, using the Dantzig selector [25] the estimate can be obtained by solving the following problem:

$$\hat{\mathbf{s}} = \min \|\mathbf{s}\|_1 \quad s.t. \|\boldsymbol{\Theta}^H(\mathbf{r} - \boldsymbol{\Theta}\mathbf{s})\|_\infty < \mu \tag{12.13}$$

The ability of the recovery algorithm to correctly detect targets depends on the number of targets and the signal-to-noise ratio (SNR). For a linear array with specific inter-element distances, it was shown in [26] that as long as the number of the targets is less than a maximal value K_{max}, and the signal to noise is larger than some minimal value SNR_{min}, the targets can be correctly detected with high probability by solving an ℓ_1-regularized least-squares problem. Explicit formulas for K_{max} and SNR_{min} as a function of the number of transmit and receive antennas and the grid size are also provided in [26].

Simulation Example 12.1

Let us consider a MIMO radar system with $M_t = 8$ TX and $N_r = 8$ RX antennas, uniformly at random located on a disk of radius 10 m, as shown in Figure 12.1. The TX antenna waveforms are taken to be Hadamard orthogonal with unit power. The received signal is corrupted by zero-mean Gaussian noise of unit power. The SNR is defined as the inverse of the power of thermal noise at a receive node and is set to 0 dB. The carrier frequency is $f = 5$ GHz. There are two targets, placed at angles θ_1 and $\theta_1 + 0.4°$. The search angle space is constructed around θ_1 as $\theta_1 + [-6° : 0.2° : 6°]$. Five hundred independent runs were conducted, with θ_1 generated at random in each run. Figure 12.4 shows the amplitude of the reflection coefficients corresponding to the angle grid points estimated in a run, that is, $\|s_i\|, i = 1, \ldots, 61$. The CS approach was compared with the Capon method (see Section 12.3.3 for details) and the matched filter method (MFM), which obtains the estimates of the target indicator vector **s** as follows:

$$\hat{\mathbf{s}} = \left| \sum_{l=1}^{N_r} \sum_{m=1}^{N_p} \mathbf{z}_{lm}^H \boldsymbol{\Psi}_{lm} \right| \qquad (12.14)$$

where $\mathbf{z}_{lm} = [z_{lm}((m-1)T + 0T_s), \ldots, z_{lm}((m-1)T + (L + \tilde{L} - 1)T_s)]^T$ is the vector of the T_s-sampled data obtained by the lth RX antenna and the mth pulse.

For the results of Figure 12.4, the length of the transmit sequences is $L = 16$, and all the three methods used all 16 fast-time samples. Figure 12.4 clearly demonstrates that among the three methods the CS method has the lowest side lobes and the narrowest main lobes. It is noted that no compression is performed here by the CS method; the goal is to just demonstrate the CS method's ability to improve angle resolution as compared with the other method for the same number of samples. For the same scenario, Figure 12.5 shows the receiver operating characteristic

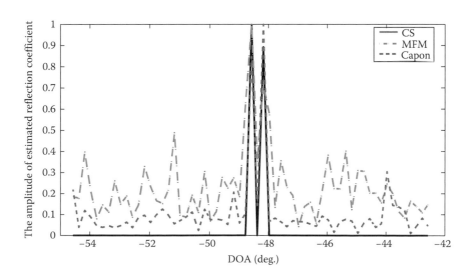

FIGURE 12.4
(See color insert.) DOA estimates of CS, Capon, and MEM using 16 received samples.

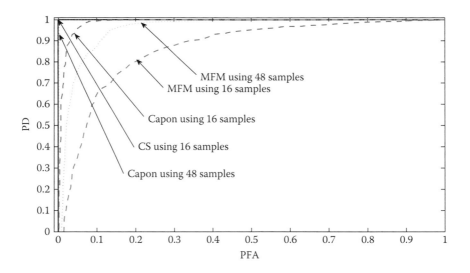

FIGURE 12.5
ROC performances for CS, Capon, and MFM methods with $M_t = N_r = 8$.

(ROC) curves of the angle estimates, produced by the aforementioned
methods, based on 500 independent runs. In each run, θ_1, the TX/RX
antenna locations and the noise signal are randomly generated. Two
cases are demonstrated in Figure 12.5, that is, $L = 16$ and $L = 48$. Again,
all the three methods use all L fast-time samples and no compression is
performed by the CS method. Here, the probability of detection (PD) is
the percentage of cases in which all the targets are detected. The probability of false alarm (PFA) is defined as the percentage of cases in which
false targets are detected. It can be seen from the figure that when all
methods use 16 fast-time samples, the CS method has the best performance. The Capon method comes close to the CS method when it uses
three times the number of samples of CS. Further, it is difficult to set
the proper threshold for the MFM when the target reflection coefficients
are unknown. However, this is not an issue for the CS approach since it
always produces a very *clean* estimate, as it can be seen in Figure 12.4.

Figure 12.6 shows the ROC curves of the angle–range estimates
obtained by the CS and MFM methods with $M_t = 8$ TX and $N_r = 5$ RX
antennas. The transmit waveforms have length $L = 30, 48$ and the sampling period is taken as $T_s = 10^{-7}$ s. The angle–range estimates have
been obtained based on 500 random and independent runs, with the
TX/RX antennas fixed at the same locations in all runs. The azimuth
angle and range of three targets are randomly generated in each run
within $[-3°, 3°]$ and $[1000\text{ m}, 1150\text{ m}]$, respectively. The search range–
angle space is $[-3°, -2.7°, \ldots, 3°] \times [1000\text{ m}, 1015\text{ m}, \ldots, 1150\text{ m}]$. The CS
receiver compresses the received signal through the Gaussian random
measurement matrix. In the presence of a single target of unit reflectivity,
the CS method performs slightly better than the MFM method with the

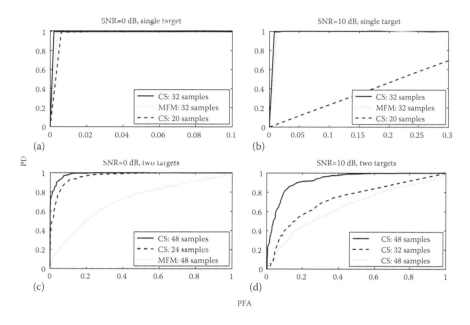

FIGURE 12.6
(See color insert.) ROC performances of range–DOA estimation for CS, Capon, and MFM methods with $M_t = 8$, $N_r = 5$. (a) SNR = 0 dB, single target, (b) SNR = − 10 dB, single target, (c) SNR = 0 dB, two targets, and (d) SNR = − 10 dB, two targets.

same number of samples as shown in Figures 12.6a and b. This is because the ripples, as seen in Figure 12.4, are not strong enough to overwhelm the single peak produced by the target. The ROC curves for the single target also demonstrate a significant performance degradation of the CS method with insufficient number of received samples when the SNR is low, that is, $M = 20$ and $SNR = −10$ dB. Figure 12.6c and d show the ROC curves produced for the scenario of two targets with reflectivities $\beta = 1, 0.7$. In this case, the estimation ripples of the MFM method mask the weaker target. The CS method performs much better than the MFM method with the same number of samples, or even with much smaller number of samples. However, the performance advantage of the CS method over the MFM method decreases as the SNR decreases as it can be seen in Figure 12.6d, and more samples are required for the former method to produce the desired performance.

12.3 Challenging Issues Associated with CS-MIMO Radars

In this section, we discuss some issues associated with CS-MIMO radars and outline possible solutions.

12.3.1 Basis Mismatch and Resolution

Like most spectral-based methods, for example, MUSIC, Capon, the CS-MIMO radars search for targets in a discretized search space. Thus, they suffer from mismatch between the mathematical model for sparsity and the physical model for sparsity, also referred to as *basis mismatch* [27]. In the presence of basis mismatch, the targets that do not fall on grid points may not be captured by the closest grid points. The mismatch error can be reduced by discretizing the search space on a finer grid, which would also help resolution. However, a fine grid leads to high correlation between the target returns from the adjacent grid points of the search space, which goes against the uniform uncertainty principle (UUP). Thus, the grid spacing must be carefully selected to balance resolution and performance requirements. In general, one can increase the number of TX and RX antennas to improve angular resolution, or the number of measurements to achieve high resolution without violating the UUP [15,17].

12.3.2 Complexity

The complexity of CS-MIMO radars depends on the CS algorithm; matching pursuit algorithms and basis pursuit algorithms, respectively, require computational cost of the order of $\mathcal{O}(N^2)$ and $\mathcal{O}(N^3)$, where N is the grid size. When the MIMO radar system needs to estimate target angle, speed, and range, a three-dimensional space needs to be discretized, resulting in a large number of grid points. This involves prohibitively high computational complexity. In addition, the requirement of a fine grid for achieving high resolution further increases the computational complexity. The complexity problem can be mitigated by successively estimating angle–range and speed in a decoupled fashion, so that the original three-dimensional problem is converted into a two-dimensional problem plus a single-dimensional problem. In particular, one can first perform an initial angle–Doppler estimation using a coarse grid and then refine the grid points around the initial estimate (as shown in Figure 12.7) [28]. Restricting the candidate angle–Doppler space reduces the samples in the angle–Doppler space that are required for constructing the basis matrix, thus reducing the complexity of CS-MIMO radars.

12.3.3 Clutter Rejection: CS-Capon

The advantages of CS-MIMO radars stem from the sparsity of targets in the target space. However, the target sparsity diminishes in the presence of clutter, as clutter is highly correlated with the returns from the target of interest. Therefore, clutter suppression needs to precede any CS recovery procedure. The problem of suppressing clutter becomes easier when some information about the clutter is available, for example, its Doppler or its angle of arrival.

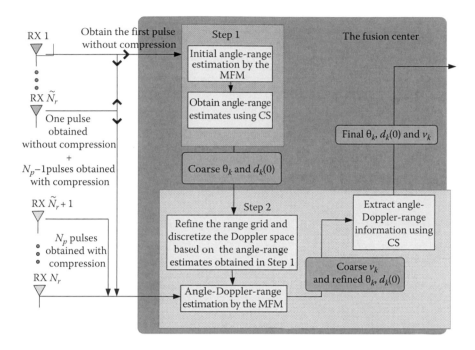

FIGURE 12.7
Schematic diagram of the decoupled scheme. (From Yu, Y. et al., CSSF MIMO radar: Compressive-sensing and step-frequency based MIMO radar, *IEEE Trans. Aerosp. Electron. Syst.*, 48(2), 1490, April 2012. With permission. © 2012 IEEE.)

For the case in which the covariance of the clutter is known, an approach, termed CS-Capon, was proposed in [29]. The approach consists of applying Capon beamforming on the compressively obtained data, followed by target estimation using CS theory. This approach is outlined next.

For simplicity, we ignore the time delays in the waveforms, which corresponds to the case in which the targets fall in the same range cell. Further, we consider stationary targets, thus we only need to estimate DOA. The received signal at N_r RX antennas can be formulated as

$$\mathbf{Z}^T = \sum_{k=1}^{K} \beta_k \mathbf{v}_r(\theta_k) \mathbf{v}^T(\theta_k) \mathbf{X}^T + \mathbf{N} \tag{12.15}$$

where
$\mathbf{Z}$ is the received signal at N_r RX antennas; its ith column contains the received signal from the ith RX antenna
$\mathbf{v}_r(\theta_k)$ is the receive steering vector at direction θ_k
$\mathbf{v}(\theta_k)$ is the transmit steering vector at direction θ_k
$\mathbf{N}$ is the clutter matrix, whose covariance matrix $\mathbf{R}_N$ is known

The Capon method yields a beamformer $\mathbf{w}$ that suppresses clutter while keeping the signal from a desired angle, θ, undistorted. In particular, the beamformer $\mathbf{w}$ can be formulated as

$$\min_{\mathbf{w}} \mathbf{w}^H \mathbf{R}_N \mathbf{w} \text{ s.t. } \mathbf{w}^H \mathbf{v}_r(\theta) = 1 \tag{12.16}$$

The solution of (12.16) is

$$\mathbf{w}(\theta) = \frac{\mathbf{R}_N^{-1} \mathbf{v}_r(\theta)}{\mathbf{v}_r^H(\theta) \mathbf{R}_N^{-1} \mathbf{v}_r(\theta)} \tag{12.17}$$

Therefore, if one applied $\mathbf{w}(\theta)$ to the compressed received signal $\mathbf{Y} = \mathbf{\Phi} \mathbf{Z}$, the clutter would be rejected while the received signal from θ would be preserved. Let the discretized angle grid be $[a_1, \ldots, a_N]$ and let $\mathbf{w}_n$ denote the Capon beamforming vector for discrete direction a_n. Then, the CS-Capon method proceeds as follows (also see Figure 12.8):

- Apply the Capon beamforming vectors $\mathbf{w}_n, n = 1, \ldots, N$ to the compressed signal $\mathbf{Y}$ to obtain

$$\mathbf{y}_n = (\mathbf{w}_n^H \mathbf{Y}^T)^T = \mathbf{Y}\mathbf{w}_n^*$$

$$= \mathbf{\Theta}_n \mathbf{s} + \mathbf{\Phi} \mathbf{Z} \mathbf{w}_n^* \quad n = 1, \ldots, N \tag{12.18}$$

where

$$\mathbf{\Theta}_n = \mathbf{\Phi} \mathbf{X} \left[\mathbf{v}(a_1)\mathbf{v}_r^T(a_1)\mathbf{w}_n^*, \ldots, \mathbf{v}(a_N)\mathbf{v}_r^T(a_N)\mathbf{w}_n^* \right],$$

and $\mathbf{s}$ is a sparse vector, the nonzero elements of which indicate the targets locations.
- Stack $\mathbf{y}_n, n = 1, \ldots, N$ into a vector and compress this long vector with matrix $\tilde{\mathbf{\Phi}}$, that is,

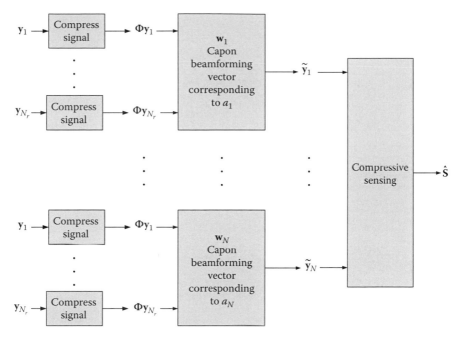

FIGURE 12.8
CS-Capon method diagram. (From Yu, Y. et al., A capon beamforming method for clutter suppression in colocated compressive sensing based MIMO radars, in: *Proceedings of SPIE 8717, Compressive Sensing II*, 87170J (May 31, 2013). doi:10.1117/12.2015635. With permission.)

$$\tilde{\mathbf{y}} = \tilde{\boldsymbol{\Phi}} \left[\tilde{\mathbf{y}}_1^T, \ldots, \tilde{\mathbf{y}}_N^T \right]^T$$

$$= \underbrace{\tilde{\boldsymbol{\Phi}} \left[\tilde{\boldsymbol{\Theta}}_1^T, \ldots, \tilde{\boldsymbol{\Theta}}_N^T \right]^T}_{\tilde{\boldsymbol{\Theta}}} \mathbf{s} + \underbrace{\tilde{\boldsymbol{\Phi}} \left[\mathbf{w}_1^H \mathbf{Z}, \ldots, \mathbf{w}_N^H \mathbf{Z} \right]^T}_{\tilde{\mathbf{Z}}}$$

$$= \tilde{\boldsymbol{\Theta}} \mathbf{s} + \tilde{\mathbf{Z}} \qquad (12.19)$$

In this step, the compression matrix $\tilde{\boldsymbol{\Phi}}$ is used for the sake of reducing the problem dimension and saving computational cost.

- Apply any CS algorithm to (12.19)

It was shown in [29] that the aforementioned scheme enables to significantly increase signal-to-clutter-plus-noise ratio (SCNR) as compared with the CS method without clutter suppression, and thus improving angle estimation in terms of ROC, in the presence of strong clutter. As in the Capon method, however, the CS-Capon method cannot yield good performance in the presence of strong clutter reflectors at the target locations.

Simulation Example 12.2

We consider the same antenna configuration as *Simulation Example 12.1*. The carrier frequency is $f = 5$ GHz. The TX antennas use Hadamard orthogonal waveforms with $L = 128$ and unit power. The received signal is corrupted by zero-mean Gaussian noise of unit power. The SNR is set to 0 dB. The clutter signal is constructed as the sum of returns reflected by 3000 reflectors, located in the DOA angle space $[-30^\circ : 60/1500^\circ : 30^\circ]$. The reflection coefficient of each reflector is 0.15. The measurement matrix, $\Phi \in \mathcal{R}^{M \times L}$ is taken to be random Gaussian with $M = 50$.

Figure 12.9 shows the averaged SCNR over 100 independent runs versus the number of RX antennas. In each run, the TX/RX antennas are randomly placed on a disk of radius 10 m. Here, we also plot the maximum and minimum SCNR for all beamformed receive vectors $\tilde{\mathbf{y}}_n$, $n = 1, \ldots, N$. In the simulations, $K = 3$ targets are generated randomly on the DOA grid $[-30^\circ : 1^\circ : 30^\circ]$ with reflection coefficients $[1, 1, 1]$. The number of TX antennas is set to $N_t = 20$. For each N_r, 100 runs are carried out and average SCNR is calculated. It can be seen in Figure 12.9 that, without beamforming, the SCNR of the received signal is around -10 dB

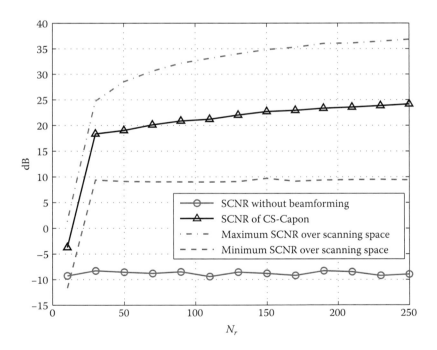

FIGURE 12.9

SCNR comparisons with/without beamforming with $N_t = 20$. (From Yu, Y. et al., A capon beamforming method for clutter suppression in colocated compressive sensing based MIMO radars, in: *Proceedings of SPIE 8717, Compressive Sensing II*, 87170J (May 31, 2013). doi:10.1117/12.2015635. With permission.)

and does not change much as N_r increases. With beamforming, the SCNR of the CS-Capon estimator is around 20 dB when $N_r = 50$ and increases as N_r increases. This indicates that the CS-Capon method could achieve significant gain in SCNR performance, that is, strong clutter suppression. We can also see in Figure 12.9 that the SCNR of the CS-Capon estimator falls in the middle between the maximum and minimum SCNR of all beamformed vectors $\tilde{y}_n$, $n = 1, \ldots, N$. This is because the SCNR of the CS-Capon estimator is averaged over the entire scanned angle space.

Figure 12.10 plots the ROC curves of the angle estimates that are obtained based on 500 random and independent runs. In each run, $K = 3$ targets are randomly generated on the angle grid of interest $[-8°, -8° + 0.2°, \ldots, 8°]$ with random reflection coefficients $[1, 0.45, 0.8]$, respectively. In the simulations, $N_t = 20$, $N_r = 30$, $L = 256$, and each RX antenna collects 20 compressed measurements. A random Gaussian matrix is used as the measurement matrix. One can see that the CS-Capon method outperforms the CS method and the Capon method. It is worth noting that if the reflection coefficients of the targets are identical, Capon performs similar to the CS-Capon method. When the reflection coefficients of three targets are different, the ripples generated by Capon may mask the targets with smaller reflectivity.

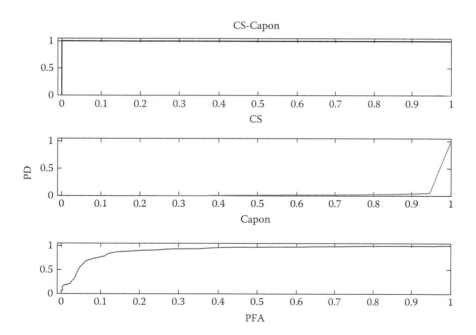

FIGURE 12.10
ROC performances for CS-Capon and Capon methods with $N_t = 20$ and $N_r = 30$. (From Yu, Y. et al., A capon beamforming method for clutter suppression in colocated compressive sensing based MIMO radars, in: *Proceedings of SPIE 8717, Compressive Sensing II*, 87170J (May 31, 2013). doi: 10.1117/12.2015635. With permission.)

12.3.4 Phase Synchronization

In a scenario in which the TX and RX antennas are not physically connected and need to communicate with the fusion center in a wireless fashion, coherent processing is hampered by carrier phase errors and frequency offsets introduced by the antennas oscillators. In such cases, synchronization needs to be conducted periodically between TX/RX antennas. Synchronization is a major issue in all distributed systems that require coherent processing.

There are two basic approaches to phase synchronization, that is, closed-loop approaches [30–32], in which the synchronization is coordinated by the fusion center while the distributed nodes are minimally involved, and open-loop approaches [33–35], in which the synchronization is implemented primarily based on information exchange between the distributed nodes. In the open-loop methods, the involvement of the fusion center is to simply broadcast an unmodulated sinusoidal beacon to the nodes.

12.4 Advanced Techniques for CS-MIMO Radars

In this section, we describe three advanced techniques that can improve the detection performance of CS-MIMO radars, namely, power allocation, waveform design, and measurement matrix design.

12.4.1 Power Allocation

A power allocation scheme distributes the system total transmit power among the transmit antennas in an optimal fashion. For widely separated CS-MIMO radars, a power allocation scheme was proposed in [13] aiming at maximizing the minimum target returns given a total transmit power, so that the probability of missing weak targets is reduced. This scheme determines the transmit power for the next set of transmit pulses by using the estimates of $\mathbf{s}$ (see (12.8)) obtained from the previously received signal. On letting p_i denote the power assigned to the ith TX antenna and the grid in the search space be of size N, the optimal $p_i, i = 1, \ldots, M_t$ are found by solving the following optimization problem

$$\max_{p_i} \min_k \sum_{i=1}^{M_t} \sum_{l=1}^{N_r} p_i |s_k^{il}|^2 \qquad (12.20)$$

under the condition that $\sum_{i=1}^{M_t} p_i = P_t$. This problem can be solved at a fusion center, and the computed p_is can be sent to the transmit antennas, to be used in the next set of pulses.

Another power allocation approach that applies to both widely separated and colocated CS-MIMO radars was proposed in [14,36,37], in which the p_is are obtained by minimizing the coherence between the target returns from different search cells (or grid points), or equivalently, the coherence of the columns of the sensing matrix (see (12.11)). Since the UUP requires the sensing matrix to be as orthogonal as possible in order to guarantee reliable performance, the power is allocated so that the difference between the Gram matrix of the sensing matrix, that is, $\Theta^H \Theta$ and an identity matrix is minimized. On letting $\mathbf{p} = [p_1, \ldots, p_{M_t}]^T$ and P_m the maximum transmit power allowed at the mth TX antenna the problem is formulated as

$$\min_{\mathbf{p}} \sum_{k \neq k'} |\mathbf{u}_{k'}^H \mathbf{u}_k|^2 + \sum_{k} |\mathbf{u}_k^H \mathbf{u}_k - P_t N_r|^2$$

$$s.t.\ \mathbf{1}^T \mathbf{p} = P_t,\ \mathbf{p} \geq \mathbf{0}_{M_t \times 1},\ \mathbf{p} \leq P_m \mathbf{1}_{M_t \times 1} \tag{12.21}$$

where $\mathbf{u}_k$ is the kth column of the sensing matrix.

By allocating the total transmit power among the TX antennas according to (12.21), the transmit antennas that offer more information on targets are assigned more transmit power while those that have little contribution to target detection may be eliminated. In this way, this scheme not only improves the accuracy of target detection but also enables reduction of the number of active transmit antennas as compared with the uniform power allocation (UPA) scheme.

In the following, we elaborate on the power allocation scheme for colocated CS-MIMO radars. For simplicity, we consider stationary targets, which are located in the same range cell, and thus we only aim at improvement in DOA estimation. Therefore, the model described in (12.3) can be reduced to

$$\mathbf{r}_l \approx \sum_{k=1}^{K} \beta_k e^{j\frac{2\pi f}{c} \eta_l^r(\theta_k)} \mathbf{\Phi X v}(\theta_k) + \mathbf{\Phi n}_l \tag{12.22}$$

Here, all RX antennas use the same measurement matrix to compress the signal, and target detection is performed using only one pulse.

Let us discretize the angle space into N discrete angles $[a_1, \ldots, a_N]$. Combining the output of N_r receive antennas, we have

$$\mathbf{r} = \Theta \mathbf{s} + \mathbf{n} \tag{12.23}$$

where Θ is the sensing matrix with its kth column equal to

$$\mathbf{u}_k = \left[e^{j2\pi(\eta_1^r(a_k)f)/c}, \ldots, e^{j2\pi(\eta_{N_r}^r(a_k)f)/c} \right]^T \otimes (\mathbf{\Phi X V}(a_k)\tilde{\mathbf{p}}) \tag{12.24}$$

where

$$\mathbf{V}(a_k) = \mathrm{diag}\{\mathbf{v}(a_k)\}, \text{ and } \tilde{\mathbf{p}} = \sqrt{\mathbf{p}}$$

It holds that

$$|\mathbf{u}_{k'}^H \mathbf{u}_k|^2 = u_{kk'} |\tilde{\mathbf{p}}^H \mathbf{B}_{kk'} \tilde{\mathbf{p}}|^2$$

$$= u_{kk'} \left[(\tilde{\mathbf{p}}^H \mathbf{Br}_{kk'} \tilde{\mathbf{p}})^2 + (\tilde{\mathbf{p}}^H \mathbf{Bi}_{kk'} \tilde{\mathbf{p}})^2 \right] \qquad (12.25)$$

where

$$u_{kk'} = \left| \sum_{l=1}^{N_r} e^{j2\pi \frac{(n_l^r(a_k) - n_l^r(a_{k'}))f}{c}} \right|^2$$

$$\mathbf{B}_{kk'} = \mathbf{V}^H(a_{k'}) \mathbf{X}^H \boldsymbol{\Phi}^H \boldsymbol{\Phi} \mathbf{X} \mathbf{V}(a_k)$$

$$\mathbf{Br}_{kk'} = \frac{\mathbf{B}_{kk'} + \mathbf{B}_{kk'}^H}{2}$$

$$\mathbf{Bi}_{kk'} = \frac{\mathbf{B}_{kk'} - \mathbf{B}_{kk'}^H}{2j}$$

It is easy to see that $\mathbf{B}_{kk'}$ is not a positive semidefinite (PSD) matrix unless $k = k'$ and thus the objective function is nonconvex. However, we can make the objective function convex via the following trick:

$$(\tilde{\mathbf{p}}^T \mathbf{Br}_{kk'} \tilde{\mathbf{p}})^2 = \left(\tilde{\mathbf{p}}^T \underbrace{\left(\mathbf{Br}_{kk'} + \frac{b}{P_t} \mathbf{I} \right)}_{\mathbf{Cr}_{kk'}} \tilde{\mathbf{p}} - b \right)^2$$

$$= (\tilde{\mathbf{p}}^T \mathbf{Cr}_{kk'} \tilde{\mathbf{p}})^2 + \tilde{\mathbf{p}}^T \underbrace{\left(-2b\mathbf{Cr}_{kk'} + \frac{d}{P_t} \mathbf{I} \right)}_{\mathbf{Dr}_{kk'}} \tilde{\mathbf{p}} + C_r \qquad (12.26)$$

where b and d are non-negative real scalars that let $\mathbf{Cr}_{kk'}$ and $\mathbf{Dr}_{kk'}$ be PSD matrices, that is, $b/P_t + \lambda_{min}(\mathbf{Br}_{kk'}) \geq 0$ and $d/P_t + \lambda_{min}(-2b\mathbf{Cr}_{kk'}) \geq 0$. C_r are constants that will not affect the objective function.

Equation 12.26 is convex since $\mathbf{Cr}_{kk'}$ and $\mathbf{Dr}_{kk'}$ are PSD matrices. By performing the same trick on $(\tilde{\mathbf{p}}^H \mathbf{Bi}_{kk'} \tilde{\mathbf{p}})^2$, we can obtain

$$(\tilde{\mathbf{p}}^T \mathbf{Bi}_{kk'} \tilde{\mathbf{p}})^2 = (\tilde{\mathbf{p}}^T \mathbf{Ci}_{kk'} \tilde{\mathbf{p}})^2 + \tilde{\mathbf{p}}^T (\mathbf{Di}_{kk'}) \tilde{\mathbf{p}} + C_i \qquad (12.27)$$

In the same way, the second term in the objective function (12.21) can be rewritten as

$$|\mathbf{u}_k^H \mathbf{u}_k - P_t N_r|^2 = N_r^2 (\tilde{\mathbf{p}}^T (\mathbf{B}_{kk} - \mathbf{I}) \tilde{\mathbf{p}})^2 = (\tilde{\mathbf{p}}^T \mathbf{C}_{kk} \tilde{\mathbf{p}})^2 + \tilde{\mathbf{p}}^T (\mathbf{D}_{kk}) \tilde{\mathbf{p}} + C \quad (12.28)$$

Then the objective function of (12.21) can be transformed into a convex function, and the problem of (12.21) becomes

$$
\min_{\tilde{\mathbf{p}}} \sum_{k \neq k'} (\tilde{\mathbf{p}}^T \mathbf{Cr}_{kk'} \tilde{\mathbf{p}})^2 + (\tilde{\mathbf{p}}^T \mathbf{Ci}_{kk'} \tilde{\mathbf{p}})^2 + \tilde{\mathbf{p}}^T (\mathbf{Dr}_{kk'} + \mathbf{Di}_{kk'}) \tilde{\mathbf{p}}
$$

$$
+ \sum_{k} (\tilde{\mathbf{p}}^T \mathbf{C}_{kk} \tilde{\mathbf{p}})^2 + \tilde{\mathbf{p}}^T (\mathbf{D}_{kk}) \tilde{\mathbf{p}}
$$

$$
s.t. \ \tilde{\mathbf{p}}^H \tilde{\mathbf{p}} = P_t, \ \tilde{\mathbf{p}} \geq \mathbf{0}_{M_t \times 1}, \ \tilde{\mathbf{p}} \leq \sqrt{P_m} \mathbf{1}_{M_t \times 1} \qquad (12.29)
$$

Looking at the constraints of (12.29), we can see that the first constraint is nonconvex. To address this issue, the solution of (12.29) can be obtained in an iterative fashion, where in each iteration, the constraint is substituted by its local affine approximation, that is,

$$
(\tilde{\mathbf{p}})^T \tilde{\mathbf{p}} \approx (\tilde{\mathbf{p}}^j)^T \tilde{\mathbf{p}}^j + 2(\tilde{\mathbf{p}}^j)^T (\tilde{\mathbf{p}} - \tilde{\mathbf{p}}^j) \qquad (12.30)
$$

where $(\tilde{\mathbf{p}}^j)$ is the estimate of $\tilde{\mathbf{p}}$ at the jth iteration.

The iteration would proceed as follows.

1. Initialize with $\mathbf{p}^{(0)} = [1, 1, \ldots, 1]^T$.
2. At the jth iteration, $\tilde{\mathbf{p}}^{(j)}$ is obtained by solving

$$
\min_{\tilde{\mathbf{p}}} \sum_{k \neq k'} |\mathbf{u}_{k'}^H \mathbf{u}_k|^2 + \sum_{k} |\mathbf{u}_k^H \mathbf{u}_k - P_t N_r|^2
$$

$$
s.t. \ (\tilde{\mathbf{p}}^{(j-1)})^H \tilde{\mathbf{p}}^{(j-1)} + 2(\tilde{\mathbf{p}}^{(j-1)})^H (\tilde{\mathbf{p}} - \tilde{\mathbf{p}}^{(j-1)}) = P_t,
$$

$$
\tilde{\mathbf{p}} \geq \mathbf{0}_{M_t \times 1}, \ \tilde{\mathbf{p}} \leq \sqrt{P_m} \mathbf{1}_{M_t \times 1}
$$

3. If $\|\tilde{\mathbf{p}}^{(j)} - \tilde{\mathbf{p}}^{(j-1)}\|_2 < \epsilon$, the iteration stops and the output is $\tilde{\mathbf{p}} = \tilde{\mathbf{p}}^{(j)}$. Otherwise, the iteration continues.

One could use a compressive receiver at each receive antenna based on measurement matrix $\mathbf{\Phi}$. The implementation of the measurement matrix may increase the complexity of the analog circuit as it is a full matrix. Therefore, we can skip the step of linear compression and directly collect a small number of samples at the RX antennas by using an identity matrix as the measurement matrix. If the TX antennas transmit orthogonal waveforms, that is, $\mathbf{X}^H\mathbf{X} = \mathbf{I}$, then $\mathbf{B}_{kk}$ is a diagonal matrix. Then (12.21) can be reduced to a simple convex problem as follows:

$$\min_{\mathbf{p}} \ \mathbf{p}^T \left(\sum_{k \neq k'} u_{kk'} \mathbf{b}_{kk'}^* \mathbf{b}_{kk'}^T \right) \mathbf{p}$$

$$s.t. \ \mathbf{1}_{M_t \times 1}^T \mathbf{p} = P_t, \ \mathbf{p} \geq \mathbf{0}, \ \mathbf{p} \leq P_m \mathbf{1}_{M_t \times 1} \qquad (12.31)$$

where $\mathbf{b}_{kk'}$ contains the diagonal elements of $\mathbf{B}_{kk'}$.

Simulation Example 12.3

We consider the same antenna configuration as *Simulation Example 12.1*. In each run, the locations of TX and RX antennas are generated uniformly at random. The carrier frequency is $f = 5$ GHz. The transmitted waveforms are orthogonal Hadamard sequences of length $L = 32$ and have unit power. The SNR is set to 0 dB. Three targets are placed uniformly at random on the angle grid $[0°: 0.1°:5°]$. The total transmitted power is set to M_t and the maximum transmit power for each antenna is 9 W. The measurement matrix Φ is taken to be the identity matrix. The Dantzig selector, as shown in (12.13), was used to extract target angle information.

Multiple independent runs are conducted; in each run the locations of TX and RX antennas are randomly generated following a uniform distribution. Figure 12.11 demonstrates the average coherence of column pairs of the sensing matrix, corresponding to the optimized power allocation (OPA) scheme, computed over 100 independent runs and for $M_t = 5, 10, 15, 20, 25, 30$. For comparison purposes, the corresponding results for UPA and random power allocation (RPA) are also shown in the same figure. It can be seen from the figure that the OPA scheme can reduce the first term in (12.21), representing the sum of the square magnitude of the cross-correlation of column pairs $\mathbf{u}_k, \mathbf{u}_{k'}$ of the sensing matrix, denoted here by *SCSM*. Figure 12.11 also shows that the SCSM can be reduced by increasing the number of RX antennas.

Figure 12.12 shows the ROC curves of the angle estimates, obtained based on 1000 independent runs. In each run, three targets are randomly generated on the angle grid. The cases of SNR $= -10$ dB, 0 dB with different numbers of TX/RX antennas are shown in Figure 12.12. One can see that the OPA scheme can improve the ROC performance as compared with the UPA scheme. Again, an increase in the number of RX antennas can improve the detection performance. Increasing the number of RX antennas cannot boost SNR and thus the performance gain comes from the improvement of the sensing matrix as shown in Figure 12.11. The performance gain with the increase of the number of TX/RX antennas is more prominent at low SNR scenarios. This is because antenna aperture would be increased by using more TX/RX antennas. In addition, an increase in the number of TX antennas can also improve the SNR of

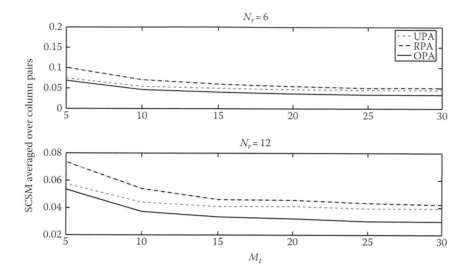

FIGURE 12.11

SCSM versus M_t under optimal (PPA), uniform (UPA), and random (RPA) power allocation in CS-based colocated MIMO radars. Case (I) $N_r = 6$; Case (II) $N_r = 12$; in both cases $L = 32$.

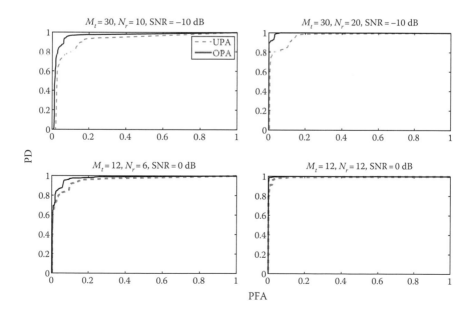

FIGURE 12.12

The ROCs of angle estimates under power allocation with $M_t = 12$ and $L = 32$.

the received signal at the RX antennas, as in our simulations the transmit power for each TX antenna is fixed.

In addition to performance improvement, the OPA scheme also reduces the number of active TX antennas. For example, in the case of $M_t = 20$ and $N_r = 12$, 6 TX antennas are allocated power less than 0.0001 W on an average of 500 runs. This indicates that the OPA scheme only requires 14 TX antennas to be active, while all 20 TX antennas are needed for the UPA scheme. Figure 12.13a shows the distribution of TX

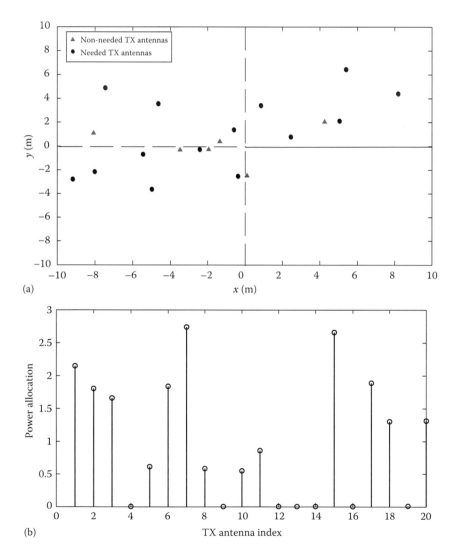

(a)

(b)

FIGURE 12.13
The TX distribution in colocated CS-MIMO radars with $M_t = 20$, $N_r = 12$, and $L = 32$. (a) TX antenna distribution and (b) power allocation.

antennas that were assigned power less than 0.0001 W in one run; the non-needed antennas are marked on the figure. Figure 12.13b illustrates the power allocation results for this case.

12.4.2 Waveform Design for Colocated CS-MIMO Radars

Along the lines of optimal power allocation, one may also design the transmit waveforms by minimizing the difference between the Gram matrix of the sensing matrix and an identity matrix, under a total transmitted power constraint. In this scenario, the kth column of the sensing matrix becomes

$$\mathbf{u}_k = \left[e^{j2\pi \frac{\eta_1^r(\theta_k)f}{c}}, \ldots, e^{j2\pi \frac{\eta_{N_r}^r(\theta_k)f}{c}} \right]^T \otimes (\mathbf{\Phi}\tilde{\mathbf{V}}(a_k)\mathbf{x}) \tag{12.32}$$

where
$\tilde{\mathbf{V}}(a_k) = \mathbf{I}_L \otimes \mathbf{v}^T(a_k)$, and $\mathbf{x} = vec(\mathbf{X}^T)$

One can formulate the following optimization problem:

$$\min_{\mathbf{x}} \sum_{k \neq k'} |\mathbf{u}_{k'}^H \mathbf{u}_k|^2 + \sum_k |\mathbf{u}_k^H \mathbf{u}_k - P_t N_r|^2$$

$$s.t. \ \mathbf{x}^H \mathbf{x} = P_t \tag{12.33}$$

which can be solved along the lines of optimal power allocation described in Section 12.4.1. However, the number of variables for waveform design is much larger than that for power allocation, that is, the former is $M_t L$, while the latter is M_t. Further, (12.33) is not convex, thus a global minimum does not exist, and the obtained waveforms depend on the initial waveforms used. However, as it will be shown in the following simulation example, this approach can still lead to performance improvement.

Simulation Example 12.4

We consider the same antenna configuration as *Simulation Example 12.1*. $M_t = 10$ TX antennas and $N_r = 10$ RX antennas are employed. The carrier frequency is $f = 5$ GHz and the SNR is set to 0 dB. Three targets are present on the angle grid $[0°, 0.2°, \ldots, 2°]$. The total transmitted power is set to M_t. The initial vector $\mathbf{x}^{(0)}$ is based on orthogonal Hadamard waveforms of length $L = 16$ and unit power. The measurement matrix $\mathbf{\Phi}$ is chosen as a Gaussian random matrix with $M = round(0.7L)$, which means each RX antenna forwards $M = 11$ samples to the fusion center. The Dantzig selector, as shown in (12.13), is used to extract target angle information. The ROC curves of the angle estimates produced by

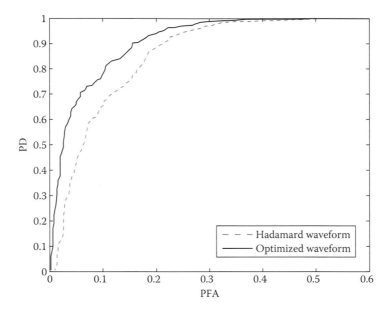

FIGURE 12.14
The ROCs of angle estimates under waveform design in CS-based colocated MIMO radars.

the proposed waveform-design method are shown in Figure 12.14, where
the performance improvement due to the designed waveform is evident. Figure 12.15 illustrates the suboptimal waveforms of 4 TX antennas
obtained from (12.33) in one run.

12.4.3 Measurement Matrix Design

In this section, we discuss the design of the measurement matrix in order to
improve the detection performance of colocated CS-MIMO radars.

Generally, there are two factors that affect the performance of CS. The
first factor is the coherence of the sensing matrix. The UUP requires low
coherence of the sensing matrix (CSM) to guarantee exact recovery of the
sparse signal. The second factor is the signal-to-interference ratio. For example, if the basis matrix obeys the UUP and the signal of interest **s** is
sufficiently sparse, then the square estimation error of the Dantzig selector
satisfies, with high probability, the following [25]:

$$\| \hat{\mathbf{s}} - \mathbf{s} \|_{\ell_2}^2 \leq C^2 2 \log N \times \left(\sigma^2 + \sum_{i}^{N} \min(s^2(i), \sigma^2) \right) \qquad (12.34)$$

where C is a constant and $\hat{\mathbf{s}}$ is the estimate of **s**. It can be easily seen from
(12.34) that an increase in the interference power degrades the performance
of the Dantzig selector.

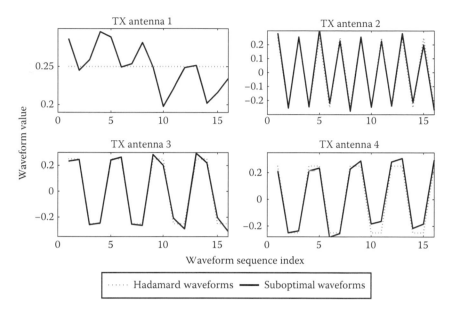

FIGURE 12.15
Local optimum waveforms and initial waveforms of 4 TX antennas in one run for CS-based colocated MIMO radars.

12.4.3.1 Measurement Matrix Design by Reducing CSM and Increasing SIR

The goal of measurement matrix design is to reduce CSM and at the same time increase SIR. The coherence of two columns of the sensing matrix, Θ, corresponding to the kth and k'th grid point is given by

$$\mu_{kk'}(\Theta) = \frac{\left| \sum_{m=1}^{N_p} \sum_{l=1}^{N_r} e^{j2\pi(q_{lmk} - q_{lmk'})} \mathbf{u}_{k'}^H \Phi^H \Phi \mathbf{u}_k \right|}{N_r N_p \sqrt{\mathbf{u}_k^H \Phi^H \Phi \mathbf{u}_k \mathbf{u}_{k'}^H \Phi^H \Phi \mathbf{u}_{k'}}} \tag{12.35}$$

where $\mathbf{u}_k = \mathbf{C}_{\lfloor 2c_k/cT_p/L \rfloor} \mathbf{X} \mathbf{v}(a_k)$ ($\mathbf{C}_{\tau_k}$ is defined in Section 12.2.1). Let the interference waveform at the lth RX antenna during the mth pulse be modeled as Gaussian distributed, that is, $n_{lm}(t) \sim \mathcal{CN}(0, \sigma^2)$. Let us also assume that the noise waveforms are independent across the RX antennas and between pulses. Then the SIR equals

$$\text{SIR} = \frac{\sum_{i=1}^K |\beta_i|^2 \mathbf{u}_{k_i}^H \Phi^H \Phi \mathbf{u}_{k_i}}{\sigma^2 \text{Tr}\{\Phi^H \Phi\}} \tag{12.36}$$

The precise manner in which CSM and SIR affect the performance of the CS approach is unknown. In [40], the measurement matrix is obtained by optimizing a linear combination of CSM and the reciprocal of SIR. The CSM can be defined in various ways. One definition of CSM is the maximum coherence produced by a pair of cross columns in the sensing matrix. This criterion works well for a uniform sensing matrix but might not capture the behavior of the sensing matrix when coherence of most of its column pairs is small [41]. In addition, the maximization of the maximum coherence would increase the coherence of some column pairs that had initially low coherence. Next, we use the sum of CSM (SCSM) as the metric to design the measurement matrix.

By minimizing the linear combination of the SCSM and SIR, we formulate the optimization problem as follows:

$$
\min_{\boldsymbol{\Phi}} \left(\sum_{k \neq k'} \mu_{kk'}^2(\boldsymbol{\Theta}) + \lambda \frac{1}{\mathrm{SIR}} \right) \tag{12.37}
$$

where λ is a positive weight, which reflects the tradeoff between the SCSM and SIR.

The problem of (12.37) is not convex. In order to obtain a solution, let us first view (12.37) as an optimization problem with respect to $\mathbf{B} = \boldsymbol{\Phi}^H \boldsymbol{\Phi}$. Furthermore, let us set the norm of the columns of the sensing matrix to 1, that is, $N_r N_p \mathbf{u}_k^H \boldsymbol{\Phi}^H \boldsymbol{\Phi} \mathbf{u}_k = 1$, $k = 1, \ldots, N$; this will significantly simplify the expression for $\mu_{kk'}(\boldsymbol{\Theta})$ and $1/\mathrm{SIR}$. Now, (12.37) can be reformulated as

$$
\min_{\mathbf{B}} \sum_{k=1}^{N-1} \sum_{k'=k+1}^{N} \left| \sum_{m=1}^{N_p} \sum_{l=1}^{N_r} e^{j2\pi(q_{lmk'} - q_{lmk})} \mathbf{u}_{k'}^H \mathbf{B} \mathbf{u}_k \right|^2 + \lambda \mathrm{Tr}\{\mathbf{B}\}
$$

$$
\text{s.t.} \quad N_r N_p \mathbf{u}_k^H \mathbf{B} \mathbf{u}_k = 1, \ k = 1, \ldots, N,
$$

$$
\mathbf{B} \geq 0 \tag{12.38}
$$

which is a convex problem with respect to $\mathbf{B}$. Once $\mathbf{B}$ is obtained, the solution of (12.37) can be obtained based on the eigendecomposition $\mathbf{B} = \mathbf{V}\boldsymbol{\Sigma}\mathbf{V}^H$ as

$$
\boldsymbol{\Phi}_{\#1} = \sqrt{\tilde{\boldsymbol{\Sigma}}} \tilde{\mathbf{V}}^H \tag{12.39}
$$

where $\tilde{\boldsymbol{\Sigma}}$ is a diagonal matrix that contains on its diagonal the nonzero eigenvalues of $\boldsymbol{\Sigma}$, and $\tilde{\mathbf{V}}$ contains as its columns the corresponding eigenvectors.

Introducing $N(N-1)/2$ auxiliary variables to (12.38), that is, $t_{kk'}$, $k = 1, \ldots, N-1$, $k' = k+1, \ldots, N$, we have

$$\min_{\mathbf{B}, t_{kk'}} \sum_{k=1}^{N-1} \sum_{k'=k+1}^{N} t_{kk'} + \lambda \mathrm{Tr}\{\mathbf{B}\}$$

$$s.t. \quad N_r N_p \mathbf{u}_k^H \mathbf{B} \mathbf{u}_k = 1, \; k = 1, \ldots, N,$$

$$\left| \mathbf{u}_{k'}^H \mathbf{B} \mathbf{u}_k \right|^2 \le t_{kk'}, \; k = 1, \ldots, N-1, \; k' = k+1, \ldots, N,$$

$$\mathbf{B} \ge 0 \qquad (12.40)$$

Then (12.40) can be recast as a semidefinite program (SDP) as follows:

$$\min_{\mathbf{t}, \mathbf{B}} \quad \mathbf{1}_{1 \times \frac{N(N-1)}{2}} \mathbf{t} + \mathbf{a}^T vec(\mathbf{B})$$

$$s.t. \quad \mathbf{A}^T vec(\mathbf{B}) = \mathbf{1}_{N \times 1}$$

$$\mathbf{F}_{kk'}(\mathbf{t}) \ge 0, \; k = 1, \ldots, N-1, \; k' = k+1, \ldots, N$$

$$\mathbf{B} \ge 0 \qquad (12.41)$$

where $\mathbf{t}$ is composed of the $N(N-1)/2$ auxiliary variables $t_{kk'}$, $\mathbf{A}$ contains $vec((\mathbf{u}_k \mathbf{u}_k^H)^T) N_r N_p$ as its kth column,

$$\mathbf{a} = [\lambda, \mathbf{0}_{1 \times (L+\tilde{L})}, \lambda, \mathbf{0}_{1 \times (L \mid \tilde{L})}, \overbrace{\cdots, \mathbf{0}_{1 \times (L+\tilde{L})}, \lambda}^{(L+\tilde{L}) \, \lambda s}]^T,$$

and

$$\mathbf{F}_{kk'}(\mathbf{t}) = \begin{bmatrix} t_{kk'} & vec((\mathbf{u}_k \mathbf{u}_{k'}^H)^T)^T vec(\mathbf{B}) \\ (vec((\mathbf{u}_k \mathbf{u}_{k'}^H)^T)^T vec(\mathbf{B}))^H & 1 \end{bmatrix},$$

$$k = 1, \ldots, N-1, \; k' = k+1, \ldots, N \qquad (12.42)$$

Multiple software packages can be used to solve (12.41), for example, Sedumi [38] and CVX [39].

12.4.3.2 Measurement Matrix Design by Improving SIR Only

Solving (12.41) requires prohibitively high computational load. Further, the solution needs to be adapted to a particular basis matrix. In order to avoid these two shortcomings, the measurement matrix can be designed targeting SIR improvement only [12]. There are two important assumptions in such design.

- A1. We assume that the Doppler shift within a pulse can be ignored, which holds if $2vf T_p / c \ll 1$. This assumption simplifies the basis matrix as it allows us to ignore the Doppler information in the basis matrix.

- A2. $1/\tilde{M} \sum_{i=1}^{\tilde{M}} a_i x_i y_i \approx E(a_i x_i y_i) = 0$ for sufficiently large $\tilde{M}$, where $x_i, y_i \sim \mathcal{N}(0, 1/\tilde{M})$ and are i.i.d., and a_i are uncorrelated with x_i, y_i, $i = 1, \ldots, \tilde{M}$. This assumption will be used to simplify the SIR expression.

As in [12], let us impose the following special structure on the measurement matrix, which reduces the dimensionality of the design problem:

$$\boldsymbol{\Phi}_{\#2} = \boldsymbol{\Phi} \mathbf{W}^H \tag{12.43}$$

where

$\boldsymbol{\Phi}$ is an $M \times \tilde{M}$ ($M \leq \tilde{M}$) zero-mean Gaussian random matrix
$\mathbf{W}$ is an $(L + \tilde{L}) \times \tilde{M}$ deterministic matrix to be determined, satisfying diag$\{\mathbf{W}^H \mathbf{W}\} = [1, \ldots, 1]^T$

The matrix $\mathbf{W}$ can be selected to improve the detection performance of the CS approach at the receiver. Matrix $\boldsymbol{\Phi}_{\#2}$ compresses the signal without increasing the CSM. As shown in [40], with the appropriate $\mathbf{W}$, $\boldsymbol{\Phi}_{\#2}$ does result in higher CSM as compared with the conventional measurement matrix. Next, we discuss the selection of $\mathbf{W}$.

Assuming that the time delay induced by the kth target follows a discrete uniform distribution, that is, $p(\tau_k = k) = 1/(\tilde{L} + 1), k = 0, \ldots, \tilde{L}$, the average SIR of the kth target return, taken over the node locations for uniformly distributed nodes, can be approximated by

$$\overline{SIR}_k = \frac{|\beta_k|^2}{\sigma^2 \tilde{M}} \sum_{\tau=0}^{\tilde{L}} \frac{1}{\tilde{L} + 1} \mathrm{Tr}\{\mathbf{W}^H \mathbf{Q}_\tau \mathbf{W}\} = \frac{|\beta_k|^2}{\sigma^2 \tilde{M}} \frac{1}{\tilde{L} + 1} \mathrm{Tr}\{\mathbf{W}^H \mathbf{C} \mathbf{W}\} \tag{12.44}$$

where

$$\mathbf{Q}_{\tau_k} = \mathbf{C}_{\tau_k} \mathbf{X} \mathbf{X}^H \mathbf{C}_{\tau_k}^H,$$

$$\mathbf{C} = \sum_{\tau=0}^{\tilde{L}} \mathbf{Q}_\tau = [\mathbf{C}_0 \mathbf{X}, \ldots, \mathbf{C}_{\tilde{L}} \mathbf{X}][\mathbf{C}_0 \mathbf{X}, \ldots, \mathbf{C}_{\tilde{L}} \mathbf{X}]^H. \tag{12.45}$$

Therefore, the optimization problem that maximizes $\overline{SIR}_k$ can be written as

$$\mathbf{W}^* = \max_{\mathbf{W}, \tilde{M}} \overline{SIR}_k$$

$$s.t. \ \mathrm{diag}\{\mathbf{W}^H \mathbf{W}\} = [1, \ldots, 1]_{\tilde{M} \times 1}^T \tag{12.46}$$

The solution $\mathbf{W}^*$ of the aforementioned problem contains as its columns the eigenvectors corresponding to the largest eigenvalue of $\mathbf{C}$. This solution may result in an ill-conditioned sensing matrix. Therefore, a feasible $\mathbf{W}$ by taking all possible delays into account is proposed as follows:

$$\mathbf{W} = [\mathbf{C}_0 \mathbf{X}, \ldots, \mathbf{C}_{\bar{L}} \mathbf{X}] \tag{12.47}$$

It was shown in [40] that $\boldsymbol{\Phi}_{\#2}$ enables SIR improvement with a negligible increase in the maximum CSM as compared with the Gaussian measurement matrix.

12.4.3.3 $\boldsymbol{\Phi}_{\#1}$ versus $\boldsymbol{\Phi}_{\#2}$

The advantages and disadvantages of $\boldsymbol{\Phi}_{\#1}$ and $\boldsymbol{\Phi}_{\#2}$ are summarized as follows:

- *Complexity*
  Solving $\boldsymbol{\Phi}_{\#1}$ involves a complex optimization problem and depends on a particular basis matrix, while $\boldsymbol{\Phi}_{\#2}$ requires only knowledge of all the possible discretized time delays. Therefore, the construction of $\boldsymbol{\Phi}_{\#1}$ involves higher computational complexity than $\boldsymbol{\Phi}_{\#2}$.
- *Performance*
  $\boldsymbol{\Phi}_{\#1}$ aims at decreasing the coherence of the sensing matrix and enhancing SIR simultaneously. The tradeoff between CSM and SIR results in $\boldsymbol{\Phi}_{\#1}$ yielding lower SIR than $\boldsymbol{\Phi}_{\#2}$. Therefore, $\boldsymbol{\Phi}_{\#1}$ is expected to perform better than $\boldsymbol{\Phi}_{\#2}$ in the case of low interference, while it should perform worse in the presence of strong interference.

Simulation Example 12.5

We consider the same antenna configuration as *Simulation Example 12.1* with $M_t = N_r = 4$. The TX antennas transmit Hadamard waveforms of length $L = 128$. A jammer is located at angle $7°$ and transmits an unknown Gaussian random waveform. Because of the high computational complexity required to obtain $\boldsymbol{\Phi}_{\#1}$, we consider a small-size grid with a limited number of grid points in the angle–range space $[0°, 1°] \times [10010 \text{ m}, 10090 \text{ m } 10, 100 \text{ m}, 10,090 \text{ m}]$. The spacing of adjacent angle–range grid points is $[0.2°, 15 \text{ m}]$. Three targets are considered, randomly distributed on this grid. The data of only one pulse are used, and thus only the angle–range estimates can be obtained.

   Figure 12.16 compares ROC performance of CS-MIMO radars using $\boldsymbol{\Phi}_{\#1}$, $\boldsymbol{\Phi}_{\#2}$ and a Gaussian random measurement matrix (GRMM), and that of MIMO radars that use MFM, based on 100 independent and random runs. The CS-based approach uses $M = 20$ samples, while the MFM-based approach uses 100 samples. Different combinations of

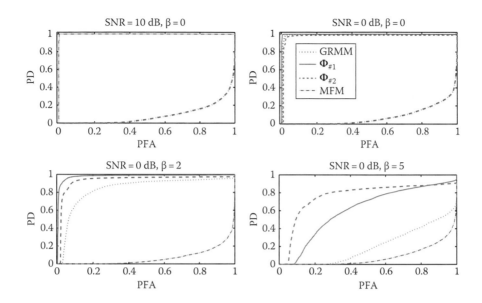

FIGURE 12.16
(See color insert.) ROC curves for CS-MIMO radars using $\Phi_{\#1}$, $\Phi_{\#2}$ and the GRMM and for MIMO radars using the MFM ($M_t = N_r = 4$ and $\lambda = 1.5$).

SNR and jammer-signal power are considered. It can be seen that $\Phi_{\#1}$ and $\Phi_{\#2}$ with Hadamard waveforms can improve detection accuracy as compared with the GRMM in the case of mild and strong interference, respectively. Note that the three measurement matrices give rise to similar performance for $SNR = 10$ dB and $\beta = 0$. This is because the interference is sufficiently small so that all the measurement matrices perform well. Again, one can see that the MFM is inferior to the CS approach, although it uses far more measurements than the CS approach.

12.5 Application to Through-the-Wall Radar

Through-the-wall radars (TWRs) enable the sensing of objects and/or humans behind obstacles or even the layout of buildings. TWRs are indispensable for situational awareness in a wide range of civilian and military applications, including surveillance or detection of survivors trapped in heaps of rubble during natural disasters.

MIMO TWRs have been studied in [42,43]. In that context, multiple spatially distributed antennas provide different points of view of the scene, allowing one to mitigate artifacts, caused by obstacles, thus enhancing target

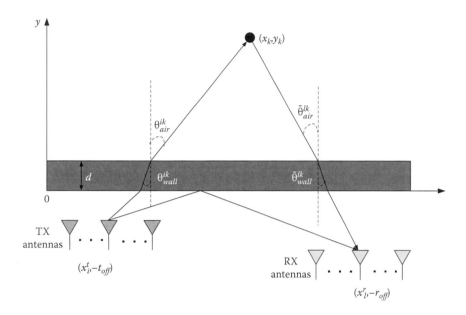

FIGURE 12.17
(See color insert.) Geometry of through-the-wall propagation and wall reflection.

detectability in indoor environments. CS-MIMO TWRs have also been proposed [44], allowing for reduction of the number of samples and data
acquisition time as compared with MIMO TWRs. In [44], the TX antennas
emit in a time division multiplexing fashion. In the following, we present
a CS-MIMO radar approach for TWRs, in which the TX antennas transmit
simultaneously different waveforms, thus allowing for further reduction of
acquisition time.

Let us consider a linear M_t-element TX array and the N_r-element RX
array, located parallel to the wall at standoff distances t_{off} and r_{off}, respectively (see Figure 12.17). Let $(x_i^t, -t_{off})$ and $(x_i^r, -r_{off})$ denote the locations
of the ith TX and RX antennas, respectively. The wall is homogeneous with
thickness d and dielectric constant ε, and is located along the x-axis. K stationary point targets are behind the wall, with their locations denoted by
$(x_k, y_k), k = 1, \ldots, K$. Wideband signals $x_i(t), i = 1, \ldots, M_t$, are transmitted
via the random step frequency approach, that is, the pulses have carrier frequency $f_n = f_0 + C_n B/(N_s), n = 1, \ldots, N_p$, where f_0 is the center frequency, B
is the bandwidth, and C_n is a random integer between 0 and $N_s - 1$.

For simplicity, and to demonstrate the concept, we only consider the
front wall reflection and a single propagation path (see Figure 12.17). Therefore, the parameters to be estimated are $x_k, y_k, k = 1, \ldots, K$. The signals
received at the RX antennas are composed of the target and front wall
returns, that is,

- *Target Returns*: The target returns in the baseband received at the lth RX antenna during the mth pulse can be expressed as

$$z_{lm}^{target}(t) = \sum_{k=1}^{K} \sum_{i}^{M_t} \beta_k x_i(t - \tau_{ilk}) e^{-j2\pi f_m \tau_{ilk}} \qquad (12.48)$$

where
β_k is the complex reflectivity of the kth target
τ_{ilk} denotes the propagation delay due to the kth target for the TX–RX antenna pair (i, l)

On letting θ_{air}^{ik} and $\tilde{\theta}_{wall}^{lk}$, respectively, denote the angle of incidence and refraction corresponding to the kth target and ith TX/RX antenna, τ_{ilk} can be expressed as

$$\tau_{ilk} = \frac{t_{off} + y_k - d}{\cos(\theta_{air}^{ik})c} + \frac{d\sqrt{\varepsilon}}{\cos(\theta_{wall}^{ik})c}$$
$$+ \frac{r_{off} + y_k - d}{\cos(\tilde{\theta}_{air}^{lk})c} + \frac{d\sqrt{\varepsilon}}{\cos(\tilde{\theta}_{wall}^{lk})c} \qquad (12.49)$$

Once the distances between the target and the antenna along the x-axis and the y-axis $(\Delta x, \Delta y)$ have been found, one can obtain the angle incidence θ_{air} and the refraction angle θ_{wall} by solving the following nonlinear equations

$$\frac{\theta_{air}}{\theta_{wall}} = \sqrt{\varepsilon} \qquad (12.50)$$

$$\Delta x = (\Delta y - d)\tan(\theta_{air}) + d\tan(\theta_{wall}) \qquad (12.51)$$

- *Wall Clutter*: The received signal due to the front wall reflection at the lth RX antenna during the mth pulse can be expressed as

$$z_{lm}^{wall}(t) = \sum_{l=1}^{N_r} \sum_{i=1}^{M_t} \beta_{wall} x_i(t - \tau_{il}^{wall}) e^{-j2\pi f_m \tau_{il}^{wall}} \qquad (12.52)$$

where
β_{wall} denotes the complex reflectivity of the front wall
τ_{il}^{wall} denotes the propagation delay for the TX–RX antenna pair (i, l)
τ_{il}^{wall} is given by [43]

$$\tau_{il}^{wall} = \frac{\sqrt{(x_i^t - x_{il})^2 + t_{off}^2}}{c} + \frac{\sqrt{(x_i^r - x_{il})^2 + r_{off}^2}}{c} \tag{12.53}$$

where

$$x_{il} = \frac{x_i^t r_{off} + x_i^r t_{off}}{t_{off} + r_{off}}. \tag{12.54}$$

Therefore, the received signal at the lth RX antenna during the mth pulse can be formulated as

$$z_{lm}(t) = z_{lm}^{target}(t) + z_{lm}^{wall}(t) + n_{lm}(t) \tag{12.55}$$

where $n_{lm}(t)$ denotes the interference.

Let us discretize the $2 - D$ imaging region into N grid points, that is, $[(x_n, y_n)], n = 1, \ldots, N$ and let s_n denote the coefficient associated with the nth grid point. The received signal $z_{lm}(t)$ can be rewritten as a linear combination of target returns reflected from all grid points, that is,

$$z_{lm}(t) = \sum_{n=1}^{N} \sum_{i=1}^{M_t} s_n x_i(t - \tau_{iln}) e^{-j2\pi f_m \tau_{iln}} + z_{lm}^{wall}(t) + n_{lm}(t) \tag{12.56}$$

where τ_{iln} is the propagation delay due to the nth grid point for the TX/RX antenna pair (i, l) and equals to

$$\tau_{iln} = \frac{t_{off} + y_n - d}{\cos\left(\theta_{air}^{in}\right)c} + \frac{d\sqrt{\varepsilon}}{\cos\left(\theta_{wall}^{in}\right)c}$$

$$+ \frac{r_{off} + y_n - d}{\cos\left(\tilde{\theta}_{air}^{ln}\right)c} + \frac{d\sqrt{\varepsilon}}{\cos\left(\tilde{\theta}_{wall}^{ln}\right)c} \tag{12.57}$$

If the kth target is located at (x_n, y_n), the coefficient s_n equals β_k; otherwise, s_n is zero.

The lth receiver collects M linearly compressed measurements during each pulse repetition interval T through the measurement matrix Φ_l. Let us stack all the compressed samples collected during the mth pulse for the lth RX antenna into a vector $\mathbf{r}_{lm}$. It holds that

$$\mathbf{r}_{lm} = \sum_{n=1}^{N} \sum_{i=1}^{M_t} s_n \mathbf{\Phi}_l \mathbf{C}_{\tau_{iln}} \mathbf{x}_i e^{-j2\pi f_m \tau_{iln}} + \mathbf{\Phi}_l \mathbf{z}_{lm}^{wall} + \mathbf{\Phi}_l \mathbf{n}_{lm}$$

$$= \mathbf{\Phi}_l \mathbf{\Psi}_{lm} \mathbf{s} + \mathbf{\Phi}_l \mathbf{z}_{lm}^{wall} + \mathbf{\Phi}_l \mathbf{n}_{lm} \tag{12.58}$$

where

$\mathbf{x}_i, i = 1, \ldots, M_t$, are the transmitted sequences of length L and time spacing T_s, and thus the pulse length is $T_p = LT_s$

$\mathbf{C}_\tau = [\mathbf{0}_{L \times \lfloor \frac{\tau}{T_s} \rfloor}, \mathbf{I}_L, \mathbf{0}_{L \times (\tilde{L} - \lfloor \frac{\tau}{T_s} \rfloor)}]^T$ is of size $(L + \tilde{L}) \times L$. $\tilde{L}T_s$ is the maximum time delay of the imaging region. Here we assume that the target returns fall completely within the sampling window of length $(L + \tilde{L})T_s$

$\mathbf{\Phi}_l$ is the $M \times (L + \tilde{L})$ measurement matrix for the lth receive node

$\mathbf{z}_{lm}^{wall} = [z_{lm}^{wall}(0T_s + (m-1)T), \ldots, z_{lm}^{wall}((L + \tilde{L} - 1)T_s + (m-1)T)]^T$

$\mathbf{n}_{lm} = [n_{lm}(0T_s + (m-1)T), \ldots, n_{lm}((L + \tilde{L} - 1)T_s + (m-1)T)]^T$

$\mathbf{s} = [s_1, \ldots, s_N]^T$ is the target indicator vector

$\mathbf{\Psi}_{lm} = \left[\sum_i^{M_t} \mathbf{C}_{\tau_{il1}} \mathbf{x}_i e^{-j2\pi f_m \tau_{il1}}, \ldots, \sum_i^{M_t} \mathbf{C}_{\tau_{ilN}} \mathbf{x}_i e^{-j2\pi f_m \tau_{ilN}} \right]$

By combining the output of N_p pulses at N_r receive antennas, the fusion center can formulate the equation

$$\mathbf{r} \stackrel{\triangle}{=} [\mathbf{r}_{11}^T, \ldots, \mathbf{r}_{1N_p}^T, \ldots, \mathbf{r}_{N_r N_p}^T]^T$$

$$= \mathbf{\Phi} \mathbf{\Psi} \mathbf{s} + \mathbf{n} \tag{12.59}$$

where

$$\mathbf{\Psi} = [(\mathbf{\Psi}_{11})^T, \ldots, (\mathbf{\Psi}_{1N_p})^T, \ldots, (\mathbf{\Psi}_{N_r N_p})^T]^T$$

$$\in \mathcal{C}^{(L+\tilde{L})N_r N_p \times N}$$

$$\mathbf{\Phi} = \text{diag}\{[\mathbf{\Phi}_1, \ldots, \mathbf{\Phi}_1, \mathbf{\Phi}_2, \ldots, \mathbf{\Phi}_2, \ldots, \mathbf{\Phi}_{N_r}, \ldots, \mathbf{\Phi}_{N_r}]\}$$

$$\in \mathcal{C}^{N_r N_p M \times (L+\tilde{L})N_r N_p} \tag{12.60}$$

and

$$\mathbf{n} = [(\mathbf{\Phi}_1(\mathbf{z}_{11}^{wall} + \mathbf{n}_{11}))^T, \ldots, (\mathbf{\Phi}_1(\mathbf{z}_{1N_p}^{wall} + \mathbf{n}_{1N_p}))^T, \ldots,$$

$$(\mathbf{\Phi}_{N_r}(\mathbf{z}_{N_r N_p}^{wall} + \mathbf{n}_{N_r N_p}))^T]^T$$

$$\in \mathcal{C}^{N_r N_p M \times 1} \tag{12.61}$$

The front wall induces very strong reflection and thus masks the targets behind the wall. Therefore, the removal or mitigation of the wall reflection is a crucial preprocessing step for TWR imaging. Typical ways to remove the wall clutter from the received signal include background subtraction [45], subspace projection [43], or time gating [44]. Then the target indicator **s** can be easily solved by applying the existing CS algorithms [19–25] to the *clean* data. The indices of the nonzero entries of **s** correspond to the indices of the grid points where the targets are located on.

Although here we have considered single-path propagation and wall reflection, the idea can be extended to a scenario in which the wall reverberations and propagation multipaths exist.

Simulation Example 12.6

Let us consider the same antenna configuration of Figure 12.17. The center carrier frequency is $f_0 = 1$ GHz, the step frequency bandwidth $B = 6$ GHz, and the number of frequency bins $N_s = 60$. The TX and RX antennas are equally spaced with spacing $1.5\lambda = 0.45$ m along the x-axis. The standoff distances of the TX and RX antenna array are 1 m and 1.5 m, respectively. The leftmost TX and RX antennas are located at the origin of the Cartesian coordinate system. The transmitted waveforms are orthogonal Hadamard sequences with $L = 8$ and $T_s = 0.5$ ns. A homogeneous wall is located along the x-axis; it has thickness $d = 0.2$ m, permittivity $\varepsilon = 4$, and reflectivity $\beta_{wall} = 50$. Three targets are located at $(0.171, 0.325)$m, $(1.89, 1.356)$m, $(1.13, 0.85)$m and of reflectivity 0.5, 1, 0.6, respectively. The imaging region, behind the wall, is of size 2 m × 2 m; it is discretized on the 2-D grid $[0, 0.02, \dots, 2]$m × $[0.2, 0.22, \dots, 2.2]$m. The wall clutter is removed by the time gating method. The measurement matrix is constructed by M equally spaced rows of the identity matrix. This type of measurement matrix enables reduction of the sampling rate without complicating the hardware implementation. The transmitted signal of a TX antenna is of unit energy during each pulse and the *SNR* is defined as the inverse of the variance of the thermal noise. The SNR is set to 0 dB here. The target indicator **s** is obtained by solving the following convex problem:

$$\min \|\mathbf{s}\|_1 \quad s.t. \|(\mathbf{r} - \boldsymbol{\Phi}\boldsymbol{\Psi}\mathbf{s})\|_2 < \sigma \qquad (12.62)$$

Figure 12.18 shows the reconstructed images in one realization produced by the CS imaging method and the MFM imaging method for MIMO TWR. The number of TX antennas is fixed to $M_t = 6$. For the MEM imaging method, the images without thresholding and with thresholding at 0.1 are shown in the second and third rows of Figure 12.18, respectively. The true image in the first subfigure is also demonstrated for reference. One can see that given the same set of parameters $(N_r = 2, N_p = 8, M = 10)$, the CS imaging method produces a clean and accurate image, while the MFM imaging method reconstructs a blurred image where the target with reflectivity 0.5 is masked by the noisy background. In this case, the postprocessing step of thresholding fails to help clean the background. As the three parameters increased to $(N_r = 4, N_p = 12, M = 35)$, the MFM imaging method performs similarly to the CS imaging method but still with an inferior resolution. Figure 12.18 perfectly demonstrates the advantages of the CS-MIMO TWR over

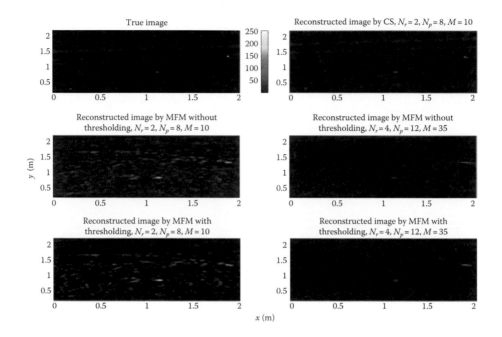

FIGURE 12.18
(See color insert.) Reconstructed image by CS and MFM.

the MFM-MIMO TWR: with fewer RX antennas, lower sampling rate (or samples), and less acquisition time (or pulses), CS-MIMO TWR achieves better performance.

Acknowledgments

This work was supported by ONR under grants ONR-N00014-07-1-0500 and ONR-N00014-12-1-0036.

References

1. E. Fishler, A. Haimovich, R. Blum, D. Chizhik, L. Cimini, and R. Valenzuela, MIMO radar: An idea whose time has come, in *Proceedings of the IEEE Radar Conference*, Philadelphia, PA, pp. 71–78, April 2004.
2. E. Fishler, A. Haimovich, R. Blum, L. Cimini, D. Chizhik, and R. Valenzuela, Performance of MIMO radar systems: Advantages of angular diversity, in *Proceedings of the 38th Asilomar Conference on Signals, Systems, and Computers*, Pacific Grove, CA, pp. 305–309, November 2004.

3. A.M. Haimovich, R.S. Blum, and L.J. Cimini, MIMO radar with widely separated antennas, *IEEE Signal Processing Magazine*, 25 (1), 116–129, January 2008.
4. P. Stoica and J. Li, MIMO radar with colocated antennas, *IEEE Signal Processing Magazine*, 24 (5), 106–114, September 2007.
5. J. Li, P. Stoica, L. Xu, and W. Roberts, On parameter identifiability of MIMO radar, *IEEE Signal Processing Letters*, 14 (12), 968–971, December 2007.
6. L. Xu, J. Li, and P. Stoica, Radar imaging via adaptive MIMO techniques, in *Proceedings of the European Signal Processing Conference*, Florence, Italy, September 2006.
7. M.I. Skolnik, *Radar Handbook*. New York: McGraw-Hill, 1990.
8. W.D. Wirth, *Radar Techniques Using Array Antennas*. London, U.K.: Institution of Electrical Engineers, 2001.
9. A.P. Petropulu, Y. Yu, and H.V. Poor, Distributed MIMO radar using compressive sampling, in *Proceedings of the 42nd Asilomar Conference on Signals, Systems, and Computers*, Pacific Grove, CA, pp. 203–207, November 2008.
10. C.Y. Chen and P.P. Vaidyanathan, Compressed sensing in MIMO radar, in *Proceedings of the 42nd Asilomar Conference on Signals, Systems, and Computers*, Pacific Grove, CA, pp. 41–44, November 2008.
11. T. Strohmer and B. Friedlander, Compressed sensing for MIMO radar— Algorithms and performance, in *Proceedings of the 43rd Asilomar Conference on Signals, Systems, and Computers*, Pacific Grove, CA, pp. 464–468, November 2009.
12. Y. Yu, A.P. Petropulu, and H.V. Poor, MIMO radar using compressive sampling, *IEEE Journal of Selected Topics in Signal Processing*, 4 (1), 146–163, February 2010.
13. S. Gogineni and A. Nehorai, Target estimation using sparse modeling for distributed MIMO radar, *IEEE Transactions on Signal Processing*, 59 (11), 5315–5325, November 2011.
14. Y. Yu and A.P. Petropulu, A study on power allocation for widely separated CS-based MIMO radar, in *Proceedings of SPIE 8365, Compressive Sensing*, 83650S (June 8, 2012). doi:10.1117/12.919734.
15. E.J. Candes, J.K. Romberg, and T. Tao, Stable signal recovery from incomplete and inaccurate measurements, *Communications on Pure and Applied Mathematics*, 59 (8), 1207–1223, August 2006.
16. D.L. Donoho, Compressed sensing, *IEEE Transactions on Information Theory*, 52 (4), 1289–1306, April 2006.
17. E. Candes and T. Tao, Decoding by linear programming, *IEEE Transactions on Information Theory*, 51 (12), 4203–4215, December 2005.
18. E.J. Candes, Compressive sampling, in *Proceedings of the International Congress of Mathematicians*, Madrid, Spain, 2006.
19. J.A. Tropp and A.C. Gilbert, Signal recovery from random measurement via orthogonal matching pursuit, *IEEE Transactions on Information Theory*, 53 (12), 4655–4666, December 2007.
20. D. Needell and R. Vershynin, Signal recovery from incomplete and inaccurate measurements via regularized orthogonal matching pursuit, *IEEE Journal of Selected Topics in Signal Processing*, 4 (2), 310–316, April 2010.
21. D. Needell and R. Vershynin, Uniform uncertainty principle and signal recovery via regularized orthogonal matching pursuit, *Foundations of Computational Mathematics*, 9 (3), 317–334, April 2009.

22. D. Needell and R. Vershynin, COSAMP: Iterative signal recovery from incomplete and inaccurate samples, *Applied and Computational Harmonic Analysis*, 26 (3), 301–321, April 2008.

23. J.A. Tropp, Just relax: Convex programming methods for identifying sparse signals in noise, *IEEE Transactions on Information Theory*, 52 (3), 1030–1051, March 2006.

24. E.J. Candes and J. Romberg, ℓ_1-MAGIC: Recovery of sparse signals via convex programming. http://www.acm.caltech.edu/l1magic/, October 2008.

25. E. Candes and T. Tao, The Dantzig selector: Statistical estimation when p is much larger than n, *Annals of Statistics*, 35 (6), 2313–2351, December 2007.

26. T. Strohmer and B. Friedlander, Analysis of sparse MIMO radar. https://www.math.ucdavis.edu/strohmer/papers/2012/sparsemimo.pdf, 2012.

27. Y. Chi, L.L. Scharf, A. Pezeshki, and A.R. Calderbank, Sensitivity to basis mismatch in compressed sensing, *IEEE Transactions on Signal Processing*, 59 (5), 2182–2195, May 2011.

28. Y. Yu, A.P. Petropulu, and H.V. Poor, CSSF MIMO radar: Compressive-sensing and step-frequency based MIMO radar, *IEEE Transactions on Aerospace and Electronic Systems*, 48 (2), 1490–1504, April 2012.

29. Y. Yu, S. Sun, and A.P. Petropulu, A capon beamforming method for clutter suppression in colocated compressive sensing based MIMO radars, in *Proceedings of SPIE 8717, Compressive Sensing II*, 87170J (May 31, 2013). doi:10.1117/12.2015635.

30. G. Barriac, R. Mudumbai, and U. Madhow, Distributed beamforming for information transfer in sensor networks, in *Proceedings of the 2004 International Symposium on Information Processing in Sensor Networks*, Berkeley, CA, pp. 81–88, April 2004.

31. R. Mudumbai, J. Hespanha, U. Madhow, and G. Barriac, Distributed transmit beamforming using feedback control, *IEEE Transactions on Information Theory*, 56 (1), 411–426, January 2010.

32. I. Thibault, A. Faridi, G.E. Corazza, A.V. Coralli, and A. Lozano, Design and analysis of deterministic distributed beamforming algorithms in the presence of noise, *IEEE Transactions on Communications*, 61 (4), 1595–1607, April 2013.

33. D.R. Brown and H.V. Poor, Time-slotted round-trip carrier synchronization for distributed beamforming, *IEEE Transactions on Signal Processing*, 56, 5630–5643, 2008.

34. R.D. Preuss and D.R. Brown, Two-way synchronization for coordinated multicell retrodirective downlink beamforming, *IEEE Transactions on Signal Processing*, 59, 5415–5427, 2011.

35. W. Hao, G. Qiang, and F. Li, Effects of carrier synchronization on link throughput of distributed beamforming, in *Proceedings of the IEEE 21st International Symposium on Personal Indoor and Mobile Radio Communications (PIMRC)*, Istanbul, Turkey, pp. 2111–2116, 2010.

36. Y. Yu and A.P. Petropulu, Power allocation for CS-based colocated MIMO radar systems, in *The Seventh IEEE Sensor Array and Multi-channel Signal Processing Workshop (SAM)*, Hoboken, NJ, June 2012.

37. Y. Yu, S. Sun, R. Madan, and A.P. Petropulu, Power allocation and waveform design for the compressive sensing based MIMO radar, to appear in *IEEE Transactions on Aerospace and Electronic Systems*.

38. SeDuMi, Optimization over symmetric cones, CORAL lab, Department of Industrial and Systems Engineering, Lehigh University. http://sedumi.ie.lehigh.edu/.
39. M. Grant and S. Boyd, CVX: Matlab software for disciplined convex programming, version 2.0 beta. http://cvxr.com/cvx, September 2012.
40. Y. Yu, A.P. Petropulu, and H.V. Poor, Measurement matrix design for compressive sensing based MIMO radar, *IEEE Transactions on Signal Processing*, 59 (11), 5338–5352, November 2011.
41. J.A. Tropp, Greed is good: Algorithmic results for sparse approximation, *IEEE Transactions on Information Theory*, 50 (10), 2231–2242, October 2004.
42. B. Boudamouz, P. Millot, and C. Pichot, Through the wall radar imaging with MIMO beamforming processing—Simulation and experimental results, *American Journal of Remote Sensing*, 7–12, January 2013. doi:10.11648/j. ajrs.20130101.12.
43. F. Ahmad and M.G. Amin, Wall clutter mitigation for MIMO radar configurations in urban sensing, in *11th International Conference on Information Sciences, Signal Processing and their Applications (ISSPA)*, Montreal, QC, Canada, pp. 1165–1170, July 2012.
44. J. Qian, F. Ahmad, and M.G. Amin, Joint localization of stationary and moving targets behind walls using sparse scene recovery, *Journal of Electronic Imaging*, 22 (2), 021002–021002, April 2013. doi:10.1117/ 1.JEI.22.2.021002.
45. F. Ahmad and M.G. Amin, Multi-location wideband synthetic aperture imaging for urban sensing applications, *Journal of the Franklin Institute*, 345 (6), 618–639, September 2008.

13

Compressive Sensing Meets Noise Radar

Mahesh C. Shastry, Ram M. Narayanan, and Muralidhar Rangaswamy

CONTENTS

ABSTRACT In this chapter, we discuss how noise radar systems are suitable for realizing practically the promises of compressive sensing in radar imaging, in general, and in urban-sensing applications, in particular. Noise radar refers to radio frequency imaging systems that employ transmit signals that are generated to resemble random noise waveforms. Noise radar has recently been successfully applied to urban sensing applications such as through-the-wall sensing (Amin 2011). Recent advances in the field of compressive sensing provide us with techniques to overcome the challenges of waveform design, sampling, and bandwidth constraints. We review existing literature related to these problems and present new results that enable us to leverage compressive sensing and sparsity to improve noise radar systems. We model compressively sampled noise radar imaging as a problem of inverting linear system with a circulant random system matrix. We demonstrate the feasibility of this model by applying it to experimental data acquired using a millimeter wave ultrawideband noise radar system. Our principal contributions lie in developing theory and algorithms for imaging and detection strategies in compressively sampled noise radar imaging. We outline an approach based on extreme value statistics that works by empirically estimating the distribution of the residue of instances of the estimation algorithm. False alarms are treated as statistically rare events for estimating event probabilities in the compressive detection problem. We extrapolate the distribution of the residue from a small number of recovery instances to calibrate compressive noise radar systems. For deploying compressively sensed noise radar systems in real applications, it is necessary to develop convenient approaches to calibrate and characterize recovery performance.

13.1 Introduction

Imaging using noise radar can be cast as a linear inverse problem. In a typical radar system, an electromagnetic wave is transmitted into free space using a transmitting antenna and the reflections are measured by a receiving antenna. Information about the targets of interest is extracted by comparing the transmitted and reflected waveforms. In this chapter, we use the terms *target scene* and *target environment* to refer to the region of space in which unknown targets are present. The terms *transmitted waveform* and *received* or *reflected* waveform refer to the time series represented by the transmitted and reflected fields, respectively. The received and transmitted waveforms are discretized for processing, and the discrete locations of targets in each scene are referred to as *cells*. We concern ourselves primarily with imaging problems that involve building range profiles using noncoherent noise radar technology.

The most basic and widespread use of radar systems is in estimating range profiles of the target scene, velocity of targets, and in performing target detection tasks. For many years, conventional radar systems used analog processing systems for imaging. One of the main reasons for this is the desire to maximize the bandwidth of the transmit signals. A high bandwidth signal is crucial to achieving a high-range resolution. For most of the last six decades, the absence of efficient sampling hardware and digital processing limited the adoption of digital processing for radar systems. Since the 1990s (Wu and Li 1998), digital radar receivers are slowly becoming pervasive in radar imaging applications. This is largely thanks to the development of efficient high-rate analog-to-digital converters (ADCs) and digital

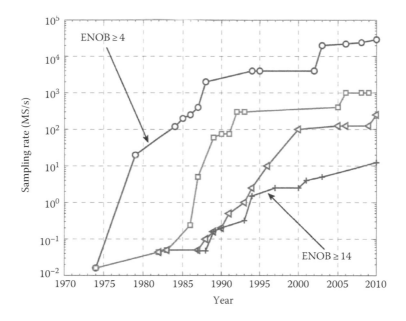

FIGURE 13.1
(See color insert.) Trends in the progress of ADC technology since 1975 to 2010, adapted from Jonsson (2010). ENOB stands for *effective number of bits* and refers to the achievable number of quantization levels. The different colors represent the progress of ADC technology for different values of ENOB. Blue line: ENOB $\geq$ 4. Red line: ENOB $\geq$ 8. Green line: ENOB $\geq$ 12. Black line: ENOB $\geq$ 14. (From Jonsson, B.E., A survey of A/D converter performance evolution, in: *17th IEEE International Conference on Electronics, Circuits, and Systems (ICECS)*, Athens, Greece, 2010, pp. 766–769. With permission.)

signal processing hardware. However, ADC technology has advanced at a much slower pace than the growth in computational capabilities. The move to digital systems has enabled us to incorporate even more advanced signal processing techniques at the receivers to improve the functionality of radar systems. The trends in the evolution of ADC technology from 1975 to 2010 are shown in Figure 13.1 (the figure has been adapted from a 2010 survey) (Jonsson 2010). Figure 13.1 is a plot of the peak sampling frequency achieved by state-of-the art ADCs for different effective number of bits (ENOBs). The figure represents the tradeoff between sampling frequency and the levels of quantization achievable by the ADC. As seen in Figure 13.1, the best 12-bit quantization ADCs available today can sample at a rate of less than 1 gigasamples per second. The development of high resolution radar technology is hindered by the cost of sampling the analog signal.

Noise radar technology involves transmitting random noise-like continuous waveforms. Traditionally, the received signals are noncoherently processed using a matched filter for range detection. The idea of using noise waveforms for radar imaging was first proposed in Horton (1959). Over the last decade (Narayanan et al. 1998), the trend has been toward the adoption of *ultrawideband* signals for noise radar. Ultrawideband waveforms are defined as signals that have a bandwidth of either 500 MHz or higher or 20% of the center frequency. In our applications, we use waveforms of 500 MHz bandwidth. The ultrawideband nature of the waveform enables us to achieve a high-range resolution. The basic ranging problem involves modeling the target scene as a linear filter with transfer function, $s(t)$, so that, for a given transmit signal $\psi(t)$ and additive noise $n(t)$, the reflected signal is simply given by

$$f(t) = \psi(t) * s(t) + n(t), \tag{13.1}$$

where $*$ represents the operation of linear convolution. In the past, noise radar signals have mainly been processed using a matched filter. Matched-filtering-based target recovery is premised on the fact that delayed versions of the transmitted signal are orthogonal to each other. If we discretize the signal model, we obtain

$$f = \Psi s + n, \tag{13.2}$$

where $f, s, n \in \mathbb{R}^N$ and $\Psi \in \mathbb{R}^{N \times N}$. With the orthogonality assumption, the recovered target is given by

$$s_{CR}^* = \Psi^{-1} f = \Psi^T f. \tag{13.3}$$

The advantage of using the matched filter is that its implementation is efficient in terms of cost and processing latency. In order to extract the target image, all we need to do is to compute the cross-correlation between the reflected waveform and the transmitted waveform.

In this chapter, we propose to use a general optimization-based approach to noise radar imaging as opposed to traditional cross-correlation processing. The discrete l_2-norm is defined for any vector $v = (v_1, v_2, \ldots, v_N)$ as $||v||_2 = \sum_{i=1}^N |v_i|^2$. The discrete l_1-norm is defined as $||v||_2 = \sum_{i=1}^N |v_i|$. The recovery of sparse solutions to underdetermined linear systems is possible in the framework of compressive sensing. We extend the signal model such that the system matrix $A = \Phi \Psi \in \mathbb{R}^{M \times N}$. Such a system matrix corresponds to sampling the received signal at a rate dictated by the information bandwidth rather than the signal bandwidth. We cast the problem as one of minimizing a least-squares-based cost function such as $\min_s ||f - \Psi s||_{l2}$ with an additional l_1 regularization term. While least-squares processing can potentially be more computationally expensive than correlation processing and more difficult to process in real time, such a model enables us to improve the functionality of the radar system. As will be shown in Section 13.2, with the least-squares formulation, we can utilize sparsity to more efficiently sample the signals. Least-squares and sparsity-based approaches are harder to analyze due to the usage of nonlinear and iterative recovery schemes. Further, target recovery involves significant computational cost. In Section 13.3, we propose to overcome these problems with the help of an efficient data-driven detection algorithm. We validate our theoretical and empirical results by analyzing experimental noise–radar data acquired using a millimeter-wave radar system operating with a bandwidth of 500 MHz.

13.1.1 State of the Art in Compressive Radar Imaging

Random noise radar (Horton 1959) involves transmitting waveforms that are generated as stochastic processes. The technology is headed toward utilizing ultrawideband transmit waveforms (Narayanan et al. 1998) with which high-range resolutions can be obtained. The use of randomly generated waveforms in noise radar makes signals immune to interception and jamming to an extent. Early stochastic waveform radars used analog processing to detect targets. Increasingly noise radar systems are using digital processing (Chen et al. 2012) for imaging in real time. Noise radar systems have been shown to be a viable technology for through-the-wall imaging in general and urban sensing in particular (Narayanan 2008). Digital noise radar systems use high rate ADC. The sampling rates of the ADC limit the maximum achievable range resolution of the system. In order to circumvent this bottleneck, we employ compressive sensing principles. The application of compressive sensing to through-the-wall imaging using stepped frequency radar was first reported in Yoon and Amin (2010). Practical compressive radar systems have been implemented for ground penetrating radar applications using stepped frequency waveforms (Gurbuz et al. 2009; Suksmono et al. 2010). Bar-Ilan and Elder (Bar-Ilan and Eldar 2014) applied the Xampling (Mishali et al. 2011) framework to develop a pulse radar prototype for joint range velocity radar imaging using Doppler focusing. Ender (Ender 2010) presented results on compressive radar imaging using pulsed chirp waveforms. Our experimental work differs from these in that we use incoherent ultrawideband continuous wave noise radar waveforms for imaging. Continuous wave noise-like waveforms have the advantage of being instantaneous wideband. Further, the waveforms are robust to additive noise, jamming, and interception (Narayanan et al. 1998).

The use of random transmit waveforms makes noise radar particularly suitable for compressive sensing. Random waveforms in compressive radar imaging were first suggested by Baraniuk and Steeghs (Baraniuk and Steeghs 2007) in the context of random demodulators. The recovery performance of compressive sensing estimators depends on the system matrix satisfying certain properties. The two most commonly studied properties are the restricted isometry property (RIP) and mutual coherence (Candes et al. 2006; Donoho 2006). The compressive noise radar imaging problem, as described in this chapter, involves a circulant random system matrix.

The suitability of circulant random matrices for compressive sensing is less well studied than the standard case of the random matrix. In the context of *random demodulators*, Romberg (Romberg 2009) showed that with specially designed circulant random system matrices recovery with probability $1 - O(n^{-1})$ is possible with $O(S \log^3 N)$ measurements. Bernoulli random circulant matrices were considered by Haupt et al. (Haupt et al. 2010) in the context of compressive channel estimation. They showed that based on the restricted isometry property of the system matrix, $O(S^2 \log N)$ measurements of the Toeplitz random matrix are sufficient for stable recovery using the Dantzig selector (Candes et al. 2006) recovery algorithm. Rauhut et al. (Rauhut et al. 2012) derived the RIP for circulant matrices, suggesting that $O\left(\max\left((S \log N)^{1.5}, S \log^2 S \log^2 N\right)\right)$ measurements guarantee stable recovery. Herman and Strohmer derived mutual coherence results for high-resolution radar imaging (Herman and Strohmer 2009) and proposed the use of Alltop sequences as transmit waveforms under the narrowband approximation. They also alluded to the effectiveness of random waveforms in high-resolution radar imaging. Noise waveforms also allow for target-matched waveform design in the context of compressive sensing (Shastry et al. 2013b). Our work extends the state of the art in considering ultrawideband random waveform radar systems that are primarily used for range imaging. An advantage of practical noise radar systems is their simplicity. Our analyses of practical issues relating to compressive radar imaging are intended to push the field toward real-world applications. Simulations indicate that the number of measurements follows the optimal compressive sensing asymptotics of $O(S \log N)$ (Shastry et al. 2010).

Portions of this chapter were presented in conference publications (Shastry et al. 2010, 2012, 2013a).

The summary of this chapter is as follows:

1. We formulate compressive sensing as a problem involving the inversion of a linear system with circulant random matrices. In Section 13.2, we use this linear system model to develop the theory of compressive noise radar imaging. We justify the circulant-matrix model by applying it to experimental noise radar data. We analyze the performance of compressive sensing through theoretical and empirical arguments. For the first time in literature, we experimentally verify the possibility of recovering targets from compressively sampled UWB noise radar. We conducted experiments using a millimeter-wave radar to validate the practicality of compressive noise radar.

2. In compressive noise radar systems, target recovery is achieved by using nonlinear convex optimization solvers. This complicates the task of target detection. First, while the target location is sparse,

the radar measurement is seldom sparse. Therefore, not accounting for this mismatch results in a performance loss. Under nonlinear recovery, it is difficult to derive theoretical closed form expressions for probabilities of detection and false alarm. The probability of false alarm and the statistics of the recovered vector are necessary to determine detection thresholds. In Section 13.3, we propose a data-driven tail estimation algorithm based on the theory of extreme value statistics. We fit a generalized Pareto distribution to the tail distribution of the detection variables to efficiently derive empirical expressions relating the probability of false alarm and the detection threshold. We test our algorithms on data acquired from experiments with real noise radar systems.

13.2 Basics of Compressive Stochastic Waveform Radar

13.2.1 Compressive Radar

The range estimation problem in radar imaging involves the inversion of a linear system. Let $\psi(t)$ be the transmit waveform, $s(t)$ denote a target scene, and $n(t)$ the additive noise in the system. The reflected waveform, $f(t)$ can be modeled as

$$f(t) = \int_{-\infty}^{\infty} \psi(\tau - t)s(\tau)d\tau + n(t). \tag{13.4}$$

The discrete form of this linear convolution problem results in the expression

$$f_i = \sum_k \psi_{i-k}s_k + \mathbf{n}_i, \tag{13.5}$$

$$f = \mathbf{\Psi}s + n, \tag{13.6}$$

where $f, s, n \in \mathbb{R}^N$ and $\Psi \in \mathbb{R}^{N \times N}$. In real situations, the vector s is typically sparse. For signals of finite duration, this equation is represented exactly by a linear system with a Toeplitz system matrix $\hat{\mathbf{\Psi}}$ so that $\hat{f} = \hat{\mathbf{\Psi}}s + n$. We approximate this as a linear system with a circulant system matrix $\mathbf{\Psi}$, in order to simplify computation and analysis. In our experiments, the substitution of the Toeplitz matrix with a circulant matrix is justified because the nonzero elements occur close to the beginning of the data record. Physically, this is due to the fact that we use ultrawideband waveforms. With a bandwidth of 500 MHz, as in our system, even a few microseconds of acquisition results in information related to a much larger range than typically useful. The error in recovering accurately at the ends of the signal is intrinsic to processing signals of finite time duration (Oppenheim et al. 1999). This type of error can only be overcome by increasing the dimensionality of our problem. Experimental justification for this approximation is illustrated by the accuracy of the experimental results described in Section 13.2.3.

In sparse target scenarios, compressive sensing theory allows us the luxury to undersample $y(t)$. We model this operation of undersampling as a premultiplication by the measurement matrix $\mathbf{\Phi} = R_\Omega \in \mathbb{R}^{M \times N}$, where R_Ω consists of the rows of the identity matrix indexed by the set $\Omega \subset \{1, 2, \ldots, N\}$. Thus, we have

$$y = R_\Omega(f + n) = R_\Omega \mathbf{\Psi} s + R_\Omega n, \tag{13.7}$$

with $y \in \mathbb{R}^M$. The performance of the compressive radar ranging problem depends on the properties of the matrix $R_\Omega \mathbf{\Psi} \triangleq A \in \mathbb{R}^{M \times N}$. We assume that the continuous target scene is discretized into N grid points, with only a small percentage of the *cells* occupied by scattering targets. We justify this assumption with experimental results. The sparsity of the anticipated solution is characterized by the assumption that $||s||_{l_0} \leq S$ with $S \ll N$. We define the quantities $\rho \triangleq S/M$ and $\delta \triangleq M/N$. If the matrix A satisfies certain properties for vectors of given sparsity, then the problem can be inverted by solving the following convex optimization problem, called basis pursuit denoising (BPDN):

$$\text{BPDN}(\rho, \delta; \sigma) : \min_{s \in \mathbb{R}^N} ||s||_1 \text{ subject to } ||y - As||_2 \leq \sigma. \tag{13.8}$$

This specific formulation of the problem is chosen over other formulations (Tropp and Wright 2010) of compressive sensing because it is useful for practical implementations. Specifically, the value σ has a natural association with the signal-to-noise ratio of the system. Practical radar systems can measure this either by aiming the radar signal at an area where targets are absent and noting the energy. When there is no opportunity to measure reflected signals in the absence of targets, this can be done by employing an SNR estimation approach (Pauluzzi and Beaulieu 2000). In either case, the radar operator has access to an approximate estimate of the value σ.

The RIP and mutual-coherence approaches to analyzing compressive sensing offer no exact results about the behavior of the residue of compressive estimation. The properties of the residue of compressive estimation are inferred via inequalities (Candes et al. 2006) that estimate the upper bound of residual error. Practical applications of compressive sensing require more detailed analyses of the residual error. With this in mind, the authors of the present chapter proposed using phase transition diagrams (Shastry et al. 2010) for characterizing radar systems.

13.2.2 Correlations in the Circulant Matrix

In practical compressive sensing systems, as we shall see in Section 13.2.4, transmit waveforms are not ideal. The assumption that transmit waveforms can be discretized into independent identically distributed (i.i.d.) random process is convenient for theoretical analysis of compressive recovery. In real systems, however, we need to characterize the effect of correlations that exist due to hardware-related nonidealities. We model the transmit waveform as the correlation of an i.i.d. random process $\tilde{psi}(t)$ and a transfer function $h(t)$ that represents the bandlimiting nonidealities so that

$$\psi(t) = \tilde{\psi}(t) * h(t) \tag{13.9}$$

$$\mathbf{\Psi} = \tilde{\mathbf{\Psi}}H. \tag{13.10}$$

In order to quantify the effect of correlations, we look at the transform point spread function (TPSF) of the system matrix. Ideally, for effective compressive signal recovery, we desire the nondiagonal elements of the normalized Gram matrix G to be as low as possible. We look at the error metric given by

$$\chi(G) = \sum_{i \neq j} |G_{i,j}|^2. \tag{13.11}$$

In Figure 13.2, we plot the values of $\chi(G)$ for various values of $G = \mathbf{\Psi}H_l$, where l denotes the width of the power spectrum of the filters. We characterize the matrix H_l with the parameter $l = ||P_h(f)||_{l0}$, where $P_h(f)$ refers to the power spectrum. Large values of l indicate that the waveform is nearly white and subsequently the sequence of random variables modeling the waveform are uncorrelated. Lower values of l correspond to narrowly filtered random processes. The narrow power spectrum corresponds to highly correlated waveforms. The interesting result from this simulation is actually that compressive sensing is fairly robust to correlations in the transmit waveform. This conforms with the experimental observations that we outline in Section 13.2.4. Even with a low-pass filter that only allows about 70% of the spectrum of the transmit waveform, we see in Figure 13.2 that χ is almost as good as the uncorrelated case.

13.2.3 Experiments

We used a millimeter-wave radar system to test the possibility of using compressive sensing for noise radar. The bandwidth of the signal used in this system is 500 MHz. A sample transmit waveform is shown in Figure 13.3. We operated the ADC at a rate of 1 gigasamples per second. The system

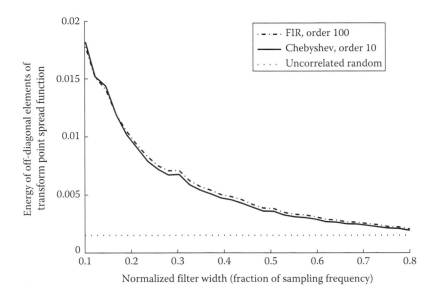

FIGURE 13.2

The effect of filtering on the TPSF. On the y-axis is $\chi(G)$. It is seen that the performance of compressive sensing is expected to deteriorate if the random transmit waveforms are highly correlated (narrow filters).

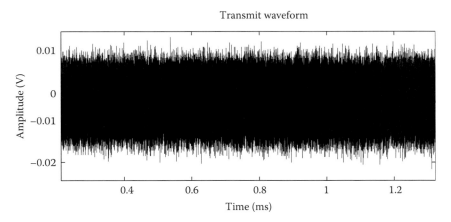

FIGURE 13.3

Time-domain plot of a transmit waveform sampled at 1 GS/s. The signal was generated using the experimental noise radar setup.

consists of two conical antennas that are used for transmitting and receiving signals. The antennas have a half-power beam width of $1°$. The conical antennas are connected to a high-power amplifier. The experiments were conducted in an outdoor setting. A photograph of the experimental setup is shown in Figure 13.4. We tested the imaging capability of the system at

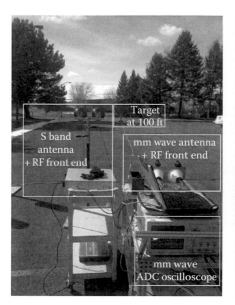 

FIGURE 13.4
(See color insert.) Photographs of radar system and target.

distances ranging from 14 m to around 33 m. We used tetrahedral *corner reflectors* and cylindrical scatterers as targets.

13.2.4 Analysis of Experimental Data

Compressive sensing recovery with circulant matrices is possible when the waveform that generates the circulant system matrix is i.i.d. random variables. Our experiments indicate that (a) waveforms in real systems are correlated as seen in Figure 13.5 which deviate from the normal distribution as seen in Figure 13.6, and (b) compressive signal recovery is tolerant to some extent to the correlations in the system matrix. The correlations in the hardware are expected to worsen the performance of compressive signal recovery. However, as we saw in Figure 13.2, the degradation may not be significant as long as the correlation is not high. This is affirmed by the accuracy of recovery seen in Figures 13.7 through 13.11.

13.2.5 Imaging Performance

In this section, we compare the performance of compressive sensing recovery with traditional correlation processing and least squares. We see evidence for the fact that the performance of compressive recovery even with just 25% of the samples compares favorably with least squares ($s*_{LS} = \Psi^{-1}f$) and correlation processing (s^*_{CR} in Equation 13.3). The scenarios and the

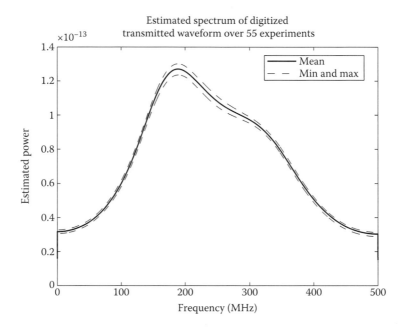

FIGURE 13.5
Power spectrum estimates (using a covariance estimator) of a millimeter-wave transmit waveform.

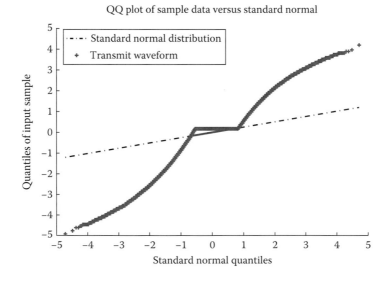

FIGURE 13.6
QQ plot of the normalized transmit waveform for millimeter-wave radar in comparison with standard normal.

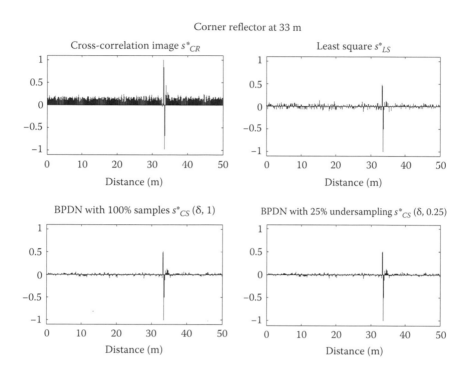

FIGURE 13.7
Corner reflector at a distance of 33 m (about 108 ft) (millimeter-wave radar).

related experimental analyses are described subsequently. In the descriptions, *target scene* refers to the sum total of all electromagnetic effects within the antenna beam.

1. The recovery of a target scene with one corner reflector placed at a distance of 33 m from the antennas is illustrated in Figure 13.7. We chose a corner reflector because most of the energy backscatter is guaranteed to be along the antenna line of sight. This scenario is intended to demonstrate the feasibility of modeling compressive noise radar imaging as the problem of inverting a circulant random matrix.

2. Two cylindrical reflectors with the first placed at around 14 m. The second cylinder is placed at 0.6 m and 0.3 m in the scenarios presented in Figures 13.8 and 13.9, respectively. We see that the resolving capabilities of compressively sampled noise radar are comparable with that of conventional cross-correlation-based processing. This favorable result is also seen in the case of a target scene consisting of two corner reflectors placed at 33 m and separated by distances of 0.3 m and 1 m in Figures 13.10 and 13.11, respectively.

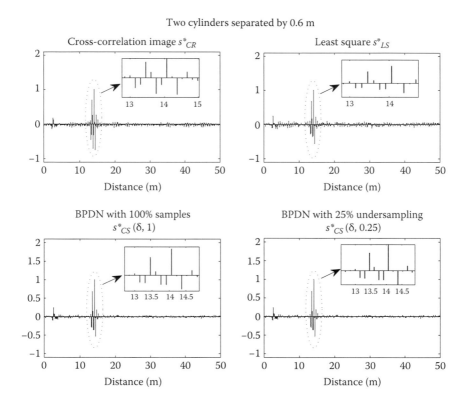

FIGURE 13.8
Two cylindrical targets at a distance of about 14 m (40 ft) from the radar and separated from each other by 0.6 m, which is the physical resolution corresponding to a 500 MHz bandwidth EM wave in free space.

13.3 Detection Strategies for Compressive Noise Radar

13.3.1 Compressive Sensing Detection

Target detection is a fundamental task in radar imaging systems. There are well-established approaches to target detection in conventional radar imaging (Richards 2005). In conventional radar systems, radar detection theory has largely been amenable to theoretical analysis. Threshold detection has proved particularly powerful in radar imaging. Improvements in the detection have primarily been achieved by optimizing objective functions constructed from expressions for the signal-to-noise ratio. We consider here the problem of detecting each element of the recovered vector in compressive radar imaging. The recovered signal is given by

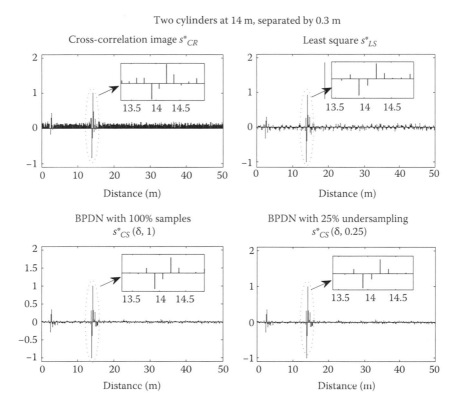

FIGURE 13.9
Two cylindrical targets at a distance of about 14 m (40 ft) from the radar and separated from each other by 0.3 m.

$$\hat{s} = \arg\min_{s \in \mathbb{R}^N} ||s||_1 \text{ s.t. } ||y - As||_{l2} \leq \sigma. \tag{13.12}$$

Our detection problem seeks the significance of each recovered pixel of the vector s. The null hypothesis is the absence of a target at each pixel, $s(k)$:

$$\mathcal{H}_0^{(k)} : s(k) = 0 \tag{13.13}$$

$$\mathcal{H}_1^{(k)} : s(k) \neq 0. \tag{13.14}$$

Relating this ground truth hypothesis with the recovered target vector $\hat{s}$, we obtain the false alarm and detection probabilities for each pixel, $\hat{s}(k)$,

$$P_D^{(k)} = \mathbb{P}[\hat{s}(k) > \xi | H_1^{(k)}] \tag{13.15}$$

$$P_{FA}^{(k)} = \mathbb{P}[\hat{s}(k) > \xi | H_0^{(k)}]. \tag{13.16}$$

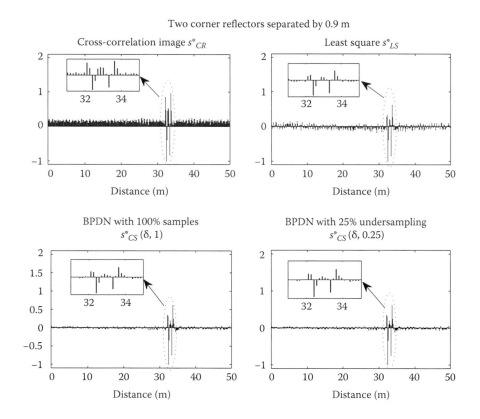

FIGURE 13.10
Two corner reflectors separated by a distance of 1 m (3 ft), located at a distance of 33 m (108 ft) from the radar.

$P_D^{(k)}$ and $P_{FA}^{(k)}$ henceforth refer to the probabilities of detection and false alarm, respectively, for each pixel indexed by k.

Detection of radar targets is accomplished by comparing different attributes of signals and images with thresholds. In our case, we look at the problem of determining the significance of each recovered pixel of the target scene. The important task of characterizing the detector requires us to derive relationships between thresholds and their corresponding probabilities of detection and false alarm. In the context of sparse radar imaging, fixing the threshold for detecting and defining significant targets is an important task. In sparse target scenes, with a large percentage of the target cells being zero, the cost of false alarms is proportionally higher.

However, the problem of casting the theory and performance of compressive radar imaging in the context of conventional radar systems remains open. The first paper to explore this problem from the perspective of signal processing was the 2010 paper on compressive signal processing in Davenport et al. (2010). Their proposal was to use a threshold detection

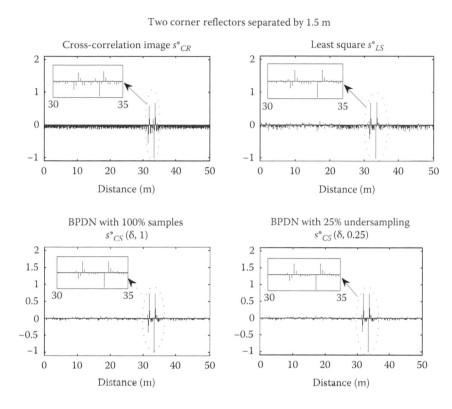

FIGURE 13.11
Two corner reflectors separated by a distance of 1.5 m (5 ft), located at a distance of 33 m from the radar.

approach based on the sufficient statistic given by $z^T (AA^T)^{-1} A\hat{s}$. This scalar detector is an ensemble quantity that is not useful for pixel-wise threshold-ing. The asymptotic results concerning the detectors of Davenport et al. were derived with the assumption that the system matrix is Gaussian random. The most recent publications related to the present section are the results in Anitori et al. (2013) on the design and analysis of detectors for com-pressive sensing. Anitori et al. propose to apply the complex-approximate message passing algorithm to arrive at a closed form for the distribution of the recovery error. However, the approximate message passing algorithm has not been shown to converge to accurate and stable solutions for problems with circulant random matrices. Thus, we develop a data-driven approach to threshold detection in compressive noise radar.

13.3.2 Statistics of the Error of Compressive Signal Recovery

There are no theoretical guarantees for the distribution of the residue of compressive sensing recovery when the system matrix is circulant random.

TABLE 13.1

Table Summarizing the Kolmogorov–Smirnov Test Comparing a
Few Instances of the Normalized Data with the Standard Normal
Distribution

Data Type	Null Hypothesis: Standard Normal	P-Value
Synthetic	Reject	0
Data set 1	Reject	0
Data set 2	Reject	0
Data set 3	Reject	0
Data set 4	Reject	0

In this section, we perform some empirical tests to study whether it is reasonable to approximate compressive signal recovery using normal distribution. We looked at those points of $\hat{s}$ that are characterized by the indices $i : s_i = 0$. This is equivalent to the instances of falsely detecting targets. In order to test the *Gaussianity*, we use the Kolmogorov–Smirnov (KS) test and quantile–quantile (QQ) plots. The KS test involves testing the null hypothesis that the distribution underlying the data is standard normal. We normalized the data by subtracting the mean and dividing by the standard deviation. Following this normalization, the empirical cumulative distribution function (CDF) of data is compared with the standard normal distribution. The results of the KS test are given in Table 13.1. The qq- plots in Figures 13.12 and 13.13 compare the quantiles of empirical data with that of theoretical estimates. Deviations

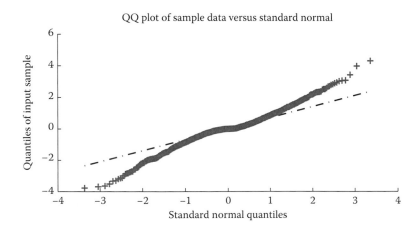

FIGURE 13.12
Quantile–quantile plot of *synthetic* normalized data. Blue markers represent the statistics of the points that satisfy $\hat{s}_i$ with i such that $s_i = 0$. Since the blue markers deviate significantly from the red line, the data cannot be modeled as Gaussian.

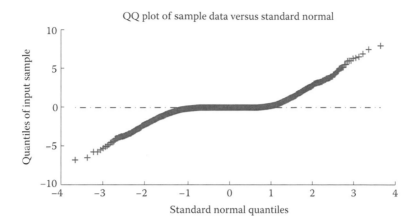

FIGURE 13.13
An example of quantile–quantile plots of recovery results from experimental data. Blue markers represent the statistics of the points that satisfy $\hat{s}_i$ with i such that $s_i = 0$. If the data are distributed according to the standard normal distribution, then the blue markers coincide with the red line.

of the empirical data from the straight line indicate non-Gaussianity. These two tests provide evidence that it is unreasonable to model the recovery residue as being drawn from Gaussian distribution. This precludes using sample mean and sample variance as estimates in computing detection thresholds and probabilities of false alarm and detection. In order to completely characterize threshold detection for compressive sensing, we need to have a fairly accurate idea of the underlying distributions.

When convex optimization recovery schemes such as l_1-based recovery are used, the nonlinear and iterative nature of the estimator algorithm coupled with the arbitrary evolution of the iterates complicate the task of deriving theoretical expressions for the distribution of the residue. The computational cost of convex optimization solvers presents a significant barrier to employing a brute force Monte Carlo approach to determine detection thresholds. For a desired probability of false alarm, P_{fa}, the number of instances of BPDN(.) should be at least of the order of $100/P_{fa}$ to compute the threshold with reasonable accuracy. This becomes a problem when we require the compressive radar system to compare favorably with conventional radar systems, where it is not unusual to see systems with probability of false alarm as low as $P_{fa} \sim 10^{-4}$.

There have been extensions to Monte Carlo simulations in the past to accommodate the occurrence of rare events (Broadwater and Chellappa 2010) in the context of radar signal processing. One such approach involves estimating probabilities using extreme value theory (Ozturk et al. 1996; Broadwater and Chellappa 2010). We propose applying to compressive noise radar, an approach based on limit theorems from extreme value theory.

Extreme value theory refers to the study of probabilities of rare events in stochastic systems. The basic results concern the statistics of the extremes of ordered random variables. In the past, it has been applied to problems in finance, climate sciences, and geophysical modeling. In electrical engineering, the utility of extreme value theory was first proposed for problems in detection theory by Ozturk et al. (Ozturk et al. 1996). In the context of compressive sensing, our proposal is to extrapolate the probabilities of rare events from a few instances of solving the convex optimization problem. A manageable number of instances of convex optimization problem are used to generate the statistics of compressive sensing for various values of (ρ, δ) and these are used to compute thresholds for small values of P_{fa} ($<10^{-4}$).

13.3.3 Threshold Estimation for Compressive Detection

We adopt a data-driven approach to estimate thresholds for compressive detection. From the sparsity assumption, it follows that a large proportion of the recovered vector will be zeros. Thus, we start with the assumption that we have access to the oracle knowledge about the location of a few zeros in the recovered vector. Determination of the exact number of nonzero coefficients is the classical model order selection problem. This remains an open problem with respect to l_1-norm optimization-based signal recovery. A subjective technique for a real radar system may, for example, involve using knowledge acquired via a visual examination of the target scene, and the realization that some locations consist of zeros. Let $Z \triangleq \{k : s(k) = 0\}$ denote the subset of $\{1, 2, \ldots, N\}$ about which we have knowledge that targets are absent. By observing the statistics of members of the set Z, across different realizations of the convex optimization solver for different ξ, we can determine the false alarm rate. Let $k_Z^{(i)}$ denote elements of subsequence of $\{1, 2, \ldots, N\}$, such that, for each $k_z \in Z$,

$$P_{FA} = \mathbb{P}[s^*(k_Z^{(i)}) > \xi].\tag{13.17}$$

Assume that the random variable $s^*(k_Z^{(i)})$ has the pdf $p_z(x)$; then we can compute P_{FA} as

$$P_{FA}(\xi) = \int_{\xi}^{\infty} p_z(x)dx.\tag{13.18}$$

If we were to use Monte Carlo simulations to estimate these probability distributions, for a given P_{FA}, we require Q, the number of realizations used for estimation to satisfy the requirement that $Q \gg 1/P_{FA}$. This requirement implies that brute-force Monte Carlo simulations are impractical when each realization of $s^*(k_z^{(i)})$ is being generated from a convex optimization solver. Following past research (Ozturk et al. 1996; Broadwater and Chellappa 2010) in this area, we use the generalized Pareto distribution (GPD) to estimate the tail distribution and thus the P_{FA}. The cumulative distribution function of GPD is given by

$$G(x) \triangleq 1 - \left(1 + \frac{\gamma x}{\zeta}\right)^{-\frac{1}{\gamma}}, \tag{13.19}$$

with

$$-\infty < \gamma < \infty, \quad \zeta > 0, \quad 0 > \gamma x \leq -\zeta. \tag{13.20}$$

GPD parametrizes numerous other distributions such as the exponential distribution when $\gamma = 0$ and the uniform distribution when $\gamma = -1$. As proved by Pickands (Pickands 1975), the following result relates the GPD to the tails of general, unspecified distributions:

$$\lim_{n \to \infty} \mathbb{P}\left[\sup_{0 \geq x < \infty} \left|\mathbb{P}[Y > y + u | Y \geq u] - 1 + G(y)\right| > \epsilon\right] = 0, \tag{13.21}$$

$$\forall \epsilon > 0. \tag{13.22}$$

Now we note that the conditional expectation given by

$$F_u(y) = \mathbb{P}[Y \leq y + u | Y > u] = \frac{F(u + y) - F(u)}{1 - F(u)}. \tag{13.23}$$

Setting $z = u + y$, we get

$$F(z) = F_u(z - u)(1 - F(u)) + F(u). \tag{13.24}$$

Further, we can define $\alpha \triangleq 1 - F(u)$ so that

$$F(z) = \alpha F_u(z - u) + (1 - \alpha). \tag{13.25}$$

We proceed by estimating the limit of $F_u(z - u)$ based on Equation 13.21 so that

$$F_u(z - u) = G(z) = 1 - \left(1 + \frac{\gamma}{\zeta}(z - u)\right)^{-\frac{1}{\gamma}} \tag{13.26}$$

and

$$F(z) = \alpha \left(1 - \left(1 + \frac{\gamma}{\zeta}(z - u)\right)^{-\frac{1}{\gamma}}\right) + (1 - \alpha) \tag{13.27}$$

$$= 1 - \alpha \left\{1 + \frac{\gamma}{\zeta}(z - u)\right\}^{-\frac{1}{\gamma}}. \tag{13.28}$$

Thus, the general strategy for applying GPD to estimate rare-event probabilities is to first set a particular α and then estimate the tail of the unknown distribution using (13.27). A typically used value is $\alpha = 0.1$, for which the value of u can be computed using Monte Carlo simulations to be the value of the threshold representing the $100 \times (1-\alpha))$th percentile of the data. The parametric function is then fitted to the given data to arrive at values for γ and ζ. We follow the proposal of Ozturk et al. (1996) and employ the Nelder–Mead algorithm for solving the maximum- likelihood formulation given by

$$\left(\hat{\zeta}, \hat{\gamma}\right) = \arg\min_{\gamma, \zeta} \left(\alpha Q \log \zeta + \left(1 + \frac{1}{\gamma}\right) \sum_{i=1}^{\alpha Q} \log\left(1 + \frac{\gamma z_i}{\zeta}\right)\right). \tag{13.29}$$

Subsequently, the relationship between the probability of false alarm and the threshold can be derived based on the GPD estimate as follows:

$$P_{FA} = \alpha \left\{1 + \frac{\hat{\gamma}}{\hat{\zeta}}(\tau - u)\right\}^{-\frac{1}{\gamma}}, \tag{13.30}$$

$$\tau = u + \frac{\hat{\zeta}}{\hat{\gamma}}\left(\left(\frac{P_{FA}}{\alpha}\right)^{-\hat{\gamma}} - 1\right). \tag{13.31}$$

13.3.4 GPD and Compressive Sensing

With the aforementioned results about GPD established, we now proceed to apply it to compressive sensing. We treat the convex optimization solver as an experiment whose reconstruction error has an unknown distribution. We wish to estimate accurate thresholds for low P_{FA}. The advantage of using the approach based on GPD is that the results so derived are to a large extent independent of the type of distributions of the target scene, s, the noise η, and residue $s - \hat{s}$. Our methodology for deriving the thresholds for compressive sensing is as follows:

The threshold is computed from the probability of false alarm as

$$\tau^{(CS)}(P_{FA}) = u + \frac{\hat{\zeta}}{\hat{\gamma}}\left(\left(\frac{P_{FA}}{\alpha}\right)^{-\hat{\gamma}} - 1\right). \tag{13.32}$$

With the relationship between P_{FA} and τ established, we can proceed to derive the probability of detection from the statistics of the nonzero values and establish the receiver operating characteristics.

13.3.5 Computational Complexity of GPD-Based Threshold Estimation

In our work, we use convex optimization to solve the compressive signal recovery problem. While there are more computationally efficient approaches, convex optimization provides the best performance in terms of recovery accuracy (Tropp and Wright 2010). The recent algorithm of approximate message passing (AMP) (Donoho et al. 2009) is an exception in that its performance is theoretically identical to convex optimization. However, this theoretical guarantee is only valid for Gaussian random system matrices. Compressive sensing is particularly sensitive to the nature of the system matrix, and it is as yet unclear how AMP-based algorithms behave for circulant matrices.

We use the spectral projected gradient algorithm-l_1 (SPGL1) (van den Berg and Friedlander 2008) for solving the basis pursuit denoising problem. Each iteration of the SPGL1 algorithm involves a matrix–vector product. In our problem, this corresponds to a computational complexity of $O(MN)$, assuming that each scalar arithmetic operation can be accomplished with complexity of $O(1)$. Multiplying with circulant system matrices is an operation that takes $O(N \log N)$ operations. While the exact estimate for the number of iterations for SPGL1 is unknown, the similar ParNes algorithm (Gu et al. 2012) requires $O(\sqrt{1/\sigma})$ iterations to converge to a solution. So we surmise that the most efficient convex-optimization-based BPDN solvers have a complexity of $O(M \log N \sqrt{1/\sigma})$.

Let us assume that we use N_{mc} number of instances of SPGL1, then the cost of constructing the empirical CDF of the residue would be $O(N_{mc}N \log N \sqrt{1/\sigma})$. Our approach to estimating the CDF seeks to achieve two objectives, computationally speaking: (a) reduce the number of instances N_{mc}, and (b) solve smaller problems on each instance so that we can lower the value of N. Let us assume that we desire a probability of false alarm of $\tilde{P}_{fa}$. In order to accurately estimate the threshold that achieves this probability, we would require around $\tilde{N}_{mc} = 10/\tilde{P}_{fa}$ instances. Thus, the cost of $\tilde{N}_{mc}$ instances of the SPGL1 algorithm would be $O(\tilde{N}_{mc}N \log N \sqrt{1/\sigma})$. Using the GPD approach for extrapolating tail distribution allows us to estimate thresholds for very low values of P_{fa} without significantly increasing N_{mc} and N.

In a real system, our goal is to compute detection thresholds with as little latency as possible. An improvement of a factor of $(1/\gamma_{mc}\gamma_s) \sim 10^4$ in the computational speed of estimating thresholds represents a significant advantage when each solution of the SPGL1 algorithm takes around 300 s for a 10,000-dimensional problem, as we see in Figure 13.14. The computational expense of convex optimization-based signal recovery presents a

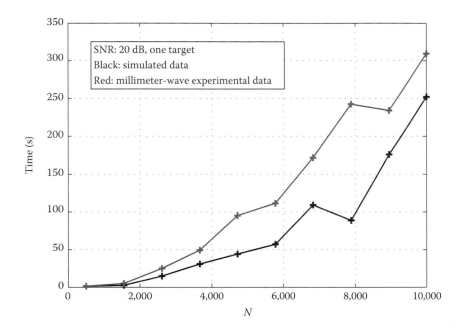

FIGURE 13.14
(See color insert.) Computational complexity of SPGL1 convex optimization signal recovery in terms of running time of algorithm on an Intel i7, 2.8 GHz with 8 GB of RAM.

significant stumbling block in running extensive Monte Carlo simulations. In Figure 13.14, we present the computational cost as a function of problem size. The numbers pertain to a MATLAB® implementation of the SPGL1 algorithm (van den Berg and Friedlander 2007) on an Intel i7, 2.8 GHz system with 8 GB RAM.

A single solution of a problem of dimension 10,000 takes about 5 min. A problem dimension of 10,000 would provide sufficient opportunities to simulate events of likelihoods of the order of 0.01. In summary, if we resort to brute-force Monte Carlo simulations, it would take us in excess of 5 min to even simulate events that occur with a probability between 10^{-3} and 10^{-2}. Thus computationally speaking, it can get prohibitively expensive to simulate events that occur with lower probabilities. The direct implication to detection theory is that it becomes impractical to estimate thresholds that can yield low false alarm rates using brute-force Monte Carlo methods.

13.3.5.1 Performance of GPD-Based Threshold Estimation

We demonstrate the effectiveness of Algorithm 13.1 to data generated from applying the SPGL1 convex optimization solver to synthesized compressive

---

Algorithm 13.1 GPD-based tail, P_{FA}, and threshold estimation

---

Estimation of P_{FA} as a function of threshold for compressive sensing using GPD.

Input: X, s.

Output: GPD parameters, $\hat{\gamma}^{(CS)}$, $\hat{\zeta}$.

    for $j \in Z$ **do**

        for $i = 1 \rightarrow Q$ **do**

            Solve the convex optimization problem given by BPDN(ρ, δ, σ);

            Use the entire recovered vector $\hat{s}$ as a training set, i.e., $T = \{1, 2, \ldots, N\}$ OR if available, choose a set of points for which the truth of the hypothesis is known, i.e. $T = \{i : s_i = 0\}$;

        end for

        Using the points $r(j) : j \in T$, construct the distribution $\hat{p}_{r(j)}(x) = \sum_{t=1}^{Q} \mathbb{1}_{r\left(k_z^{(t)}\right)}(x)$;

    end for

Set $\alpha = 0.1$, and select $u = r(b)$ such that $\#\{t : r(t) > r(b)\} = \lfloor \alpha Q \rfloor$, and let $T \triangleq \{t : r(t) > r(b)\}$;

Let $z^{(u)}$ be a sequence such that $\forall j, z_j^{(u)} \in T$;

Estimate GPD parameters μ and ζ by applying the Nelder–Mead solver to solve the *maximum likelihood* function optimization problem given by $\min_{\gamma, \zeta} \alpha Q \log \zeta + \left(1 + \frac{1}{\gamma}\right) \sum_{i=1}^{\alpha Q} \log\left(1 + \frac{\gamma}{\zeta}(z_i^{(u)} - u)\right)$. Call this solution $\hat{\gamma}^{(CS)}$ and $\hat{\zeta}^{(CS)}$;

---

noise radar problems. In Figure 13.15, we plot the complete real CDF of the absolute value of the residual data with the relatively small value of $Q = 5 \times 10^4$ for the training data. The parameters for the GPD-based estimate of the P_{FA} are derived from the data. The value of α has to be chosen carefully based on the number of reliable nonzero values in the training data. A high value of α will mean that the training data contain too many samples too far from the tail. If α is too low, then there will be too few samples in the training data. This GPD-based extrapolation is seen to conform with both training and test data.

    The GPD estimates are obtained for 50 independent realizations of Algorithm 13.1. From these estimates, the threshold is computed for various values of the probability of false alarm from Equation 13.32. The 50 realizations are plotted as smoothed probability density function in Figure 13.16, estimated as

$$p_\tau(x) = \sum_{i=1}^{50} \mathbb{1}_{\hat{T}}(x),$$

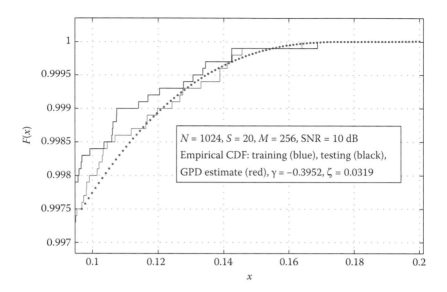

FIGURE 13.15

(See color insert.) Plot showing application of Algorithm 13.1 for GPD-based extrapolation of the cumulative distribution of empirical residue data. Compressive recovery with $N = 1024$.

where $\hat{T}$ represents the set of all numbers larger than $\hat{\tau}$, which is the estimated threshold value. The indicator function $\mathbb{1}_{\hat{\tau}}(x)$ takes the value of 1 when $x \in \hat{T}$ and 0 otherwise. In the simulations, we used the `ksdensity` function in MATLAB. The median threshold for each P_{FA} is indicated in the plots. As the desired P_{FA} is lower, the unknown threshold values are farther away from the training data. Thus, the uncertainty in the estimated threshold value increases, as indicated by the increasing variance of the parameter estimates. The reliable median allows us to extract a meaningful estimate even when the desired probability of false alarm is as low as 10^{-8}.

We applied our threshold estimation algorithm on experimental data that was acquired as described in Section 13.2. The fully sampled (1 gigasamples per second) received waveform and transmitted waveform were a record of 10^5 data samples. We divided the entire record into smaller sets of length $N = 4096$. For one of these sets, we solved the l_1-minimization-based compressive recovery algorithm to extract the radar target image. We then utilized partial knowledge of the locations of the nonzero values to construct a set that represented the points $\hat{s}_i : s_i = 0$. For the subset of locations where targets are absent, we estimated the tail using Algorithm 13.1. We verify the performance of this algorithm by comparing it with the estimated cumulative distribution functions of the empirical data. We combined the results of several such experiments to empirically construct the extended cumulative distribution function. The accurate reconstruction of these values is seen in Figure 13.17.

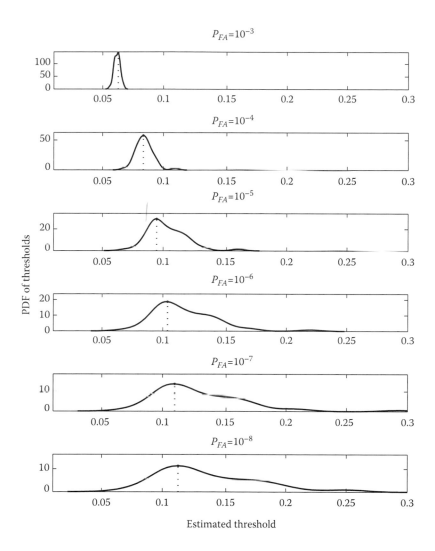

FIGURE 13.16
The probability distribution of the estimates of the threshold value for various desired values of P_{FA}. Compressive recovery was performed with $S = 10, M = 256, N = 1024$, and SNR = 10 dB. The estimate of the P_{FA} was done using $\alpha = 0.01$. The pdf of the threshold estimation itself was generated using 50 instances of Algorithm 13.1.

13.4 Conclusions and Future Work

13.4.1 Compressive Noise Radar Imaging and Detection

In this chapter, we showed through theoretical arguments, numerical simulations, and experiments, the suitability of stochastic waveforms for

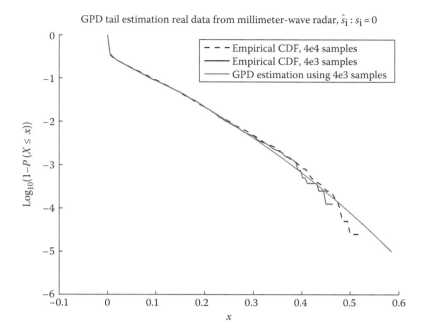

FIGURE 13.17
(See color insert.) Tail estimation for real data. We observe that the empirical CDF corresponds closely to the GPD estimate for the tail. The target scenario involved one corner reflector target at 100 ft imaged multiple times using the millimeter-wave radar described in Section 13.2.

designing practical compressive radar systems. In spite of anticipated non-idealities, the performance of compressively sampled radar compares favorably with conventional radar imaging systems. We developed an approach based on extreme value theory to estimate the tail of compressive sensing recovery residues. The closed form of the distribution of the tail thus obtained belongs to the family of GPD. We successfully tested our algorithms on experimental noise radar data proving that compressive noise radar imaging is a feasible technology that could replace or augment conventional noise radar systems.

13.4.2 Open Problems

Going forward, we have identified two important open problems related to compressive noise radar imaging:

1. Developing an extension of the approximate message passing state evolution (Bayati and Montanari 2011) framework to circulant-matrix based compressive sensing. This provides us an approach to derive expressions for detection statistics.

2. Designing a sampling system that is suitable for high-throughput compressive sampling applications such as ultrawideband noise radar. The current system can be used to achieve higher resolutions however, for a generally applicable system, it is necessary to design a hardware system for undersampling signals. The principal challenge is in keeping track of the index and temporal location of each acquired sample.

References

Amin, M. G. 2011. *Through-the-Wall Radar Imaging*. Boca Raton, FL: CRC Press.

Anitori, L. et al. 2013. Design and analysis of compressed sensing radar detectors. *IEEE Transactions on Signal Processing* 61 (4): 813–827.

Bar-Ilan, O. and Y. C. Eldar. 2014. Sub-Nyquist radar via Doppler focusing. *IEEE Transactions on Signal Processing* 62 (7): 1796–1811.

Baraniuk, R. and P. Steeghs. 2007. Compressive radar imaging. In *2007 IEEE Radar Conference*, Waltham, MA, pp. 128–133.

Bayati, M. and A. Montanari. 2011. Dynamics of message passing on dense graphs, with applications to compressed sensing. *IEEE Transactions on Information Theory* 57 (2): 764–785.

van den Berg, E. and M. P. Friedlander. 2007. SPGL1: A solver for large-scale sparse reconstruction. http://www.cs.ubc.ca/labs/scl/spgl1.

van den Berg, E. and M. P. Friedlander. 2008. Probing the Pareto frontier for basis pursuit solutions. *SIAM Journal on Scientific Computing* 31 (2): 890–912.

Broadwater, J. B. and R. Chellappa. 2010. Adaptive threshold estimation via extreme value theory. *IEEE Transactions on Signal Processing* 58 (2): 490–500.

Candes, E. J., J. Romberg, and T. Tao. 2006. Robust uncertainty principles: Exact signal reconstruction from highly incomplete frequency information. *IEEE Transactions on Information Theory* 52 (2): 489–509.

Chen, P.-H. et al. 2012. A portable real-time digital noise radar system for through-the-wall imaging. *IEEE Transactions on Geosciences and Remote Sensing* 50 (10): 4123–4134.

Davenport, M. A. et al. 2010. Signal processing with compressive measurements. *IEEE Journal of Selected Topics in Signal Processing* 4 (2): 445–460.

Donoho, D. L. 2006. Compressed sensing. *IEEE Transactions on Information Theory* 52 (4): 1289–1306.

Donoho, D. L., A. Maleki, and A. Montanari. 2009. Message-passing algorithms for compressive sensing. *Proceedings of the National Academy of Sciences* 106 (45): 18914–18919.

Ender, J. H. G. 2010. On compressive sensing applied to radar. *Signal Processing* 90 (5): 1402–1414.

Gu, M., L.-H. Lim, and C. J. Wu. 2012. PARNES: A rapidly convergent algorithm for accurate recovery of sparse and approximately sparse signals. *Numerical Algorithms* 64 (2): 1–27.

Gurbuz, A. C., J. H. McClellan, and W. R. Scott. 2009. A compressive sensing data acquisition and imaging method for stepped frequency GPRs. *IEEE Transactions on Signal Processing* 57 (7): 2640–2650.

Haupt, J. et al. 2010. Toeplitz compressed sensing matrices with applications to sparse channel estimation. *IEEE Transactions on Information Theory* 56 (11): 5862–5875.

Herman, M. and T. Strohmer. 2009. High-resolution radar via compressed sensing. *IEEE Transactions on Signal Processing* 57 (6): 2275–2284.

Horton, B. M. 1959. Noise modulated distance measuring system. *Proceedings of the IRE* 47 (5): 821–828.

Jonsson, B. E. 2010. A survey of A/D-converter performance evolution. In *17th IEEE International Conference on Electronics, Circuits, and Systems (ICECS)*, Athens, Greece, pp. 766–769.

Mishali, M. et al. 2011. Xampling: Analog to digital at sub-Nyquist rates. *IET Circuits, Devices & Systems* 5 (1): 8–20.

Narayanan, R. M. et al. 1998. Design, performance, and applications of a coherent ultra-wideband random noise radar. *Optical Engineering* 37 (6): 1855–1869.

Narayanan, R. M. 2008. Through-wall radar imaging using UWB noise waveforms. *Journal of the Franklin Institute* 345 (6): 659–678.

Oppenheim, A. V., R. W. Schafer, and J. R. Buck. 1999. *Discrete-time Signal Processing*, 2nd edn. Upper Saddle River, NJ: Prentice-Hall, Inc.

Ozturk, A., P. R. Chakravarthi, and D. D. Weiner. 1996. On determining the radar threshold for non-Gaussian processes from experimental data. *IEEE Transactions on Information Theory* 42 (4): 1310–1316.

Pauluzzi, D. R. and N. C. Beaulieu. 2000. A comparison of SNR estimation techniques for the AWGN channel. *IEEE Transactions on Communications*, 48 (10): 1681–1691.

Pickands III, J. 1975. Statistical inference using extreme order statistics. *The Annals of Statistics* 3 (1): 119–131.

Rauhut, H., J. Romberg, and J. A. Tropp. 2012. Restricted isometries for partial random circulant matrices. *Applied and Computational Harmonic Analysis* 32 (2): 242–254.

Richards, M. A. 2005. *Fundamentals of Radar Signal Processing*. New York: McGraw-Hill.

Romberg, J. 2009. Compressive sensing by random convolution. *SIAM Journal on Imaging Sciences* 2 (4): 1098–1128.

Shastry, M. C., R. M. Narayanan, and M. Rangaswamy. 2010. Compressive radar imaging using white stochastic waveforms. In *2010 International Waveform Diversity and Design Conference*, Niagara Falls, Canada.

Shastry, M. C., R. M. Narayanan, and M. Rangaswamy. 2013a. Characterizing detection thresholds using extreme value theory in compressive noise radar imaging. In *SPIE Defense, Security, and Sensing Conference*, Baltimore, MD.

Shastry, M. C., R. M. Narayanan, and M. Rangaswamy. 2013b. Waveform design for compressively sampled ultrawideband radar. *Journal of Electronic Imaging* 22 (2): 021011–021011.

Shastry, M. C. et al. 2012. Analysis and design of algorithms for compressive sensing based noise radar systems. In *2012 IEEE Seventh Sensor Array and Multichannel Signal Processing Workshop (SAM)*, Hoboken, NJ, pp. 333–336.

Suksmono, A. B. et al. 2010. Compressive stepped-frequency continuous-wave ground-penetrating radar. *IEEE Geoscience and Remote Sensing Letters* 7 (4): 665–669.

Tropp, J. A. and S. J. Wright. 2010. Computational methods for sparse solution of linear inverse problems. *Proceedings of the IEEE* 98 (6): 948–958.

Wu, Y. and J. Li. 1998. The design of digital radar receivers. *IEEE Aerospace and Electronic Systems Magazine* 13 (1): 35–41.

Yoon, Y.-S. and M. G. Amin. 2010. Through-the-wall radar imaging using compressive sensing along temporal frequency domain. In *2010 IEEE International Conference on Acoustics Speech and Signal Processing (ICASSP)*, Dallas, TX, pp. 2806–2809.

Index

Printed and bound by CPI Group (UK) Ltd, Croydon, CR0 4YY

18/10/2024

01776271-0011